INTRODUCTION TO
BIOTECHNOLOGY

INTRODUCTION TO
BIOTECHNOLOGY

THIRD EDITION

William J. Thieman

Ventura College, Emeritus

Michael A. Palladino

Monmouth University

PEARSON

Boston   Columbus   Indianapolis   New York   San Francisco   Upper Saddle River
Amsterdam   Cape Town   Dubai   London   Madrid   Milan   Munich   Paris   Montreal   Toronto
Delhi   Mexico City   São Paulo   Sydney   Hong Kong   Seoul   Singapore   Taipei   Tokyo

Editor-in-Chief: Beth Wilbur
Senior Acquisitions Editor: Michael Gillespie
Executive Director of Development: Deborah Gale
Project Editor/Development Editor: Leata Holloway
Assistant Editor: Leslie Allen
Executive Marketing Manager: Lauren Harp
Director of Production: Erin Gregg
Managing Editor: Michael Early
Production Project Manager: Lori Newman
Production Management: Cenveo Publisher Services/Nesbitt Graphics, Inc.
Copyeditor: Heidi Thaens
Compositor: Cenveo Publisher Services/Nesbitt Graphics, Inc.
Interior Designer: Cenveo Publisher Services/Nesbitt Graphics, Inc.
Cover Designer: Jodi Notowitz
Illustrators: Cenveo Publisher Services/Nesbitt Graphics, Inc.
Photo Researcher: Bill Smith Group
Image Lead: Donna Kalal
Manufacturing Buyer: Michael Penne
Cover & Interior Printer: Edwards Brothers

Cover Photo Credits: Plush Studios/Chris Gramly/Getty Images; Charles Smith/Corbis; Danace2000/Dreamstime; Martin Shields/Getty Images; Cutcaster

Credits and acknowledgments borrowed from other sources and reproduced, with permission, in this textbook appear on the appropriate page within the text [or on p. C-1].

1 2 3 4 5 6 7 8 9 10—EDB—15 14 13 12 11

ISBN 10: 0-321-81892-X
ISBN 13: 978-0-321-81892-8

To my wife, Billye, the love of my life,
and to the hundreds of biotechnology graduates
who are now doing good science at biotechnology companies
and loving every minute of it.

W. J. T.

To my 10 nieces and nephews, Jordan, Eric, Hannah, Vincent,
Noah, Sara, Andrew, Sofia, David and Amelia, you are unique
and wonderful treasures of our family!

M. A. P.

About the Authors

Authors Michael Palladino and Bill Thieman

William J. Thieman taught biology at Ventura College for 40 years and biotechnology for 11 years before retiring from full time teaching in 2005. He continues to teach the biotechnology course part-time. He received his B.A. in biology from California State University at Northridge in 1966 and his M.A. degree in Zoology in 1969 at UCLA. In 1993, he started a biotechnician training program at Ventura College. In 1995, he added laboratory skills components to the course and articulated it as a state-approved vocational program.

Mr. Thieman has taught a broad range of undergraduate courses including general, human, and cancer biology. He received the Outstanding Teaching Award from the National Biology Teachers Association in 1996 and the 1997 and 2000 Student Success Award from the California Community Colleges Chancellor's Office. The Economic Development Association presented its 1998 Program for Economic Development Award to the biotechnology training program at Ventura College for its work with local biotechnology companies. His success in acquiring grants to support the program was recognized at the 2007 Conference of the National Center for Resource Development.

Michael A. Palladino is Dean of the School of Science and an Associate Professor of Biology at Monmouth University in West Long Branch, New Jersey. He received his B.S. degree in Biology from Trenton State College (now known as The College of New Jersey) in 1987 and his Ph.D. in Anatomy and Cell Biology from the University of Virginia in 1994. He joined the Monmouth faculty in 1999.

Dr. Palladino has taught a wide range of undergraduate courses. He has received several awards for research and teaching, including the Distinguished Teacher Award from Monmouth University, the Caring Heart Award from the New Jersey Association for Biomedical Research, and the Young Investigator Award of the American Society of Andrology. At Monmouth, he has an active lab of undergraduate students involved in research on the cell and the molecular biology of male reproductive organs. He is founder and director of the New Jersey Biotechnology Educators Consortium, a statewide association of biotechnology teachers.

Dr. Palladino is author of *Understanding the Human Genome Project,* the first volume in the *Benjamin Cummings Special Topics in Biology Series,* for which he also serves as series editor. Dr. Palladino is also coauthor on the writing team of W. S. Klug, M. R. Cummings, and C. A. Spencer for the textbooks *Concepts of Genetics* and *Essentials of Genetics,* both published by Pearson Education.

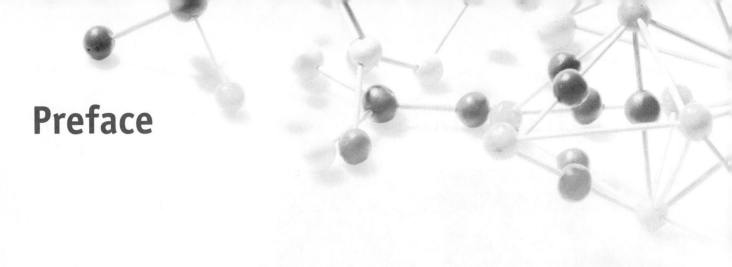

Preface

It is hard to imagine a more exciting time to be studying biotechnology. Advances are occurring at a dizzying pace, and biotechnology has made an impact on many aspects of our everyday lives. Now in its third edition, *Introduction to Biotechnology* remains the first biotechnology textbook written specifically for the diverse backgrounds of undergraduate students. Appropriate for students at both 2- and 4-year and vocational technical schools, *Introduction to Biotechnology* provides students with the tools for practical success in the biotechnology industry through its balanced coverage of molecular biology, details on contemporary techniques and applications, integration of ethical issues, and career guidance.

Introduction to Biotechnology was designed with several major goals in mind. The text aims to provide:

- An engaging and easy-to-understand narrative that is appropriate for a diverse student audience with varying levels of scientific knowledge.

- Assistance to instructors teaching all major areas of biotechnology and help to students learning fundamental scientific concepts without overwhelming and excessive detail.

- An overview of historic applications while emphasizing modern, cutting-edge, and emerging areas of biotechnology.

- Insights on how biotechnology applications can provide some of the tools to solve important scientific and societal problems for the benefit of humankind and the environment.

- Inspiration for students to ponder the many ethical issues associated with biotechnology.

Introduction to Biotechnology provides broad coverage of topics including molecular biology, bioinformatics, genomics, and proteomics. We have striven to incorporate balanced coverage of basic molecular biology with practical and contemporary applications of biotechnology to provide students with the tools and knowledge they need to understand the field.

In our effort to introduce students to the cutting-edge techniques and applications of biotechnology, we have dedicated specific chapters to such emerging areas as agricultural biotechnology (Chapter 6), forensic biotechnology (Chapter 8), bioremediation (Chapter 9), and aquatic biotechnology (Chapter 10). Consideration of the many regulatory agencies and issues that affect the biotechnology industry are discussed in Chapter 12. In addition to the ethical issues included in each chapter as **You Decide** boxes, a separate chapter (Chapter 13) is dedicated to ethics and biotechnology.

New Features of the Third Edition

The third edition of *Introduction to Biotechnology* includes several new instructor resources and exciting features:

- **Forecasting the Future** briefly highlights exciting new areas of biotechnology that the authors predict will be worth watching in the future.

- **Making a Difference** at the end of each chapter spotlights particularly beneficial aspects of biotechnology applications that have had major impacts in improving the quality of life.

- More end-of-chapter **Questions & Activities**, including more Internet-based exercises.

- A computerized test bank with multiple-choice questions for each chapter; electronic files for all images in the textbook; and PowerPoint Lecture Outline slides conveniently located on the Instructor's Resource Center, www.pearsonhighered.com/educator.

- New **You Decide** entries have been added to stimulate student interest in controversial areas of biotechnology.

In addition, each chapter has been thoroughly revised and updated to provide students with current information on emerging areas of biotechnology. Of special note are the following changes:

- **Chapter 1: The Biotechnology Century and Its Workforce.** Includes updated content on the current state of the biotechnology industry, company mergers, biotechnology and pharmaceutical company revenues, funding sources for starting a biotechnology company, and investigational new drugs.

- **Chapter 2: An Introduction to Genes and Genomes.** Includes a new section titled "Revealing the Epigenome," which provides an introduction to epigenetics and its importance to genetic diseases and disease treatments.

- **Chapter 3: Recombinant DNA Technology and Genomics.** Includes updated content on the Human Genome Project, a new section called "10 Years after the Human Genome Project," and new coverage of the Genome 10K Plan, Human Microbiome Project, and personalized genomics. Major content updates have been made to DNA sequencing technologies, including a new section and figure titled "Next-Generation Sequencing."

- **Chapter 4: Proteins as Products.** Includes examples of delivery vehicles for proteins that overcome the difficulty of administering protein drugs; progress on designing protein structures based on the 1,200 superfamilies that have been discovered; the increased emphasis on discovery of biomarker proteins that can indicate disease at earlier stages; and the design of nanoparticles that deliver proteins designed to attach to sites on cancer cells to destroy them.

- **Chapter 5: Microbial Biotechnology,** includes new content on metagenomics, the Human Microbiome Project, a new section on synthetic genomes, and a new section on microbes for making biofuels.

- **Chapter 6: Plant Biotechnology** Includes the work of nonprofit research groups in developing new varieties of transgenic plants; the shift in emphasis to plant transgenic crops in developing countries, which are now the majority users; the addition of new transgenic crops and the expansion of others to include stacks of traits; new emphasis on different biofuels from plant wastes and algae; progress in biopharming, edible vaccines, and their importance; and current ways of dealing with resistant insects and weeds that have developed from the use of these crops.

- **Chapter 7: Animal Biotechnology.** Includes a shift in direction from drugs to vaccines for humans of all ages and the rationale behind it; the significance of animal testing for drugs toward treatments for animal diseases; the benefits of cell-culture testing before animal testing for regulatory approval; the first approval of a drug produced in a transgenic goat to treat a type of stroke; new method for creating animals with gene knockouts and knock-ins; and the importance of a national project to determine the function of all the genes in a rat by using knockout technology.

- **Chapter 8: DNA Fingerprinting and Forensic Analysis.** Includes the progress in utilizing personal DNA sequencing as a precursor to diagnosis; the shift from RFLP to PCR fingerprinting and the reasoning behind it; new examples of DNA fingerprint comparisons using the CODIS sites; new examples of DNA sequences to identify certified products; and new examples of nonhuman DNA comparisons.

- **Chapter 9: Bioremediation.** Includes updated content of GM species for bioremediation and a new section highlighting the roles of bioremediation in cleanup at the *Deepwater Horizon* oil spill in the Gulf of Mexico.

- **Chapter 10: Aquatic Biotechnology.** Includes revised content on aquaculture, bioprospecting, and biotechnology products from aquatic organisms.

- **Chapter 11: Medical Biotechnology.** Includes reorganized and revised content on the Human Genome Project and genetic testing along with a new section on direct-to-consumer genetic tests, updated content on gene therapy technologies, and new and updated content on induced-pluripotent stem cells and stem cell regulations.

- **Chapter 12: Biotechnology Regulations.** Includes a discussion of the uses and dangers of synthetic genomes and potential regulations; the potential for faster drug approvals through the sharing of information on drug failures in trials; effect of FDA publication of long-term effects of previously approved drugs; the effect of the USPTO's decision that gene sequences for diagnostic purposes are not patentable; new FDA policies designed to protect participants in drug trials; and the importance of current training of biotech company employees based on examples.

- **Chapter 13: Ethics and Biotechnology.** Includes reorganized content and an abbreviated chapter format, new information on risk assessments, and a new **You Decide** on field trials of GM insects.

Returning Features

Introduction to Biotechnology is specifically designed to provide several key elements that will help students enjoy learning about biotechnology and prepare them for a career in biotechnology.

Learning Objectives

Each chapter begins with a short list of learning objectives presenting key concepts that students should understand after studying the chapter.

Abundant Illustrations

Approximately 200 figures and photographs provide comprehensive coverage to support chapter content. Illustrations, instructional diagrams, tables, and flowcharts present step-by-step explanations that give students visual help to learn about the laboratory techniques and complex processes that are important in biotechnology.

Career Profiles

A special box at the end of each chapter introduces students to different job options and career paths in the biotechnology industry and provides detailed information on job functions, salaries, and guidance for preparing to enter the workforce. Experts currently working in the biotechnology industry have contributed information to many of these **Career Profile** boxes. We strongly encourage students to refer to these profiles if they are interested in learning more about careers in the industry.

You Decide

From genetically modified foods to stem cell research, there are an endless number of topics in biotechnology that provoke strong ethical, legal, and social questions and dilemmas. **You Decide** boxes stimulate discussion in each chapter by presenting students with information that relates to the social and ethical implications of biotechnology, followed by a set of questions for them to consider. The goal of these boxes is to help them understand *how* to consider ethical issues and to formulate their own informed decisions.

Tools of the Trade

Biotechnology is based on the application of various laboratory techniques or tools in molecular biology, biochemistry, bioinformatics, genetics, mathematics, engineering, computer science, chemistry, and other disciplines. **Tools of the Trade** boxes in each chapter present modern techniques and technologies related to the content of each chapter to help students learn about the techniques and laboratory methods that are the essence of biotechnology.

Questions & Activities

Questions are included at the conclusion of each chapter to reinforce student understanding of concepts. Activities frequently include Internet assignments that ask students to explore a cutting-edge topic. Answers to these questions are provided at the end of the text.

Glossary

Like any technical discipline, biotechnology has a lexicon of terms and definitions that are routinely used in discussing processes, concepts, and applications. The most important terms are shown in **boldface type** throughout the book and are defined as they appear in the text. Definitions of these key terms are included in a glossary at the end of the book.

Supplemental Learning Aids

Introduction to Biotechnology **Companion Website**
(www.pearsonhighered.com/biotechnology)

The Companion Website is designed to help students study for their exams and deepen their understanding of the text's content. Each chapter contains learning objectives, content reviews, flashcards, glossary terms, jpegs of figures, and an extensive collection of Keeping Current Web Links which explore related topics in other areas. In addition, **References and Further Readings** and **Q&A Boxes** (both formerly in the text) now appear as part of the Companion Website.

Instructor Resource Center (IRC)

The Instructor Resource Center, www.pearsonhighered.com/educator, is designed to support instructors teaching biotechnology. The IRC is an online resource that supports and augments material in the textbook. New instructor supplements available for download include:

- **Computerized Test Bank:** 10 to 20 multiple-choice test questions per chapter
- **Jpeg Art Files:** electronic files of all text tables, line drawings, and photos
- **PowerPoint Lecture Outlines:** a set of PowerPoint presentations consisting of lecture outlines for each chapter augmented by key text illustrations

Instructors using *Introduction to Biotechnology* can contact their Benjamin Cummings sales representative to access the Instructor Resource Center free of charge.

Benjamin Cummings Special Topics in Biology Series

The *Benjamin Cummings Special Topics in Biology Series* is a series of booklets designed for undergraduate students. These small booklets present the basic scientific facts and social and ethical issues surrounding current hot topics. Booklets in the series include:

- *Alzheimer's Disease* (ISBN 0-1318-3834-2)
- *Biology of Cancer, 2/e* (ISBN TBD) NEW EDITION!
- *Biological Terrorism* (ISBN 0-8053-4868-9)
- *Emerging Infectious Diseases* (ISBN 0-8053-3955-8)

- *Genetic Testimony* (ISBN 0-1314-2338-X)
- *Gene Therapy* (ISBN 0-8053-3819-5)
- *HIV and AIDS* (ISBN 0-8053-3956-6)
- *Mad Cows and Cannibals* (ISBN 0-1314-2339-8)
- *Stem Cells and Cloning, 2/e* (ISBN 0-3215-9002-3)
- *Understanding the Human Genome Project, 2/e* (ISBN 0-8053-4877-8)

Contact your local Benjamin Cummings sales representative about bundling booklets in this series with *Introduction to Biotechnology* or visit www.pearsonhighered.com for more information.

Acknowledgments

A textbook is the collaborative result of hard work by many dedicated individuals, including students, colleagues, editors and editorial staff, graphics experts, and many others. First, we thank our family and friends for their support and encouragement while we spent countless hours on this project. Without your understanding and patience, this book would not have been possible.

We gratefully acknowledge the help of many talented people at Benjamin Cummings, particularly the editorial staff. We thank Becky Ruden for believing in the book's mission and for her vision in supporting and improving the text. We thank Project Editor Leata Holloway for keeping us on schedule and for her attention to detail, patience, enthusiasm, editorial suggestions, and great energy for the project. Senior Acquisitions Editor Michael Gillespie has been a stabilizing force in moving the project forward at a critical time.

We thank Production Supervisor Lori Newman for guiding us through the production process, and Caroline Cummins for her help in identifying and securing key art and suggesting alternatives. We thank Judith Bucci, Production Manager at Cenveo Publisher Services, for her expert work. Derek Bacchus delivered a fresh, innovative cover design and we appreciate Derek's artistic skills!

Undergraduate students at Monmouth University read many drafts of the manuscript for the first and second editions and critiqued art scraps for clarity and content. We thank former students in BY 201, "Introduction to Biotechnology," for their honest reviews, error finding, and suggestions from a student's perspective. We also thank Monmouth University graduate Robert Sexton for contributing to the Career Profile section of Chapter 3. Our students inspire us to strive for better ways to teach and help them understand the wonders of biotechnology. We applaud you for your help in creating what we hope future students will deem to be a student-friendly textbook.

We also thank Mr. Peter Kim for his suggestions on the business organization of a biotechnology company, Dr. Daniel Rudolph for contributing to the Career Profile in Chapter 5 ("Microbial Biotechnology")

and Gef Flimlin for contributing to the Career Profile in Chapter 10 ("Aquatic Biotechnology").

Finally, *Introduction to Biotechnology* has greatly benefited from the valued input of many colleagues and instructors who helped us in aiming for the highest levels of scientific accuracy, clarity, and pedagogical insight, offering suggestions for improvements in each chapter. The many instructors who have developed biotechnology courses and programs, and enthusiastically teach majors and nonmajors about biotechnology, provided reviews of the text and art that have been invaluable in shaping this textbook. Your constructive criticism helped us to revise drafts of each chapter, and your words of praise helped to inspire us to move ahead. All errors or omissions in the text are our responsibility. We thank you all and look forward to your continued feedback. Reviewers of *Introduction to Biotechnology*, include:

For the third edition:

James Crowder *Brookdale Community College*
Mary Colavito *Santa Monica College*
Craig Fenn *Housatonic Community College*
John Goudie *Kalamazoo Area Math and Science Center*
James Hewlett *Finger Lakes Community College*
Kevin Lampe *Montgomery County Community College*
Melanie Lenahan *Raritan Valley Community College*
Timothy Metz *Campbell University*
Stan Metzenberg *California State University, Northridge*
Melissa Rowland-Goldsmith *Santiago Canyon College*
Salvatore Sparace *Clemson University*
Lisa Werner *Pima Community College*
Dave Westenberg *Missouri University of Science and Technology*
Angela Wheeler *Austin Community College*

Past edition reviewers:

D. Derek Aday *Ohio State University*
Marcie Baer *Shippensburg University*
Joan Barber *Delaware Tech & Community College*

Theresa Beaty   *LeMoyne College*
Steve Benson   *California State University East Bay*
Peta Bonham-Smith   *University of Saskatchewan*
Krista Broten   *University of Saskatchewan*
Heather Cavenagh   *Charles Sturt University*
Ming-Mei Chang   *SUNY Genesco*
Jim Cheaney   *Iowa State University*
Wesley Chun   *University of Idaho*
Peter Eden   *Marywood University*
Mary A. Farwell   *East Carolina University*
Timothy S. Finco   *Agnes Scott College*
Mark Flood   *Fairmont State College*
Kathryn Paxton George   *University of Idaho*
Joseph Gindhart   *University of Massachusetts, Boston*
Jean Hardwick   *Ithaca College*
George Hegeman   *Indiana University, Bloomington*
Anne Helmsley   *Antelope Valley College*
David Hildebrand   *University of Kentucky*
Paul Horgen   *University of Toronto at Mississauga*
James Humphreys   *Seneca College of Applied Arts &
   Technology*
James T. Hsu   *Lehigh University*
Tom Ingebritsen   *Iowa State University*
Lisa Johansen   *University of Colorado, Denver*
Ken Kubo   *American River College*
Michael Lawton   *Rutgers University*
Theodore Lee   *SUNY Fredonia*
Edith Leonhardt   *San Francisco City College*
Lisa Lorenzen Dahl   *Iowa State University*
Caroline Mackintosh   *University of Saint Mary*
Keith McKenney   *George Mason University*
Toby Mapes   *A B Tech Community College*
Patricia Phelps   *Austin Community College*
Robert Pinette   *University of Maine*
Ronald Raab   *James Madison University*
Lisa Rapp   *Springfield Technical Community College*
Melody Ricci   *Victor Valley College*
Stephen Rood   *Fairmont State College*
Bill Sciarrapa   *Rutgers University*
Alice Sessions   *Austin Community College*

Carl Sillman   *Pennsylvania State University*
Teresa Singleton   *Delaware State University*
Salvatore A. Sparace   *Clemson University*
Sharon Thoma   *Edgewood College*
Danielle Tilley   *Seattle Community College*
Janice Toyoshima   *Evergreen Valley College*
Jagan Valluri   *Marshall University*
Dennis Walsh   *Massachusetts Bay Community
   College*
Lianna Wong   *Santa Clara University*
Brooke Yool   *Ohlone College*
Mike Zeller   *Iowa State University*

Whether you are a student or instructor, we invite your comments and suggestions for improving the next edition of *Introduction to Biotechnology*. Please write to us at the following addresses below or contact us via e-mail at bc.feedback@pearson.com.

Bill Thieman
Ventura College
Department of Biology
4667 Telegraph
Ventura, CA 93003
BThieman@vcccd.edu

Michael Palladino
Monmouth University
School of Science
Department of Biology
400 Cedar Avenue
West Long Branch, NJ 07764
mpalladi@monmouth.edu
www.monmouth.edu/mpalladi

As students ourselves, we too continue to learn about biotechnology every day. We wish you great success in your explorations of biotechnology!

W. J. T.
M. A. P.

Contents

CHAPTER 4
Proteins as Products 100

CHAPTER 5
Microbial Biotechnology 122

CHAPTER 6
Plant Biotechnology 158

CHAPTER 7
Animal Biotechnology   175

CHAPTER 8
DNA Fingerprinting and Forensic Analysis   193

CHAPTER 9
Bioremediation   209

CHAPTER **13**
Ethics and Biotechnology   321

The Biotechnology Century and Its Workforce

After completing this chapter, you should be able to:

- Define biotechnology and understand the many scientific disciplines that contribute to biotechnology.

- Provide examples of historic and current applications of biotechnology and its products.

- List and describe different types of biotechnology and their applications.

- Provide examples of potential advances in biotechnology.

- Discuss how medical diagnosis will change as a result of biotechnology and provide examples of how data from the Human Genome Project will be used to diagnose and treat human disease conditions.

- Understand that there are pros and cons to biotechnology and many controversial issues in this field.

- Describe career categories and options in biotechnology and ways to explore them.

- Evaluate the specific skills needed to fill an opening in a biotechnology company.

- Understand the basics of how a biotechnology company is started, funded, and valued, and describe the organizational structure of a typical biotech company.

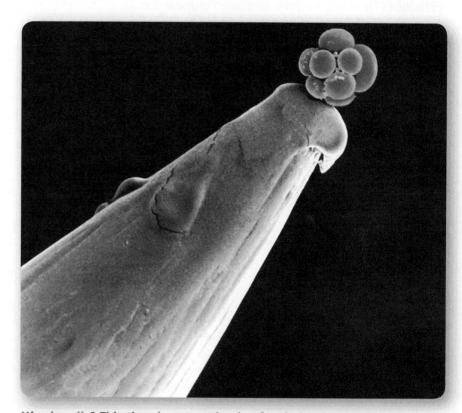

Miracle cells? This tiny cluster on the tip of a pin is a human embryo approximately three days after fertilization. Some scientists believe that stem cells contained within embryos may have the potential for treating and curing a range of diseases in humans through biotechnology. Use of these cells is also one of the most controversial topics in biotechnology.

If you have ever eaten a corn chip, you may have been affected by biotechnology. Don't eat chips? How about sour cream, yogurt, cheese, or milk? In this century, more and more of the foods we eat will be produced by organisms that have been genetically altered through biotechnology. Such **genetically modified (GM) foods** have become a controversial topic over the last few years, as have human embryos such as the one shown in the opening photo. This chapter was designed to provide you with a basic introduction to the incredible range of biotechnology topics that you will read about in this book. As you will see, biotechnology is a multidisciplinary science with many powerful applications and great potential for future discoveries.

In this chapter, we do not offer a comprehensive review of the history of biotechnology and its current applications. Instead, we present a brief introduction and overview of many topics that we discuss in greater detail in future chapters. We begin by defining biotechnology and presenting an overview of the many scientific disciplines that contribute to this field. We highlight both historic and modern applications and define the different types of biotechnology that you will study in this book. At the end of the chapter, we discuss aspects of the biotechnology workforce and skills required to work in the industry. Be sure that you are familiar with the different types of biotechnology and the key terms presented in this chapter, because they will form the foundation for your future studies.

FORECASTING THE FUTURE

Advancements in biotechnology have progressed significantly since sequencing of the human genome was completed in 2001, as you will read in the "Forecasting the Future" boxes at the beginning of each chapter. First, we'll see how genomes are being analyzed and the roles genomic studies have in biotechnology (Chapter 2). We'll explore how recombined genes hold the potential to create new diagnostics and treatments for diseases (Chapter 3) and how protein drugs that are hard to administer and assimilate can be absorbed through the skin when packaged in nanovesicles specifically designed for timed release (Chapter 4). Bacteria can be used to make biofuels and ways to use synthetic genomes to create bacteria with novel properties will be discussed (Chapter 5). We'll discuss how collaboration between nonprofits and biotech companies is accelerating the development and public acceptance of transgenic plants (Chapter 6). Vaccine development is one area of animal biotechnology that is likely to reduce the cost of health care for humans and animals (Chapter 7). We'll see how DNA is used for identification in a number of ways including disease, paternity, and security (Chapter 8). The creation of organisms to convert toxic materials to harmless substrates and other bioremediation techniques has received a boost from studies of the bacteria degrading the oil released from the 2010 Gulf of Mexico spill, as we'll see (Chapter 9). Because we live on a planet that is largely water, aquatic biotechnology is becoming the source of new products, drugs, and methods (Chapter 10). New drug delivery methods deserve careful reading (Chapter 11). Revised federal regulations have accelerated approvals of new biotech discoveries, but they have also resulted in tighter safeguards (Chapter 12). Finally, we'll see how ethical thought and meaningful dialogue is a necessity for biotech improvements (Chapter 13).

1.1 What Is Biotechnology and What Does It Mean to You?

Have you ever eaten a Flavr Savr tomato, been treated with a monoclonal antibody, received tissue grown from embryonic stem cells, or seen a "knockout" mouse? Have you ever had a flu shot, known a person with diabetes who requires injections of insulin, taken a home pregnancy test, used an antibiotic to treat a bacterial infection, sipped a glass of wine, eaten cheese, or made bread? Although you may not have experienced any of the scenarios on the first list, at least one of the items on the second list must be familiar to you. If so, you have experienced the benefits of biotechnology.

Can you imagine a world free of serious diseases, where food is abundant for everyone and the environment is free of pollution? These scenarios are exactly what many people in the biotechnology industry envision as they dedicate their lives to this exciting science. Although you may not fully understand the range of disciplines and scientific details of biotechnology, you have experienced biotechnology firsthand. **Biotechnology** is broadly defined as the science of using living organisms, or the products of living organisms, for human benefit (or to benefit human surroundings)—that is, to make a product or solve a problem. Remember this definition. As you learn more about biotechnology, we will expand and refine this definition with historical examples and modern applications from everyday life and look ahead to the biotechnology future.

You would be correct in thinking that biotechnology is a relatively new discipline that is only recently getting a lot of attention; however, it may surprise you to know that in many ways this science involves several ancient practices. As we discuss in the next section, old and new practices in biotechnology make this field one of the most rapidly changing and exciting areas of science. It affects our everyday lives and will become even more important during this century—what some have called the "century of biotechnology."

A Brief History of Biotechnology

If you asked your friends and family to define biotechnology, their answers might surprise you. They may have no idea of what biotechnology is. Perhaps they might tell you that biotechnology involves serious-looking scientists in white lab coats secretively carrying out sophisticated gene-cloning experiments in expensive laboratories. When pressed for details, however, your friends probably will not be able to tell you how these "experiments" are done, what information is gained from such work, and how this knowledge is used. Although DNA cloning and the genetic manipulation of organisms are exciting modern-day techniques, biotechnology is not a new science. In fact, many applications represent old practices with new methodologies. Humans have been using other biological organisms for their benefit in many processes for several thousand years. Historical accounts have shown that the Chinese, Greeks, Romans, Babylonians, and Egyptians, among many others, have been involved in biotechnology since about 2000 B.C.

Biotechnology does not mean hunting and gathering animals and plants for food; however, the domestication of animals such as sheep and cattle for use as livestock is a classic example of biotechnology. Our early ancestors also took advantage of microorganisms and used **fermentation** to make breads, cheeses, yogurts, and alcoholic beverages such as beer and wine. During fermentation, some strains of yeast decompose sugars to derive energy, and in the process they produce ethanol (alcohol) as a waste product. When bread dough is being made, yeast (*Saccharomyces cerevisiae*, commonly called baker's yeast) is added to make the dough rise. This occurs because the yeast ferments sugar-releasing carbon dioxide, which causes the dough to rise and creates holes in the bread. Alcohol produced by the yeast evaporates when the bread is cooked. If you make bread or pizza dough at home, you have probably added store-bought *S. cerevisiae* from an envelope or jar to your dough mix. Similar processes are very valuable for the production of yogurts, cheeses, and beverages.

For thousands of years, humans have used **selective breeding** as a biotechnology application to improve production of crops and livestock used for food purposes. In selective breeding, organisms with desirable features are purposely mated to produce offspring with the same desirable characteristics. For example, crossbreeding plants that produce the largest, sweetest, and most tender ears of corn is a good way for farmers to maximize their land to produce the most desirable crops (**Figure 1.1a**). Similar breeding techniques are used with farm animals, including turkeys (to breed birds producing the largest and most tender breast

(a)

(b)

(c)

FIGURE 1.1 Selective Breeding Is an Old Example of Biotechnology That Is Still Common Today (a) Corn grown by selective breeding. From left to right is teosinte (*Zea canina*), selectively bred hybrids, and modern corn (*Zea mays*). (b) A normal zebrafish and (c) "Casper," a transparent zebrafish produced by selective breeding.

meat), cows, chickens, and pigs. Other examples include breeding wild species of plants, such as lettuces and cabbage, over many generations to produce modern plants that are cultivated for human consumption. Many of these approaches are really genetic applications of biotechnology. Without realizing it—and without expensive labs, sophisticated equipment, PhD-trained scientists, and well-planned experiments—humans have been manipulating genes for hundreds of years.

By selecting plants and animals with desirable characteristics, humans are choosing organisms with useful genes and taking advantage of their genetic potential for human benefit. Scientists at the Children's Hospital of Boston produced a transparent zebrafish named Casper (Figure 1.1c). Casper was created by mating a zebrafish mutant that lacked reflective pigment with a zebrafish that lacked black pigment. Zebrafish are important experimental **model organisms**, and scientists believe that Casper will be important for drug testing and in vivo studies of stem cells and cancer. Casper has already proven to be valuable for studying how cancer cells spread: scientists injected fluorescent tumor cells into the fish's abdominal cavity and were able to track the migration of those cells to specific locations in the body.

One of the most widespread and commonly understood applications of biotechnology is the use of antibiotics. In 1928, Alexander Fleming discovered that the mold *Penicillium* inhibited the growth of a bacterium called *Staphylococcus aureus,* which causes skin disease in humans. Subsequent work by Fleming led to the discovery and purification of the **antibiotic** penicillin. Antibiotics are substances produced by microorganisms that will inhibit the growth of other microorganisms. In the 1940s, penicillin became widely available for medicinal use to treat bacterial infections in humans. In the 1950s and 1960s, advances in biochemistry and cell biology made it possible to purify large amounts of antibiotics from many different strains of bacteria. **Batch (large-scale) processes**—in which scientists can grow bacteria and other cells in large amounts and harvest useful products in large batches—were developed to isolate commercially important molecules from microorganisms (explained further in Chapter 4).

Since the 1960s, rapid development of our understanding of genetics and molecular biology has led to exciting new innovations and applications in biotechnology. As we have begun to unravel the secrets of DNA structure and function, new technologies have led to **gene cloning,** the ability to identify and reproduce a gene of interest, and **genetic engineering,** manipulating the DNA of an organism. Through genetic engineering, scientists are able to combine DNA from different sources. This process, called **recombinant DNA technology,** is used to produce many proteins of medical importance, including insulin, human growth hormone, and blood-clotting factors. From its inception, recombinant DNA technology has dominated many important areas of biotechnology, and as you will soon learn, many credit recombinant DNA technology with starting modern biotechnology as an industry. You will learn that recombinant DNA technology has led to hundreds of applications, including the development of disease-resistant plants, food crops that produce greater yields, "golden rice" engineered to be more nutritious, and genetically engineered bacteria capable of degrading environmental pollutants.

Gene cloning and recombinant DNA technology have had a tremendous impact on human health through the identification of thousands of genes involved in human genetic disease conditions. The ultimate gene cloning project, the **Human Genome Project,** was an international effort that began in 1990. A primary goal of the Human Genome Project was to identify all genes—the **genome**—contained in the DNA of human cells and to map their locations to each of the 24 human chromosomes (chromosomes 1 to 22 and the X and Y chromosomes). The Human Genome Project has provided unlimited potential for the development of new diagnostic approaches for detecting disease and molecular approaches for treating and curing human genetic disease conditions.

Just imagine the possibilities. The Human Genome Project can tell us the chromosomal location and code of *every* human gene, from genes that control normal cellular processes and determine characteristics such as hair color, eye color, height, and weight to the myriad of genes that cause human genetic diseases (**Figure 1.2**). As a result of the Human Genome Project, new knowledge about human genetics will have tremendous and wide-ranging effects on basic science and medicine in the near future. The 1000 Genomes Project has already contributed significantly to the understanding of human genetics by comparison with other organisms' genomes (see Chapter 3). In many ways, in our efforts to understand the functions of all human genes, we are unraveling one of the great unsolved mysteries in biology. We explore the mysteries of the genome in several chapters.

As you have just learned, biotechnology has a long and rich history. Future chapters are dedicated to exploring advances in biotechnology and looking ahead to what the future holds. As you study biotechnology, you will be introduced to what may seem to be an overwhelming number of terms and definitions. Be sure to use the index and glossary at the end of the book to help you find and define important terms.

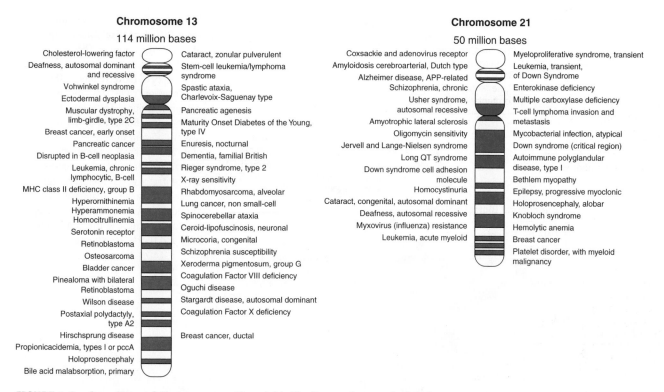

Chromosome 13

114 million bases

Cholesterol-lowering factor
Deafness, autosomal dominant
and recessive
Vohwinkel syndrome
Ectodermal dysplasia
Muscular dystrophy,
limb-girdle, type 2C
Breast cancer, early onset
Pancreatic cancer
Disrupted in B-cell neoplasia
Leukemia, chronic
lymphocytic, B-cell
MHC class II deficiency, group B
Hyperornithinemia
Hyperammonemia
Homocitrullinemia
Serotonin receptor
Retinoblastoma
Osteosarcoma
Bladder cancer
Pinealoma with bilateral
Retinoblastoma
Wilson disease
Postaxial polydactyly,
type A2
Hirschsprung disease
Propionicacidemia, types I or pccA
Holoprosencephaly
Bile acid malabsorption, primary

Cataract, zonular pulverulent
Stem-cell leukemia/lymphoma
syndrome
Spastic ataxia,
Charlevoix-Saguenay type
Pancreatic agenesis
Maturity Onset Diabetes of the Young,
type IV
Enuresis, nocturnal
Dementia, familial British
Rieger syndrome, type 2
X-ray sensitivity
Rhabdomyosarcoma, alveolar
Lung cancer, non small-cell
Spinocerebellar ataxia
Ceroid-lipofuscinosis, neuronal
Microcoria, congenital
Schizophrenia susceptibility
Xeroderma pigmentosum, group G
Coagulation Factor VIII deficiency
Oguchi disease
Stargardt disease, autosomal dominant
Coagulation Factor X deficiency

Breast cancer, ductal

Chromosome 21

50 million bases

Coxsackie and adenovirus receptor
Amyloidosis cerebroarterial, Dutch type
Alzheimer disease, APP-related
Schizophrenia, chronic
Usher syndrome,
autosomal recessive
Amyotrophic lateral sclerosis
Oligomycin sensitivity
Jervell and Lange-Nielsen syndrome
Long QT syndrome
Down syndrome cell adhesion
molecule
Homocystinuria
Cataract, congenital, autosomal dominant
Deafness, autosomal recessive
Myxovirus (influenza) resistance
Leukemia, acute myeloid

Myeloproliferative syndrome, transient
Leukemia, transient,
of Down Syndrome
Enterokinase deficiency
Multiple carboxylase deficiency
T-cell lymphoma invasion and
metastasis
Mycobacterial infection, atypical
Down syndrome (critical region)
Autoimmune polyglandular
disease, type I
Bethlem myopathy
Epilepsy, progressive myoclonic
Holoprosencephaly, alobar
Knobloch syndrome
Hemolytic anemia
Breast cancer
Platelet disorder, with myeloid
malignancy

FIGURE 1.2   Gene Maps of Chromosomes 13 and 21 The Human Genome Project has led to the identification of nearly all human genes and has mapped their location on each chromosome. The maps of chromosomes 13 and 21 indicate genes known to be involved in human genetic disease conditions. Identifying these genes is an important first step toward developing treatments for many genetic diseases.

Biotechnology: A Science of Many Disciplines

One of the many challenges you will encounter as you study biotechnology will be trying to piece together complex information from many different scientific disciplines. It is impossible to talk about biotechnology without considering the important contributions of the different fields of science. Although a major focus of biotechnology involves molecular biology techniques, biotechnology is not a single, narrow discipline of study. It is an expansive, *interdisciplinary* field that absolutely relies on biology, chemistry, mathematics, computer science, and engineering in addition to other disciplines such as philosophy and economics. Later in this chapter, we consider how biotechnology provides a wealth of employment opportunities for people who have been trained in diverse fields.

Figure 1.3, on the next page, provides a diagrammatic view of the many disciplines that contribute to biotechnology. Notice that the "roots" are primarily formed by work in the **basic sciences**—research into

fundamental processes of living organisms at the biochemical, molecular, and genetic levels. When pieced together, basic science research from many areas, with the help of computer science, can lead to genetic engineering approaches. At the top of the tree, applications of genetic engineering can be put to work to create a product or process to help humans or our living environment. Many of these processes have yet to be developed and await the intuitive participation of people working in biotechnology today.

A simplified example of the interdisciplinary nature of biotechnology can be summarized as follows. At the basic science level, scientists conducting research in microbiology at a college, university, government agency, or public or private company may discover a gene or gene product in bacteria that shows promise as an agent for treating a disease condition. Typically, biochemical, molecular, and genetic techniques would be used to determine the role of this gene. This process also involves using computer science in sophisticated ways to study the sequence of a gene and analyze the structure of the protein produced by the gene (part of a field called **bioinformatics**).

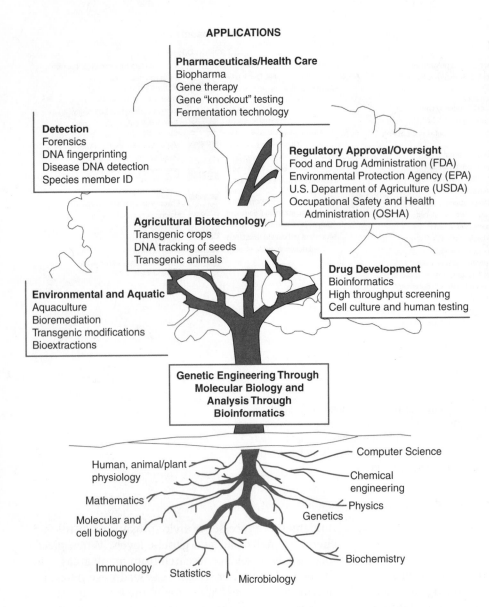

APPLICATIONS

Pharmaceuticals/Health Care
Biopharma
Gene therapy
Gene "knockout" testing
Fermentation technology

Detection
Forensics
DNA fingerprinting
Disease DNA detection
Species member ID

Regulatory Approval/Oversight
Food and Drug Administration (FDA)
Environmental Protection Agency (EPA)
U.S. Department of Agriculture (USDA)
Occupational Safety and Health
Administration (OSHA)

Agricultural Biotechnology
Transgenic crops
DNA tracking of seeds
Transgenic animals

Drug Development
Bioinformatics
High throughput screening
Cell culture and human testing

Environmental and Aquatic
Aquaculture
Bioremediation
Transgenic modifications
Bioextractions

**Genetic Engineering Through
Molecular Biology and
Analysis Through
Bioinformatics**

Computer Science

Chemical
engineering

Human, animal/plant
physiology

Physics

Mathematics

Genetics

Molecular and
cell biology

Biochemistry

Immunology

Statistics

Microbiology

FIGURE 1.3 The Biotechnology Tree: Different Disciplines Contribute to Biotechnology The basic sciences are the foundation or "roots" of all aspects of biotechnology. The central focus or "trunk" for most biotechnological applications is genetic engineering. Branches of the tree represent different organisms, technologies, and applications that "stem" from genetic engineering and bioinformatics, central aspects of most biotechnological approaches. Regulation of biotechnology occurs through governmental agencies like the FDA, USDA, EPA, and OSHA (see Chapter 12).

Once basic research has arrived at a detailed understanding of this gene, the gene may then be used in a variety of ways, including drug development, agricultural biotechnology, and environmental and marine applications (Figure 1.3). The many applications of biotechnology will become much clearer as we cover each area. At this point keep in mind that biotechnology is a science that requires skills from many disciplines.

Products of Modern Biotechnology

Throughout the book, we consider many cutting-edge and innovative products and applications of biotechnology. We look not only at products for human use but also at biotechnology applications of microbiology, marine biology, and plant biology, among other disciplines. The multitude of biotechnology products currently available are far too numerous to mention in this introductory chapter; however, many products reflect the current needs of humans—for example, pharmaceutical production, creating drugs for the treatment of human health conditions. In fact, more than 65 percent of biotechnology companies in the United States are involved in pharmaceutical production. In 1982, the California biotechnology company **Genentech,** widely regarded as the world's first biotech company, received approval for recombinant insulin, used for the treatment of diabetes, as the first biotechnology product for human benefit (**Figure 1.4**). There are now several hundred drugs, vaccines, and diagnostics on the market with more than 300 biotechnology medicines in development targeting over 200 diseases. As shown in **Figure 1.5**, drug development by the biotechnology industry is focused

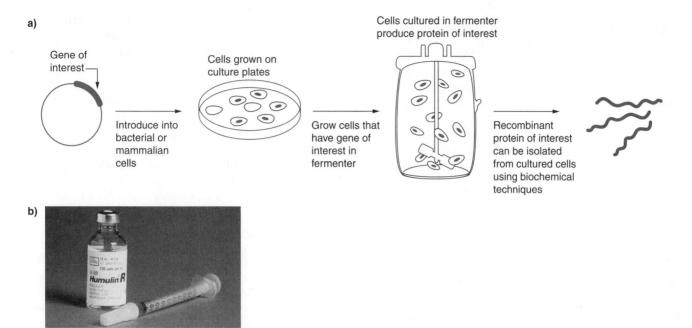

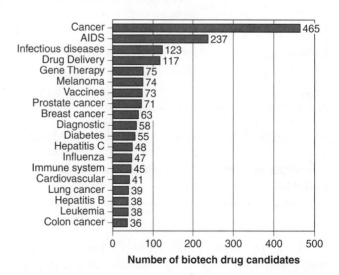

FIGURE 1.4   Using Genetically Modified Cultured Cells to Make a Protein of Interest   Genes of interest can be introduced into bacterial or mammalian cells. Such cells can be grown using cell culture techniques. Recombinant proteins isolated from these cells are used in hundreds of different biotechnology applications. In this example, mammalian cells are shown, but this process is also commonly carried out using bacteria.

on combating major diseases that affect humans, and over half of the new drugs in the development "pipeline" are designed to treat cancer. Table 1.1 on the next page provides a brief list of some of the top-selling biotechnology drugs and the companies that developed them. Diagnosis and/or treatment of a variety of human diseases and disorders—including acquired immunodeficiency syndrome (AIDS), stroke, diabetes, and cancer—make up the bulk of biotechnology products on the market.

Many of the most widely used products of biotechnology are proteins created by gene cloning (Table 1.2, next page). These proteins are called **recombinant proteins** because they are produced by gene-cloning techniques. For example, the majority of these proteins are produced from human genes inserted into bacteria to make the recombinant proteins used to treat human disease conditions.

How genes are cloned and used to produce proteins of interest is discussed in great detail later (Chapter 3). As an introduction to this idea, consider the diagram shown in Figure 1.4. As you will soon learn, scientists can identify a gene of interest and put it into bacterial or mammalian cells that are grown by a technique called **cell culture.** In cell culture, cells are grown in dishes or flasks within liquid culture media designed to provide the nutrients necessary for cell growth. Large culture

containers, called **fermenters** or **bioreactors,** are used to mass produce cells containing the DNA of interest. Scientists can harvest the protein produced by the gene of interest from these cells and use it in applications such as those described in Table 1.2.

FIGURE 1.5   Investigational Biotechnology Drugs by Disease Category   The production of drugs to combat cancer dominates the biotechnology industry's interest in treating human disease, with AIDS-related research and treatment of infectious diseases such as flu also near the top of this list.

TABLE 1.1   TOP 10 BIOTECHNOLOGY DRUGS (WITH SALES OVER $1 BILLION)

Drug	Developer	Function (Treatment of Human Disease Conditions)
Enbrel	Amgen & Wyeth	Rheumatoid arthritis
Remicade	Johnson & Johnson	Rheumatoid arthritis
Rituxan	Roche	Non-Hodgkin's lymphoma
Avastin	Roche	Colon cancer
Herceptin	Roche	Breast cancer
Humira	Abbott Labs	Rheumatoid arthritis
Levenox	sanofi-aventis	Blood clots
Lantus	sanofi-aventis	Diabetes
Aranesp	Amgen	Anemia

TABLE 1.2   EXAMPLES OF PROTEINS MANUFACTURED FROM CLONED GENES

Product	Application
Blood factor VIII (clotting factor)	Treat hemophilia
Epidermal growth factor	Stimulate antibody production in patients with immune system disorders
Growth hormone	Correct pituitary deficiencies and short stature in humans; other forms are used in cows to increase milk production
Insulin	Treat diabetes
Interferons	Treat cancer and viral infections
Interleukins	Treat cancer and stimulate antibody production
Monoclonal antibodies	Diagnose and treat a variety of diseases including arthritis and cancer
Tissue plasminogen activator	Treat heart attacks and stroke

If Figure 1.5 and Tables 1.1 and 1.2 have not provided you with convincing examples of the importance of biotechnology for human health, consider that, in the near future, genes may be routinely introduced into human cells as **gene therapy** approaches are employed to treat and cure human disease conditions. Genetics and tissue engineering

may lead to the ability to grow organs for transplantation that would only rarely be rejected by their recipients. New biotechnology products from marine organisms are being used to treat cancers, strokes, and arthritis. Specialized proteins, needed in quantity, will continue to come from additional gene transfer to animals (like ATryn from transgenic goats; see Chapter 7). Modern advances in medicine, driven by new knowledge from the Human Genome Project, will result in healthier lives and potentially increase the human life span.

Ethics and Biotechnology

Just as in any other type of technology, the powerful applications and potential promise of biotechnology applications raises many ethical concerns, and it should be no surprise to you that not everyone is a fan of biotechnology. A wide range of ethical, legal, and social implications of biotechnology are a cause of great debate and discussion among scientists, the general public, clergy, politicians, lawyers, and many others around the world (**Figure 1.6**). Throughout our discussion, we present ethical, legal, and social issues for you to consider. Increasingly, you will be faced with ethical issues of biotechnology that may influence you directly.

For instance, now that organism cloning has been accomplished in mammals such as sheep, cows, and monkeys, some have suggested that human cloning be permitted. How do you feel about this? If, in the future, you and your spouse were unable to have children by any other means, would you want the opportunity to create a baby by cloning a replica of yourself? As another example, the introduction of synthetic genomes into organisms and the production of synthetic proteins will undoubtedly add additional controversy to the risks and benefits of these techniques and place a greater emphasis on a knowledgeable public that is expected to decide these issues (see Chapter 5 for a discussion of synthetic organisms).

If you choose to work in biotechnology, you will need to develop teamworking skills that allow for differences in opinion on many ethical issues, necessitating an understanding of the basis for the arguments supported by some of your colleagues. Look for the "You Decide" boxes in each chapter, where we present scenarios or ethical dilemmas for you to consider. Realize that there are pros and cons and controversial issues associated with almost every application in biotechnology. Our goal is not to tell you *what* to think but to empower you with knowledge you can use to make your own decisions.

FIGURE 1.6   Biotechnology Is a Controversial Science That Presents Many Ethical Dilemmas

1.2 Types of Biotechnology

Now that you have learned about the many areas of science that contribute to biotechnology, you should recognize that there are many different types of biotechnology. Consider this section an introduction to what you will learn in greater detail in the chapters that follow.

Microbial Biotechnology

In Chapter 5, we explore the many ways in which microbial biotechnology affects society. As we discussed previously, the use of yeast for making beer and wine is one of the oldest applications of biotechnology. By manipulating microorganisms such as bacteria and yeast, microbial biotechnology has created better enzymes and organisms for making many foods, simplifying manufacturing and production processes, and making decontamination processes for the removal of industrial waste products more efficient. Microbes are used to make vaccines and to clone and produce batch amounts of important proteins used in human medicine, including insulin and growth hormone. We will also explore strategies used to detect microbes for diagnostic purposes in humans, food samples, and other sources and approaches to detect and combat microbes as possible bioweapons.

Agricultural Biotechnology

Chapter 6 is dedicated to plant biotechnology and agricultural applications of biotechnology. In "ag biotech," we examine a range of topics from genetically engineered, pest-resistant plants that do not need to

be sprayed with pesticides to foods with higher protein or vitamin content and drugs developed and grown as plant products. Agricultural biotechnology is already a big business that is rapidly expanding. The United Nations Food and Agriculture Organization has predicted that feeding a world population of 9.1 billion people in 2050 will require raising overall food production by some 70 percent (nearly 100 percent in the developing countries). Agricultural biotechnology provides solutions for today's farmers in the form of plants that are more environmentally friendly while yielding more per acre, resisting diseases and insect pests, and reducing farmers' production costs.

Genetic manipulation of plants has been used for over 20 years to produce genetically engineered plants with altered growth characteristics such as drought resistance, tolerance to cold temperature, and greater food yields. Research conducted during the past 10 years clearly demonstrates that plants can be engineered to produce a wide range of pharmaceutical proteins in a broad array of crop species and tissues. The use of plants as sources of pharmaceutical products is an application of agricultural biotechnology commonly called **molecular pharming.** For example, tobacco is a nonfood crop that has been the subject of many years of breeding and agronomic research. Tobacco plants have been engineered to produce recombinant proteins in their leaves, and these plants can be grown in large fields for molecular pharming But gene transfer to nontarget plants has already occurred and some varieties of "super weeds" have been documented, resulting in changes in the use of some bioengineered plants. The bioethics of these solutions has already created strong opinions on both sides about the continued development and use of GM plants.

The Presidential Advanced Energy Initiative of 2007 to allow biofuels to ease the "addiction" of the United States to foreign oil has been interpreted by advocates as meaning that 25 percent of U.S. energy would have to come from arable land by 2025. This goal will require significant advances in biotechnology to provide bioethanol sources other than the corn kernel, since this is not an efficient energy source. Agricultural waste, prairie grass, and other high-cellulose sources (including corn by-products) will have to become efficient sources of energy through new decomposition and fermentation methods developed by biotechnology. These challenges are well under way.

Animal Biotechnology

In Chapter 7, we examine many areas of animal biotechnology, one of the most rapidly changing and exciting areas of biotechnology. Animals can be used as "bioreactors" to produce important products. For

YOU DECIDE

Genetically Modified Foods: To Eat or Not to Eat?

Many experts believe that genetically modified foods are safe and that they will provide significant benefits in the future. But public opinion on the use and safety of GM foods is mixed. About one third of Americans polled believe that using scientific methods—such as recombinant DNA technology—to enhance the flavor, color, nutrition, or freshness of foods is wrong. Other polls indicate that opposition to the use of GM foods may be as high as 50 percent. Skeptics frequently comment that "GM foods are against nature," and some people worry about potential health effects such as food allergies.

But it appears that Americans expect possible benefits in the future. In the United States, 45 percent of respondents in a 2010 poll indicated that they would accept GM foods; but the percentages of those who would accept them are much lower in Japan and Europe, where the controversy is greater. If given a choice, many people have indicated that they would look for another product rather than choose food labeled as genetically modified. This attitude raises another controversy that we will consider later in the book, which is whether GM foods should be labeled as such.

Current U.S. regulations require labeling only if GM foods pose a health risk or if the product's nutritional value has changed. A 2004 report by the National Academies, for example, found that "biotech crops do not pose any more health risks than do crops created by other techniques, and that food safety evaluations should be based on the resulting food product, not the technique used to create it."

"THE LOWER-PRICED ITEMS CONTAIN GENETICALLY-MODIFIED FOODS NOT YET APPROVED FOR HUMAN USE."

■ What do you think about the use of GM foods?
■ Would you be likely to buy GM foods if they were engineered to require fewer pesticide applications than "natural" foods?
■ What if GM foods stayed fresher longer?
■ What if they were more nutritious and less expensive?
■ How much risk should consumers be willing to take to reap the benefits of GM foods?

Consider making a list of the questions you would want answered before you took your first bite of a GM food product. GM foods—to eat or not to eat; you decide.

---

example goats, cattle, sheep, and chickens are being used as sources of medically valuable proteins such as **antibodies**—protective proteins that recognize and help body cells to destroy foreign materials. Antibody treatments are being used to help improve immunity in patients with immune system disorders. Many other human therapeutic proteins produced from animals are in use, yet most of these proteins are needed in quantities that exceed hundreds of kilograms. To achieve this large-scale production, scientists can create female **transgenic animals** that express therapeutic proteins in their milk. Transgenic animals contain genes from another source. For instance, human genes for clotting proteins can be introduced into goats for the production of these proteins in their milk.

Animals are also very important in basic research as **model organisms**. For instance, gene "knockout"

experiments, in which one or more genes are disrupted, can be helpful for learning about the function of a gene. The idea behind a knockout is to disrupt a gene and then, by looking at what functions are affected in an animal as a result of the loss of a particular gene, determine the role and importance of that gene. Because many of the genes found in animals (including mice and rats) are also present in humans, learning about gene function in animals can lead to a greater understanding of gene function in humans. Similarly, the design and testing of drug and genetic therapies in animals often leads to novel treatment strategies in humans.

In 1997, scientists and the general public expressed surprise, excitement, and reservations about the announcement that scientists at the Roslin Institute in Scotland had cloned the now-famous sheep called Dolly

FIGURE 1.7 Dolly, the First Mammal Produced by Nuclear Transfer Cloning Dolly poses with her surrogate mother. Dolly was created by cloning technologies that may result in promising new techniques for improving livestock and cloning commercially valuable animals such as those containing organs for human transplantation. Unfortunately, Dolly developed early complications and was euthanized in February 2003.

(**Figure 1.7**). Dolly was the first mammal created by a cell nucleus transfer process (see Chapter 7). Many other animals have been cloned since Dolly. In 2009, the FDA approved the first drug (an anticlotting protein) produced in animals. Although animal cloning has elicited fears and concern about the potential for human cloning, scientists, for a number of reasons, are generally excited about the techniques used to clone animals. For instance, these techniques may lead to the cloning of animals containing genetically engineered organs that can be transplanted into humans without fear of tissue rejection. Does a ready supply of donor organs of all types for all people who need organ transplants sound like a good plan to you? If so, not everyone agrees. Animal cloning and the controversies surrounding organism cloning are important subjects (discussed in Chapter 7).

Forensic Biotechnology

In Chapter 8, we discuss forensic biotechnology. **DNA fingerprinting**— a collection of methods for detecting an organism's unique DNA pattern—is a primary tool used in forensic biotechnology (**Figure 1.8**). Forensic biotechnology is a powerful tool for law enforcement that can lead to the inclusion or exclusion of a person from suspicion, based on DNA evidence. DNA fingerprinting can be accomplished using trace amounts of tissue, hair, blood, or body fluids left behind at a crime scene. It was first used in 1987 to convict a rapist in England but is now routinely introduced as evidence in

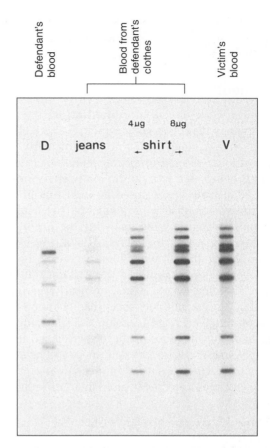

FIGURE 1.8 DNA Fingerprinting for a Murder Case This photo shows the results of DNA fingerprinting techniques (which you will learn about in Chapter 8) comparing the DNA from blood stains on a defendant's clothes to the DNA fingerprints of a victim's blood. DNA fingerprinting cannot always be used to determine definitely that an accused person has committed a crime. In this case, DNA fingerprinting provided evidence that the defendant could be linked to the crime scene, although it does not mean that the defendant was guilty of the murder.

court cases throughout the world to convict criminals as well as to free those wrongly accused of a crime.

DNA fingerprinting has many other applications, including use in paternity cases for pinpointing a child's father and identifying human remains. Another application is the DNA fingerprinting of endangered species. This has already reduced poaching and led to convictions of criminals by analyzing the DNA fingerprints of their "catches." Scientists also use DNA fingerprinting to track and confirm organisms that spread disease, such as *Escherichia coli* in contaminated meat, and to track diseases such as AIDS, meningitis, tuberculosis, Lyme disease, and the West Nile virus. Recently a French company even developed a gene expression test designed to determine if expensive food products contain cheap, substitute, mystery meats from species such as cats and eels. The need to develop tests for valuable species that can be used to fingerprint DNA has created a larger demand for small biotechnology companies that can develop these testing methods.

Bioremediation

In Chapter 9, we discuss **bioremediation,** the use of biotechnology to process and degrade a variety of natural and human-made substances, particularly those that contribute to environmental pollution. Bioremediation is being used to clean up many environmental hazards that have been caused by industrial progress. One of the most publicized examples of bioremediation in action occurred in 1989 following the *Exxon Valdez* oil spill in Prince William Sound, Alaska (**Figure 1.9**). By stimulating the growth of oil-degrading bacteria, which were already present in the Alaskan soil, many miles of shoreline were cleaned up nearly three times faster than they would have been had chemical cleaning agents alone been used. As you will learn, the rapid degradation by microbes of the dispersed oil droplets from the *Deep Water Horizon* spill in 2010 has already enabled research into natural oil-degrading organisms and the enzymes that may be used in a future spill.

Aquatic Biotechnology

In Chapter 10, we explore the vast biotechnology possibilities offered by water—the medium that covers the majority of our planet. One of the oldest applications of aquatic biotechnology is **aquaculture,** raising finfish or shellfish in controlled conditions for use as food sources. Trout, salmon, and catfish are among many important aquaculture species in the United States.

FIGURE 1.9 **Bioremediation in Action** Strains of the bacterium *Pseudomonas* were used to help clean Alaskan beaches following the *Exxon Valdez* oil spill. Scientists on this Alaskan beach are applying nutrients that will stimulate the growth of *Pseudomonas* to help speed up the bioremediation process.

FIGURE 1.10 **Aquatic Biotechnology Is an Emerging Science** From using aquaculture to raise shellfish and finfish for human consumption to isolating biologically valuable molecules from marine organisms for medical applications, aquatic biotechnology has the potential for an incredible range of applications. Shown here is a genetically engineered salmon (top) bred to grow to adult size for market sale in half the time of a normal salmon (bottom).

Aquaculture is growing in popularity throughout the world, especially in developing countries. It has recently been estimated that close to 50 percent of all fish consumed by humans worldwide are now produced by aquaculture.

In recent years, a wide range of fascinating new developments in aquatic biotechnology have emerged. These include the use of genetic engineering to produce disease-resistant strains of oysters and vaccines against viruses that infect salmon and other finfish. Transgenic salmon have been created that overproduce growth hormone, leading to extraordinary growth rates over short growing periods and thus decreasing the time and expense required to grow salmon for market sale (**Figure 1.10**).

The uniqueness of many aquatic organisms is another attraction for biotechnologists. In our oceans, marine bacteria, algae, shellfish, finfish, and countless other organisms live under some of the harshest conditions in the world. Extreme cold, pressure from living at great depths, high salinity, and other environmental constraints are hardly a barrier because aquatic organisms have adapted to their difficult environments. As a result, such organisms are thought to be rich and valuable sources of new genes, proteins, and metabolic processes that may have important human applications and benefits. **Bioprospecting** efforts are ongoing around the world to identify aquatic organisms with novel properties that may be exploited for commercial purposes. For instance, certain species of marine plankton and snails have been found to be rich sources of antitumor and anticancer molecules. Intensive research efforts are under way to better understand the wealth of potential biotechnology applications that our aquatic environments may harbor.

Medical Biotechnology

In Section 1.1, we introduced the concept that many recombinant proteins are being manufactured for human medical applications; however, this is just one example of **medical biotechnology.** Chapter 11 covers a range of different applications of medical biotechnology. From preventative medicine to the diagnosis of health and illness to the treatment of human disease conditions, medical biotechnology has resulted in an amazing array of applications designed to improve human health. Over 325 million people worldwide have been helped by drugs and vaccines developed through biotechnology. Although many powerful applications have already been designed and are currently being applied, the biotechnology century will see some of the greatest advances in medical biotechnology in history.

It seems as though hardly a week goes by without news of a genetic breakthrough such as the discovery of a human gene involved in a disease process. Television, newspapers, and popular magazines all report important discoveries of new genes and other headlines involving DNA (**Figure 1.11**). Every day, new information from the Human Genome Project is helping scientists identify defective genes and decipher the details of genetic diseases such as sickle cell anemia, Tay-Sachs disease, cystic fibrosis, and cancer as well as and forms of infertility, to give just a few examples. The Human Genome Project has resulted in new techniques for genetic testing to identify defective genes and genetic disorders, and we explore many of these techniques in this book. The 1000 Genomes Project has already identified over 20,000 genetic variations between 629 humans whose DNA was sequenced for genes that they shared. These variations are been extensively researched as possibly beneficial in protecting us from disease.

Gene therapy approaches, in which genetic disease conditions can be treated by inserting normal genes into a patient or replacing diseased genes with normal genes, are being pioneered. In the near future, these technologies are expected to become increasingly more common. **Stem cell** technologies are some of the newest, most promising aspects of medical biotechnology, but they are also among the most controversial topics in all of science. Stem cells are immature cells that have the potential to develop and specialize into nerve cells, blood cells, muscle cells, and virtually

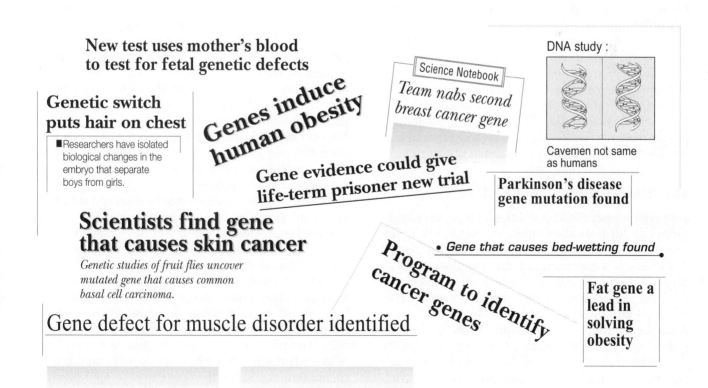

FIGURE 1.11 Genes Are Headline News Items Television, newspaper, and magazine headlines frequently report the discovery of genes involved in human disease conditions and many other new developments involving DNA.

any other type of cell in the body. Stem cells can be grown in a laboratory and, when treated with different types of chemicals, can be coaxed to develop into different types of human tissue that might be used in transplantation to replace damaged tissue. There are many exciting potential applications for stem cells, but, as we discuss in the next section and in Chapters 11 and 13, many complex scientific, ethical, and legal issues surround their use.

Biotechnology Regulations

An essential aspect of the biotechnology business involves the regulatory processes that govern the industry. In much the same way as pharmaceutical companies must evaluate their drugs based on specific guidelines designed to maximize the safety and effectiveness of a product, most biotechnology products must also be carefully examined before they are available for use. Although the FDA sets the standards for biotechnology products, many times the U.S. Department of Agriculture and the Environmental Protection Agency are involved. In fact, it has been said that biotechnology is one of the most heavily regulated industries. Two important aspects of the regulatory process include **quality assurance (QA)** and **quality control (QC).** QA measures include all activities involved in regulating the final quality of a product, whereas QC procedures are the part of the QA process, involving lab testing and monitoring of processes and applications to ensure consistent product standards. From QA and QC procedures designed to ensure that biotechnology products meet strict standards for purity and performance to issues associated with granting patents, and abiding by the regulatory processes required for clinical trials of biotechnology products in human patients, we consider these and other important biotechnology regulatory issues in Chapter 12.

The Biotechnology "Big Picture"

Although we have described different types of biotechnology as distinct disciplines, do not think about biotechnology as a field with separate and unrelated disciplines. It is important to remember that almost all areas of biotechnology are closely interrelated. For example, applications of bioremediation are heavily based on using microbes (microbial biotechnology) to clean up environmental conditions. Even medical biotechnology relies on the use of microbes to produce recombinant proteins, and all branches of biotechnology are regulated. A true appreciation of biotechnology involves understanding the biotechnology "big picture"—how biotechnology involves many different areas of science and how different types of biotechnol-

ogy depend on each other. This interdependence of many areas of science will be put to the test in solving important problems in the twenty-first century.

1.3 Biological Challenges of the Twenty-First Century

Numerous problems and challenges have the potential to be solved by biotechnology. For many of the greatest challenges—such as curing life-threatening human diseases—the barriers to overcoming these challenges are not insurmountable. Answers lie in our ability to better understand biological processes and to design and adapt biotechnological solutions. Rather than speculating about all of the ways that biotechnology may affect society in this century (an impossible task!), in this section we entice you with a few ideas on how medical biotechnology in particular will change our lives in the years ahead. In future chapters, we explore these and other ideas in much more detail.

What Will the New Biotechnology Century Look Like?

History will show that 2001 was a landmark in the biotechnology time line. In February 2001, some of the world's best-known molecular biologists gathered at a press conference to announce the publication of the rough draft of the human genome, a major accomplishment of the Human Genome Project. The DNA sequence—read as the letters A, G, C, and T—of human chromosomes was almost complete. One great surprise from this gathering was the announcement that the human genome consists of far fewer than the 100,000 genes that had been expected. As you will learn in other chapters, the Human Genome Project was completed in 2003 and has led to exciting new advances in biotechnology.

Identifying the chromosomal location and sequencing of all genes in the human genome has greatly increased our understanding of the complexity of human genetics. Basic research on the molecular biology and functions of human genes and controlling factors that regulate genes is providing immeasurable insight into how genes direct the activities of living cells, how normal genes function, and how defective genes are the molecular basis of many human disease conditions. An understanding of human genes will also allow us to study the genes of other organisms; in finding biochemical methods that they have used to fight disease, we may discover applications to human diseases.

An advanced understanding of human genetic disease conditions will also transform medicine as it is currently practiced. A new era of medicine is on the horizon. But is the Human Genome Project a quick and simple way to find defective genes so that we can quickly and simply cure human disease? If you think so, then you are overlooking the complexity of biology. The human genome is not the "biological crystal ball" that will immediately solve all of our medical problems. Identifying all human genes is just the tip of the iceberg for understanding how genes determine our health and susceptibility to disease. Scientists continue to work on unlocking the secrets of how all human genes function and how genes and proteins cause disease. A better understanding of human disease will require that we understand the structures and functions of the proteins that genes encode, the **proteome,** the collection of proteins responsible for human cells. But neither the genome nor the proteome is a software program that predetermines our health and our lives. Unlocking the mysteries of the human genome and human proteome alone makes the twenty-first century a most exciting time in which to be part of the scientific discovery process.

A Scenario in the Future: How Might We Benefit from the Human Genome Project?

Imagine the following scene in the year 2020 or so. A man seeks advice at a local pharmacy. He recently switched from one major drug to another, and the current drug is not working any better than the first for his arthritis. He tells his pharmacist, "This drug is so expensive and doesn't work any better for me than the last one, but I don't want to waste it or throw it out." "Well, sometimes the drugs don't work for everyone," says the pharmacist. This exchange represents one difficulty inherent in current health care strategies. Some drugs work for only some patients. How will the biotechnology century help this patient? The Human Genome Project might change medicine as we now know it and help patients such as this.

Many people currently experience the same problem that the man at the pharmacy encountered. The standard over-the-counter or routinely prescribed treatments available for arthritis and a host of other medical problems rarely work in the same way for everyone. Genome information has and will continue to result in the rapid, sensitive, and early detection and diagnosis of genetic disease conditions in humans of all ages, from unborn children to the elderly. In the case of arthritis, we know that there are different forms of this disease which have similar symptoms. Recent genetic studies have revealed that these different forms of arthritis are caused by different genes. Increased knowledge about genetic disease conditions such as arthritis will lead to preventive medicine approaches

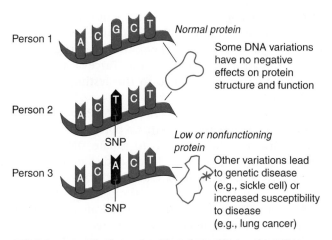

FIGURE 1.12 Single Nucleotide Polymorphisms A small piece of a gene sequence for three different individuals is represented. For simplicity, only one strand of a DNA molecule is shown. Notice how person 2 has a SNP in this gene, which has no effect on protein structure and function. Person 3, however, has a different SNP in the same gene. This subtle genetic change may affect how this person responds to a medical drug, or it may influence the likelihood that person 3 will develop a genetic disease.

designed to foster healthier lifestyles and to new, safer, and more effective treatment strategies to cure disease.

Let us consider how identifying the genes causing arthritis in our imaginary patient might help him. From its inception, the Human Genome Project yielded immediate dividends in our ability to identify and diagnose disease conditions. The identification of disease genes has enabled scientists and physicians to screen for a wide range of genetic diseases. This screening ability will continue to grow in the future. One area expected to be a great aid in the diagnosis of genetic disease conditions will comprise applications involving **single nucleotide polymorphisms (SNPs;** pronounced "snips"). SNPs are single nucleotide changes or **mutations** in DNA sequences that vary from individual to individual (**Figure 1.12**). These subtle changes represent one of the most common examples of genetic variation in humans.

SNPs are the cause of some genetic diseases, such as sickle cell anemia. Most scientists believe that SNPs will help them identify some of the genes involved in medical conditions such as arthritis, stroke, cancer, heart disease, diabetes, and behavioral and emotional illnesses as well as a host of other disorders. On average, each person is found to carry approximately 250 to 300 loss-of-function variants in studied genes and 50 to 100 variants in genes causing inherited disorders. The significance of this can be seen in the drugs that have been developed for human breast cancer. The two well-known breast cancer genes, *BRCA1* and

BRCA2, have about 1,700 variants worldwide. Before these genes were known, all breast cancer was treated with the same cell-toxic chemotherapy drugs. Now Herceptin, Rituxan, Gleevec, and Tarceva are available drug treatments linked to the testing for these two genes.

Testing one's DNA for different SNPs is one way to identify the disease genes that a person may be carrying. One way to do this is to isolate DNA from a small amount of a patient's blood and then apply this sample to a **DNA microarray,** also called a **gene chip.** Microarrays contain thousands of DNA sequences (Chapter 3). Using sophisticated computer analysis, scientists can compare patterns of DNA binding between a patient's DNA and the DNA on the microarray to reveal a patient's SNP patterns. For instance, researchers can use microarrays to screen a patient's DNA for a pattern of genes that might be expressed in a disease condition such as arthritis.

DNA chips are also being explored for their potential as DNA-based computers. Scientists postulate that DNA molecules on chips can function as logic gates, much as silicon-based chips are used to power traditional computers. DNA in a computer? Stay tuned.

The discovery of SNPs is partially responsible for the emergence of a field called **pharmacogenomics,** a field still in its infancy. Pharmacogenomics is customized medicine. It involves tailor-designing drug therapy and treatment strategies based on the genetic profile of a patient—that is, using his or her genetic information to determine the most effective and specific treatment approach (refer to Figure 11.7). Can pharmacogenomics solve some of our medical problems? Right now, doctors and physicians can only make guesses. Someday, however, with the right tools and human gene information, this will change.

Pharmacogenomics might help our arthritis patient in the pharmacy in 2020. We know that arthritis is a disease that shows familial inheritance for some individuals and, as mentioned earlier, a number of different genes are involved in different forms of arthritis. In many other cases of arthritis, a clear mode of inheritance is not seen. Perhaps there may be additional genes or nongenetic factors at work in these cases. A simple blood test from our patient could be used to prepare DNA for SNP and microarray analysis. SNP and microarray data could be used to determine which genes are involved in the form of arthritis that this man has. Armed with this genetic information, a physician could design a drug-treatment strategy—based on the genes involved—that would be *specific* and *most effective* against this man's type of arthritis. A second man with a different genetic profile for his particular type of arthritis might undergo a different treatment than the first. This is the power of pharma-

cogenomics in action. It is predicted that eventually everyone will have a whole-genome scan to provide information for useful and specific treatment. Of course, such screenings for genes that are related to medical conditions must be done in an ethical fashion, with proper security and integration into the health care delivery system.

The same principles of pharmacogenomics will also be applied to a host of other human diseases, such as cancer. As you probably already know, many drugs currently used to treat different types of cancer through **chemotherapy** may be effective against cancerous cells but may also affect normal cells. Hair loss, dry skin, changes in blood cell counts, and nausea are all conditions related to the effects of chemotherapy on normal cells. But what if drugs that are effective against cancer cells could be designed so that they had no effect on normal cells in other tissues? This may become possible as the genetic basis of cancer is better understood and drugs can be designed based on the genetics of different types of cancer. In addition, SNP and microarray information could also be used to figure out a person's risk of developing a particular type of cancer long before he or she would otherwise begin to show signs of the disease, especially when there was a family history. Such information might be used to help that person develop changes in lifestyle, such as diet and exercise habits, that might be important for preventing disease.

Another example of the benefits of studying differences in human genotypes has been in the area of **metabolomics,** a biochemical snapshot of the small molecules—such as glucose, cholesterol, ATP, and signaling molecules that result from a cellular change—produced during cellular metabolism. This snapshot directly reflects physiologic status and can be used to monitor drug effects on disease states. The exact number of human metabolites is unknown, but estimates of between 2,000 and 10,000 have been published. The use of this tool can distinguish between disease process and physiologic adaptation, and it can save time and money when it is incorporated into early-stage drug discovery. For example, a major drug company recently funded a study in which groups of mice were fed a diet designed to increase cholesterol. Then, their lipids in plasma, adipose tissue, and liver were characterized at intervals over a number of weeks. Over 500 "unusual" lipids were identified as the response. One group of mice was susceptible to atherosclerosis, thus providing a good measure of differential response due to this disease physiology.

In Chapter 11 we also discuss examples of **nanotechnology,** or applications that incorporate extremely small devices (a "nano" scale). Nanotechnology is an

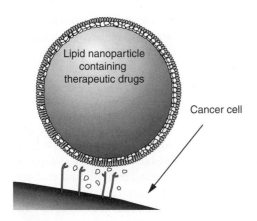

FIGURE 1.13 **Nanobiotechnology in Action** Nanoparticles containing chemotherapy agents can be specifically directed to target cancer cells by a coat of tumor-specific proteins bound to the target cells. In this way, chemotherapy agents that cannot pass through the cell membrane can be released inside these target cancer cells.

entirely new field that is rapidly emerging as a major research area. One promising application of nanotechnology has been the development of small particles that can be used to deliver drugs to cells (**Figure 1.13**).

In addition to advances in drug treatment, gene therapy represents one of the ultimate strategies for combating genetic disease. Gene therapy technologies involve replacing or augmenting defective genes with normal copies of them. Think about the potential power of this approach. Scientists are working on a variety of ways to deliver healthy genes into humans, such as using viruses to carry healthy genes into human cells. Promising techniques have been developed for treating some blood disorders and diseases of the nervous system, such as Parkinson's disease (Figure 1.13). However, many barriers must be overcome before gene therapy becomes a safe, practical, effective, and well-established approach to treating disease.

Obstacles currently prevent gene therapy from being widely used in humans. For example, how can normal genes be delivered to virtually all cells in the body? What are the long-term effects of introducing extra genes into humans? What must be done to be sure that the normal protein is properly made after the extra genes are delivered into the body? Gene therapy applications have come under increased scrutiny following complications, including the tragic deaths of several patients (see Chapter 11). Another exciting new technology with the potential for modifying a genetic defect by silencing a gene is being aggressively pursued using **small interfering RNA (siRNA;** see Chapter 3).

Stem cell technologies are expected to provide powerful tools for treating and curing disease. Stem cells are immature cells that can grow and divide to produce different types of cells, such as skin, kidney, and blood cells. Most stem cells are obtained from embryos **(embryonic stem cells, or ESCs),** and because this process involves the death of the embryo they are controversial. Scientists have successfully isolated stem cells from adult tissues **(adult-derived stem cells,** or **ASCs)** that are being compared with their embryonic counterparts for their potential for regenerating nervous tissue and other tissues lost to disease or damage. You will also learn about **induced-pluripotent stem cells (iPSCs)**, which may eventually be a great source of stem cells that can be acquired without destroying an embryo.

Stem cells can be coaxed to form almost any tissue of interest depending on how they are treated. Imagine growing skin cells, blood cells, and even whole organs in the lab and using these to replace damaged tissue or failing organs such as the liver, pancreas, and retina (**Figure 1.14**). **Regenerative medicine** is the phrase used to describe this approach. In the future, scientists may be able to collect stem cells from patients with genetic disorders, genetically manipulate these cells by gene therapy, and reinsert them into the patient from whom they were collected to help treat

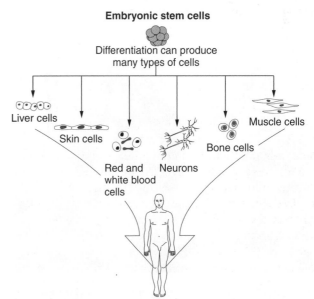

FIGURE 1.14 **Embryonic Stem Cells Can Give Rise to Many Types of Differentiated Cells** Embryonic stem cells (ESCs) are derived from embryos or early-stage fetuses; they are immature cells that can be stimulated to differentiate into a variety of cell types.

genetic disease conditions. Some of this work is already possible, and these technologies will be optimized in the near future.

We hope that the examples in this section demonstrated how the future is indeed bright for marvelous advances in medical biotechnology. Pharmacogenomics, gene therapy, and stem cell technologies are not the answers to all our genetic problems, but with continued rapid advances in genetic technology, many seemingly impossible problems may not be so insurmountable in the future. Here we have presented basic examples of medical applications in the biotechnology century, but in future chapters you will learn about exciting applications from other areas of biotechnology that will potentially change our lives for the better. We conclude our introduction to the world of biotechnology by discussing career opportunities in the industry.

1.4 The Biotechnology Workforce

How will the world prepare for the biotechnology century? Recent advances have created a range of new opportunities for biotechnology companies and individuals seeking employment in the biotechnology industry (**Figure 1.15**).

Ultimately, biotechnology companies are looking for people who are comfortable analyzing complex data and sharing their expertise with others in team-

FIGURE 1.15   The Biotechnology Industry Provides Exciting Opportunities for Many Types of Scientists From biologists and chemists to engineers, information technologists, and salespeople, the biotechnology industry offers a great range of high-tech employment opportunities. Shown here is a senior undergraduate student working on a biotechnology research project. Gaining research experience as an undergraduate is an excellent way to prepare for a career in biotechnology.

oriented, problem-solving working environments. The biotechnology workforce depends on important contributions from talented people in many different disciplines of science.

The Business of Biotechnology

In 1976, Genentech Inc., a small company near San Francisco, California, was founded. Genentech is generally recognized as the first biotechnology company, and its success ushered in the birth of this exciting industry with the release of human insulin—offering the first opportunity for diabetics to receive this human protein. Today, biotechnology is a global industry with hundreds of products on the market generating more than $63 billion in worldwide revenues, including $40 billion in sales of biological drugs (such as enzymes, antibodies, growth factors, vaccines, and hormones) in the United States. Many biotechnology companies are working on cures for cancer, in part because in the United States alone, nearly 40 percent of Americans will receive a diagnosis of cancer in their lifetimes. Cancer is the second leading cause of death in the United States, behind heart disease. Over 350 biotechnology products are currently in development targeting cancers, diabetes, heart disease, Alzheimer's and Parkinson's diseases, arthritis, AIDS, and many other diseases.

North America, Europe, and Japan account for approximately 95 percent of biotechnology companies, but biotechnology firms are found throughout the world, with over 4,900 companies in 54 countries. Countries without a traditional history in **research and development (R&D)** worldwide are turning to biotechnology for high-tech innovations. For example, biotechnology is a rapidly developing industry in India and China. Still, many of the world's leading biotechnology companies remain located in the United States (see **Figure 1.16**). There are currently around 1,500 biotechnology companies in the United States, many of which are often closely associated with colleges and universities or located near major universities where basic science ideas for biotechnological applications are generated. Visit the Biotechnology Industry Organization Website listed on the Companion Website for excellent information on biotechnology centers around the nation. At this site, you can find biotechnology companies located near you and learn about their current products.

What Is a Biotechnology Company?

By now you may be wondering "what is the difference between a biotechnology company and a pharmaceutical company?" Most people can name phar-

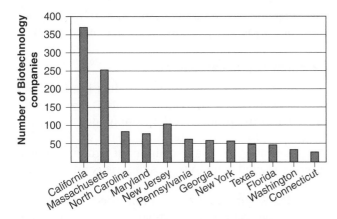

FIGURE 1.16 Distribution of U.S. Biotechnology Companies Public and private biotechnology companies are located throughout the United States.

maceutical companies such as Merck, Johnson & Johnson, or Pfizer because they or a family member may have used one or more of their products, but most people cannot name a biotechnology company (Table 1.3) or explain why a biotech company is different from a pharmaceutical company. The large pharmaceutical companies are commonly referred to as "big pharma." Generally speaking, **pharmaceutical companies** are involved in drug development by chemically synthesizing or purifying compounds used to make the drug—products such as aspirin, antacids, and cold medicines. Pharmaceutical companies typically do not use living organisms to grow or produce a product (such as a recombinant protein), as is the focus of biotechnology companies. But these days the distinctions between the two are blurring, because many large pharmaceutical companies are often involved in biotechnology-related research and product development either directly or indirectly by partnering with a biotechnology company. Also remember that biotechnology involves much more than drug development. There are many different companies of varying sizes dedicated to working on specific areas of biotechnology.

Biotechnology companies vary in size from small companies of less than 50 employees to large companies with over 300 employees. Historically many a biotechnology company began as a small **startup company** formed by a small team of scientists who believe that they might have a promising product to make (such as a recombinant protein to treat disease). The team must typically then seek investors to fund their company so that they can buy or rent lab facilities, buy equipment and supplies, and continue the research and development necessary to make their product. But starting a biotechnology company

is risky business; on average, about 40 percent of startup companies close without providing any return to investors.

Biotechnology startup companies rely on financial investments in the company, such as **venture capital (VC)** funds provided at an early stage to startup companies with a potential for success. Sources of VC funds can be individuals, financial institutions and other companies for example. Venture capital funds make money by owning equity in startup companies that have a promising technology to develop. **Angel investors**—affluent individuals who provides VC capital for a startup in exchange for company ownership—are key to providing startup biotechnology companies with the funds needed to carry out the research and testing necessary to make a product.

VC investments are the essential pipeline of funds that supports biotech companies. During the recent global financial crisis, as biotechnology VC investments experienced a 46 percent decrease between 2007 and 2008, many biotechnology companies had only enough cash flow for 12 months or less.

Eventually, if a startup biotechnology company is successful in bringing a product to the market (a process that takes around 10 years on average at a cost, in the United States, of over $1 billion for a medical drug!), many startups are often bought by larger, well-established companies. Bringing a promising

TABLE 1.3	**TOP FIVE BIOTECHNOLOGY COMPANIES AND TOP FIVE PHARMACEUTICAL COMPANIES BY REVENUE**
Biotech Companies	**Revenue (Millions)**
Amgen	$14,268
Genentech (now part of Roche)	$11,724
Genzyme	$ 3,187
UCB	$ 3,169
Gilead	$ 3,026
Pharma Companies	
Johnson & Johnson	$61,897
Pfizer	$50,009
Roche	$49,051
GlaxoSmithKline	$45,830
Novartis	$44,267

Adapted from: Ernst and Young, *Beyond Borders: Global Biotechnology Report 2010* (www.ey.com/beyondborders). Revenue based on preliminary results reported by companies.

product close to market ultimately creates value for a company, which may enable it to file for an **initial public offering (IPO),** which means that it is available for the public to purchase shares of company stock.

There are similarities between how pharmaceutical companies and biotechnology companies are organized (**Figure 1.17**). We discuss many aspects of this in the next section, in which we describe different job opportunities in each area of a biotechnology company.

Jobs in Biotechnology

The biotechnology industry in the United States employs over 200,000 people. Biotechnology offers numerous employment choices, such as laboratory technicians involved in basic research and development, computer programmers, laboratory directors, and sales and marketing personnel. All are essential to

the biotechnology industry. In this section, we consider some of the job categories available in biotechnology.

Research and development

Development of a new biotech product is a long and expensive process. Individuals in R&D are directly involved in the process of developing ideas and running experiments to determine if a promising idea (for example, using a recombinant protein from a recently cloned gene to treat a disease condition) can actually be developed into a product. It requires a great deal of trial and error. From the largest to the smallest biotechnology companies, all have some staff dedicated to R&D. On average, biotechnology companies invest at least four times more on R&D than any other high-tech industry. For some companies, the R&D budget is close to 50 percent of the operating budget. R&D is the lifeblood of most companies—without new discoveries, companies cannot make new products.

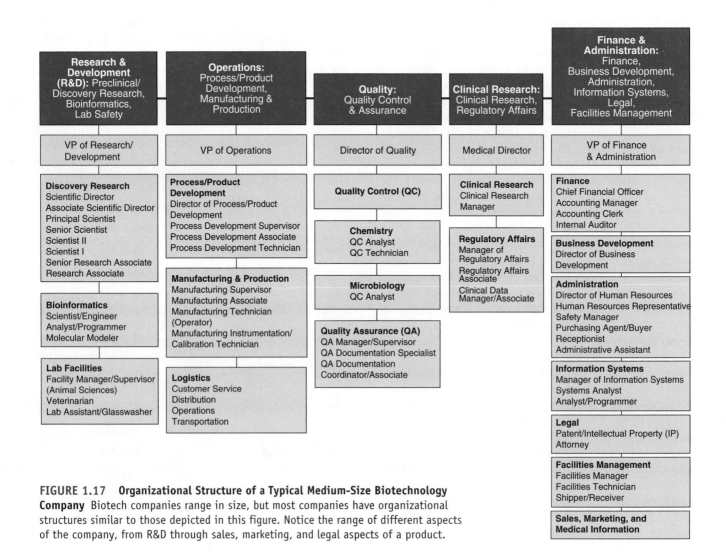

FIGURE 1.17 Organizational Structure of a Typical Medium-Size Biotechnology Company Biotech companies range in size, but most companies have organizational structures similar to those depicted in this figure. Notice the range of different aspects of the company, from R&D through sales, marketing, and legal aspects of a product.

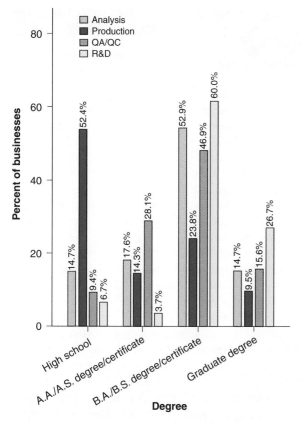

FIGURE 1.18 Minimal Level of Education Required of Entry-Level Technicians The Resource Group performed a survey of entry-level education requirements for 69 biotechnology companies in the three-county area of central coast California. Results are comparable with other areas of the United States and indicate that R&D generally requires a greater amount of training than other job areas.

The majority of positions in R&D usually require a bachelor's or associate's degree in chemistry, biology, or biochemistry (**Figure 1.18**). **Laboratory technicians** are responsible for duties such as cleaning and maintaining equipment used by scientists and keeping labs stocked with supplies. Technician positions usually require a B.A. in science or a B.S in biology or chemistry. **Research assistants** or **research associates** carry out experiments under the direct supervision of established and experienced scientists. These positions require a B.S. or M.S. degree in biology or chemistry. Research assistants and associates are considered "bench" scientists, carrying out research experiments under the direction of one or more principal or senior scientists. Assistants and associates perform research in collaboration with others. Involved in the design, execution, and interpretation of experiments and results, they may also be required to review scientific literature and prepare technical reports, lab protocols, and data summaries.

Principal or **senior scientists** usually have a PhD and considerable practical experience in research and management skills for directing other scientists. These individuals are considered the scientific leaders of a company. Responsibilities include planning and executing research priorities of the company, acting as spokespeople on company research and development at conferences, participating in patent applications, writing progress reports, applying for grants, and serving as advisers to the top financial managers of the company. The job titles and descriptions we have given can vary from company to company; however, if you are interested in making new scientific discoveries, then R&D might be an exciting career option for you.

The rapidly expanding field of bioinformatics, the use of computers to analyze and store DNA and protein data, requires an understanding of computer programming, statistics, and biology. Until recently, many experts in bioinformatics were computer scientists who had trained themselves in molecular biology or molecular biologists self-trained in computer science, database analysis, and mathematics. Today, people with computer science interests are being encouraged to take classes in biotechnology, and biotechnology students are being encouraged to take computer science classes. In addition, specific programs in bioinformatics have been developed at four-year colleges and universities, technical colleges, and community colleges to train people to become **bioinformaticists.**

Many speculate that the massive amount of data from the Human Genome Project will result in a merger of biotechnology and information technology. Bioinformaticists are needed to analyze, organize, and share DNA and protein sequence information. The human genome alone contains over 3 billion base pairs, and hundreds of thousands of bases of sequence data from other species are added to databases around the world each day. Sophisticated programs are required to analyze this information. How will biotechnology companies keep from drowning in the ever-increasing sea of data that has inundated biology and chemistry? Robust data-mining and data-warehousing systems are just beginning to enter the bioinformatics market. To put this in perspective, a financial database for a major bank might have 100 columns representing different customers, with 1 million rows of data. A major pharmaceutical database, in contrast, may contain 30,000 columns for genes and only about 40 rows of patients. Computer modeling tools, such as neural networks and decision trees, are also widely used by bioinformaticists to identify patterns between SNP markers and a disease status. If you are interested in merging an understanding of biology with computer science skills, then bioinformatics may be a good career option for you.

Operations, biomanufacturing, and production

Operations, biomanufacturing, and *production* are terms that describe the divisions of a biotechnology company that oversee specific details of product development, such as the equipment and laboratory processes involved in producing a product. This often includes **scale-up processes**, in which cultured cells making up a product must be grown on a large scale. This is not a trivial task. As a simple analogy, scaling-up is the difference between cooking a meal for yourself versus preparing a full-course Thanksgiving dinner for 50 people. Biomanufacturing and production units maintain and monitor the large-scale and large-volume equipment used during production, and they ensure that the company is following proper procedures and maintaining appropriate records for the product. Biomanufacturing job details are specific to the particular product a company is manufacturing. Entry-level jobs include material handlers, manufacturing assistants, and manufacturing associates. Supervisory and management-level jobs usually require a bachelor's or master's degree in biology or chemistry and several years of experience in manufacturing the products or type of product being produced by that company. Manufacturing and production also involve many different types of engineers, including those trained in chemical, electrical, environmental, or industrial engineering. Engineering positions usually require a B.A. degree in engineering or a M.S. degree in biology or an area of engineering.

Quality assurance and quality control

Most products from biotechnology are highly regulated by such federal agencies as the U.S. FDA, Environmental Protection Agency (EPA), and U.S. Department of Agriculture (USDA) (see Chapter 12). These federal agencies require that manufacturing follow exact methods approved by regulatory officials. As discussed previously, the overall purpose of quality assurance is to guarantee the final quality of all products. Quality control efforts are designed to ensure that products meet stringent regulations mandated by federal agencies. In addition to guaranteeing that components of the product manufacturing process meet the proper specifications, QA and QC workers are also responsible for monitoring equipment, facilities, and personnel, maintaining correct documentation, testing product samples, and

YOU DECIDE

Generic Biotech Drugs?

As the biotechnology industry has aged, many of the earlier biotech products that received patent approval are set to lose patent protection in the next few years. Patents for about 20 products with annual sales of over $10 billion expired in 2006. Patents are designed to provide a monopoly right for the developer of an invention, and in the United States patents can last for up to 20 years. There is controversy in the industry over whether many of these products will receive approval to become **generic drugs**. You may already know that generic drugs are copies of brand-name products that generally have the same effectiveness, safety, and quality as the original but are produced at a cheaper cost to the consumer than the brand-name drugs. One way that generics can cut costs is that they are often approved for use without having to undergo the same expensive safety and effects studies required for name drugs.

Many biotechnology companies are fighting the production of generic biotech drugs, claiming that the higher costs of making a biological product such as an antibody compared with a pharmaceutically-produced drug earns them a right to manufacture named drugs and make a profit without competition from generics. Some also doubt whether a generic, called a **biosimilar drug** when referring to therapeutic recombinant proteins and other biologically produced proteins such as antibodies, could be made at a greatly reduced price given that it is generally still more expensive to produce a biotech product than a pharmaceutical product and because biosimilars will be very difficult to replicate exactly. Although, the Institute for One World Health recently received funding from the Bill and Melinda Gate Foundation to develop drugs for developing countries as a nonprofit organization contracting with a biomanufacturing company. Similar questions about profit can be raised regarding drug costs in developing countries, where many of the people who need drugs cannot afford them, although many companies sell drugs in the developed world at a higher price and use some of these profits to provide drugs at low or no cost to developing nations. Should biotechnology companies be forced to produce generic drugs? You decide.

addressing customer inquiries and complaints, along with other responsibilities. Entry-level jobs in QC and QA include validation technician, documentation clerk, and QC inspector. Jobs usually require at least a B.S. degree in biology, and managerial or supervisory positions require more education. **Customer relation specialists** or **product complaint specialists** often work in the QA divisions of a company. One function of such specialists is to investigate consumer complaints about a problem with a product and to follow up with the consumer to provide an appropriate response or solution to the problem encountered.

Clinical research and regulatory affairs

In the United States, developing a drug product is a long and expensive process of testing the new drug candidate in volunteer subjects to ultimately receive new drug approval from the FDA (see Chapter 12). The clinical trial process, along with many other clinical and nonclinical areas of biotechnology, is regulated by a number of different agencies. As a result, every biotechnology company has staff monitoring regulatory compliance to make sure that proper regulatory procedures are in place and are being followed. Biotechnology companies involved in developing drugs for humans often have very large clinical research divisions with science and nonscience personnel that conduct and oversee clinical trials.

Marketing, sales, finance, and legal divisions

Marketing and selling a variety of biotechnology products, from medical instruments to drugs, is a critical area of biotechnology. Most people employed in biotechnology marketing and sales have a B.S. degree in the sciences and familiarity with scientific processes in biotechnology, perhaps combined with course-work in business or even a B.A. degree in marketing. **Sales representatives** work with medical doctors, hospitals, and medical institutions to promote a company's products. **Marketing specialists** devise advertising campaigns and promotional materials to target customer needs for the products a company sells. Representatives and specialists frequently attend trade shows and conferences. An understanding of science is important because the ability to answer end-user questions is an essential skill in marketing and sales. **Finance divisions** of a biotechnology company are typically run by vice presidents or chief financial officers who oversee company finances and are also often involved in raising funds from partners or venture capitalists seeking investments in technology companies. **Legal specialists** in biotechnology companies typically work on legal issues associated with product development and marketing, such as copyrights, naming rights, and obtaining patents. These are essential issues for protecting the ideas and products that a biotechnology company works so hard to develop (see Chapter 12 for further discussion). Staff in this area will also address legal circumstances that may arise if there are problems with a product or litigation from a user of a product.

Salaries in Biotechnology

People working in the biotechnology industry are making groundbreaking discoveries that fight disease, improve food production, clean up the environment, and make manufacturing more efficient and profitable. Although the process of using living organisms to improve life is an ancient practice, the biotechnology industry has been around for only about 25 years. As an emerging industry, biotechnology offers competitive salaries and benefits, and employees at almost all levels report high job satisfaction.

Salaries for life scientists who work in the commercial sector are generally higher than those paid to scientists in academia (colleges and universities). Scientists working in the biotechnology industry are among the most highly paid of those in the professional sciences. In 2006, in California alone, the biotechnology industry generated about $20 billion in personal wages and salary. In this same year, the top five biotechnology companies in the world spent an average of $93,400 on each employee.

According to a survey of more than 400 biotechnology companies conducted recently by the Radford Division of AON Consulting, a PhD in biology, chemistry, and molecular biology with no work experience was starting at an average annual salary of $55,700, with senior scientists earning in excess of $120,000 a year. For individuals with an M.A. degree in the same fields, the average salary was $40,600 annually, with a range from $60,000 to $70,000 per year for research associates and $32,500 annually for those with a B.A. degree, with a range of $52,000 to $62,000 per year for research associates. Visit the Radford Biotechnology Compensation Report, the Commission on Professionals in Science and Technology, and the U.S. Office of Personnel Management on the Web for updates on the surveys used for the salary figures described in this section. Biotechnology salary reports websites are listed in the Keeping Current: Web Links at the Companion Website.

Based on a national survey, 56 percent of the college students entering biotechnology training programs had little or no science background. If you have the proper background in biology and good lab skills, many good positions are available at many different levels, but increasingly educational training at the community college, technical college, four-year college, or university level is becoming a requirement for employment in biotechnology.

Hiring Trends in the Biotechnology Industry

Career prospects in biotechnology are very good. The industry has more than tripled in size since 1992, and worldwide company revenues increased from approximately $8 billion in 1992 to nearly $28 billion in 2001. But the industry is not immune to changes in the economic climate. Mergers and acquisitions have dominated the industry since 2006, following the downturn in the global economy. Major layoffs occurred among small biotech companies and biotech divisions in large pharmaceutical companies. In the United States alone, biotechnology companies reduced their workforce by about 38 percent. In the four years between 1995 and 1999, the U.S. biotechnology industry increased its employee workforce by 48.5 percent. The competition created by generic drugs has caused big pharmaceutical companies to spend more than $110 billion for more than a dozen biotechnology companies since 2006. Drugs produced by genetic engineering are harder to copy by generic competitors, making these biotechnology companies (like Genentech, and Genzyme) a good buy for the big pharmaceutical companies (like Roche, sanofi-aventis, and others). Some biotechnology firms and research labs have also found that they are better off filling skilled technician-level jobs with people who have more specialized training than pursuing their more traditional practice of attempting to find people with M.A. and Ph.D. degrees. Many human resources departments and recruitment firms indicate that there is a tremendous increase in the number of open positions in bioinformatics, proteomics, and genome studies as well as for experienced technicians.

There is currently also a hot job market for scientists in drug discovery. Larger biotechnology companies, no matter in which region, report that they are growing rapidly and find that almost every career choice is in demand. In particular, the most sought-after jobs more often include work that requires team interaction, both inside and outside the company. Partnering has become the landscape of drug development, and skills in this area are required for any person's career in the industry.

Another trend that has reached a stage of critical importance is the need for people with multiple skill areas. For instance, an individual with a degree in molecular biology or biochemistry, a minor in information technology, and course work in mathematics

CAREER PROFILE

Finding a Biotechnology Job That Appeals to You

Throughout this book we will use this career profile feature to highlight potential career options, including educational requirements, job descriptions, salary, and related information. A number of websites are outstanding resources for biotechnology career information. Links to each of these are available through the Companion Website:

- Visit the Biotech Career Center, a very good site for career materials, links to job resources, a wealth of information on over 600 biotechnology companies, and much more.

- Visit the Biotechnology Industry Organization website and access one of the biotechnology company sites.

- Visit the Access Excellence Careers in Biotechnology website for job descriptions and excellent links to resources for careers in biotechnology.

- Visit the Bio-Link website to find useful biotechnology workforce resources. It has several sections of career information, job descriptions and educational requirements, job posting sites, and state-by-state listings of biotechnology companies, among many other resources. From this site, access "Careers in Biotechnology: A Counselor's Guide to the Best Jobs in the United States" by Gina Frierman-Hunt and Julie Solberg.

- The California State University Program for Education and Research in Biotechnology (CSUPERB) is a great resource for educational and career materials in biotechnology. In particular, visit the "career site" and "job links" pages.

- Visit the Massachusetts Biotechnology Industry Organization, and follow the "careers" link to one of the most comprehensive listing of job descriptions in the biotechnology industry, from vice president of research and development to glasswasher positions (yes, this person does what the title says!).

Search these sites for a biotechnology job that appeals to you. Next rewrite your résumé to fit this job description, or identify the course work or experience you would need to apply successfully for this position.

can potentially have a great advantage in the job market, especially with companies seeking people with unique skill combinations.

In addition to good technical skills and an understanding of business and finance practices, companies also emphasize the importance of "soft skills," which they consider nearly as important at technical skills. These include skills in writing and communication, presenting information to different audiences, and teamwork. Employment prospects in biotechnology are exciting indeed. Opportunities are excellent for individuals with solid scientific training and good verbal and written communication skills coupled with a strong ability to work as part of a team in a collaborative environment.

MAKING A DIFFERENCE

There are many ways to be part of biotechnology and to influence the lives of others, as we will explore at the end of each chapter in the *Making a Difference* feature. Here is a look at what's in store in the coming chapters. Early success in the use of cell-based beads that release the chemical levodopa into the brains of Parkinson's patients sets the stage for other cell based therapies (Chapter 2). Recombinant therapeutic proteins such as insulin have improved the quality of life for many humans (Chapter 3). Trapping biomarker proteins that reflect the presence of a disease could boost the search for other treatments, bringing personalized medicine using purified proteins one step closer (Chapter 4). Antibiotics isolated from bacteria are one of the most successful examples of biotech products (Chapter 5). The "Green Revolution" may occur more quickly thanks to biotech methods that select algae that produce the most petroleum; such methods have been used in the past to enhance the natural capacities of organisms with selected traits (Chapter 6). Studying the function of genes in the mouse genome will help us to understand the human genome and could lead to the development of new drugs (Chapter 7). The 1000 Genomes Project will sequence the DNA of 1000 individuals for comparison and has revealed over 2,300 genes that were missed by the Human Genome Project; this new genome project will lead to new treatments and more personalized care (Chapter 8). Successful cleanup of groundwater pollution is one example of a bioremediation success story (Chapter 9). A test developed from horseshoe crab blood is a key technique for checking the sterility of medical devices (Chapter 10). Bone marrow transplantations are a successful example of stem cell technologies (Chapter 11). Compliance with regulations that protect companies and employees is a necessary part of biotech and will continue to provide safe products and a safe working environment (Chapter 12). Finally, we'll explore the purpose of the Genetic Information Nondiscrimination Act (Chapter 13). The opportunity to be part of a science that truly makes a difference awaits you in the coming chapters!

QUESTIONS & ACTIVITIES

Answers can be found in Appendix 1.

1. Provide two examples of historical and current applications of biotechnology.

2. Pick an example of a biotechnology application and describe how it has affected your everyday life.

3. Which area of biotechnology involves using living organisms to clean up the environment?

4. Describe how pharmacogenomics will influence the treatment of human diseases.

5. Distinguish between QC and QA, and explain why both are important for biotechnology companies.

6. Access the library of the National Center for Biological Information (www.ncbinlm.nih.org) and search for new information on adult-derived stem cells. This free source will provide abstracts and titles for full-text articles that can be obtained at other libraries.

7. Visit the "About Biotech" section of the Access Excellence website at www.accessexcellence.org/AB. This section provides an outstanding overview of historical and current applications of biotechnology. Survey the different topics presented at this site. Many interesting aspects and examples of biotechnology are described in student-friendly terms. Find a biotechnology topic that fascinates you, print out the information you are interested in, and share your newly discovered knowledge with a friend or family member who is unfamiliar with biotechnology.

8. Interacting with others in a group setting is an essential skill in most areas of science. As you learned in this chapter, biotechnology involves groups of scientists, mathematicians, and computing experts with different backgrounds collaborating to solve a problem or achieve a common goal. Discussing biology with other people is fun and beneficial to everyone working on the same problem in a company, and working with other students is an excellent way to help you learn and enjoy your studies in biotechnology. Teaching

someone else is a great way to test your knowledge. Analyze your ability to work in a group by forming a study group for the next test. Assign a group leader who will be responsible for organizing meetings and keeping the study group focused on helping each person to learn. Make sure that everyone has a topic to present to the group and that all of you offer constructive criticism to their suggestions. If you cannot get together with your classmates in one room, share your thoughts via e-mail or ask your professor to set up an electronic bulletin board for discussion purposes.

9. The National Library of Medicine is a worldwide database of biology research publications in scientific journals and it can be accessed for free at http://www.ncbi.nlm.nih.gov. Access PubMed and conduct a search on any topic of biotechnology that may be of interest to you to find recent papers on the latest new research developments in biotechnology.

10. Organism cloning has led to concerns that humans may be cloned. At least one group has even claimed to have already done so. Run a Google search with the terms *Raelians* and *cloning*.

At the time this book was published, no claims of human cloning had been proven to be valid, and most scientists are adamantly against human cloning. What do you think? Should humans be cloned?

11. Explain how biotechnology companies get started. Include in your answer a description of how a startup company is funded.

12. Describe differences and similarities between the pharmaceutical industry and the biotechnology industry.

13. Visit the FierceBiotech website (http://www.fiercebiotech.com) and sign up to receive free daily alerts on biotech industry news items. This is a great way to learn about exciting new discoveries in biotechnology, biotech company updates, career information and more.

Visit www.pearsonhighered.com/biotechnology

To download learning objectives, chapter summary, "Keeping Current" web links, glossary, flashcards, and jpegs of figures from this chapter.

An Introduction to Genes and Genomes

After completing this chapter, you should be able to:

- Compare and contrast prokaryotic and eukaryotic cells.
- Discuss experiments that determined that DNA is the inherited genetic material of living organisms.
- Describe the structure of a nucleotide and explain how they form double-helical DNA molecules.
- Describe the process of DNA replication and discuss the role of key proteins involved.
- Understand what genomes are and why biologists study them.
- Describe the process of transcription and understand how mRNA creates a functional mRNA molecule.
- Describe the process of translation, and the roles of mRNA, tRNA, and rRNA.
- Explain why gene expression regulation is important.
- Explain what the epigenome is and understand why epigenetics is important.
- Discuss the role of operons in regulating gene expression in bacteria.
- Name different types of mutations and give examples of the consequences of mutations.

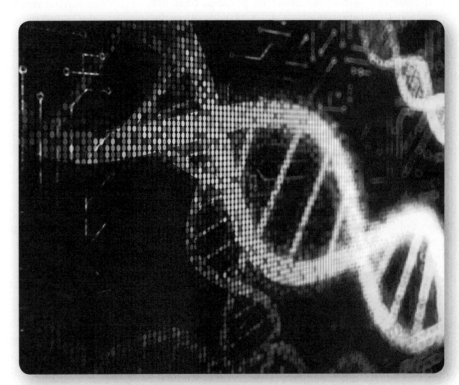

Encoded within DNA are genes that provide instructions controlling the activities of all cells. Genes influence our behavior; determine our physical appearances such as skin, hair, and eye color; and affect our susceptibility to genetic disease conditions.

Central to the study of biotechnology is an understanding of the structure of DNA as the molecule of life—the inherited genetic material. Later, we will consider how extraordinary techniques in molecular biology enable biologists to clone and engineer DNA—manipulations that are essential for many applications in biotechnology (Chapter 3). In this chapter, we review DNA structure and replication, discuss how genes code for proteins, provide an overview of genomics, and consider causes and consequences of mutations.

FORECASTING THE FUTURE

Although much of what we will discuss in this chapter describes basic information about cell structure and genes that scientists have known about for decades, there are many exciting potential future developments especially related to genomics. An active area of research is the creation of "artificial cells" in which synthetic genomes are assembled in a lab and introduced into cells to create cells with novel properties based on the inserted genome. Similarly, another area of intense investigation is the creation of novel proteins by combining DNA sequences. These approaches have not yet advanced far enough to produce valuable applications. But there are signs that synthetic genome strategies for making cells with novel properties can work. Fundamental information about cells, gene expression, and genomes is playing a major role in the development of cell-based treatment strategies such as using stem cell to create tissues and organs (which we will discuss in Chapter 11). Advances in genetic testing and novel therapies designed to target human genes will continue to be an area of focus for biotechnology companies.

2.1 A Review of Cell Structure

Cells are the structural and functional units of life. Organisms such as bacteria consist of a single cell, whereas humans have approximately 75 trillion, including over 200 different types that vary in appearance and function. Cells vary greatly in size and complexity, from tiny bacterial cells to human neurons that may stretch for more than 3 feet from the spinal cord to muscles in the toes. Virtually all cells share a common component, genetic information in the form of **deoxyribonucleic acid (DNA).** Genes control numerous activities in cells by directing the synthesis of proteins. Genes influence our behavior; determine our physical appearance, such as skin, hair, and eye color; and affect our susceptibility to genetic disease conditions. Before we begin our study of genes and genomes, we will review basic aspects of cell structure and function and briefly compare different types of cells.

Prokaryotic Cells

Cells are complex entities with specialized structures that determine cell function. Generally, every cell has a **plasma (cell) membrane,** a double-layered structure of primarily lipids and proteins that surrounds its outer surface; **cytoplasm,** the inner contents of the cell between the nucleus and the plasma membrane; and **organelles** ("little organs"), structures in the cell that perform specific functions. Throughout this book, we not only consider how plant and animal cells play important roles in biotechnology but also cover many biotechnology applications involving bacteria, yeasts, and other microorganisms. Bacteria are referred to as **prokaryotic cells,** or simply prokaryotes, from the Greek words meaning "before nucleus," because they do not have a **nucleus,** an organelle that contains DNA in animal and plant cells. Prokaryotes include true bacteria (eubacteria) and cyanobacteria, a type of blue-green algae (Table 2.1), and members of the domain Archaea (ancient bacteria with some eukaryotic characteristics).

Bacteria have a relatively simple structure (**Figure 2.1**). Their outer boundary is defined by the plasma membrane, which is surrounded by a rigid cell wall that protects the cell. Except for ribosomes, which are used for protein synthesis, bacteria have few organelles. The cytoplasm contains DNA, usually in the form of a single circular molecule, which is attached to the plasma membrane and located in an area called the nucleoid region (Figure 2.1). Some bacteria also have a tail-like structure called a flagellum, which is used for locomotion.

Eukaryotic Cells

Plant and animal cells are considered **eukaryotic cells**, from the Greek words meaning "true nucleus," because they contain a membrane-enclosed nucleus and many organelles. Eukaryotes also include fungi and single-celled organisms called protists, which include most

TABLE 2.1 PROKARYOTIC AND EUKARYOTIC CELLS

	Prokaryotic Cells	Eukaryotic Cells
Cell types	True bacteria (eubacteria) Archaebacteria	Protists, fungi, plant, animal cells
Size	100 nm–10 μm	10–100 μm
Structure	No nucleus; DNA located in the cytoplasm. No organelles.	DNA enclosed in a membrane-bound cytoplasm. Nucleus. Many organelles.

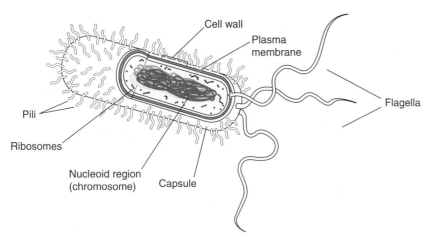

FIGURE 2.1   Prokaryotic Cell Structure
Bacteria are prokaryotes. Shown here is a drawing of structures contained in a typical rod-shaped bacterium.

algae. Diagrams of plant and animal cells are shown in **Figure 2.2**. The plasma membrane is a fluid, highly dynamic, complex double-layered barrier composed of lipids, proteins, and carbohydrates. The membrane performs essential roles in cell adhesion, cell-to-cell communication, and cell shape, and it is essential for transporting molecules into and out of the cell. The membrane also serves an important role as a selectively permeable barrier, because it contains proteins involved in complex transport processes that control which molecules can enter and leave the cell. For example, hormones such as insulin are released from the cell in a process called secretion; other molecules, such as glucose, can be taken into the cell and, within mitochondria, be converted into energy in the form of a molecule called **adenosine triphosphate (ATP).** Membranes also enclose or comprise many organelles.

The cytoplasm of eukaryotes consists of **cytosol**, a nutrient-rich, gel-like fluid, and many organelles. The cytoplasm of prokaryotes also contains cytosol, but few organelles. Think of the organelle as the compartment in which chemical reactions and cellular processes occur. Organelles allow cells to carry out thousands of different complex reactions simultaneously. Each organelle is responsible for specific biochemical reactions. For instance, lysosomes break down foreign materials and old organelles; organelles such as the endoplasmic reticulum and Golgi apparatus synthesize proteins, lipids, and carbohydrates (sugars). By compartmentalizing reactions, cells can carry out a multitude of reactions in a highly coordinated fashion simultaneously without interference. Be sure to familiarize yourself with the functions of organelles presented in Figure 2.2 and Table 2.2.

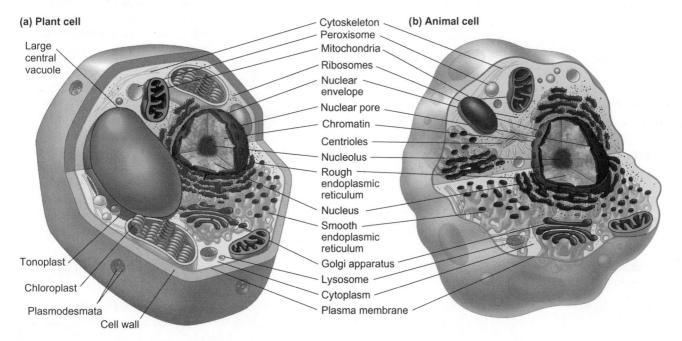

FIGURE 2.2   Eukaryotic Cell Structure Sketches of common structures present in plant (a) and animal cells (b).

TABLE 2.2   STRUCTURE AND FUNCTION OF EUKARYOTIC CELLS

Cell Part	Structure	Functions
Plasma membrane	Membrane made of a double layer of lipids (primarily phospholipids and cholesterol) within which proteins are embedded; proteins may extend entirely through the lipid bilayer or protrude on only one face; externally facing proteins and some lipids have attached sugar groups	Serves as an external cell barrier; acts in the transport of substances into or out of the cell; maintains a resting potential that is essential for the functioning of excitable cells; externally facing proteins act as receptors (for hormones, neurotransmitters, and so on) and in cell-to-cell recognition
Cytoplasm	Cellular region between the nuclear and plasma membranes; consists of fluid cytosol (containing dissolved solutes), inclusions (stored nutrients, secretory products, pigment granules), and organelles (the metabolic machinery of the cytoplasm)	
Cytoplasmic organelles		
▪ Mitochondria	Rod-like double-membrane structures; the inner membrane is folded into projections called cristae	Site of adenosine triphosphate (ATP) synthesis; powerhouse of the cell
▪ Ribosomes	Dense particles consisting of two subunits, each composed of ribosomal RNA and protein; free or attached to rough endoplasmic reticulum	The sites of protein synthesis
▪ Rough endoplasmic reticulum	Membrane system enclosing a cavity (the cisterna) and coiling through the cytoplasm; externally studded with ribosomes	Sugar groups are attached to proteins within the cisternae; proteins are bound in vesicles for transport to the Golgi apparatus and other sites; external face synthesizes phospholipids and cholesterol
▪ Smooth endoplasmic reticulum	Membranous system of sacs and tubules; free of ribosomes	Site of lipid and steroid synthesis, lipid metabolism, and drug detoxification
▪ Golgi apparatus	A stack of smooth membrane sacs and associated vesicles close to the nucleus	Packages, modifies, and segregates proteins for secretion from the cell, inclusion in lysosomes, and incorporation into the plasma membrane
▪ Lysosomes	Membranous sacs containing hydrolases (digestive enzymes)	Sites of intracellular digestion
▪ Peroxisomes	Membranous sacs of oxidase enzymes	The enzymes detoxify a number of toxic substances; the most important enzyme, catalase, breaks down hydrogen peroxide
▪ Microtubules	Cylindrical structures made of tubulin proteins	Support the cell and give it shape; involved in intracellular and cellular movements; form centrioles
▪ Microfilaments	Fine filaments of the contractile protein actin	Involved in muscle contraction and other types of intracellular movement; help form the cell's cytoskeleton
▪ Intermediate filaments	Protein fibers; composition varies	The stable cytoskeletal elements; resist mechanical forces acting on the cell
▪ Centrioles	Paired cylindrical bodies, each composed of nine triplets of microtubules	Organize a microtubule network during mitosis to form the spindle and asters; form the bases of cilia and flagella

TABLE 2.2   *(CONTINUED)*

Cell Part	Structure	Functions
■ Cilia	Short, cell surface projections; each cilium is composed of nine pairs of microtubules surrounding a central pair	Move in unison, creating a unidirectional current that propels substances across cell surfaces
■ Flagella	Like cilia but longer; the only example in humans is the sperm tail	Propels the cell
Nucleus	Largest organelle, surrounded by the nuclear envelope; contains fluid nucleoplasm, nucleoli, and chromatin	Control center of the cell; responsible for transmitting genetic information and providing the instructions for protein synthesis
■ Nuclear envelope	Double-membrane structure pierced by the pores; outer membrane continuous with the cytoplasmic endoplasmic reticulum	Separates the nucleoplasm from the cytoplasm and regulates the passage of substances to and from the nucleus
■ Nucleoli	Dense spherical (non-membrane-bound) bodies composed of ribosomal RNA and proteins	Site of ribosome subunit manufacture
■ Chromatin	Granular, thread-like material composed of DNA and histone proteins	DNA contains genes
Central vacuole (plant cells)	Large membrane-enclosed compartment	Used to store ions, waste products, pigments, protective compounds
Chloroplasts (plant cells)	Membrane-enclosed organelle containing stacked structures (grana) of chlorophyll-containing membrane sacs called thylakoids surrounded by an inner fluid (stroma)	Site of photosynthesis

The nucleus contains DNA. This organelle is a spherical structure enclosed by a double-layered membrane, the **nuclear envelope,** and is typically the largest structure in a eukaryotic cell. Nearly 6 feet of DNA is coiled into the nucleus of every human cell; if the DNA in all human cells were connected end to end, there would be enough to stretch to the sun and back about 500 times. Although the majority of DNA in a eukaryotic cell is contained within the nucleus, mitochondria and chloroplasts also contain small circular DNA molecules.

2.2 The Molecule of Life

Every high school or college biology course involves some discussion of DNA, and DNA is routinely manipulated by students in college biology laboratories and many high school classes. With the wealth of information available about many detailed aspects of DNA and genes, the study of biology in the twenty-first century might give you the impression that the structural details of DNA were always understood. However, the structure of DNA—and its function as genetic material—was not always well known. Many extraordinary researchers and incredible discoveries have contributed to our modern-day understanding of DNA

structure and function. We begin this section with a brief overview, highlighting the evidence for DNA as the genetic material; then we discuss DNA's structure.

Evidence That DNA Is the Inherited Genetic Material

In 1869, Swiss biologist Friedrich Miescher identified a cellular substance from the nucleus that he called "nuclein." Miescher purified nuclein from white blood cells and found that it could not be broken down (degraded) by protein-digesting enzymes called **proteases.** This discovery suggested that nuclein was not made only of proteins. Subsequent studies determined that this material had acidic properties, which led nuclein to be renamed "nucleic acids." DNA and **ribonucleic acid (RNA)** are the two major types of **nucleic acids.** While biochemists worked to identify the different components of nucleic acids, British microbiologist Frederick Griffith's experiments (1928) with two strains of *Streptococcus pneumoniae* provided evidence that DNA is the inherited genetic material of cells.

Griffith was studying two strains of the bacterium *Streptococcus pneumoniae*, a microbe that causes pneumonia. At the time of these studies, this strain was called *Diplococcus pneumoniae*. Griffith worked with a virulent (disease-causing) variety called the smooth

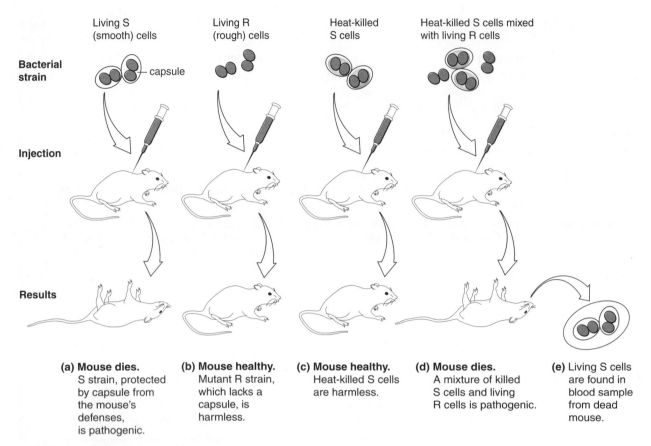

Bacterial strain

Living S (smooth) cells — capsule   Living R (rough) cells   Heat-killed S cells   Heat-killed S cells mixed with living R cells

Injection

Results

(a) Mouse dies. S strain, protected by capsule from the mouse's defenses, is pathogenic.

(b) Mouse healthy. Mutant R strain, which lacks a capsule, is harmless.

(c) Mouse healthy. Heat-killed S cells are harmless.

(d) Mouse dies. A mixture of killed S cells and living R cells is pathogenic.

(e) Living S cells are found in blood sample from dead mouse.

FIGURE 2.3   Griffith's Transformation Experiment The S strain of *S. pneumoniae* kills mice (a); the R strain is harmless (b). Heat-killed S cells are harmless (c). Mice injected with heat-killed S cells mixed together with live R cells died (d), and live S cells could be detected in the blood of dead mice (e). This result is a demonstration of transformation. Living R cells took in DNA from dead S cells, transforming the R cells into S cells.

strain (S cells) along with a harmless strain called rough cells (R cells). S cells are surrounded by a capsule (smooth coat) of proteins and sugars, whereas R cells lack this coat. When Griffith injected mice with living S cells, the mice died, and Griffith found live S cells in the blood of the dead mice (**Figure 2.3**). When live R cells were injected into mice, the mice lived and showed no living R cells in their blood. These experiments suggested that the protein coat was responsible for the death of the mice. To test this idea, Griffith then killed S cells by heating them, which destroys proteins in the coat. Not surprisingly, mice injected with heat-killed S cells lived, with no signs of live S cells in their blood. But when Griffith mixed heat-killed S cells with living R cells in a tube and injected this mixture into mice, the mice died and living S cells were found in the blood of the dead mice. Where did the live S cells come from?

This experiment provided evidence that the genetic material from heat-killed S cells had transformed (changed) or converted R cells into S cells. Griffith's experiments demonstrated **transformation,** the uptake

of DNA by bacteria. Heat treatment broke open S cells, which released their DNA into the tube. Living R cells took up S cell DNA, which transformed the properties of the R cells so that they became virulent, resembling S cells. As you will learn, transformation is a very powerful technique in molecular biology and is routinely used to introduce genes into bacteria for DNA cloning, protein production, and other important purposes. Although Griffith hypothesized that some genetic factor was responsible for the transformation results he saw, he didn't actually identify DNA as the "transforming factor." However, his experiments were instrumental in leading others in search of this factor.

In 1944, Oswald Avery, Colin MacLeod, and Maclyn McCarty purified DNA from large batches of *S. pneumoniae* grown in liquid culture. Their experiments provided definitive evidence that DNA is the genetic material and proved that DNA was the transforming factor in the Griffith experiments. In their now famous experiment, Avery, MacLeod, and McCarty ground up (homogenized) mixtures of bacterial cells from *S. pneumoniae* and treated these extracts with proteases, RNA-degrading enzymes

(RNase), or DNA-degrading enzymes **(DNase).** They subsequently carried out transformation experiments using these treated extracts. Extracts from killed S cells were mixed with living R cells and injected into mice. This experiment demonstrated that DNase-treated extracts could not transform R cells to S cells because DNA in these mixtures was degraded by DNase. Extracts treated with protease or RNase still maintained their transforming ability because DNA in these mixtures remained intact. Although other studies with viruses were essential for determining the role of DNA, the work of Avery, MacLeod, and McCarty provided definitive evidence for DNA as the genetic material causing transformation.

DNA Structure

While evidence supporting DNA as hereditary material was building, a significant question still remained: What is the structure of DNA? Erwin Chargaff provided some insight to this question by isolating DNA from a variety of different species and revealing that the percentage of DNA bases called adenine was proportional to the percentage of bases called thymines, and that the percentage of cytosine bases in an organism's DNA were roughly proportional to the percentage of guanine. This valuable observation suggested that the bases adenine, thymine, cytosine, and guanine were somehow intricately related components of DNA structure—an important principle to remember because, as we explore next, these bases are essential components of DNA.

The building block of DNA is the **nucleotide** (**Figure 2.4**). Each nucleotide is composed of a (five-carbon) **pentose sugar** called deoxyribose, a phosphate molecule, and a **nitrogenous base.** The bases are interchangeable components of a nucleotide. Each nucleotide contains one base, either **adenine (A), thymine (T), guanine (G),** or **cytosine (C)**—the so-called As, Ts, Gs, and Cs of DNA.

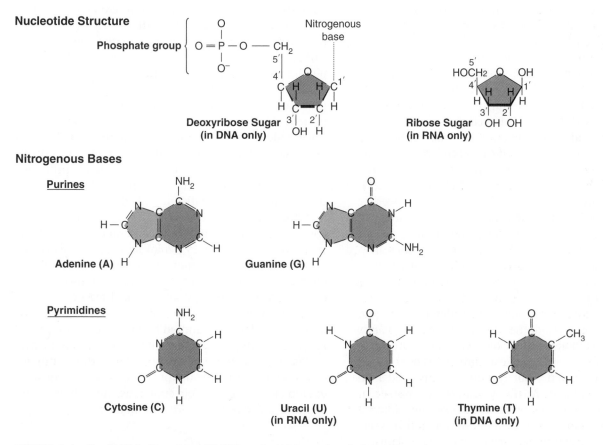

FIGURE 2.4   Nucleotide Structure All DNA nucleotides consist of a nitrogenous base, A, C, G, or T; a sugar; and a phosphate group. The sugar in DNA is deoxyribose. RNA molecules contain a pentose sugar called ribose. The pentose sugar in DNA is called deoxyribose because it lacks an oxygen at carbon number 2 (2′) of the sugar compared with the ribose sugar in RNA. A base is attached to carbon number 1 (1′) of the sugar; the phosphate group is attached to carbon number 5 (5′) of the sugar. Because of their structure, adenine and guanine belong to a group of bases called purines, whereas cytosine, thymine, and uracil belong to a group called pyrimidines.

Nucleotides are the building blocks of DNA, but how are these structures arranged to form a DNA molecule? Many scientists have contributed to the answer to this question, but the definitive structure of DNA was finally revealed by James Watson and Francis Crick, working at the Cavendish Laboratories in Cambridge, England. Chemists Rosalind Franklin and Maurice Wilkins, of University College, London, used x-ray crystallography to provide Watson and Crick with invaluable data on the structure of DNA. By firing an x-ray beam onto crystals of DNA, Franklin and Wilkins revealed a model of DNA indicating that its structure could be helical. From these data, Chargaff's findings, and other studies, Watson and Crick assembled a wire model of DNA.

Watson and Crick published "The Molecular Structure of Nucleic Acids: A Structure for Deoxyribose Nucleic Acid" in the prestigious journal *Nature* on April 25, 1953. The first paragraph of this paper reads, "We wish to suggest a structure for the salt of deoxyribose nucleic acid (D.N.A.). This structure has novel features which are of considerable biological interest." Given the importance of DNA and what we have learned about DNA structure over the last 50 years, this description might be one of the greatest understatements ever made in a published scientific paper. The significance of this discovery was appropriately recognized in 1962, when Watson, Crick, and Wilkins received the Nobel Prize in Medicine.

Watson and Crick determined that nucleotides form long strands of DNA and that each DNA molecule consists of two strands that join together and wrap around each other to form a double helix (**Figure 2.5**). A strand of DNA is a string of nucleotides held together by **phosphodiester bonds** that connect the sugar of one nucleotide to the phosphate group of an adjacent nucleotide (Figure 2.5). The sequence of bases in a strand can vary. For instance, a nucleotide containing a C can be connected to a nucleotide containing an A, T, G or another nucleotide containing a C.

Each strand of nucleotides has a **polarity** to it; there is a *5' end* and a *3' end* to the strand (Figure 2.5). *Polarity* refers to the carbons of the deoxyribose sugar. At the 5' end of a strand, the phosphate at carbon 5 is not bonded to another nucleotide, but carbon 3 is involved in a phosphodiester bond. At the 3' end, the phosphate at carbon 5 is bonded to another nucleotide, but carbon 3 is not joined to another nucleotide. Although this aspect of nucleotide structure may seem trivial, the polarity of DNA is important for replication and for the routine manipulation of DNA in the laboratory.

Watson and Crick determined that DNA molecules consist of two interconnecting strands that wrap around each other to form a right-handed double helix, perhaps the most famous molecular model in all of biology (Figure 2.5). The two strands are joined together by hydrogen bonds between **complementary** **base pairs** in opposite strands (Figure 2.5). Adenine base pairs with thymine, and guanine base pairs with cytosine. From this model, Chargaff's observations are easily understood. The proportions of A's and T's are equivalent in an organism's DNA, as are the proportions of G's and C's, because they pair with each other in a DNA molecule.

The two strands of nucleotides in a double helix are **antiparallel**—the polarity of each strand is reversed relative to the other (Figure 2.5). This orientation is necessary for the complementary base pairs to align and for hydrogen bonds to align with one another. The double helix resembles a twisted ladder of sorts. The rungs of this ladder consist of the complementary base pairs, and the sides of the ladder consist of sugar and phosphate molecules, creating the "backbone" of DNA.

What Is a Gene?

Genes are often described as units of inheritance, but what exactly is a gene? A **gene** is a sequence of nucleotides that provides cells with the instructions to synthesize a specific protein or a particular type of RNA. Not all genes are used to produce a protein. For example, genes for transfer RNA (tRNA) are used to make tRNA molecules, and while tRNAs are required for protein synthesis, they are not translated to produce a protein. Most genes are approximately 1,000 to 4,000 nucleotides (nt) long, although many smaller and larger genes have been identified. Largely by controlling the proteins produced by a cell, genes influence how cells, tissues, and organs appear, both through the microscope and with the naked eye. These inherited appearances are called **traits.** Through the DNA contained in your cells, you have inherited traits from your parents, such as eye color and skin color. Genes influence not only cell metabolism and behavioral and cognitive abilities such as intelligence but also our susceptibility to certain types of genetic diseases.

Some traits are controlled by a single gene, but many others are determined by multiple genes, which produce many proteins that interact in complex ways. In Section 2.4, we explore how genes direct protein synthesis in cells. Throughout this book, we consider examples of genes, their functions, and their many applications in different areas of biotechnology.

2.3 Chromosome Structure, DNA Replication, and Genomes

Before we consider how genes function, it is important that you understand how and why DNA is organized into chromosomes and how DNA is replicated in cells.

(a) Sugar-Phosphate Backbone

(b)

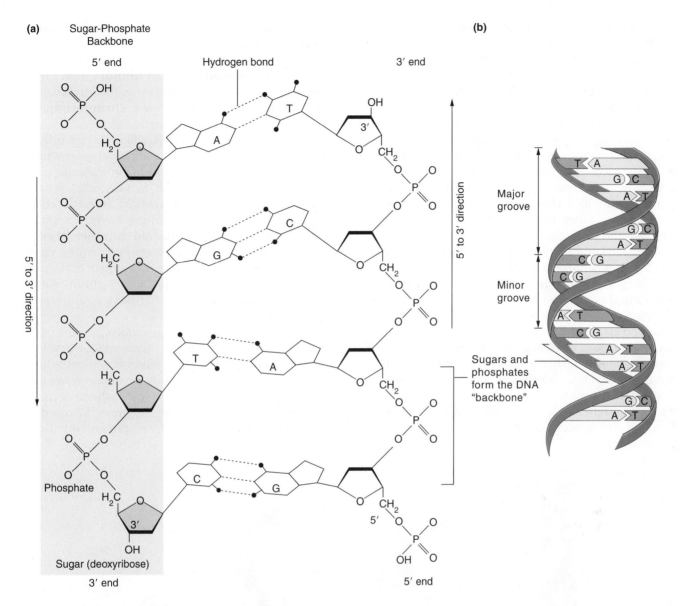

FIGURE 2.5 **DNA Is a Double-Stranded Helix** (a) Two strands of nucleotides are joined together by hydrogen bonds between complementary base pairs. Adenine bases (A) always base pair with thymine bases (T), and cytosine (C) base pairs with guanine (G). (b) The two strands wrap around each other so that the overall structure of DNA is a double-stranded helix with a sugar-phosphate "backbone," in which the bases are aligned in the center of the helix.

Chromosome Structure

Suppose you were presented with a challenge. If you solved it, you would earn free tuition for the rest of your undergraduate courses. You are given a basket containing 46 packages of different-colored yarn all unraveled and intertwined. Your challenge is to sort the yarn into 46 even balls. How would you solve this challenge? If you started to cut the tangled pile of yarn randomly, you probably would not succeed. Of course, if you painstakingly unraveled the yarn and wound each color of yarn

into a ball, you would eventually sort it into 46 even balls. This analogy provides a highly simplified view of the challenge presented to a human cell when it has to divide and sort its DNA into even packages.

The 3 billion base pairs (bp) of DNA in every human cell would stretch to around 6 feet if unraveled—an amazing amount of material packed into the tiny nucleus of each cell. This DNA must be separated evenly when a cell divides; otherwise the loss of DNA can have devastating consequences. Fortunately such mistakes in DNA separation are rare, in part because cells effectively separate and package DNA into **chromosomes.**

Inside the nucleus, DNA exists in a relatively unraveled state. This does not mean that the DNA is *uncoiled* from its double-helical structure; rather, the DNA is somewhat loosely arranged and not fully compacted into tightly coiled chromosomes, although all of the DNA in a chromosome remains together within the nucleus. When a cell is not dividing, chromosomes in the nucleus exist as an intricate combination of DNA and DNA-binding proteins called **histones** to form strings called **chromatin.** During cell division, chromatin is coiled into tight fibers that eventually wrap around each other, so that chromosomes become highly coiled and tightly condensed packages of DNA and histones (**Figure 2.6**).

The size and number of chromosomes vary from species to species. Most bacteria have a single circular chromosome, in the size range of several hundred thousand base pairs, which contains a few thousand genes. Eukaryotes typically contain one or more sets of chromosomes, which have a linear shape, and often these chromosomes are several million base pairs in size. Most human cells have two sets (pairs) of 23 chromosomes each, for a total of 46 chromosomes. Through the process of fertilization, you inherited 23 chromosomes from your mother (**maternal chromosomes**) and 23 chromosomes from your father (**paternal chromosomes**). These chromosome pairs are called **homologous pairs,** or **homologues.** Chromosomes 1 through 22 are known as the **autosomes;** the 23rd pair are called the **sex chromosomes—** consisting of X and Y chromosomes.

Human egg and sperm cells, called the sex cells or **gametes,** contain a single set of 23 chromosomes, called the **haploid number (*n*)** of chromosomes. All other cells of the body—such as skin cells, muscle cells, and liver cells—are known as **somatic cells**. Somatic cells from many organisms have two sets of chromosomes, called the **diploid number (2*n*)** of chromosomes. Human somatic cells contain 46 chromosomes. Somatic cells of a normal human male have 22 pairs of autosomes and an X and Y chromosome; cells of a normal female have 22 pairs of autosomes and two X chromosomes.

Sex chromosomes were so named because they contain genes that influence sex traits and the development of reproductive organs, whereas the autosomes were originally thought primarily to contain genes that affect body features unrelated to sex, such as skin color and eye color. Although there are genes involved in sex organ determination that are present on the Y

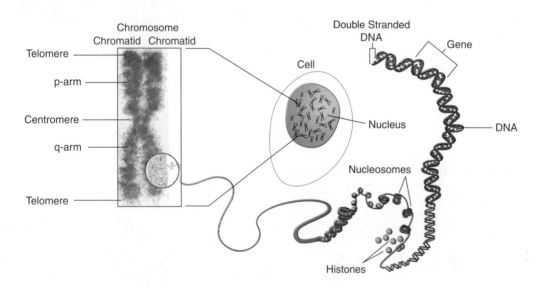

FIGURE 2.6 Chromosome Organization In a nondividing cell, chromosomal DNA exists in an unraveled state called *chromatin*. Histone proteins serve as particles around which DNA becomes tightly wound to give a "beads on a string" appearance when viewed with an electron microscope. When cells divide, chromatin is further compacted into tight fibers and supercoiled looped structures. Ultimately these supercoiled loops are tightly packed together with the assistance of other proteins to create an entire chromosome, a highly compact assembly of DNA. Each chromosome consists of two sister chromatids attached by a centromere. Chromosome arms are the portions of the chromatid on one side of the centromere, labeled as the *p* and *q arms*. The ends of a chromosome are called *telomeres*.

chromosome, there are other genes involved in sex determination that are present on autosomes, and the majority of genes on the X chromosome are not required for development of the reproductive organs.

Several characteristics are common to most eukaryotic chromosomes. Each chromosome consists of two thin, rod-like structures of DNA called **sister chromatids** (Figure 2.6). The sister chromatids are exact replicas of each other, copied during DNA synthesis. During cell division, sister chromatids are separated so that newly forming cells receive the same amount of DNA as the original cell from which they arose. Each eukaryotic chromosome has a single **centromere,** a constricted region of the chromosome consisting of intertwined DNA and proteins that join the two sister chromatids to each other. This region of a chromosome also contains proteins that attach chromosomes to organelles called microtubules, which play an essential role in moving chromosomes and separating sister chromatids during cell division.

The centromere delineates each sister chromatid into two arms—the short arm, called the **p arm,** and the long arm, or **q arm.** Each arm of a chromosome ends with a segment called a **telomere** (Figure 2.6). Telomeres are highly conserved repetitive sequences of nucleotides that are important for attaching chromosomes to the nuclear envelope. Telomeres are a subject of intense research. Changes in telomere length are believed to play a role in the aging process and in the development of certain types of cancers

Karyotype analysis for studying chromosomes

One most common way to study chromosome number and basic aspects of chromosome structure is to prepare a **karyotype.** In karyotype analysis, cells are spread on a microscope slide and then treated with chemicals to release and stain the chromosomes. For example, G-banding, in which chromosomes are treated with a DNA-binding dye called Giemsa stain, creates a series of alternating light and dark bands in stained chromosomes. Each stained chromosome shows a unique and reproducible banding pattern that can be used to identify different chromosomes. Chromosomes can be aligned and paired based on their staining pattern and size (**Figure 2.7**). In humans, chromosome 1 is the largest chromosome

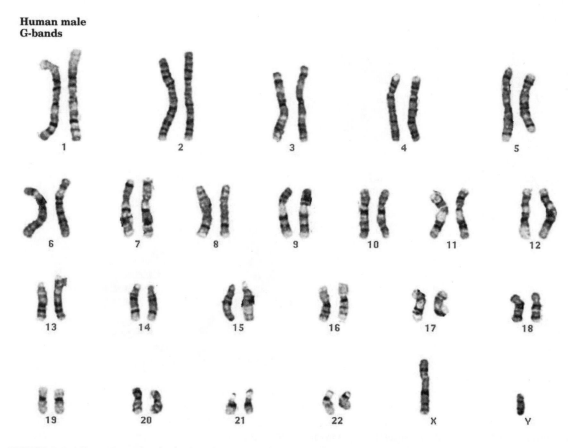

Human male G-bands

FIGURE 2.7 Karyotype Analysis In a karyotype, dividing cells are spread out onto a glass microscope slide to release their chromosomes. Chromosomes are stained and aligned based on their overall size, the position of the centromere, and their staining pattern to create a karyotype.

and chromosome 21 is the smallest. Karyotypes are very valuable for studying and comparing chromosome structure. A more modern method for karyotype analysis (called **spectral karyotyping**) incorporates specific probes and techniques to colorize chromosomes; it provides a more detailed analysis of chromosome structure than traditional karyotypes, such as the one shown in Figure 2.7. Karyotype analysis can be used to identify human genetic disease conditions associated with abnormalities in chromosomal structure and number.

DNA Replication

When a cell divides, it is essential that the newly created cells contain equal copies of replicated DNA. Somatic cells divide by a process called **mitosis,** wherein one cell divides to produce daughter cells, each of which contains an identical copy of the DNA of the original (parent) cell. For instance, a human skin cell divides to produce two daughter cells, each containing 23 pairs of chromosomes. Gametes are formed by a process called **meiosis**, wherein a parent cell divides to create up to four daughter cells, which can be either sperm or egg cells. During meiosis, the chromosome number in daughter cells is cut in half to the haploid number. Sperm and egg cells each contain a single set of 23 chromosomes. Through sexual reproduction, a fertilized egg, called the **zygote,** is formed. The zygote, which divides by mitosis to form an embryo and eventually a complete human, contains 46 chromosomes: 23 paternal chromosomes and 23 maternal chromosomes.

Prior to cell division by either mitosis or meiosis, DNA must be replicated in the cell. Replication occurs by a process called **semiconservative replication**. **Figure 2.8** shows an overview of this process. Before replication begins, the two complementary strands of the double helix must be pulled apart into single strands. Once separated, these strands serve as templates for copying two new strands of DNA. At the end of this process, two new double helices are formed. Each helix contains one original DNA (parental) strand and one newly synthesized strand, thus the term *semiconservative.*

DNA replication occurs in a series of stages involving a number of different proteins. Because prokary-

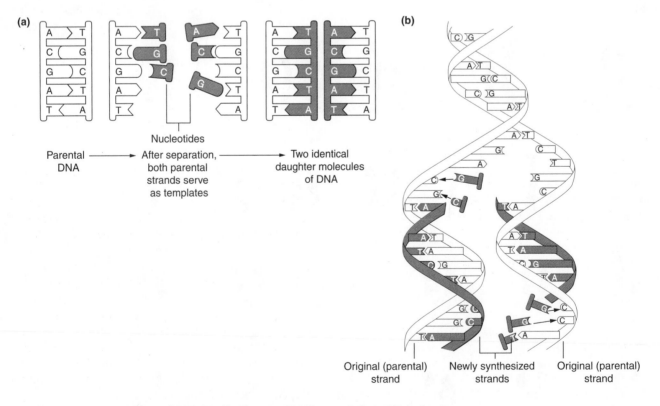

(a)

Parental DNA → After separation, both parental strands serve as templates → Two identical daughter molecules of DNA

Nucleotides

(b)

Original (parental) strand   Newly synthesized strands   Original (parental) strand

FIGURE 2.8   An Overview of DNA Replication Nucleotide strands in a DNA molecule must first be separated (a). Each strand serves as a template for the synthesis of new strands, producing two DNA molecules, each containing one original strand and one newly synthesized strand (b).

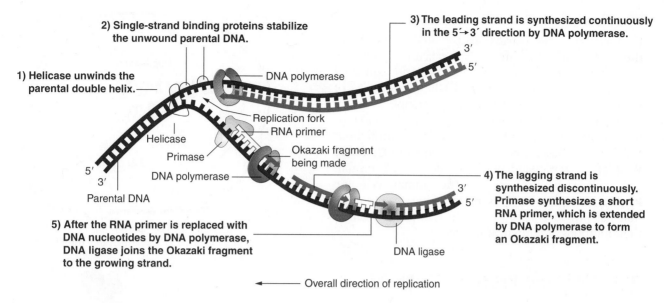

1) Helicase unwinds the parental double helix.

2) Single-strand binding proteins stabilize the unwound parental DNA.

DNA polymerase

Helicase

Primase

DNA polymerase

5′
3′

Parental DNA

Replication fork

RNA primer

Okazaki fragment being made

3) The leading strand is synthesized continuously in the 5′→3′ direction by DNA polymerase.

3′
5′

4) The lagging strand is synthesized discontinuously. Primase synthesizes a short RNA primer, which is extended by DNA polymerase to form an Okazaki fragment.

3′
5′

5) After the RNA primer is replaced with DNA nucleotides by DNA polymerase, DNA ligase joins the Okazaki fragment to the growing strand.

DNA ligase

← Overall direction of replication

FIGURE 2.9 Semiconservative Replication of DNA

otes contain circular chromosomes, DNA replication in prokaryotes is slightly different from that in eukaryotes. Here we consider the main players and key concepts in DNA replication overall. But keep in mind that there are subtle differences distinguishing the replication process in prokaryotes from that in eukaryotes. Replication is initiated by **DNA helicase,** an enzyme that separates the two strands of nucleotides, literally "unzipping" the DNA by breaking hydrogen bonds between complementary base pairs (**Figure 2.9**). The separated strands form a replication fork. As helicase unwinds the DNA and **single-strand binding proteins** attach to each strand and prevent them from base pairing and reforming a double helix. This step is important because the DNA strands must be held apart during DNA replication. Strand separation occurs at sites called **origins of replication.** Bacterial chromosomes have a single origin. Because of their large size, eukaryotic chromosomes have multiple origins. Starting DNA replication at multiple origins allows eukaryotic chromosomes to be copied rapidly.

The next step in DNA replication involves the addition of short segments of RNA approximately 10 to 15 nucleotides long. These sequences, called **RNA primers**, are synthesized by an enzyme called **primase** (in eukaryotes, a form of DNA polymerase called α [alpha] acts as the primase). Primers start the process of DNA replication because they serve as binding sites for **DNA polymerases,** the key enzymes that synthesize new strands of DNA. Several different forms of DNA polymerase are involved in copying DNA. In bacteria, the enzyme **DNA polymerase III**

(called DNA polymerase δ [delta] in eukaryotes) binds to each single strand, moving along the strand and using it as a template to copy a new strand of DNA. During this process, DNA polymerase uses nucleotides present in the cell to synthesize complementary strands of DNA. DNA polymerase always works in one direction, synthesizing new strands in a 5′-to-3′ orientation and adding nucleotides to the 3′ end of a newly synthesized strand (Figure 2.9) by forming phosphodiester bonds between the phosphate of one nucleotide and the sugar in the previous nucleotide.

Because DNA polymerase proceeds only in a 5′-to 3′-direction, replication along one strand, the **leading strand**, occurs in a continuous fashion (Figure 2.9). Synthesis on the opposite strand, the **lagging strand,** occurs in a discontinuous fashion because DNA polymerase must wait for the replication fork to open. On the lagging strand, short pieces of DNA, called Okazaki fragments (named after Reiji and Tuneko Okazaki, the scientists who discovered these fragments), are synthesized as the DNA polymerase works its way out of the replication fork. Covalent bonds between Okazaki fragments in the lagging strand are formed by **DNA ligase** to ensure that there are no gaps in the phosphodiester backbone. Finally, the RNA primers are removed and these gaps are filled by DNA polymerase.

Remember the functions of enzymes involved in DNA synthesis. As we will discuss, DNA polymerase and DNA ligase are routinely used in the lab in working with cloned DNA (Chapter 3).

What is a genome?

DNA contains the instructions for life—genes. All of the DNA in an organism's cells is called the **genome.** Contained in the human genome are approximately 20,000 genes scattered among 3 billion base pairs of DNA. The study of genomes, a discipline called **genomics,** is currently one of the most active and rapidly advancing areas of biological science. Throughout this book, we discuss aspects of the **Human Genome Project,** a worldwide effort to identify all human genes on each chromosome. The Human Genome Project was an enormous undertaking in genomics that has provided exciting insight into human genes, their locations, and their functions.

2.4 RNA and Protein Synthesis

Genes govern the activities and functions within a cell by directing the synthesis of proteins. Some of the myriad functions of these essential molecules are as follows:

- Proteins are necessary for cell structure, for example, as important components of membranes, organelles, and the cytoplasm.

- Proteins carry out essential reactions in the cell as enzymes.

- Proteins perform critical roles as hormones and other "signaling" molecules that cells use to communicate with one another.

- Receptor proteins bind to other molecules, such as hormones, and transport proteins, enabling molecules to enter and leave cells.

- Proteins as antibodies recognize and destroy foreign materials in the body.

Quite simply, cells cannot function without proteins. How does DNA make proteins? Actually, DNA does not make proteins directly. To synthesize proteins, genes are first copied into molecules called **messenger RNA (mRNA) (Figure 2.10)**. RNA synthesis is called **transcription,** because genes are literally transcribed (copied) from a DNA code into an RNA code. In turn, mRNA molecules, which are exact copies of genes, contain information that is deciphered into instructions for making a protein through a process known as **translation.**

RNA molecules are single-stranded, not double-stranded like DNA, but the chemical composition of RNA is very similar to that of DNA. The bases of RNA are also very similar to those of DNA. One key difference is that RNA contains a base called uracil (U) instead of thymine (T) (see Figure 2.4). The other primary difference is that RNA contains a pentose sugar

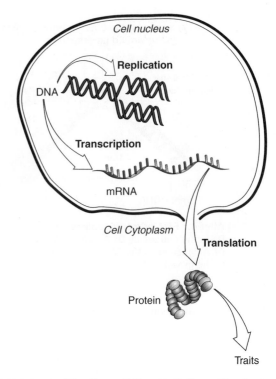

FIGURE 2.10   The Flow of Genetic Information in Cells DNA is copied into RNA during the process of transcription. RNA directs the synthesis of proteins during translation. Through proteins, genes control the metabolic and physical properties or traits of an organism.

called ribose, which has a slightly different structure than the deoxyribose sugar contained in DNA.

An easy way to remember the difference between transcription and translation is to remember that translation involves a change in code from RNA to protein, much like translating one language to another. Through the production of mRNA and the synthesis of proteins, DNA controls the properties of a cell and its traits (**Figure 2.11**). This process of transcription and translation directs the flow of genetic information in cells, controlling a cell's activities and properties. Here we study basic principles of transcription and translation and aspects of how gene expression can be controlled by cells.

Copying the Code: Transcription

How is DNA used as a template to make RNA? **RNA polymerase** is a key enzyme for transcription. Inside the nucleus, RNA polymerase unwinds the DNA helix and then copies one strand of DNA into RNA. Unlike DNA replication, in which the entire DNA molecule is copied, transcription occurs only in segments of a chromosome that contain genes. How does RNA polymerase know where to begin transcription?

TOOLS OF THE TRADE

Enzymes Involved in DNA Replication Have Valuable and Essential Roles in Molecular Biology Research

This chapter introduces important principles of DNA structure, replication, transcription, and translation and provides an essential background on genes and genomes. These principles are important not only for understanding what genes are and how they function but also many of the components involved in processes such as DNA replication, which have become essential tools for research in molecular biology.

In the next chapter, you will learn how scientists can identify, clone, and analyze genes—a fascinating aspect of molecular biology research and biotechnology. The technology for cloning and studying genes became possible only as scientists learned more about the enzymes involved in processes such as DNA replica-

tion. For instance, DNA polymerase, the key enzyme that synthesizes DNA during semiconservative replication, is widely used in molecular biology labs to copy DNA. Similarly, RNA polymerase is also widely used in molecular research. In addition, many DNA cloning procedures rely on DNA ligase to join together DNA fragments from different sources—a process called recombinant DNA technology.

Because most of these enzymes are relatively inexpensive and readily available from supply companies, their use has become commonplace in most molecular biology labs. Without our understanding of the enzymes that act on DNA and RNA and their functions, many modern applications of biotechnology and a majority of techniques routinely used in molecular biology research would be impossible.

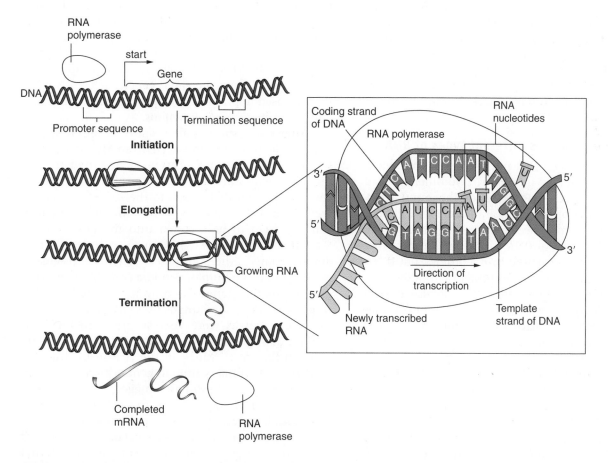

FIGURE 2.11 Transcription During transcription, RNA polymerase binds to DNA at a promoter region adjacent to a gene sequence and unwinds the DNA. RNA polymerase moves along the DNA template, copying one strand into a molecule of RNA. When RNA polymerase reaches a termination sequence, it releases from the DNA and transcription ends.

Adjacent to most genes is a **promoter,** specific sequences of nucleotides that allow RNA polymerase to bind at specific locations next to genes. Proteins called **transcription factors** help RNA polymerase find the promoter and bind to DNA; sequences called **enhancers** can also play important roles in transcription.

After RNA polymerase binds to a promoter, it unwinds a region of the DNA to separate the two strands. Only one of the strands, called the **template strand** (the opposite strand is called the *coding strand*), is copied by RNA polymerase. The presence of a promoter determines which strand of DNA will be transcribed. Once properly oriented, RNA polymerase proceeds along the DNA template strand to copy a complementary strand of RNA by forming phosphodiester bonds between ribonucleotides in a 5'-to-3' direction (Figure 2.11). When RNA polymerase reaches the end of a gene, it encounters a termination sequence. These sequences bind either specific proteins or base pairs to create loops at the end of the RNA. As a result, the RNA polymerase and newly formed RNA are released from the DNA molecule. Unlike DNA replication, in which the DNA is copied only once each time a cell divides, multiple copies of mRNA are transcribed from each gene during transcription. Sometimes a cell transcribes thousands of copies of mRNA from a gene. Later in this section, you will see that cells with a high requirement for a particular protein generally produce large amounts of mRNA to encode that protein.

Transcription produces different types of RNA

We have already seen that mRNA is produced when many genes are copied into RNA. Two other types of RNA, **transfer RNA (tRNA)** and **ribosomal RNA (rRNA),** are also produced by transcription. Different RNA polymerases produce each type of RNA. As we will soon learn, only mRNA carries information that directly codes for the synthesis of a protein, but tRNA and rRNA are also essential for protein synthesis.

It is only relatively recently that scientists discovered a new class of non-protein-coding RNA molecules called **microRNAs (miRNAs).** MicroRNAs are part of a rapidly growing family of small RNA molecules about 20 to 25 nucleotides in size that play novel roles in regulating gene expression. Later we'll briefly discuss the role of miRNAs.

mRNA processing

In eukaryotic cells, the initial mRNA copied from a gene is called a **primary transcript (pre-mRNA)**. This mRNA is immature and not fully functional. Primary transcripts undergo a series of modifications, collectively called mRNA processing, before they are ready for protein synthesis. One modification involves **RNA splicing** (**Figure 2.12**). When the details of transcription were first worked out, scientists were surprised to learn that genes are interrupted by stretches of DNA that do not contain protein-coding information, called **introns.** Introns are interspersed between **exons,** or protein-coding sequences of a gene. Introns and exons are copied during transcription of mRNA. Before mRNA can be used to make a protein, the exons must be spliced together. As a simple analogy, think of introns as randomly inserted letters in a sentence that must be removed before the sentence can make sense. In the splicing process, introns are cut out of the primary transcript and adjacent exons are spliced to form a fully functional mRNA molecule with no introns.

Splicing provides flexibility in the types of proteins that can ultimately be produced from a single gene. When genes were first discovered, scientists thought that a single gene could produce only one protein. But as the process of splicing was revealed, it became clear that when a gene contains several exons, splicing doesn't always occur in the same way. As a result, multiple proteins can be produced from a single gene. In a complex process called **alternative splicing,** splicing can sometimes join together certain exons and *cut out* other exons, essentially treating them as introns (Figure 2.12b). This process creates multiple mRNAs of different sizes from the same gene. Each mRNA can then be used to produce different proteins with different, sometimes unique functions.

Alternative splicing allows several different protein products to be produced from the same gene sequence. For instance, certain genes used to produce antibodies are alternatively spliced to produce some antibodies that attach to the surface of cells as well as other antibodies with different structures causing them to be secreted into the bloodstream. Similar splicing occurs with neurotransmitter genes, among many others. In fact, scientists originally believed that the human genome contained approximately 100,000 genes based on a predicted number of proteins made by human cells. As you will learn, genome scientists were very surprised to find that the human genome contains only about 20,000 genes (Chapter 3). Much of the reason for this discrepancy is that many human genes can be spliced in different ways, and estimates suggest that over 90% of human genes may use alternative splicing to create multiple different RNAs and proteins.

Another type of processing occurs at the 5' end of mRNA, where a guanine base containing a methyl group is added (Figure 2.12). Known as a 5' cap, this structure plays a role in ribosome recognition of the 5' end of the mRNA molecule during translation.

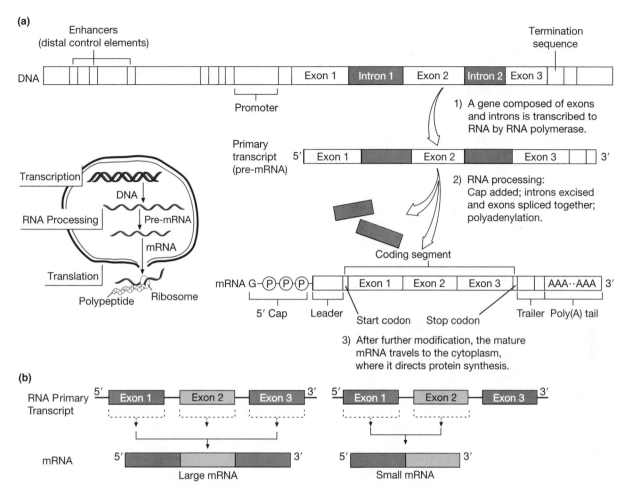

FIGURE 2.12 **A Eukaryotic Gene and mRNA Processing** (a) Transcription of a eukaryotic gene produces a primary transcript or pre-mRNA, which undergoes processing through RNA splicing, the addition of a 5' cap, and polyadenylation. After processing, the final, mature mRNA is ready for export to the cytoplasm, where it will be translated into a protein. (b) Alternative splicing can produce different mRNAs and protein products from the same gene. Notice that the larger mRNA on the left contains three exons spliced together but that the shorter mRNA on the right contains only two exons spliced together.

Last, in a process called **polyadenylation**, a string of adenine nucleotides around 100 to 300 nucleotides in length is added to the 3' end of the mRNA, creating a poly(A) "tail" (Figure 2.12). This tail protects mRNA from RNA-degrading enzymes in the cytoplasm, increasing its stability and availability for translation. Following processing, a mature, processed mRNA molecule leaves the nucleus and enters the cytoplasm, where it is now ready for translation (Figure 2.12).

Translating the Code: Protein Synthesis

The ultimate function of a protein-coding gene is to produce a protein. Here we look at a brief overview of translation, using information in mRNA to synthesize a protein from amino acids. Translation occurs in the cytoplasm of cells as a multistep process that involves several different types of RNA molecules. It will be much easier for you to understand the details of translation if you are familiar with important functions of each type of RNA. These are the major components of translation:

- Messenger RNA (mRNA)—a copy of a gene. Acts as a "messenger" by carrying the genetic code, encoded by DNA, from the nucleus to the cytoplasm, where this information can be read to produce a protein. Messenger RNAs vary in length from approximately 1,000 to several thousand nucleotides.

- Ribosomal RNA (rRNA)—short single-stranded molecules around 1,500 to 4,700 nucleotides long.

Ribosomal RNAs are important components of **ribosomes**, organelles that are essential for protein synthesis. Ribosomes recognize and bind to mRNA and "read" the mRNA during translation.

- Transfer RNA (tRNA)—molecules that transport amino acids to the ribosome during protein synthesis. Transfer RNA molecules are approximately 75 to 90 nucleotides in length.

We have already discussed the details of mRNA structure, but before we explore the details of translation, you need to be familiar with the genetic code of mRNA and the specific structures of ribosomes and tRNA.

The genetic code

What is the **"genetic code"** contained within mRNA? As you will learn shortly, ribosomes read the code and then produce proteins, which are formed by joining the building blocks called **amino acids** (see Appendix 2). A chain of amino acids linked together by covalent bonds is a **polypeptide.** Some proteins consist of a single polypeptide chain; others contain several polypeptide chains that must wrap and fold around each other to form complicated three-dimensional structures (see Figure 4.3). Proteins can contain combinations of up to 20 different amino acids, yet there are

only four bases in mRNA molecules. So how does this code work? What information is the ribosome decoding to tell a cell what amino acids belong in a protein? If there are only four nucleotides in mRNA, how can mRNA provide information coding for 20 different amino acids?

The answers to these questions lie in the genetic code, a universal language of genetics used by virtually all living organisms. The code works in three-nucleotide units called **codons,** which are contained within mRNA molecules. Each codon codes for a single amino acid (Table 2.3). For instance, notice that the codon UAC codes for the amino acid tyrosine, and the codon UGC codes for the amino acid cysteine. Although each codon codes for one amino acid, there is flexibility in the genetic code. There are 64 different potential codons corresponding to all possible combinations of the four possible bases assembled into three nucleotide codons (4^3). But because there are only 20 amino acids, most amino acids may be coded for by more than one codon. For example, notice in Table 2.3 that the amino acid lysine may be coded for by AAA and AAG. Having a redundancy of codons increases the efficiency of translation. Some codons are present in mRNAs with greater frequency than others, just as some words in the English language are preferred over others with identical meanings.

TABLE 2.3 THE GENETIC CODE

First Position (5' End)	Second Position: U	Second Position: C	Second Position: A	Second Position: G	Third Position (3' End)
U	UUU UUC Phenylalanine (Phe) UUA UUG Leucine (Leu)	UCU UCC UCA UCG Serine (Ser)	UAU UAC Tyrosine (Tyr) UAA Stop UAG Stop	UGU UGC Cysteine (Cys) UGA Stop UGG Tryptophan	U C A G
C	CUU CUC CUA CUG Leucine (Leu)	CCU CCC CCA CCG Proline (Pro)	CAU CAC Histidine (His) CAA CAG Glutamine (Gln)	CGU CGC CGA CGG Arginine (Arg)	U C A G
A	AUU AUC AUA Isoleucine (Ile) AUG Methionine (Met) START	ACU ACC ACA ACG Threonine (Thr)	AAU AAC Asparagine (Asn) AAA AAG Lysine (Lys)	AGU AGC Serine (Ser) AGA AGG Arginine (Arg)	U C A G
G	GUU GUC GUA GUG Valine (Val)	GCU GCC GCA GCG Alanine (Ala)	GAU GAC Aspartic Acid (Asp) GAA GAG Glutamic Acid (Glu)	GGU GGC GGA GGG Glycine (Gly)	U C A G

Also contained in the genetic code are codons that tell ribosomes where to begin or end translation. The start codon, AUG, codes for the amino acid methionine and signals the starting point for mRNA translation. As a result, the first amino acid in many proteins is methionine, although this amino acid is removed shortly after translation in some proteins. Stop codons terminate translation. UGA is a commonly used stop codon, but UAA and UAG are other stop codons (Table 2.3). Stop codons do not code for amino acids; they simply signal the end of translation.

The genetic code is universal. It is used by cells in humans, bacteria, plants, earthworms, fruit flies, and all other species. There are some subtle differences to the code in certain species, but at the basic level it operates the same way throughout biology. Because of this, biologists can use techniques called recombinant DNA technology to clone a human gene such as the insulin gene and insert it into bacteria so that bacterial cells transcribe and translate insulin—a

protein they normally do not produce. Another helpful aspect of the universal genetic code is that it enables scientists to clone a gene in one species, such as a mouse, and then use sequence information from the mouse gene to identify a similar gene in humans. Because different species share a common genetic code, this approach is a very common strategy for identifying human genes, including many involved in disease processes.

Ribosomes and tRNA molecules

Ribosomes are complex structures consisting of aggregates of rRNA and proteins that form structures called subunits. Each ribosome contains two subunits, the large and small subunits. These subunits associate to form two grooves, called the **A (aminoacyl) site** and the **P (peptidyl) site,** into which tRNA molecules can bind, and an **E site** through which tRNA molecules leave the ribosome (**Figure 2.13**).

Transfer RNAs are small molecules less than 100 nucleotides long. Transfer RNAs fold in intricate ways,

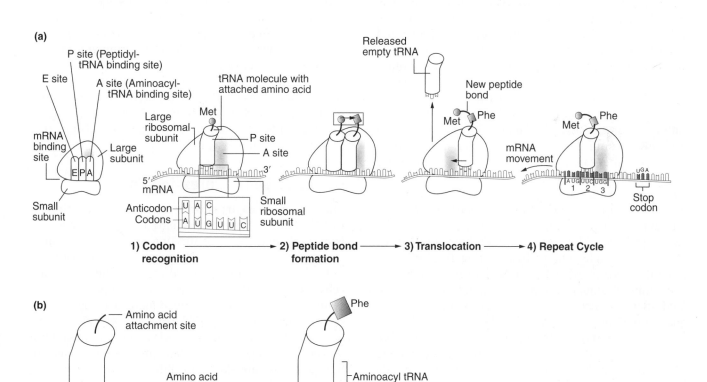

FIGURE 2.13 Stages of Protein Synthesis (a) Each ribosome contains a large and a small subunit. Shown here is a ribosome attached to mRNA; abbreviated steps of translation are shown in steps 1 to 4. (b) Diagrammatic example of the tRNA symbol used in this book. At one end of each tRNA is an amino acid binding site and at the opposite end is a three-nucleotide anticodon sequence.

and certain nucleotides in a tRNA base pair with each other. As a result, a tRNA assumes a structure called a cloverleaf, because as regions of the molecule base pair, other unpaired segments create loops. At one end of each tRNA is an amino acid attachment site (Figure 2.13b). Enzymes in the cytoplasm called aminoacyl tRNA synthetases attach a single amino acid to each tRNA molecule, creating what is known as an **aminoacyl transfer RNA (tRNA)** or "charged" tRNA. Aminoacyl tRNA molecules carry their amino acids to the ribosome and bind within grooves of the ribosomes at the A site. At the opposite end of each tRNA molecule is a three-nucleotide sequence called an **anticodon.** Different amino acids have different anticodon sequences. Anticodons are designed to complementary base pair with codons in mRNA. Now that you know the "players" of translation—mRNA, ribosomes, and tRNA—we will examine how these components come together to produce a protein.

Stages of translation

There are some fundamental differences between translation in prokaryotes and eukaryotes. Here we provide an overview of basic aspects of the three major stages of translation in eukaryotes: initiation, elongation, and termination. The beginning of translation is called *initiation*. During initiation, the small ribosomal subunit binds to the 5′ end of the mRNA molecule by recognizing the 5′ cap of the mRNA. Other proteins, called initiation factors, are also involved in guiding the small subunit to the mRNA. The small subunit moves along the mRNA until it encounters the start codon AUG. Pausing at the start codon, the small subunit waits for the correct tRNA, called the initiator tRNA, to come along (Figure 2.13). This tRNA has the amino acid methionine (met) attached to it (remember that most proteins begin with this amino acid) and contains the anticodon UAC. The UAC anticodon binds to the start codon by complementary base pairing (Figure 2.13); then the large ribosomal subunit binds to this complex containing the small subunit, initiation factors, mRNA, and initiator tRNA. Once these components are in place, the ribosome can start translating a protein.

The next step of translation is called *elongation*, because during this phase additional tRNAs enter the ribosome, one at a time, and a growing polypeptide chain is elongated. The ribosome, paused at the second codon, waits for the (second) tRNA to enter the A site. In Figure 2.13, notice that the second codon is UUC, which codes for the amino acid phenylalanine (phe). The phe-tRNA enters the A site of the ribosome and the anticodon (AAG) base pairs with the codon. After two tRNAs are attached to the ribosome, an enzyme in the ribosome called **peptidyl transferase** catalyzes the formation of a peptide bond between the amino acids

(attached to their tRNAs). Peptide bonds join together amino acids to form a polypeptide chain.

After the amino acids are attached to each other, the initiator tRNA, without methionine attached, pauses briefly in the E site and then is released from the ribosome. Released "empty" tRNAs are recycled by the cell. A new amino acid is attached to the tRNA to create an aminoacyl tRNA, which can be used again for translation. The newly forming polypeptide remains attached to the tRNA in the A site. During a phase called *translocation*, the ribosome shifts so the tRNA and growing protein move into the P site of the ribosome. The tRNA with a growing polypeptide chain attached is called a peptidyl tRNA. The A site of the ribosome is now aligned with the third codon in sequence (UGG, which codes for tryptophan), and the ribosome waits for the proper aminoacyl tRNA to enter the A site. The cycle continues as described to attach the next amino acid (tryptophan) to the growing protein and repeats itself as the ribosome moves along the mRNA.

Elongation cycles continue to form a new protein until the ribosome encounters a stop codon (for instance, UGA). This signals the third stage of translation called *termination*. Remember that stop codons do not code for an amino acid. Proteins called *releasing factors* interact with the stop codon to terminate translation. The ribosomal subunits come apart and release from the mRNA, and the newly synthesized protein is released into the cell. Ribosomes do recycle and subsequently can bind to any other mRNA molecule (not just the mRNA for one particular gene) and start the process of translation again.

Basics of Gene Expression Control

The term **gene expression** refers to the production of mRNA (and sometimes protein) by a cell. Cells are exquisitely effective at controlling gene expression and translation to accommodate their needs. All genes are not transcribed and translated at the same rate in all cells. Gene expression is a dynamic process. Individual genes can be actively transcribed for a particular period of time, followed by periods of relatively little transcription depending on the needs of a cell. In addition, all cells of an organism contain the same genome, so how and why are skin cells different from brain cells or liver cells? Different cell types have different properties and carry out different functions because cells can *regulate* or control the genes they express. At any given time in a cell, only certain genes are "turned on" or expressed to produce proteins, while many other genes are silenced or repressed. These genes may be expressed by cells only at certain times, in response to specific cues from inside or outside of the cell, to make proteins as needed. These cues can be environmental

YOU DECIDE

Access to Biotechnology Products for Everyone?

Later, you will learn how genes can be identified, cloned, and studied in great detail (Chapter 3). One benefit of gene cloning has been the identification of genes involved in human diseases. As a result, it is possible to make many gene products in the laboratory and use them for medical purposes. For instance, when the gene for insulin was cloned in bacteria, it became possible to produce large amounts of insulin for treating people with diabetes. Similarly, cloning of the gene for human growth hormone (hGH), which stimulates growth of bones and muscles during childhood, provided a readily available source of this hormone. Legally available by prescription only, hGH is used widely and effectively to treat children with certain forms of short stature or dwarfism. Illegal use of hGH by professional athletes has received much attention in recent years. Dwarfism is generally defined as a condition that results in an adult height of 4 feet, 10 inches or shorter. The availability of hGH and other products of biotechnology raises an ethical question. Should hGH be available to everyone who wants taller children or only those children with dwarfism? Suppose parents wanted their average-size son to be taller so that he would have a better chance of making his high school varsity basketball team. Should these parents be able to give their son hGH simply to enhance his height? You decide.

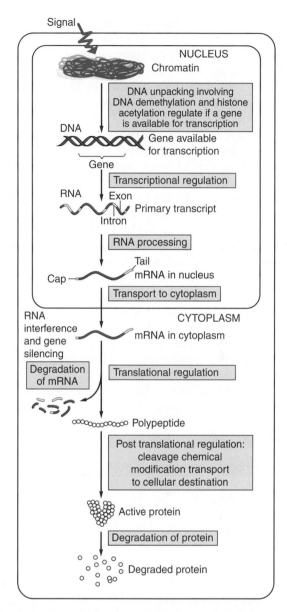

FIGURE 2.14   Levels of Gene Expression Regulation
Prokaryotic and eukaryotic cells can regulate gene expression in a variety of complex ways. This figure summarizes the primary ways in which gene expression can be regulated in eukaryotic cells. Notice that gene expression regulation can occur at many different "levels," (highlighted in blue boxes here) beginning with how chromatin is folded or chemically modified to controlling degradation or turnover of a protein once it is made. Transcriptional regulation is a commonly used control mechanism in both prokaryotic and eukaryotic cells.

signals such as temperature changes, nutrients in the external environment, hormones, or other complex chemical signals exchanged by cells.

How can genes be turned on and off in response to different signals? Biologists call this process **gene regulation.** Prokaryotic cells and eukaryotic cells regulate gene expression in a variety of complex ways (**Figure 2.14**). One common mechanism used by both types of cells is called **transcriptional regulation—** controlling the amount of mRNA transcribed from a particular gene as a way to turn genes on or off. Here we provide an introduction to transcriptional regulation and consider basic examples of this process in eukaryotes and prokaryotes.

Transcriptional regulation of gene expression

Because the amount of protein translated by a cell is often directly related to the amount of mRNA in the cell, cells can regulate the amount of mRNA produced for any given gene to control indirectly the amount of protein a cell produces. How do cells know which genes to turn on and which to shut off? To understand transcriptional regulation, we must look at the role of promoter sequences more closely.

Promoters are found "upstream" of gene sequences, meaning that they are found at the 5′ end of a gene. Genes in prokaryotes and eukaryotes do not all use the same promoter sequences. In eukaryotes, common promoter sequences found upstream of many genes include a **TATA box** (TATAAAA), located about 30 nucleotides (−30) upstream of the start site of a gene, and a **CAAT box** (GGCCAATCT), located about 80 nucleotides (−80) upstream of a gene (**Figure 2.15**).

Earlier in this chapter we discussed how RNA polymerase initiates transcription by binding to promoter sequences adjacent to genes. For most eukaryotic genes, RNA polymerase cannot recognize and bind to a promoter unless transcription factors are present at the promoter. Transcription factors are DNA-binding proteins that can bind promoters and interact with RNA polymerase to stimulate transcription of a gene (Figure 2.15). In eukaryotes and prokaryotes, common transcription factors interact with promoters for many genes; however, both types of cells also use specific transcription factors that interact only with certain promoters. Transcription of some genes also depends on the binding of specific transcription factors to regulatory sequences adjacent to the promoter. In addition, many genes that are tightly regulated by cells also contain regulatory sequences called **enhancers.**

Enhancer sequences are usually located around 50 or more base pairs upstream of the promoter, but they can also be located downstream of a gene. Enhancer sequences bind regulatory proteins, generally referred to as *activators*. Activator molecules interact with transcription factors and RNA polymerase, forming a complex that stimulates (activates) transcription of a gene. Cells use a wide variety of different activator molecules. Each activator binds to a particular enhancer sequence. Some activator molecules can be hormones. For instance, the male sex steroid testosterone can be an activator to stimulate gene expression. You may know that testosterone stimulates cellular activities such as muscle and hair growth in developing boys, but how does testosterone work? This hormone binds to a receptor protein inside cells. The testosterone-receptor protein complex acts as an activator to bind to a specific enhancer element in DNA called an *androgen-response element* (5′-TGTTCT-3′). These elements are usually found close to a promoter. In turn, testosterone and its receptor stimulate gene expression. The female sex steroid estrogen works in a similar manner.

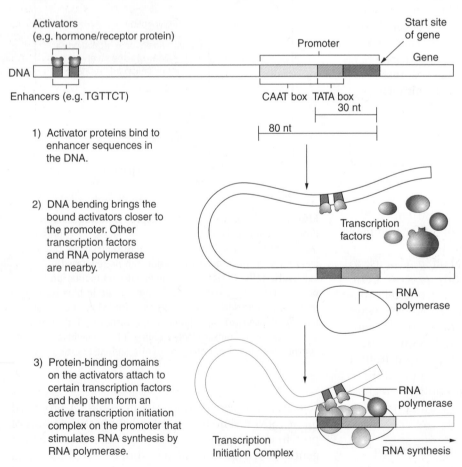

FIGURE 2.15 Promoters, Transcription Factors, and Enhancers

But testosterone and other activators don't stimulate expression of all genes in all cells. Activators act only on those genes that contain enhancer sequences to which they can bind. Testosterone stimulates expression of genes involved in muscle and hair growth because these genes contain androgen-response elements. Transcription of other genes without androgen-response elements are not directly affected by the hormone. Incidentally, steroid abuse by bodybuilders and athletes looking to increase muscle mass and tone can cause serious long-term health effects, in part because steroids abnormally stimulate gene expression for prolonged periods of time.

Through activators and enhancers, cells can use transcriptional control to regulate gene expression and control cellular activities. Some genes even contain repressor sequences that decrease transcription. Because different cells produce different transcription factors and activator molecules, genes can be turned on in some tissues and not others. Skin cells turn on different genes than muscle cells do, so each cell type produces different proteins, giving each cell type different functions. Consequently tissue- and cell-specific gene expression is one way for cells to control the proteins they express even though all body cells contain the same genome. These important control mechanisms are part of why different cells have different functions.

In addition, identifying the promoters, enhancer sequences, and transcription factors that bind to regulatory regions of a gene is important for making biotechnology products. For instance, the identification of transcription factors that stimulate the expression of proteins needed for bone growth and development is helping scientists develop new drugs that can be used to stimulate bone growth in people suffering from forms of arthritis when their cells no longer produce the factors that stimulate bone growth.

Bacteria use operons to regulate gene expression

Bacteria are very important organisms for many applications of biotechnology, such as the production of human proteins. Many initial studies on gene regulation were carried out in bacteria. Bacteria and other microorganisms can and must rapidly control gene expression in response to environmental stimuli such as growth nutrients and changes in temperature and light intensity. One interesting aspect of gene expression and regulation in bacteria is that many bacterial genes are organized in arrangements called **operons**. Operons are essentially clusters of several related genes located together and controlled by a single promoter.

The genes of an operon can be regulated in response to changes within the cell. Operons can be stimulated (induced) or inhibited (repressed) depending on the needs of a cell. Many genes controlling nutrient metabolism by bacteria are organized as operons. Bacteria can use operons to tightly regulate gene expression in response to their nutrient requirements. Here we present a well-studied classic example of gene regulation in bacteria by describing the *lac* operon (**Figure 2.16**).

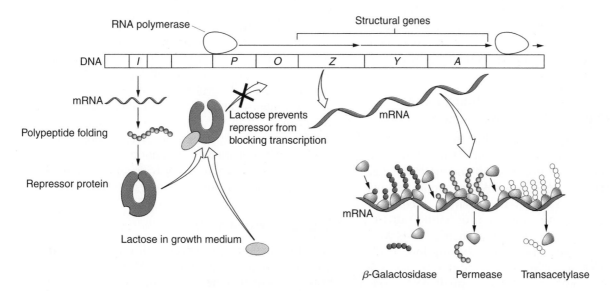

FIGURE 2.16   The *lac* Operon By controlling the *lac* operon, bacterial cells can regulate gene expression in response to availability of the sugar lactose. In the absence of lactose, the *lac* repressor binds to the operator, blocking its transcription. In the presence of lactose, lactose binds and inactivates the repressor, allowing transcription of the operon to occur.

The *lac* operon consists of the following three genes:

- *lacZ*, encoding the enzyme β-*galactosidase*
- *lacY*, encoding the enzyme *permease*
- *lacA*, encoding the enzyme *acetylase*

Together, these three enzymes are necessary for the transport and breakdown of lactose by bacterial cells. Lactose, a sugar present in milk, is an important energy source for many bacteria. For bacteria to metabolize lactose, the sugar must be transported into cells by permease and then degraded into glucose and galactose by β-galactosidase. The function of acetylase is not clear, although it may play a role in protecting cells from the toxic products of lactose degradation. The *lac* operon is regulated by a protein called the **lac repressor,** which is encoded by a separate gene called the *lacI* gene. When bacteria are grown in the absence of lactose, the repressor protein uses helix-turn-helix motifs to bind a sequence within the *lac* operon promoter (p) called the **operator** (o). By binding to the operator, the repressor blocks RNA polymerase from binding to the promoter and blocks transcription of the *Z*, *Y*, and *A* genes in the operon (Figure 2.16). This is a nice way for bacteria to control their metabolism. Why expend energy transcribing genes and translating proteins if there is no lactose available for producing energy?

Conversely, in the presence of lactose, the sugars act as inducer molecules that stimulate transcription of the *lac* operon. Lactose binds to the *lac* repressor, changing the shape of the repressor protein and preventing it from binding to the operator (Figure 2.16). With no repressor in its way, RNA polymerase can bind to the *lac* promoter and stimulate transcription of the operon. Transcribed mRNA from the operon is translated to produce the enzymes required by the cell to metabolize lactose.

MicroRNAs silence genes revealing a recent discovery of gene expression regulation

In 1998, scientists studying the roundworm *Caenorhabditis elegans* discovered small (21 or 22 nt) double-stranded pieces of non-protein-coding RNA called **short interfering RNA (siRNA),** so named because they were shown to bind to mRNA and subsequently block or interfere with translation of bound mRNAs. For a short time it appeared that perhaps siRNAs were restricted to *C. elegans*, but while looking for siRNAs in other species, scientists discovered that microRNAs (miRNAs) are another category of small RNA molecules that do not encode proteins. Genes for miRNAs have been identified in all multicellular organisms studied to date.

Like siRNAs, miRNAs are regulatory molecules that regulate gene expression by "silencing" gene expression through blocking translation of mRNA or by causing degradation of mRNA. MicroRNA genes produce RNA transcripts that are processed by enzymes (Drosha and Dicer) into short single-stranded pieces 21 to 22 nucleotides in length (**Figure 2.17**). The processed miRNAs then bind to a complex of proteins (called RISCs), which allows them to bind to mRNAs that are complementary to the miRNA sequence. Binding of miRNA to an mRNA sequence can either block translation by ribosomes or trigger enzymatic degradation of an mRNA. Both mechanisms inhibit or silence expression by preventing the mRNA from being translated into a protein.

RNA-based mechanisms of gene silencing are generally referred to as **RNA interference (RNAi).** The identification of human genes that are silenced by miRNAs is currently a very active area of research. Scientists are working on exploiting RNAi as a potentially promising way to use small RNA molecules to selectively turn off or silence genes involved in human disease conditions (discussed in Chapters 3 and 11). As evidence of how valuable RNAi is as a research technique, the 2006 Nobel Prize in Physiology or Medicine was awarded to Andrew Fire of Stanford University and Craig Mello of the University of the Massachusetts School of Medicine for their discovery of this method for switching genes off.

2.5 Mutations: Causes and Consequences

We conclude with a brief discussion of how genes can be affected by **mutations,** or changes in the nucleotide sequences of DNA. Mutations are a major cause of genetic diversity. For instance, the underlying basis of the evolution of species to develop and acquire new characteristics is governed by mutations of genes over time. Mutations can also be detrimental. Mutation of a gene can result in the production of an altered protein that functions poorly or in some cases no longer encodes a functional protein. Such mutations can cause genetic diseases. In this section, we provide an overview of different types of mutations and their consequences.

Types of Mutations

Sometimes mutations occur through spontaneous events such as errors during DNA replication. For instance, DNA polymerase can insert the wrong nucleotide into a newly synthesized strand of DNA, say inserting a T where a C belongs. Even though enzymes in cells work to detect and correct mistakes, errors occasionally occur during DNA replication. Mutations can also be induced by environmental causes. For

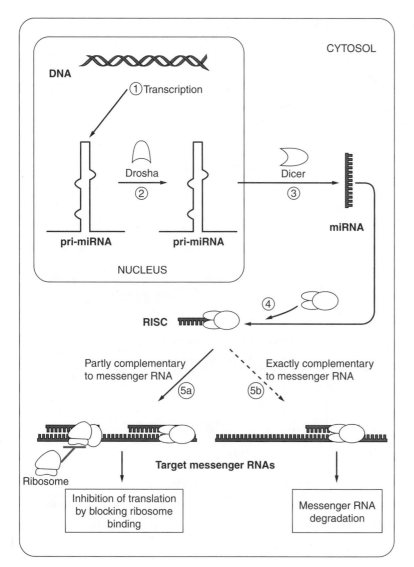

FIGURE 2.17 **MicroRNAs Silence Gene Expression** Gene silencing by miRNAs is a multistep process. (1) MicroRNA genes transcribe a primary transcript called pri-RNA that (2) folds into a hairpin loop similar to the folding patterns of tRNA molecules. (3) The enzyme Dicer cleaves pri-mRNAs into single-stranded miRNAs that are 21 to 22 nucleotides long. (4) MicroRNAs bind to a cluster of proteins to form a ribonucleoprotein complex. (5) This miRNA complex can then bind to mRNA molecules by complementary base pairing with sequences that are an exact match or a close match to complementary sequences in the mRNA and miRNA. (5a) miRNA binding then silences gene expression by blocking translation or by targeting the bound mRNA for degradation by enzymes in the cell.

example, chemicals called **mutagens,** many of which mimic the structure of nucleotides, can mistakenly be introduced into DNA and change DNA structure. Exposure to x-rays or ultraviolet light from the sun can also mutate DNA (That glowing tan you may enjoy during the summer is not as healthy as you think).

Regardless of how mutations arise, depending on the type of mutation, they may have no effect on protein production or they can dramatically change protein production and protein structure and function. Mutations can involve large changes in genetic information or single nucleotide changes in a gene, such as changing an A to a C or a G to a T. The most common mutations in a genome are single-nucleotide changes (or a few nucleotide mutations) called **point mutations.** Such mutations are commonly referred to as **single-nucleotide polymorphisms (SNPs,** pronounced "snips"), and they represent one major

type of genetic variation in the genomes of different humans. We will discuss SNPs in several chapters. Point mutations often involve base-pair substitutions, in which a base pair is replaced by a different base pair; insertions, in which a nucleotide is inserted into a gene sequence; or deletions, the removal of a base pair (**Figure 2.18** on the next page).

Mutations ultimately exert their effects on a cell by changing the properties of a protein, which in turn can affect traits.

Gene mutation can cause:

↓

Changes in protein structure and function; synthesis of a nonfunctional protein or no protein synthesized can lead to:

↓

Change or loss of a trait

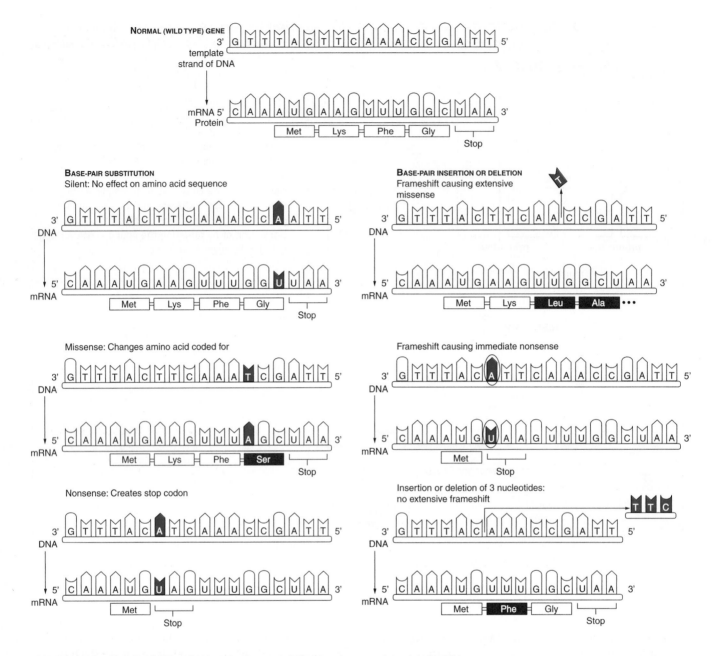

FIGURE 2.18 **Types of Mutations** Mutations can influence the genetic code of mRNA and resulting proteins translated from a gene. Shown here is a portion of mRNA copied from a gene, but mutations generally occur within DNA. Mutations in a gene can have different effects on the protein that is translated.

Proteins are large, complicated molecules. To work correctly, most proteins must be folded into complex three-dimensional shapes. Changing one or two amino acids in a protein can alter its overall shape, dramatically disrupting its function or in some cases preventing it from functioning at all. A mutation may have no effect on a protein if it changes the codon sequence of a gene to another codon that codes

for the same amino acid (Figure 2.18). This is a **silent mutation** because it has no effect on the structure and function of the protein. Similarly, a mutation can change a codon so that a different amino acid is coded for. Such **missense mutations** are also considered "silent" if the new amino acid coded for doesn't change the protein's structure and function. However, if the newly coded amino acid changes the structure

CAREER PROFILE

Careers in Genomics

It is an incredibly exciting time to consider a career in genomics. Never before have there been greater opportunities or a wider variety of career options for anyone interested in genomes. In addition to studying the human genome, scientists are actively involved in studying the genomes of many other species, including model organisms such as mice, fruit flies, zebrafish, agriculturally important crop plants and plant pests, disease-causing microbes, and marine organisms. As enormous amounts of genomic information become available, decades of work will be necessary to study what different genes do. The process of deciphering the secrets contained in genomes will involve the combined efforts of many people.

Career opportunities in genomics primarily fall into four major categories: laboratory scientists, clinical doctors, genetic counselors, and bioinformatics experts. Laboratory scientists are the so-called bench scientists, because they conduct experiments at the lab bench daily. Lab scientists are often involved in carrying out experiments to discover and clone genes and study their functions. Entry-level lab scientist opportunities exist as lab technicians for those with associate's and bachelor's degrees; higher-level scientist positions such as laboratory director positions require a master's degree or a doctor of philosophy (PhD.) degree. Clinical doctors are M.D.-trained physicians who conduct research and interact with patients as members of research teams. Clinical doctors may be involved in research studies to treat patients with new genetics-based treatments such as gene therapy protocols. Physicians are also important because genomics is having profound effects on medicine and will continue to do so in the future. There is an increased emphasis on *translational medicine* designed to bring genetics research from the lab bench to the bedside by creating new genetics-based treatments of human diseases. Genetic counselors help people understand genomics information such as how genes affect disease susceptibility. Counselor positions usually require a M.S. degree with training in psychology, biology, and genetics.

Bioinformatics is a discipline that merges biology with computer science. Bioinformatics involves storing, analyzing, and sharing gene and protein data. The tremendous amount of DNA sequence data being generated by genome projects has made bioinformatics a rapidly emerging area that is absolutely essential to genomics. Bioinformaticians generally have solid training in both biology and computer science, and they work together with bench scientists to analyze genome data. Most positions in bioinformatics require a B.S., M.S., or PhD. degree.

Visit the Human Genome Program Information page in Keeping Current: Web Links at the Companion Website for an outstanding view of career possibilities in genetics; this site also includes links to many other valuable resources.

---

of the protein, its function may be significantly altered. Later, we consider a dramatic example of how a SNP is responsible for the human genetic condition called sickle cell disease.

Sometimes mutations called **nonsense mutations** change a codon for an amino acid into a stop codon, which causes an abnormally shortened protein to be translated, usually creating a nonfunctional protein. Insertions or deletions can also dramatically affect the protein produced from a gene by creating a **frameshift mutation.** As shown in Figure 2.18, the insertion of a nucleotide causes the reading frame of the codons to be shifted to the right of the insertion, changing the protein encoded by the mRNA. Frameshifts often create nonfunctional proteins.

Mutations Can Be Inherited or Acquired

It is important to realize that not all mutations have the same effect on body cells. The effects of a mutation depend not only on the type of mutation s but also on the cell type in which it occurs. Gene mutations can be inherited or acquired. **Inherited mutations** are mutations passed to offspring through gametes—sperm or egg cells. As a result, the mutation is present in the genome of all of the offspring's cells. Inherited mutations can cause birth defects or inherited genetic diseases.

Acquired mutations occur in the genome of somatic cells and are not passed along to offspring. Although not inherited, acquired mutations can cause abnormalities in cell growth, leading to the formation of cancerous tumors as well as metabolic disorders and other conditions. For instance, prolonged exposure to ultraviolet light can cause acquired mutations in skin cells, leading to skin cancer. The effort to understand the genetic basis of cancer and many other human diseases is a major area of biotechnology research.

Mutations are the basis of variation in genomes and a cause of human genetic diseases

Mutations form the molecular basis of many human genetic diseases. Sickle cell disease was the first genetic disease discovered; its cause was pinpointed to a particular mutation (**Figure 2.19**). Sickle cell disease is

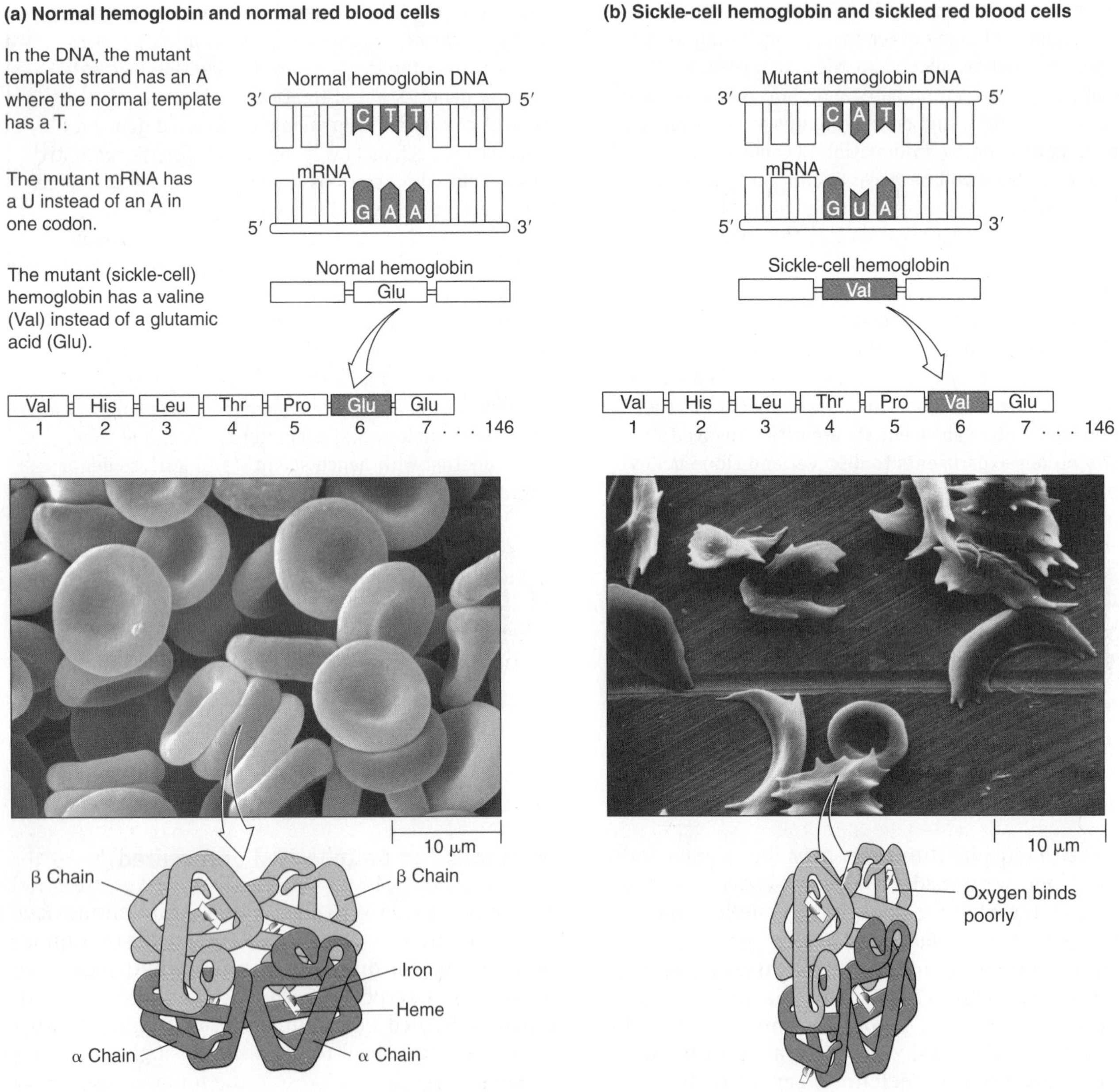

(a) Normal hemoglobin and normal red blood cells

In the DNA, the mutant template strand has an A where the normal template has a T.

The mutant mRNA has a U instead of an A in one codon.

The mutant (sickle-cell) hemoglobin has a valine (Val) instead of a glutamic acid (Glu).

(b) Sickle-cell hemoglobin and sickled red blood cells

FIGURE 2.19 Molecular Basis of Sickle-Cell Disease (a) Hemoglobin, the oxygen-binding protein of red blood cells, consists of four polypeptide chains. A portion of the normal hemoglobin gene is shown here along with its transcribed mRNA and the first seven amino acids for one of the hemoglobin polypeptides, which contains a total of 146 amino acids. (b) The defective gene that causes sickle cell disease contains a single base-pair change in its sequence that alters translation of the hemoglobin protein. This subtle change alters the shape of red blood cells, causing them to sickle.

created by a single nucleotide change, a base-pair substitution, in the gene coding for one of the polypeptides in the protein **hemoglobin**, the oxygen-binding protein in red blood cells. Red blood cells contain millions of hemoglobin molecules, each of which consists of four polypeptide chains (Figure 2.19). A point mutation in the β-globin gene changes the genetic code so that the gene codes for a different amino acid at position 6 in one of the hemoglobin polypeptides. As a result, the amino acid valine, instead of glutamic acid, is inserted into sickle cell hemoglobin. Individuals with two defective copies of the hemoglobin gene suffer the effects of sickle cell disease. This subtle mutation alters the oxygen-transporting ability of hemoglobin and dramatically changes the shape of red blood cells to an abnormal sickled shape. Sickled cells block blood vessels, and patients suffer from poor oxygen delivery to tissues, causing joint pain and other symptoms. Sickle cell disease is one of the best-understood inherited genetic disorders.

As scientists have learned more about the human genome, they have discovered that DNA sequences from people of different backgrounds around the world are very similar. Regardless of ethnicity, human genomes are approximately 99.9 percent identical. In other words, your DNA sequence is about 99.9 percent the same as DNA sequences from President Barack Obama, LeBron James, Britney Spears, Julia Roberts, Justin Bieber—or virtually any other human.

But because there is about a 0.1 percent difference in DNA between one individual and another, or around one base out of every thousand, this means that there are roughly 3 million differences between different individuals. Many of these variations are created by mutation, and most of them have no obvious effects; however, other mutations strongly influence cell function, behavior, and susceptibility to genetic diseases. These variations are the basis of differences in all inherited traits among people, from height and eye color to personality, intelligence, and life span.

Most genetic variation between human genomes is created by SNPs. For instance, at a particular sequence, a certain base in a given region of DNA sequence from President Barack Obama may read "A," Britney Spear's sequence may read "T," LeBron's sequence may read "C," and the same site in your sequence may read "G." Most SNPs are harmless because they occur in intron regions of DNA; however, when they occur in exons, they can affect the structure and function of a protein, which can influence cell function and result in disease. Sickle cell disease is caused by an SNP. Refer to Figure 1.11, which shows a comparison of a gene sequence from three different people. An SNP in this gene in person 2 may have no effect on protein structure and function if it is

FIGURE 2.20   Biotechnology Is Being Used to Detect and Correct Mutations

a silent mutation. Other SNPs (person 3) cause disease if they change protein structure and function.

Later, we'll consider several genetic disease conditions, discuss how defective genes can be detected, and examine how scientists are working on gene therapy approaches to cure these diseases (**Figure 2.20**; see also Chapter 11 for more on genetic disease conditions).

2.6 Revealing the Epigenome

Our study of the genomes of humans and other species has provided great insight on the **epigenome** and its influences on gene expression. The term *epigenome* refers to modifications in chromatin structure, which *do not* involve mutations in DNA sequence. Some epigenetic modifications are inherited, while others vary from generation to generation. Some epigenetic modifications may be reversible, while others are long-lasting. Epigenetic changes can differ between cell types in the body and in normal and diseased tissues. Diet and environmental conditions can influence the epigenome. We also know that epigenetics plays a major role in clarifying how patterns of gene expression can vary during embryonic development.

The more we study the epigenome, the more we learn about how complex it is. We are only now beginning to truly appreciate the importance of the epigenome. So what kinds of chromatin modifications contribute to the epigenome? Epigenetic modifications affect both DNA and histones. For example, the addition of methyl groups (-CH3) to DNA, known as *methylation*, commonly occurs on cytosine bases.

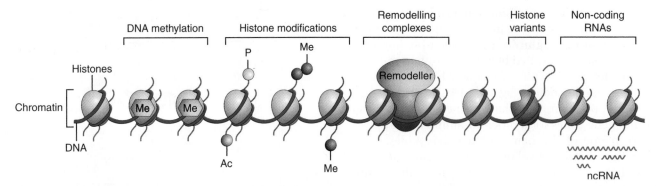

FIGURE 2.21 **Chromatin Modifications Involved in Epigenetics** The chromatin modifications shown here are responsible for many epigenetic influences on gene expression. These include DNA methylation (Me); histone modification by acetylation (Ac), phosphorylation (P), and methylation; remodeling of histones and other DNA-binding proteins; histone variations; and noncoding RNAs (ncRNAs).

Methylation of histone proteins also occurs. Other examples of histone modifications include the addition of acetyl groups (CO CH₃), or *acetylation*, and the addition of phosphate groups (PO₄), or *phosphorylation* (**Figure 2.21**).

The extent to which methylation, acetylation, and phosphorylation affect a genome can vary greatly from one region of a chromosome to another. These chemical modifications work in complex ways to affect how DNA loops and folds in a chromosome. Such structural effects can ultimately influence how easily gene expression occurs in a particular region of a genome because epigenetic modifications affect how "accessible" a segment of DNA is to proteins involved in transcription. For instance, heavily acetylated regions of a genome cause the DNA to maintain a more open and less compact arrangement that allows for transcription. Removing acetyl groups from such a section of the genome typically inhibits transcription. Other epigenetic modifications that affect transcription include remodeling of histones and other DNA-binding proteins which can mask or activate regions of a genome, variations in the types of histone proteins found in DNA, and the presence or absence of noncoding RNAs that can coat DNA and influence gene expression (Figure 2.21).

Because the epigenome regulates gene expression, almost every major pharmaceutical company and many biotech companies are working on epigenetics projects. We know that the epigenome is involved in many diseases, including cancer. Inhibiting or enhancing epigenetic modifications to target epigenetic changes that can affect gene expression for disease treatment is a very hot area of research. For example, a relatively new category of drugs called histone methyltransferases (HMTs), which affect gene expression by adding methyl groups to histone proteins, have shown promise in treating specific malignant tumors and leukemia. Keep in mind this brief introduction to the epigenome as you learn more about genomes and ways in which the biotechnology industry is working on treatments for genetic diseases.

MAKING A DIFFERENCE

Scientists around the world are working on a variety of cell therapy approaches to the treatment of human disease. For example, researchers at Rush University Medical Center and Emory University are investigating applications of a novel cell therapy using retinal pigmented epithelial cells (RPE cells) attached to tiny gelatin beads. In early-stage studies with human patients, these beads have been implanted in the brains of patients with Parkinson's disease (PD). Patients treated so far are showing improvements in parkinsonian symptoms such as tremors, rigidity, slowness of movements, and impaired balance and coordination. This cell therapy approach holds great promise, as some of the treated patients continue to show improvements 6 years or more after treatment. The RPE cells produce levodopa, a precursor to the neurotransmitter dopamine, produced by neurons in the brain. Loss of dopamine-producing neurons is a major cause of PD. In this approach, RPE cells are cultured in a lab and attached to gelatin beads, which are necessary for the cells to survive in the brain. Then the beads are implanted into patients. The implanted cells provide a source of levodopa, which can be used by local neurons to produce dopamine. Only time will tell if this approach will become a viable treatment option or cure for PD, but without question there are many cell-based strategies under way to treat disease that will ultimately "make a difference" and alleviate pain and suffering.

QUESTIONS & ACTIVITIES

Answers can be found in Appendix 1.

1. Compare and contrast genes and chromosomes and describe their roles in the cell.

2. If the sequence of one strand of a DNA molecule is 5'-AGCCCGACTCTATTC-3', what is the sequence of the complementary strand?

3. What does the phrase "gene expression" mean?

4. Suppose you identified a new strain of bacteria. If the DNA content of this organism's cells is 13% adenine, approximately what percentage of this organism's genome consists of guanine? Explain your answer.

5. Provide at least three important differences between DNA and RNA.

6. Consider the following sequence of mRNA: 5'-AGCACCAUGCCCCGAACCUCAAAGUGAAA-CAAAAA-3'. How many codons are included in this mRNA? How many amino acids are coded for by this sequence? Use Table 2.3 to determine the amino acid sequence encoded by this mRNA. Note: Remember that mRNA molecules are actually much larger than the very short sequence shown here.

7. Consider the following sequence of DNA:

    5'-TTTATGGG TTGGCCCGGGTCATGATT- 3'
    3'-AAATCCCAACCGGGCCCAGT ACTAA- 5'

    a. Transcribe each of these sequences into mRNA. Which DNA sequence (top or bottom strand) produces a functional mRNA containing a start codon? What is the amino acid sequence of the polypeptide produced from this mRNA?

    b. For the DNA strand producing a functional mRNA, number each base from left to right

with the first base numbered "1." Insert a "T" between bases 10 and 11, representing a base insertion mutation. Transcribe an mRNA from this new strand and translate it into a protein. Compare the amino acid sequence of this protein to the one translated in (a). What happened? Explain.

8. Name the three types of RNA involved in protein synthesis and describe the function of each.

9. What is an operon? How do bacteria use operons to regulate gene expression?

10. What is gene expression regulation? Why is it important? Describe examples of ways in which gene expression can be regulated.

11. Diagram the process of semiconservative replication of DNA by drawing a replication fork and indicating important enzymes, proteins, and other components involved in this process. Provide a *one-sentence* description of the function of each component.

12. Why do some mutations affect changes in protein structure and function that can result in disease whereas other mutations have no significant effects on protein structure and function?

13. In a nucleotide of DNA, which carbon in the deoxyribose sugar is bonded to the nitrogenous base?

14. What is the epigenome and why are biotechnology and pharmaceutical companies interested in the epigenome?

Visit www.pearsonhighered.com/biotechnology

To download learning objectives, chapter summary, "Keeping Current" web links, glossary, flashcards, and jpegs of figures from this chapter.

Recombinant DNA Technology and Genomics

After completing this chapter you should be able to:

- Define recombinant DNA technology and explain how it is used to clone genes and manipulate DNA.

- Compare and contrast different types of cloning vectors; describe their practical features and applications.

- Discuss how DNA libraries are created and screened to identify cloned genes of interest.

- Describe how agarose gel electrophoresis, polymerase chain reaction (PCR), DNA sequencing, and other molecular techniques are used to study gene structure, function, and expression.

- Describe how whole-genome shotgun sequencing and high-throughput sequencing techniques enable scientists to rapidly analyze genomes.

- Understand major findings of the Human Genome Project, potential scientific and medical applications of knowledge about the human genome, and associated ethical, legal, and social issues.

- Explain why genomics-related "omics" disciplines are developing rapidly as areas of research.

- Provide examples of how bioinformatics can be used to analyze nucleic acid and protein sequences and structures.

Undergraduate biology students working on a recombinant DNA experiment.

As you have learned, biotechnology is not a new science. However, the modern era of biotechnology began when DNA cloning techniques were developed. Since the 1970s and continuing today, amazing and rapidly developing laboratory methods in **recombinant DNA technology** and **genetic engineering** have changed molecular biology, basic science, and medical research forever. In this chapter, we present an overview of recombinant DNA technology. We then take a look at an amazing range of techniques that scientists use to clone genes and study gene structure and function. The chapter concludes with an introduction to genomics and bioinformatics, which provide methods for studying genomes.

FORECASTING THE FUTURE

There are many exciting potential future directions in recombinant DNA and genomics research. Without question one area of substantial progress will be the completion of thousands of genome projects for different plants, animals, bacteria, viruses, and other organisms. Scientists continue to prospect genomes from organisms around the world to look for new genes with potential applications in biotechnology. Following on the success of the Human Genome Project, the use of genomic information for new applications—including novel methods for the early detection of disease genes and novel gene-based treatment strategies for diseases such as cancer and many other human health conditions—holds tremendous promise for alleviating pain and suffering through biotechnology. And on the near horizon is the sequencing (personal genomics) of individual genomes.

3.1 Introduction to Recombinant DNA Technology and DNA Cloning

When scientists James Watson and Francis Crick discovered that the structure of DNA is a double-helical molecule, they hinted at the potential importance and impact of this discovery. However, not even these Nobel Prize winners could have imagined the astonishing pace at which molecular biology would advance over the next half century.

In the years before and after Watson and Crick's discovery, many other scientists contributed to our understanding of DNA as the genetic material of living cells. A number of researchers studied DNA structure and replication in bacteria and in **bacteriophages.** Bacteriophages, often simply called *phages*, are viruses

that infect bacterial cells. Much of what we know about DNA replication and DNA-synthesizing enzymes has been learned from studying bacteria and phages. For example, recall that DNA ligase joins together adjacent DNA fragments (Okazaki fragments) during DNA replication (Chapter 2). As you will soon learn, DNA ligase is an important enzyme in recombinant DNA technology.

In the late 1960s, many scientists were interested in gene cloning; they speculated that it might be possible to clone DNA by cutting and pasting DNA from different sources (recombinant DNA technology). It may seem that the terms *gene cloning, recombinant DNA technology*, and *genetic engineering* describe the same process. Recombinant DNA technology is commonly used to make gene cloning possible, whereas genetic engineering often relies on recombinant DNA technology and gene cloning to modify an organism's genome. However, the terms *recombinant DNA technology* and *genetic engineering* are frequently used interchangeably. *Clone* is derived from a Greek word that describes a cutting (of a twig) that is used to propagate or copy a plant. A modern biological definition of a clone is a molecule, cell, or organism produced from another single entity. The laboratory methods required for gene cloning as described in this chapter are different from the techniques used to clone whole organisms (Chapter 7).

Restriction Enzymes and Plasmid DNA Vectors

In the early 1970s, gene cloning became a reality. Many nearly simultaneous discoveries and collaborative efforts among several researchers led to the discovery of two essential components that made gene cloning and recombinant DNA techniques possible—**restriction enzymes** and **plasmids (plasmid DNA).** Restriction enzymes are DNA-cutting enzymes, and plasmid DNA is a circular form of self-replicating DNA that scientists can manipulate to carry and clone other pieces of DNA.

Microbiologists in the 1960s discovered that some bacteria are protected from destruction by bacteriophages because they can *restrict* phage replication. Swiss scientist Werner Arber proposed that restricted growth of phages occurred because some bacteria contained enzymes that could cut viral DNA into small pieces, thus preventing viral replication. Because of this ability, these enzymes were called *restriction enzymes*. In 1970, working with the bacterium *Haemophilus influenzae*, Johns Hopkins University researcher Hamilton Smith isolated *Hin*dIII, the first restriction enzyme to be well characterized and used for DNA cloning. Restriction enzymes are also called restriction endonucleases (*endo,* "within"; *nuclease,* "nucleic acid–cutting enzyme") because they cut *within* DNA sequences as opposed to enzymes that cut from the ends of DNA sequences (exonucleases). Smith demonstrated that *Hin*dIII could be used to cut or digest DNA into small fragments. In 1978, Smith shared a Nobel Prize with Werner Arber and Daniel Nathans for their discoveries on restriction enzymes and their applications.

Restriction enzymes are primarily found in bacteria, and they are given abbreviated names based on the genus and species names of the bacteria from which they are isolated. For example, one of the first restriction enzymes to be isolated, *Eco*RI, is so named because it was discovered in the *E. coli* strain called RY13. Restriction enzymes cut DNA by cleaving the phosphodiester bond (in the sugar-phosphate backbone) that joins adjacent nucleotides in a DNA strand. However, restriction enzymes do not just randomly cut DNA, nor do all restriction enzymes cut DNA at the same locations. Like other enzymes, restriction enzymes show specificity for certain **substrates.** For these enzymes, the substrate is DNA. As shown in **Figure 3.1a**, restriction enzymes bind to, recognize, and cut (digest) DNA within specific sequences of bases called **restriction sites.** Why don't restriction enzymes digest DNA in bacterial

cells? Bacteria protect their DNA from restriction enzyme digestion because some of the nucleotides in their DNA contain methyl groups that block restriction enzymes from digestion (Figure 3.1b).

Restriction enzymes are commonly referred to as four- or six-base-pair cutters because they typically recognize restriction sites with a sequence of four or six nucleotides. Eight-base-pair cutters have also been identified. Each restriction site is a **palindrome**— the arrangement of nucleotides reads the same forward and backward on opposite strands of the DNA molecule. (Remember the word *madam* or the phrase *a Toyota* as examples of palindromes.) Some restriction enzymes, such as *Eco*RI, cut DNA to create DNA fragments with overhanging single-stranded ends called "sticky" or **cohesive ends** (see Figure 3.1a); other enzymes generate fragments with double-stranded ends called **blunt ends.** Table 3.1 shows some common restriction enzymes, their source microorganisms, and their restriction sites. Notice that the first three enzymes in the table are six-base-pair cutters that produce DNA molecules with cohesive ends. The fourth enzyme (*Taq*I) is a four-base-pair cutter that produces cohesive ends, and the lower three enzymes produce blunt-ended DNA fragments. Enzymes that produce cohesive ends are often

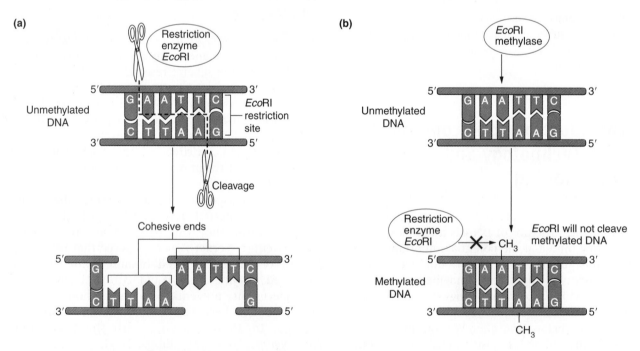

FIGURE 3.1 Restriction Sites and Restriction Enzyme Action (a) Digestion of DNA by *Eco*RI produces DNA fragments with cohesive ends. (b) Methylation of the *Eco*RI restriction site by the enzyme *Eco*RI methylase blocks DNA cleavage by *Eco*RI. Note: Methylation of the *Eco*RI restriction site occurs on an adenine nucleotide, but methylation more frequently occurs on cytosine nucleotides.

TABLE 3.1 COMMON RESTRICTION ENZYMES

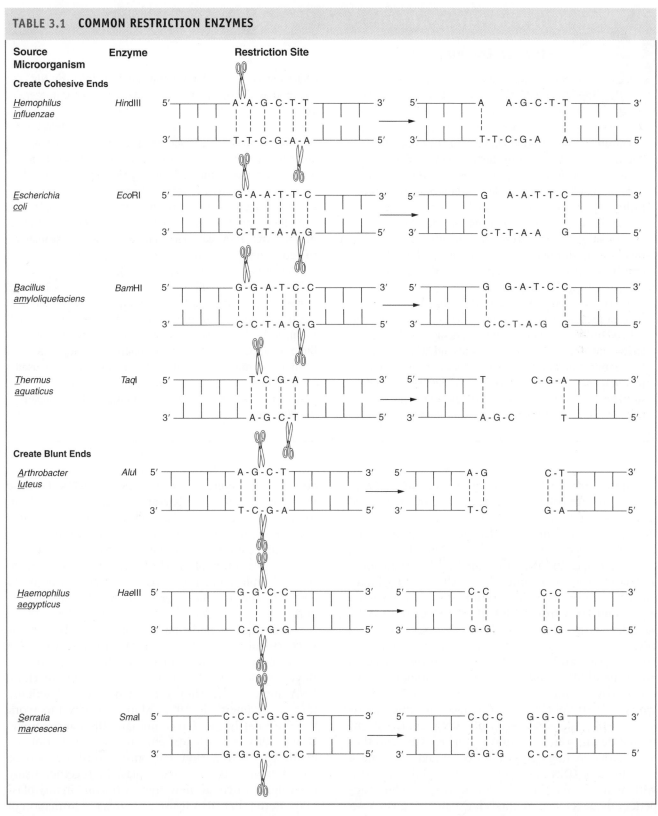

Source Microorganism	Enzyme	Restriction Site
Create Cohesive Ends		
Hemophilus influenzae	*Hind*III	
Escherichia coli	*Eco*RI	
Bacillus amyloliquefaciens	*Bam*HI	
Thermus aquaticus	*Taq*I	
Create Blunt Ends		
Arthrobacter luteus	*Alu*I	
Haemophilus aegypticus	*Hae*III	
Serratia marcescens	*Sma*I	

favored over blunt-end cutters for many cloning experiments because DNA fragments with cohesive ends can easily be joined together. DNA from *any* source—such as bacteria, humans, dogs, cats, frogs, dinosaurs, or ancient human remains—can be digested by a particular restriction enzyme as long as the DNA has a restriction site for that enzyme. In the simplest sense, the discovery of restriction enzymes provided molecular biologists with the "scissors" needed to carry out gene cloning.

TOOLS OF THE TRADE

Restriction Enzymes

Restriction enzymes are sophisticated "scissors" that molecular biologists use to manipulate DNA. Working with restriction enzymes has become easier over the 30 years, since Hamilton Smith and others pioneered their use. Over 300 restriction enzymes are commercially available rather inexpensively. Many enzymes are readily available because they have been cloned using recombinant DNA technology, so they are made and isolated in large quantities. Commercially prepared enzymes come in conveniently sized prepackages with buffer solutions that provide all the components necessary for optimal enzyme activity. If researchers must work with an enzyme with which they are unfamiliar, they can use the restriction enzyme database REBASE, an outstanding tool for locating enzyme suppliers and enzyme specifics.

In addition, a variety of software packages and websites are available to assist scientists who work with restriction enzymes and DNA sequences. For example, imagine that you are a molecular biologist who just cloned and sequenced a 7,200-bp piece of DNA and you want to see if there is an enzyme that will cut your gene to create a 250-bp piece of DNA for a probe you want to make. Not too long ago, if you had a lot of enzymes in your freezer, you could digest this DNA and run gels to see if you could get a 250-bp piece, but this imprecise approach took a lot of time and resources. If you had sequenced your gene, you could scan the sequence with your eyes, looking for a restriction site of interest—a very time-consuming and eye-straining effort! The Internet makes this task much easier, because many websites function as online tools for analyzing restriction enzyme cutting sites. For example, in Webcutter, DNA sequences can be entered and searched to determine restriction enzyme cutting patterns (see problem 8 in "Questions & Activities").

The widespread application of recombinant DNA techniques in many areas of biological and medical research has led to hundreds of technique books, websites, and journals. In *Biotechniques* (a popular monthly journal), biologists publish and share information on molecular cloning techniques. Several sites that are commonly used for designing PCR primers, and other applications are provided in Keeping Current: Web Links, on the Companion Website.

In the early 1970s, Paul Berg, Herbert Boyer, Stanley Cohen, and colleagues at Stanford University used gene cloning to change molecular biology forever. Berg founded recombinant DNA technology when he created a piece of recombinant DNA by joining together (splicing) DNA from the *E. coli* chromosome and DNA from a primate virus called SV40 (simian virus 40). Berg first isolated chromosomal DNA from *E. coli* and DNA from SV40. Then he cut both DNA samples with *Eco*RI, added *E. coli* DNA and viral DNA fragments to a reaction tube with the enzyme DNA ligase, and succeeded in creating a hybrid molecule of SV40 and *E. coli* DNA. The importance of this discovery was fully recognized when Paul Berg won the 1980 Nobel Prize in Chemistry for this experiment, which demonstrated that DNA could be cut from different sources with the same enzyme and that the restriction fragments could be joined to create a recombinant DNA molecule. Berg shared this prize with Walter Gilbert and Frederick Sanger, who independently developed methods for sequencing DNA—a process we discuss later in this chapter.

Berg worked with chromosomal DNA; Cohen was interested in the molecular biology of small circular pieces of DNA known as **plasmids**. Plasmid DNA is primarily found in bacteria. Plasmids are primarily found, and are considered *extrachromosomal* DNA because they are present in the bacterial cytoplasm in addition to the bacterial chromosome. Plasmids are small; most average approximately 1,000 to 4,000 base pairs (bp) in size and they are self-replicating; that is, they duplicate independently of the chromosome. Cohen studied plasmid replication, the transfer of plasmids between bacteria, and mechanisms of bacterial resistance to antibiotics.

Cohen postulated that plasmids could be used as **vectors**—pieces of DNA that can accept, carry, and replicate (clone) other pieces of DNA. Cohen and Boyer worked with two bacterial plasmids to clone DNA successfully. They published results describing these experiments in 1973. Many consider this work the informal birth of recombinant DNA technology. Using *Eco*RI, a restriction enzyme previously isolated by Herbert Boyer, they cut both plasmids and then joined fragments from each plasmid together using DNA ligase to create new hybrid (recombinant) plasmids. Recall that DNA ligase catalyzes the formation of phosphodiester bonds between nucleotides. Ligase can join together DNA with cohesive ends as well as blunt-ended fragments. As a result of these and other experiments, Cohen and Boyer produced the first plasmid vector for cloning purposes, called pSC101 and named

"SC" for Stanley Cohen. pSC101 contained a gene for tetracycline resistance and restriction sites for several enzymes including *Eco*RI and *Hin*dIII. In subsequent work they used similar experiments to clone DNA from the South African claw-toed frog *Xenopus laevis* (another important model organism in genetics and developmental biology) into the *Eco*RI site of pSC101. **Figure 3.2** illustrates how recombinant DNA can be formed in a process similar to that used in the Cohen and Boyer experiments.

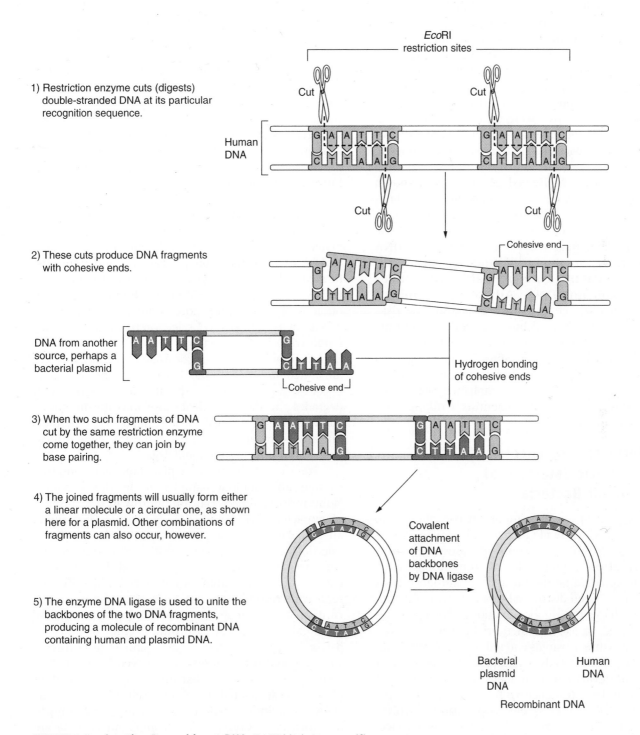

1) Restriction enzyme cuts (digests) double-stranded DNA at its particular recognition sequence.

2) These cuts produce DNA fragments with cohesive ends.

DNA from another source, perhaps a bacterial plasmid

3) When two such fragments of DNA cut by the same restriction enzyme come together, they can join by base pairing.

4) The joined fragments will usually form either a linear molecule or a circular one, as shown here for a plasmid. Other combinations of fragments can also occur, however.

5) The enzyme DNA ligase is used to unite the backbones of the two DNA fragments, producing a molecule of recombinant DNA containing human and plasmid DNA.

Hydrogen bonding of cohesive ends

Covalent attachment of DNA backbones by DNA ligase

Bacterial plasmid DNA

Human DNA

Recombinant DNA

FIGURE 3.2 Creating Recombinant DNA *Eco*RI binds to a specific sequence (5'-GAATTC-3') and then cleaves the DNA backbone, producing DNA fragments. The single-stranded ends of the DNA fragments can form hydrogen bonds with each other because they have complementary base pairs. DNA ligase can then catalyze the formation of covalent bonds in the DNA backbones of the fragments to create a piece of recombinant DNA.

Cohen and Boyer had created the first DNA cloning vector—a vehicle for the insertion and replication of DNA—and in 1980 were awarded patents for pSC101 and for the gene splicing and cloning techniques they had developed. These experiments ushered in the birth of modern biotechnology, because many of the current techniques used for gene cloning and gene manipulations are based on these fundamental methods of recombinant DNA technology.

In 1974, as a direct result of the Berg, Cohen, and Boyer experiments, gene cloning pioneers and critics voiced concerns about the safety of genetically modified organisms. Scientists were concerned about what might happen if recombinant bacteria were to leave the lab or if such bacteria could transfer their genes to other cells or survive in other organisms, including humans. In 1975, an invited group of well-known molecular biologists, virologists, microbiologists, lawyers, and journalists gathered at the Asilomar Conference Center in Pacific Grove, California, to discuss the benefits and potential hazards of recombinant DNA technology. As a result of the historic Asilomar meeting, the **National Institutes of Health (NIH)** formed the **Recombinant DNA Advisory Committee (RAC),** which was charged with evaluating the risks of recombinant DNA technology and establishing guidelines for recombinant DNA research. In 1976, the RAC published a set of guidelines for working with recombinant organisms. The RAC continues to oversee gene cloning research, and compliance with RAC guidelines is mandatory for scientists working with recombinant organisms.

Transformation of Bacterial Cells and Antibiotic Selection of Recombinant Bacteria

Cohen also made another important contribution to gene cloning, which made the pSC101 cloning experiments possible. His laboratory demonstrated how **transformation,** a process for inserting foreign DNA into bacteria, could be used to reliably introduce DNA into bacteria. Cohen discovered that if he treated bacterial cells with calcium chloride solutions, added plasmids to cells chilled on ice, and then briefly heated the cell and DNA mixture, plasmids entered bacterial cells. Once inside bacteria, plasmids replicate and express their genes. Transformation techniques are explained in greater detail later (Chapter 5). A more modern transformation method, called **electroporation,** involves applying a brief (millisecond) pulse of high-voltage electricity to create tiny holes in the bacterial cell wall that allow DNA to enter. Electroporation can also be used to introduce DNA into mammalian cells and to transform plant cells.

Ligation of DNA fragments and transformation by any method are somewhat inefficient. During ligation, some of the digested plasmid ligates back to itself to create recircularized plasmid that lacks foreign DNA. During transformation, a majority of cells do not take up DNA.

Now that you have learned how DNA can be inserted into a vector and introduced into bacterial cells, we consider how recombinant bacteria—those transformed with a recombinant plasmid—can be distinguished from a large number of nontransformed bacteria and bacterial cells that contain plasmid DNA without foreign DNA. This screening process is called **selection** because it is designed to facilitate the identification of (selecting *for*) recombinant bacteria while preventing the growth of (selecting *against*) nontransformed bacteria and bacteria that contain plasmid without foreign DNA.

Cohen and Boyer used **antibiotic selection,** a technique in which transformed bacterial cells are plated on agar plates with different antibiotics, as a way to identify recombinant bacteria and nontransformed cells. For many years antibiotic selection was a widely used approach. Modern cloning techniques often incorporate other more popular selection strategies, such as "blue-white" selection (the reason for this name will soon be obvious). In blue-white selection, DNA is cloned into a restriction site in the *lacZ* gene, as illustrated in **Figure 3.3.** Recall that the *lacZ* gene encodes β-galactosidase (β-gal), an enzyme that degrades the disaccharide lactose into the monosaccharides glucose and galactose (Chapter 2). When it is interrupted by an inserted gene, the *lacZ* gene is incapable of producing functional β-gal.

Transformed bacteria are plated on agar plates that contain an antibiotic—ampicillin, in this example. Nontransformed bacteria cannot grow in the presence of ampicillin because they lack plasmids containing an ampicillin resistance gene (ampR). But antibiotic selection alone does not distinguish transformed bacteria with nonrecombinant plasmid that has recircularized from recombinant plasmids. To identify bacteria with recombinant plasmids, the agar must also contain a chromogenic (color-producing) substrate for β-gal called X-gal (5-bromo-4-chloro-3-indolyl-β-D-galactopyranoside). X-gal is similar to lactose in structure and turns blue when cleaved by β-gal. As a result, nonrecombinant bacteria—those that contain plasmid that ligated back to itself without insert DNA—contain a functional *lacZ* gene, produce β-gal, and turn blue. Conversely, recombinant bacteria are identified as white colonies. Because these cells contain plasmid with foreign DNA inserted into the *lacZ* gene, β-gal is not produced, and these cells cannot metabolize X-gal. Therefore, through blue-white selection,

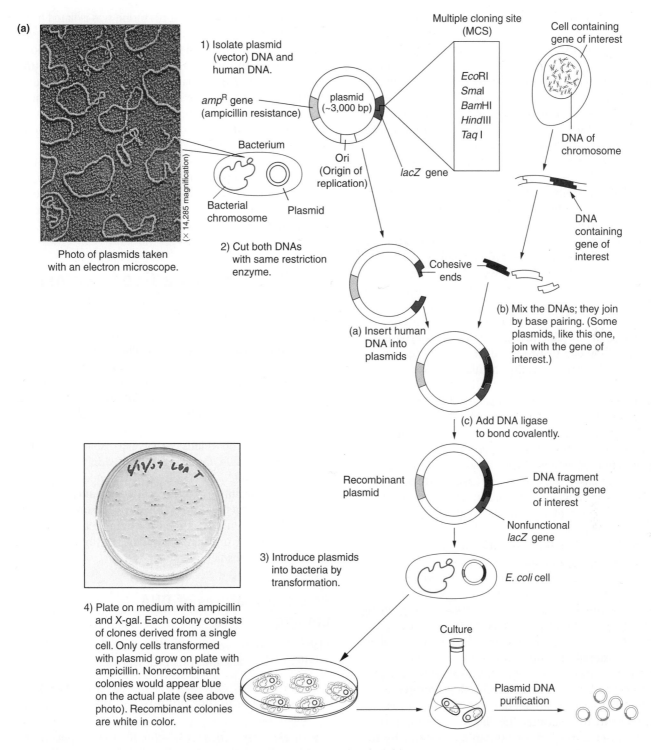

(a)

Photo of plasmids taken with an electron microscope.

(× 14,285 magnification)

1) Isolate plasmid (vector) DNA and human DNA.

amp^R gene (ampicillin resistance)

plasmid (~3,000 bp)

Ori (Origin of replication)

Bacterium

Bacterial chromosome

Plasmid

Multiple cloning site (MCS)

*Eco*RI
*Sma*I
*Bam*HI
*Hind*III
Taq I

lacZ gene

Cell containing gene of interest

DNA of chromosome

DNA containing gene of interest

2) Cut both DNAs with same restriction enzyme.

Cohesive ends

(a) Insert human DNA into plasmids

(b) Mix the DNAs; they join by base pairing. (Some plasmids, like this one, join with the gene of interest.)

(c) Add DNA ligase to bond covalently.

Recombinant plasmid

DNA fragment containing gene of interest

Nonfunctional *lacZ* gene

3) Introduce plasmids into bacteria by transformation.

E. coli cell

4) Plate on medium with ampicillin and X-gal. Each colony consists of clones derived from a single cell. Only cells transformed with plasmid grow on plate with ampicillin. Nonrecombinant colonies would appear blue on the actual plate (see above photo). Recombinant colonies are white in color.

Culture

Plasmid DNA purification

FIGURE 3.3   Cloning a Gene in a Plasmid and Blue-White Selection

nontransformed and nonrecombinant bacteria are *selected against* and white colonies are identified or *selected for* as the desired colonies containing recombinant plasmids. Colonies containing recombinant plasmids are **clones**—genetically identical bacterial cells each containing copies of the recombinant plasmids.

Introduction to Human Gene Cloning

Restriction enzymes, DNA ligase, and plasmids are major tools of molecular biologists for manipulating and cloning genes from virtually any source. With transformation, scientists had a way to introduce recombinant

DNA into bacterial cells. Recombinant DNA technology made it possible to cut and join together DNA fragments, insert DNA into a plasmid (DNA cloned into a plasmid is commonly called *insert* DNA), and produce large amounts of the insert DNA by allowing bacteria to be the workhorses for replicating recombinant DNA.

If the cloned DNA fragment is a gene that encodes a protein product, bacterial cells could be used to synthesize the protein product of the cloned gene. We call this *expressing* a protein. Molecular biologists recognized that if human genes could be cloned and expressed, recombinant DNA technology would become an invaluable tool with powerful and exciting applications in research and medicine. Because bacteria can be grown in large-scale preparations, scientists can produce large amounts of the cloned DNA and isolate quantities of protein that would normally be very difficult or expensive to purify without cloning (these processes are described in more detail in Chapters 4 and 5). Because of recombinant DNA technology, a wide range of valuable proteins that are otherwise difficult to obtain can be produced from cloned genes.

The first commercially available human gene product of recombinant DNA technology was human **insulin,** a peptide hormone produced by cells in the pancreas called beta cells. When blood glucose rises—for example, after eating a sugar-rich meal—insulin lowers blood glucose by stimulating glucose storage in liver and muscle cells as long chains of glucose called glycogen. Individuals with **type I,** or **insulin-dependent, diabetes mellitus** do not produce insulin on their own. As a result of this insulin deficiency, diabetics experience excessively high blood sugar levels (hyperglycemia), which, over time, can lead to serious damage to many bodily organs.

In 1977, the insulin gene was cloned into plasmids, expressed in bacterial cells, and isolated by scientists at **Genentech** (named for *gen*etic *en*gineering *tech*nology), the San Francisco biotechnology company cofounded in 1976 by Herbert Boyer and Robert Swanson (details of the techniques used for cloning insulin are examined in Chapter 5). Genentech is generally regarded as the first biotechnology company.

In 1982, the recombinant form of human insulin, called **Humulin,** became the first recombinant DNA product to be approved for human applications by the **U.S. Food and Drug Administration (FDA).** Shortly after insulin became available, growth hormone—used to treat children who suffer from a form of dwarfism—was cloned, and because of recombinant DNA technology, a wide variety of other medically important proteins that were once difficult to obtain in adequate amounts became readily available.

Prior to recombinant DNA technology, important hormones like insulin and growth hormone had to be isolated from tissues. Growth hormone was isolated from the pituitary glands of human cadavers. Not only was this process expensive and inefficient, but these isolations also carried with them the risk of unknowingly co-purifying viruses and other pathogens as contaminants that could be passed to people receiving the hormone. There are now several hundred products of recombinant DNA technology on the market with widespread applications in basic research, medicine, and agriculture.

With a basic understanding of the techniques involved in manipulating a piece of DNA, in the next section we go on to examine some important aspects of DNA vectors and how different vectors are chosen and used depending on what is to be accomplished.

3.2 What Makes a Good Vector?

The number of different DNA vectors, vector functions, and applications has increased substantially since Stanley Cohen constructed pSC101. Plasmids are still the most commonly used cloning vectors because they allow for the routine cloning and manipulation of small pieces of DNA that form the foundation for many techniques used daily in a molecular biology laboratory. In addition, it is fairly simple to transform bacterial cells with plasmids and relatively easy to isolate plasmids from bacterial cells. One of the first widely used plasmid vectors, called pBR322, was designed to include genes for ampicillin and tetracycline resistance and several useful restriction sites. However, plasmid cloning vectors have been engineered over the years to incorporate a number of other important features that have made pBR322 obsolete.

Practical Features of DNA Cloning Vectors

Plasmid cloning vectors usually include most of the following desirable and practical features:

- *Size*—They should be small enough to be easily separated from the chromosomal DNA of the host bacteria.

- *Origin of replication* (ori)—Site for DNA replication that allows plasmids to replicate separately from the host cell's chromosome. The number of plasmids in a cell is called **copy number.** The normal copy number of plasmids in most bacterial cells is small (usually less than 12 plasmids per cell); however, many of the most desirable cloning plasmids are known as high-copy-number plasmids because they replicate to create hundreds or thousands of plasmid copies per cell.

- *Multiple cloning site* (MCS)—The MCS is a segment of DNA with recognition sites for different common restriction enzymes (see Figure 3.3a). These sites are engineered into the plasmid so that digestion of the plasmid with restriction enzymes does not result in the loss of a fragment of DNA. Rather, the circular plasmid simply becomes linearized when it is digested with a restriction enzyme. An MCS provides for great flexibility in the range of DNA fragments that can be cloned into a plasmid because it is possible to insert DNA fragments generated by cutting with many different enzymes.

- *Selectable marker genes*—These genes allow for the selection and identification of bacteria that have been transformed with a recombinant plasmid. Some of the most common selectable markers are genes for ampicillin resistance (ampR), tetracycline resistance (tetR), and the *lacZ* gene used for blue-white selection.

- *RNA polymerase promoter sequences*—These sequences are used for the transcription of RNA in vivo and in vitro by RNA polymerase. Recall that RNA polymerase copies DNA into RNA during transcription (Chapter 2). In vivo, these sequences allow bacterial cells to make RNA from cloned genes, which in turn leads to protein synthesis. In vitro, transcribed RNA can to synthesize RNA "probes," which are useful in studying gene expression (Section 3.4).

- *DNA sequencing primer sequences*—These sequences permit nucleotide sequencing of cloned DNA fragments that have been inserted into the plasmid (Section 3.4).

Types of Vectors

Just as one screwdriver cannot be used for all sizes and types of screws, bacterial plasmid vectors cannot be used for all applications in biotechnology. There are limitations to how plasmids can be used in cloning. One primary limitation is the size of the DNA fragment that can be inserted into a plasmid. Insert size usually cannot exceed approximately 6 to 7 kilobases (1 kb = 1,000 bp). In addition, sometimes bacteria express proteins from eukaryotic genes poorly. Because of these limitations, molecular biologists have developed many other types of DNA vectors, each of which has particular benefits depending on the cloning application. Table 3.2 compares the important features, sources, and applications of different types of cloning vectors.

TABLE 3.2 A COMPARISON OF DNA VECTORS AND THEIR APPLICATIONS

Vector Type	Maximum Insert Size (kb)	Applications	Limitations
Bacterial plasmid vectors (circular)	~ 6–12	DNA cloning, protein expression, subcloning, direct sequencing of insert DNA	Restricted insert size; limited expression of proteins; copy number problems; replication restricted to bacteria
Bacteriophage vectors (linear)	~ 25	cDNA, genomic and expression libraries	Packaging limits DNA insert size; host replication problems
Cosmid (circular)	~ 35	cDNA and genomic libraries, cloning large DNA fragments	Phage packaging restrictions; not ideal for protein expression; cannot be replicated in mammalian cells
Bacterial artificial chromosome (BAC, circular)	~ 300	Genomic libraries, cloning large DNA fragments	Replication restricted to bacteria; cannot be used for protein expression
Yeast artificial chromosome (YAC, circular)	200–2,000	Genomic libraries, cloning large DNA fragments	Must be grown in yeast; cannot be used in bacteria
Ti vector (circular)	Varies depending on type of Ti vector used	Gene transfer in plants	Limited to use in plant cells only; number of restriction sites randomly distributed; large size of vector not easily manipulated

Bacteriophage vectors

DNA from bacteriophage lambda (λ) was one of the first phage vectors used for cloning. The λ chromosome is a linear structure approximately 49 kb in size. Cloned DNA is inserted into restriction sites in the center of the λ chromosome. Recombinant chromosomes are then packaged into viral particles in vitro, and these phages are used to infect *E. coli* growing as a lawn (a continuous layer covering the plate). At each end of the λ chromosome are 12 nucleotide sequences called cohesive sites (COS), which can base pair with each other. When λ infects *E. coli* as a host, the λ chromosome uses COS sites to circularize and then replicate. Bacteriophage λl replicates through a process known as a **lytic cycle.** As λ replicates to create more viral particles, infected *E. coli* are lysed (to lyse means to split or rupture) by λ, creating zones of dead bacteria called **plaques,** which appear as cleared spots on the bacterial lawn. Each plaque contains millions of recombinant phage particles. An advantage of these vectors is that they allow for the cloning of larger DNA fragments (up to approximately 25 kb) than plasmids. Phage vectors are no longer very widely used.

Cosmid vectors

Cosmid vectors contain COS ends of λ DNA, a plasmid origin of replication, and genes for antibiotic resistance, but most of the viral genes have been removed. DNA is cloned into a restriction site and the cosmid is packaged into viral particles, as is done with bacteriophage vectors, and used to infect *E. coli*, wherein cosmids replicate as a low-copy-number plasmid. Bacterial colonies are formed on a plate, and recombinants can be screened by antibiotic selection. One advantage of cosmids is that they allow for the cloning of DNA fragments in the 20- to 45-kb range. But much like phage vectors, because of the development of other types of vectors, cosmid vectors are no longer very widely used and have limited applications.

Expression vectors

Protein **expression vectors** allow for the high-level synthesis (expression) of eukaryotic proteins within bacterial cells because they contain a prokaryotic promoter sequence adjacent to the site where DNA is inserted into the plasmid. Bacterial RNA polymerase can bind to the promoter and synthesize large amounts of RNA (for the insert), which is then translated into protein. Protein may then be isolated using biochemical techniques (described in Chapter 4). However, it is not always possible to express a functional protein in bacteria. For example, bacterial ribosomes sometimes cannot translate eukaryotic mRNA sequences. If a protein is produced, it may not fold and be processed correctly, as occurs in eukaryotic cells that use organelles to fold and modify proteins. Also, making some recombinant products in bacteria can be a problem because *E. coli* often does not secrete proteins; therefore expression vectors are often used in *Bacillus subtilis*, a strain more suitable for protein secretion.

In some cases, the host bacteria can recognize recombinant proteins as foreign and degrade the protein, whereas in others the expressed protein is lethal to the host bacterial cells. Certain viruses, such as SV40, can be used to deliver expression vectors into mammalian cells. Typically SV40-derived vectors contain a strong (viral) promoter sequence for high-level transcription and a poly(A) addition signal for adding a poly(A) tail to the 3' end of synthesized mRNAs. Variations of such vectors have been used for human gene therapy.

Bacterial artificial chromosomes

Bacterial artificial chromosomes (BACs) are large low-copy-number plasmids present as one to two copies in bacterial cells and they containing genes that encode the F-factor (a unit of genes controlling bacterial replication). BACs can accept DNA inserts in the 100- to 300-kb range. BACs were widely used in the Human Genome Project to clone and sequence large pieces of human chromosomes.

Yeast artificial chromosomes

Yeast artificial chromosomes (YACs) are small plasmids grown in *E. coli* and introduced into yeast cells (such as *Saccharomyces cerevisiae*). A YAC is a miniature version of a eukaryotic chromosome. YACs contain an origin of replication, selectable markers, two telomeres, and a centromere that allows for replication of the YAC and segregation into daughter cells during cell division. Foreign DNA fragments are cloned into a restriction site in the center of the YAC. YACs are particularly useful for cloning large fragments of DNA from 200 kb to approximately 2 megabases (mb = 1 million bases) in size. Like BACs, YACs played an important role in the cloning efforts of the Human Genome Project.

Ti vectors

Ti vectors are naturally occurring plasmids (around 200 kb in size) isolated from the bacterium *Rhizobium radiobacter* (formerly called *Agrobacterium tumefaciens* and recently renamed on the basis of genome data) a soil-borne plant pathogen that causes a condition in plants called crown gall disease. When *R. radiobacter* enters host plants, a piece of DNA (T-DNA) from the Ti plasmid (Ti stands for tumor-inducing) inserts into the host chromosome. T-DNA encodes for the synthesis of a hormone called auxin, which weakens the host's cell wall. Infected plant cells divide and enlarge to form a tumor (gall). Plant geneticists recognized that if they could remove auxin and other detrimental genes from the Ti plasmid, the resulting vector could be used to deliver genes into plant cells. Ti vectors are widely used to transfer genes into plants as you will learn in Chapter 6.

Now that we have examined different types of vectors and their applications, in the next section we will turn our attention to how scientists can use recombinant DNA technology to identify and clone genes of interest.

3.3 How Do You Identify and Clone a Gene of Interest?

Cutting and pasting different pieces of DNA to produce a recombinant DNA molecule has become a routine technique in molecular biology. But the types of cloning experiments we have described so far allow for the *random* cloning of DNA fragments based on restriction enzyme cutting sites, not precise cloning of a single gene or particular piece of DNA of interest. For example, if you were interested in cloning the insulin gene and you simply took DNA from the pancreas, cut it with enzymes, and then ligated digested DNA into plasmids, you would create hundreds of thousands of recombinant plasmids and not just recombinant plasmids that contain only the insulin gene.

Molecular biologists call this approach "shotgun" cloning, because many fragments are randomly cloned at once and no individual gene is specifically targeted for cloning. How would you know which recombinant plasmid contained the insulin gene? Moreover, if the insulin gene (or adjacent sequences) does not have restriction sites for the restriction enzyme you used, you might not have any recombinant plasmids containing the insulin gene. Even if you did create plasmids with the insulin gene, how would you separate these from the other recombinant plasmids? So how do you find a particular gene of interest and clone only the DNA sequence that you want to study? These questions can often be answered by a cloning approach involving DNA libraries.

Creating DNA Libraries: Building a Collection of Cloned Genes

Many cloning strategies begin by preparing a **DNA library**—a collection of cloned DNA fragments from a particular organism contained within bacteria or viruses as the hosts. Libraries can be saved for relatively long periods of time and "screened" to pick out different genes of interest. Two types of libraries are typically used for cloning, **genomic DNA libraries** and **complementary DNA libraries (cDNA libraries).** **Figure 3.4** on the next page shows how genomic libraries and cDNA libraries are constructed.

Genomic versus cDNA libraries

In a genomic library, chromosomal DNA from the tissue of interest is isolated and then digested with a restriction enzyme (see Figure 3.4a). This process produces fragments of DNA that include the organism's entire genome. A plasmid, BAC, YAC, or bacteriophage vector is digested with the same enzyme, and DNA ligase is used to ligate genomic DNA pieces and vector DNA randomly. In theory, all DNA fragments in the genome will be cloned into a vector. Recombinant vectors are then used to transform bacteria, and each bacterial cell clone will contain recombinant vector with a plasmid containing a genomic DNA fragment. Consider each clone a "book" in this "library" of DNA fragments. One disadvantage of creating this type of library for eukaryotic genes is that non-protein-coding pieces of DNA, called introns, are cloned in addition to protein-coding sequences (exons). Because a majority of DNA in any eukaryotic organism consists of introns, many of the clones in a genomic library will contain non-protein-coding pieces of DNA. Another limitation of genomic libraries is that many organisms, including humans, have such large genomes that searching for a gene of interest would be like searching for a needle in a haystack.

In a cDNA library, mRNA from the tissue of interest is isolated and used for making the library. However, mRNA cannot be cut directly with restriction enzymes, so it must be converted to a double-stranded DNA molecule. An enzyme called **reverse transcriptase (RT)** is used to catalyze the synthesis of single-stranded DNA from the mRNA (see Figure 3.4b). This enzyme is made by viruses called **retroviruses**—so named because they are exceptions to the usual flow of genetic information. Instead of having a DNA genome that can be used to make RNA, retroviruses have an RNA genome. After infecting host cells, they use RT to convert RNA into DNA, so that they can replicate. Human immunodeficiency virus (HIV), the causative agent of acquired immunodeficiency syndrome (AIDS), is a retrovirus. As we will discuss, retroviruses also have important applications in biotechnology as gene therapy vectors (Chapter 11). Because RT synthesizes DNA that is an exact copy of mRNA, it is called **complementary DNA (cDNA).** The mRNA is degraded by treatment with an alkaline

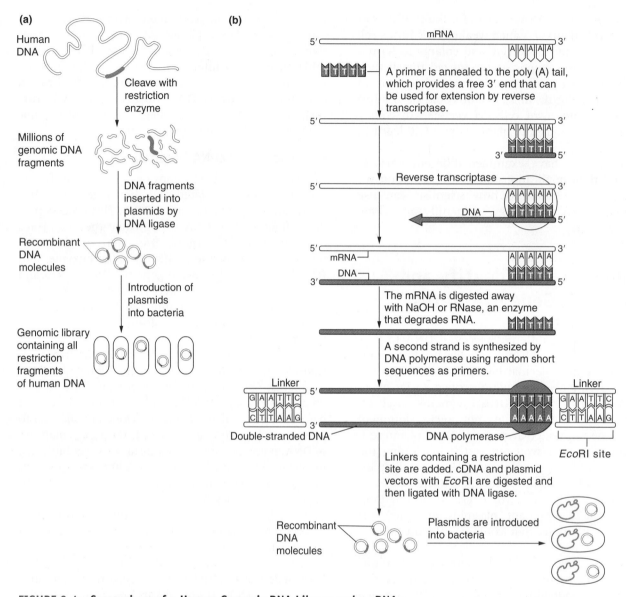

FIGURE 3.4 Comparison of a Human Genomic DNA Library and a cDNA Library (a) Human DNA is cleaved with a restriction enzyme to create a series of smaller fragments that are cloned into plasmids or other vectors. A (human) genomic library consists of a collection of bacteria each containing a different fragment of human DNA. In theory, all DNA fragments from the genome will be represented in the library. (b) In a cDNA library, mRNA is converted into cDNA by the enzyme reverse transcriptase. Linkers containing a restriction site are added to the cDNA to create cohesive ends. The cDNA can now be cloned into a plasmid for subsequent replication in bacteria.

solution or enzymatically digested; then DNA polymerase is used to synthesize a second strand to create double-stranded cDNA.

Because cDNA sequences do not necessarily have a convenient restriction site at each end, short, double-stranded DNA sequences called *linker sequences* are often enzymatically added to the ends of the cDNA. Linkers contain restriction sites. Different linkers for different restriction sites are commercially available.

By adding linkers, cDNA can now be ligated into a convenient restriction site in a vector of choice, often a plasmid. The recombinant plasmid is then used to transform bacteria.

One primary advantage of cDNA libraries over genomic libraries is that they are a collection of actively *expressed* genes in the cells or tissue from which the mRNA was isolated. Also, introns are not cloned in a cDNA library. By contrast, when cloned genomic DNA, containing introns and exons is inserted into bacteria, the cells cannot splice mRNA transcribed from this DNA and remove introns. For this reason, cDNA libraries are typically preferred over genomic libraries when attempting to clone and express a gene of interest.

Another advantage of cDNA libraries is that they can be created and screened to isolate genes that are primarily expressed only under certain conditions in a tissue. For example, if a gene is expressed only in a tissue stimulated by a hormone, researchers make libraries from hormone-stimulated cells to increase the likelihood of cloning hormone-sensitive genes. Libraries have become such a routine aspect of molecular biology that many companies sell libraries prepared from a range of tissues from different species. One disadvantage is that cDNA libraries can be difficult to create and screen if a source tissue with an abundant amount of mRNA for the gene is not available. But as you will learn, a technique called the *polymerase chain reaction* (*PCR*) can frequently solve this problem.

Library screening

Once either a genomic library or a cDNA library is created, it must be *screened* to identify the genes of interest. One of the most common library screening techniques is called **colony hybridization (Figure 3.5)**. In colony hybridization, bacterial colonies from the library containing recombinant DNA are grown on an agar plate. A nylon or nitrocellulose membrane is placed over the plate, and some of the bacterial cells attach to the membrane at the same location where they are found on the plate. If bacteriophage vectors are used, phages are transferred onto the nylon. The nylon is treated with an alkaline solution to lyse bacteria and denature their DNA, which binds to the nylon as single-stranded molecules. Typically, the nylon is then incubated with a DNA **probe,** a single-stranded DNA fragment that is complementary to the gene of interest because it can base pair by hydrogen bonding to the target DNA to be cloned. Probes are "tagged" or labeled using either a radioactive nucleotide or, more commonly, a fluorescent dye or other compounds that can be used to catalyze light-releasing reactions called **chemiluminescence.** The dye makes it possible to track the probe to determine where it binds. The probe binds to complementary sequences on the nylon—a process is called **hybridization.**

The nylon is then washed to remove excess unbound probe and exposed to photographic film in a process called **autoradiography.** Anywhere the probe has bound to the filter, radioactivity from radioactive probes or released light (fluorescence or chemiluminescence) from nonradioactive probes exposes silver grains in the film. Depending on the abundance

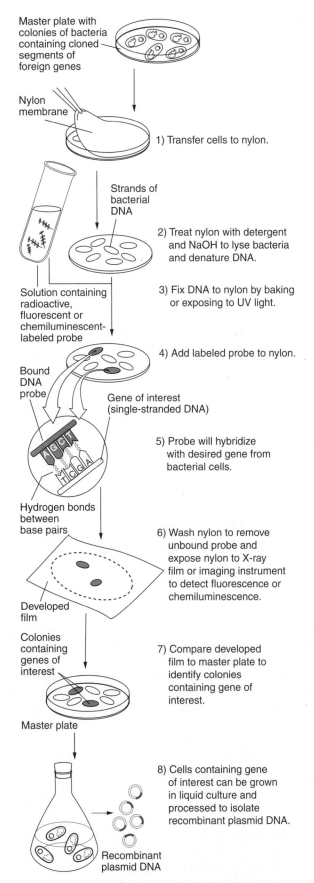

Master plate with colonies of bacteria containing cloned segments of foreign genes

Nylon membrane

1) Transfer cells to nylon.

Strands of bacterial DNA

2) Treat nylon with detergent and NaOH to lyse bacteria and denature DNA.

3) Fix DNA to nylon by baking or exposing to UV light.

Solution containing radioactive, fluorescent or chemiluminescent-labeled probe

4) Add labeled probe to nylon.

Bound DNA probe

Gene of interest (single-stranded DNA)

5) Probe will hybridize with desired gene from bacterial cells.

Hydrogen bonds between base pairs

6) Wash nylon to remove unbound probe and expose nylon to X-ray film or imaging instrument to detect fluorescence or chemiluminescence.

Developed film

Colonies containing genes of interest

7) Compare developed film to master plate to identify colonies containing gene of interest.

Master plate

8) Cells containing gene of interest can be grown in liquid culture and processed to isolate recombinant plasmid DNA.

Recombinant plasmid DNA

FIGURE 3.5 Colony Hybridization: Library Screening with a DNA Probe to Identify a Cloned Gene of Interest

of the gene of interest, there may be only a few colonies (or plaques) on the filter that hybridize to the probe. Film is developed to create a permanent record, called an autoradiogram (or autoradiograph), which is then compared with the original plate of bacterial colonies to identify which colonies contained recombinant plasmid with the gene of interest. These colonies can now be grown on a larger scale to isolate the cloned DNA. Often, when hybridization is done using fluorescent or chemiluminescent probes, a digital imaging instrument can be used to detect probe binding and then a photograph aligned with the bacterial plate to identify colonies of interest.

The type of probe used for library screening often depends on what is already known about the gene of interest. For example, the screening probe is frequently a gene cloned from another species. A cDNA clone of a gene from a rat or mouse is often a very effective probe for screening a human genomic or cDNA library, because many gene sequences in rats and mice are similar to those found in human genes. If the gene of interest has not been cloned in another species but some information is available about the protein sequence, a series of chemically synthesized **oligonucleotides** can be made based on a prediction of codons that can code for the known protein sequence. If some partial amino acid sequence is known for a protein encoded by a gene to be cloned, it is possible to "work backward" and design oligonucleotides based on the predicted nucleotides that coded for the amino acid sequence. In addition, if an antibody is available for the protein encoded by the gene of interest, an expression library, which results in protein expression in bacteria, can be used, and the library can be screened with the antibody to detect colonies expressing the recombinant protein.

Library screening rarely results in the isolation of clones that contain full-length genes. It is more common to obtain clones with small pieces of the gene of interest (one reason why this occurs with cDNA libraries is because it may be difficult to isolate full-length mRNA or synthesize full-length cDNA for the gene of interest). When small pieces of a gene are cloned, scientists sequence these pieces and look for sequence overlaps. Overlapping fragments of DNA can then be pieced together like a puzzle in an attempt to reconstruct the full-length gene. This often requires screening the library several times, with large numbers of bacteria being plated and used for colony hybridization. Looking for start and stop codons in the sequenced pieces is one way to predict if the entire gene has been pieced together. Through this process, overlapping fragments can be pieced together to assemble an entire gene.

Later in this chapter we will discuss how whole-genome shotgun sequencing strategies can enable scientists to sequence entire genomes. Because of genomic studies, libraries are becoming a less common way to identify and clone genes. Instead of using a library to identify one or a few genes at a time, genomics enables scientists to identify sequences for all genes in a genome.

Polymerase Chain Reaction

Although libraries are very effective and commonly used for cloning and identifying a gene of interest, the **polymerase chain reaction (PCR)** is a much more rapid approach to cloning than building and screening a library. PCR is often the technique of choice when cloning genes but it also has many other applications. Developed in the mid-1980s by Kary Mullis, PCR turned out to be a revolutionary technique that has had an impact on many areas of molecular biology. In 1993, Mullis won the Nobel Prize in Chemistry for his invention. PCR is a technique for making copies or amplifying a *specific sequence* of DNA in a short period of time. The concept behind a PCR reaction is remarkably simple. Target DNA to be amplified is added to a thin-walled tube and mixed with deoxyribonucleotides (dATP, dCTP, dGTP, dTTP), buffer, and DNA polymerase. A paired set of **primers** is added to the mixture. Primers are short single-stranded DNA oligonucleotides usually around 20 to 30 nucleotides long. These primers are complementary to nucleotides flanking opposite ends of the target DNA to be amplified (**Figure 3.6**).

The reaction tube is then placed in a *thermal cycler*. In the simplest sense, a thermal cycler is a sophisticated heating block that is capable of rapidly changing temperature over very short time intervals. The thermal cycler takes the sample through a series of reactions called a PCR cycle (Figure 3.6). Each cycle consists of three stages. In the first stage, called *denaturation*, the reaction tube is heated to approximately 94°C to 96°C, causing separation of the target DNA into single strands. In the second stage, called *hybridization* (or *annealing*), the tube is then cooled slightly to between 55°C and 65°C, which allows the primers to hydrogen bond to complementary bases at opposite ends of the target sequence. During extension (or elongation), the last stage of a PCR cycle, the temperature is usually raised slightly (to about 70°C to 75°C), and DNA polymerase copies the target DNA by binding to the 3' ends of each primer and using the primers as templates. DNA polymerase adds nucleotides to the 3' end of each primer to synthesize a complementary strand.

At the end of one complete cycle, the amount of target DNA has been doubled. The thermal cycler

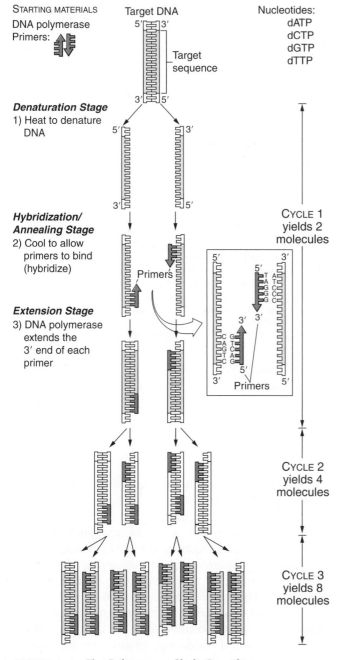

STARTING MATERIALS
DNA polymerase
Primers:

Target DNA
5' 3'

Target
sequence

Nucleotides:
dATP
dCTP
dGTP
dTTP

Denaturation Stage
1) Heat to denature
DNA

3' 5'

5' 3'

3' 5'

**Hybridization/
Annealing Stage**
2) Cool to allow
primers to bind
(hybridize)

Primers

CYCLE 1
yields 2
molecules

Extension Stage
3) DNA polymerase
extends the
3' end of each
primer

Primers

CYCLE 2
yields 4
molecules

CYCLE 3
yields 8
molecules

FIGURE 3.6   The Polymerase Chain Reaction

produced from a reaction starting with one molecule of target DNA.

One key to PCR is the type of DNA polymerase used in the reaction. The repeated heating and cooling required for PCR would denature and destroy most DNA polymerases after just a few cycles. Several sources of PCR-suitable DNA polymerases are available. One of the first and most popular enzymes for PCR is known as ***Taq* DNA polymerase. Taq** is isolated from the domain Archaea called *Thermus aquaticus*, a species that thrives in hot springs. Because *T. aquaticus* is adapted to live in hot water (it was first discovered in the hot springs of Yellowstone National Park), it has evolved a DNA polymerase that can withstand high temperatures. Because *Taq* is stable at high temperatures, it can withstand the temperature changes necessary for PCR without being denatured. Thermostable polymerases such as *Taq* are essential for PCR.

There are many variations and different applications in PCR technology. For example, **real-time** or **quantitative PCR (qPCR),** uses specialized thermal cyclers that enable researchers to quantify amplification reactions as they occur. We will discuss qPCR later in this chapter. An excellent tutorial on PCR can be viewed at the Cold Spring Harbor DNA Learning Center website listed at the Companion Website.

PCR has widespread applications in research and medicine, such as making DNA probes, studying gene expression, amplifying minute amounts of DNA to detect viral pathogens and bacterial infections, amplifying DNA to diagnose genetic conditions, detecting trace amounts of DNA from tissues at a crime scene, and even amplifying ancient DNA from fossilized dinosaur tissue (**Figure 3.7**). Many PCR applications are described throughout the book.

Cloning PCR products

PCR is often used instead of library screening approaches for cloning a gene because it is rapid and effective (**Figure 3.8**). A disadvantage of PCR cloning is that, to design primers, you need to know something about the DNA sequences that flank your gene of interest. Cloning by PCR is easiest if the gene has already been cloned in another species—for instance, using primers for a gene cloned previously from mice to clone the equivalent gene from humans.

There are many ways to clone a gene using PCR. One modern approach to PCR cloning takes advantage of an interesting quirk of thermostable polymerases. As DNA is copied, *Taq* and other polymerases used for PCR normally add a single adenine nucleotide to the 3' end of all PCR products (Figure 3.8). After amplifying a

repeats these three stages again according to the total number of cycles determined by the researcher, usually 20 or 30 cycles. The greatest advantage of PCR is its ability to amplify millions of copies of target DNA from a very small amount of starting material in a short period of time. Because the target DNA is doubled after every round of PCR, after 20 cycles of PCR, approximately 1 million copies (2^{20}) of target DNA are

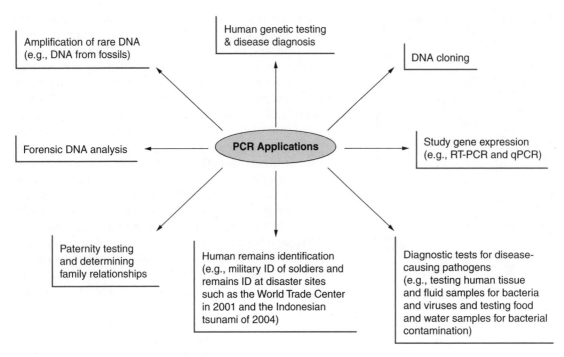

FIGURE 3.7   PCR Applications   The amplification of DNA by PCR has become an essential technique in molecular biology with a wide range of different applications. Examples of some of the more common biotechnology-related applications are represented in this figure.

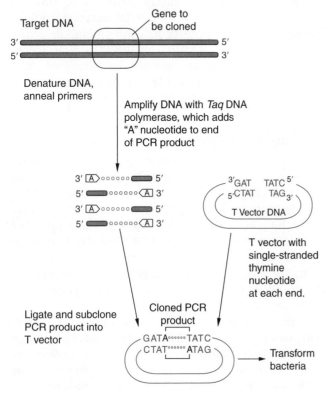

FIGURE 3.8   Cloning a Gene by PCR

target gene, cloned PCR products can be ligated into plasmids called T vectors. T vectors contain a single-stranded thymine nucleotide at each end that can complementarily base pair with overhanging adenine nucleotides in PCR products. Once ligated into a T vector, the recombinant plasmid containing the cloned PCR product can be introduced into bacteria and its nucleotide sequence can be determined.

Now that you have learned some of the most common strategies used to clone genes, in the next section we will consider a wide range of different approaches that scientists use to study cloned genes.

3.4 Laboratory Techniques and Applications of Recombinant DNA Technology

Why clone DNA? What can you do with a cloned gene? There are numerous applications of gene cloning and recombinant DNA technology. **Figure 3.9** summarizes common gene-cloning applications, many of which are discussed further in other chapters. In this section, we present some important routine molecular biology laboratory techniques and basic applications of gene cloning.

Agarose Gel Electrophoresis

Agarose gel electrophoresis is one of the most common laboratory techniques used in working with DNA because it allows one to separate and visualize DNA fragments based on size (**Figure 3.10** on the next page). Agarose is a material that is isolated from seaweed, melted in a buffer solution, and poured into a plastic tray. As the agarose cools, it solidifies to form a horizontal semisolid gel containing small holes or pores through which DNA fragments will travel. The percentage of agarose used to create the gel determines its ability to resolve DNA fragments of different sizes. Most applications generally involve gels that contain 0.5% to 2% agarose. A gel with a high percentage of agarose (say 2%) is better suited for separating small DNA fragments because they will snake

their way through the pores more easily than large fragments, which do not separate through the dense gel very well. A lower percentage of agarose is better suited for resolving large DNA fragments.

To run a gel, it is submerged in a buffer solution that will conduct electricity. DNA samples are loaded into small depressions called *wells*, and an electric current is applied through electrodes at opposite ends of the gel. Separating DNA by electrophoresis is based on the fact that DNA migrates through a gel according to its charge and size (in base pairs). The sugar-phosphate backbone renders DNA negatively charged; therefore, when DNA is placed in an electrical field, it migrates toward the anode (positive pole) and is repelled by the cathode (negative pole). Because all DNA is negatively charged regardless of the length or source, the rate of DNA migration and separation through an agarose gel depends on the *size* of a DNA molecule. Migration distance is inversely proportional to the size of a DNA fragment, so large DNA fragments migrate short distances

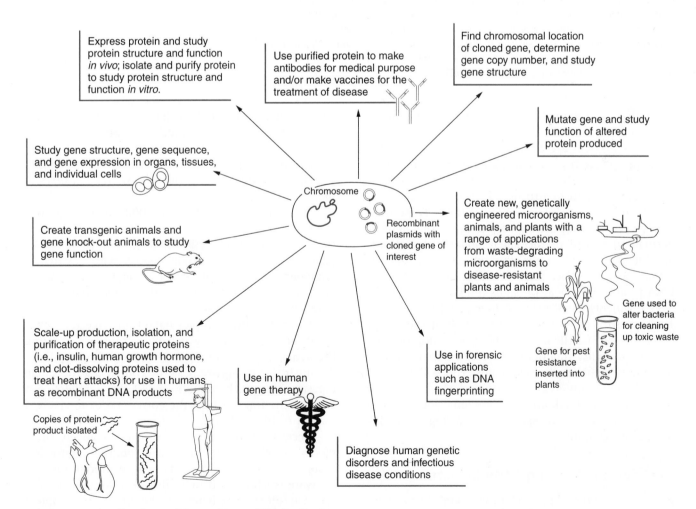

FIGURE 3.9   Applications of Recombinant DNA Technology

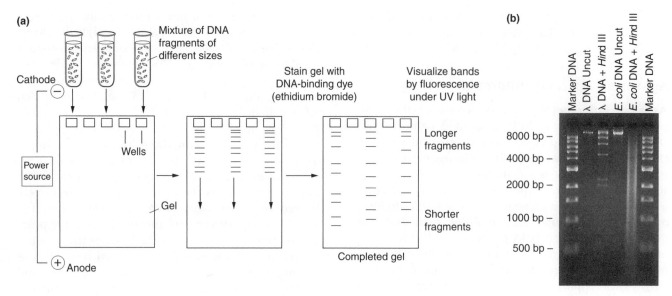

FIGURE 3.10   Agarose Gel Electrophoresis (a) DNA fragments can be separated and visualized by agarose gel electrophoresis. (b) Photograph of an agarose gel stained with ethidium bromide. Lanes labeled as "Marker DNA" were loaded with commercially prepared DNA size standards. These serve as a ladder of fragments of known size that are used to determine the size of experimental samples of DNA being analyzed. The lane labeled "λ DNA uncut" shows high-molecular-weight uncut chromosomal DNA from phage λ; "λ DNA + *Hin*dIII" shows a series of discrete fragments created when λ DNA is digested with the restriction enzyme *Hin*dIII. The lane labeled "*E. coli* DNA uncut" contains undigested chromosomal DNA, and the adjacent lane shows *E. coli* chromosomal DNA digested with *Hin*dIII (*E. coli* DNA + *Hin*dIII).

through a gel relatively slowly and small fragments migrate faster.

Tracking dyes are added to monitor DNA migration during electrophoresis. After the desired time of electrophoresis, DNA in the gel is stained using dyes such as **ethidium bromide,** which intercalate (penetrate) in between the base pairs of DNA. These dyes fluoresce when they are exposed to ultraviolet light. A permanent record of the gel is obtained by photographing the gel while it is exposed to ultraviolet light (Figure 3.10b). Notice how the *Hin*dIII-digested *E. coli* DNA produces a smear of bands unlike the set of discrete fragments visualized with *Hin*dIII-digested λ DNA. This smearing is due to the large size of the *E. coli* chromosome and the large number of cutting sites for *Hin*dIII; so many fragments are created that it is not possible to visualize discrete bands. You will encounter techniques involving agarose gel electrophoresis throughout this book and, as a student, you are very likely to learn to run agarose gels during your biotechnology training.

Restriction Mapping Gene Structure

In the early days of gene cloning, soon after a gene was cloned, typically a type of physical map of the gene would be created to determine which restriction enzymes cut the cloned gene and to pinpoint the location of these cutting sites. Knowing the **restriction map** of a gene was very useful for making clones of small pieces of the gene (called *subcloning*) and manipulating many relatively small pieces of DNA (for example, 100 to 1,000 bp) to sequence DNA and prepare DNA probes to study gene expression. To create a restriction map, cloned DNA is subjected to a series of single digests with restriction enzymes as well as double digests with combined enzymes; the digested DNA is then separated by agarose gel electrophoresis.

Once the DNA samples have been digested, separated, and visualized by gel electrophoresis, creating the actual restriction map is like assembling a puzzle. As illustrated in **Figure 3.11**, by comparing the single digests with each double digest, researchers can arrange the fragments in the correct order to create a map of restriction sites.

Because DNA sequencing of even relatively small pieces of DNA has become fairly common, restriction mapping is now typically done using bioinformatics software (such as Webcutter, described in problem 8 in the "Questions & Activities" at the end of this chapter) to identify restriction cutting sites in a DNA sequence

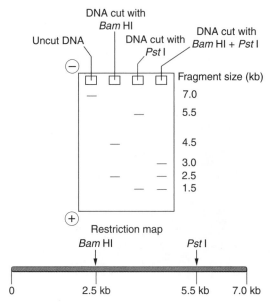

FIGURE 3.11 Restriction Mapping The location of restriction sites for *Bam*HI and *Pst*I is determined in a DNA fragment that is 7 kb long. Digestion with *Bam*HI cleaves the DNA into two fragments measuring 2.5 and 4.5 kb, indicating that the DNA was cut at a single site located 2.5 kb from one end. Digestion with *Pst*I cleaves the DNA into two fragments at 1.5 and 5.5 kb, indicating that the DNA was cut at a single site located 1.5 kb from one end. A double digest with both enzymes cleaves the DNA into three fragments, 3.0 kb, 2.5 kb, and 1.5 kb. Because the 3.0-kb fragment is not created by either *Bam*HI or *Pst*I digestion alone, it must represent the DNA located between the *Bam*HI and *Pst*I cutting sites. Once this "puzzle" of fragments has been arranged, it becomes clear that the restriction map at the bottom of the figure is the only map consistent with the pattern of fragments created by the digests in this experiment.

without having to actually digest DNA and create a map experimentally. However, because of the historical importance of restriction mapping and its occasional use today, it is still of value to be aware of this technique.

DNA Sequencing

After a gene is cloned, it is important to determine the nucleotide sequence of the gene—its exact order of As, Gs, Ts, and Cs. Knowing the DNA sequence of a gene can be helpful (1) to deduce the amino acid sequence of a protein encoded by a cloned gene, (2) to determine the exact structure of gene, (3) to identify regulatory elements such as promoter sequences, (4) to identify differences in genes created by gene splicing, and (5) to identify genetic mutations among other reasons.

Different methods of **DNA sequencing** are available including techniques for PCR "cycle" sequencing and computer-automated DNA sequencing, and these technologies are rapidly improving each year. Initially, the most widely used sequencing approach was chain-termination sequencing, a manual method developed in 1977 by Frederick Sanger and often referred to as the Sanger method. In this technique, a DNA primer was hybridized to denatured template DNA, such as a recombinant plasmid to be sequenced, in a reaction tube containing deoxyribonucleotides and DNA polymerase. Because many modern plasmids are designed with sequencing primer binding sites adjacent to the multiple cloning site, DNA polymerase can be used to extend a complementary strand from the 3′ end of primers hybridized to the plasmid. The original approach utilized radioactively labeled primer sequences.

A small amount of a modified nucleotide called a **dideoxyribonucleotide (ddNTP)** was mixed in with the vector, primer, polymerase, and deoxyribonucleotides. A ddNTP differs from a normal deoxyribonucleotide (dNTP) because it has a hydrogen group attached to the 3′ carbon of the deoxyribose sugar instead of a hydroxyl group-OH (see **Figure 3.12a** on the next page). When a ddNTP is incorporated into a chain of DNA, the chain cannot be extended because the absence of a 3′-OH prevents the formation of a phosphodiester bond with a new nucleotide; hence, the chain is "terminated."

In the original Sanger approach, four separate reaction tubes were set up. Each tube contained vector, primer, and all four dNTPs, but each tube also contained a small amount of one ddNTP. As synthesis of a new DNA strand from the primer begins, DNA polymerase randomly inserts a ddNTP into the sequence instead of a normal dNTP, preventing further synthesis of a complementary strand. Over time, a ddNTP will be incorporated at all positions in the newly synthesized strands, creating a series of fragments of varying lengths that are terminated at dideoxy residues. For the original Sanger technique, the DNA strands were separated on a thin polyacrylamide gel, which can separate sequences that differ in length by a single nucleotide. Autoradiography was used to detect the radioactive sequencing fragments shown in Figure 3.12c. The sequence determined from the autoradiogram is "read" from the bottom to the top as individual nucleotides. As shown in Figure 3.12c, the sequence determined from the autoradiogram is *complementary* to the sequence on the template strand in the vector.

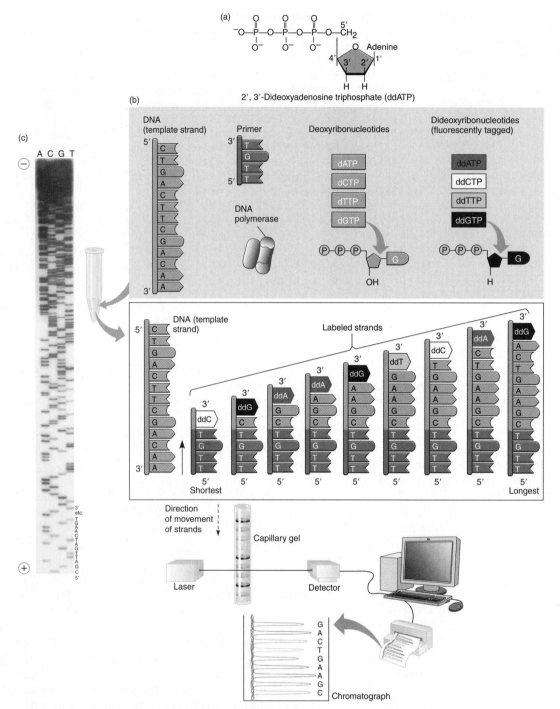

FIGURE 3.12   Computer-Automated DNA Sequencing (a) The structure of a dideoxynucleotide (ddNTP). Note that the 3′ group attached to the carbon is a hydrogen rather than a hydroxyl group (OH). Because another nucleotide cannot be attached at the 3′ end of a ddNTP, these nucleotides are the key to DNA sequencing by the Sanger method as shown (b). Computer-automated approaches separate DNA fragments using a capillary gel. A laser and detector are used to detect fluorescence of each dideoxynucleotide. (c) An autoradiogram from an original dideoxy sequencing reaction in which radioactive ddNTPs or primers were used and four separate reactions carried out. The letters over the lanes (A, C, G, and T) correspond to the particular ddNTP used in the sequencing reaction analyzed in the lane. Analyzing Sanger sequencing by autoradiography is largely obsolete, having been replaced by computer-automated sequencing approaches as shown in (b) and next-generation sequencing technologies.

YOU DECIDE

To Patent or Not?

The Human Genome Project was completed ahead of schedule in part because of competition between publicly funded genome centers and privately funded companies such as Celera Genomics, originally directed by former NIH researcher Craig Venter. While at The Institute for Genomic Research (TIGR), Venter and colleagues were the first scientists to completely sequence the genome of a living organism, the bacterium *Haemophilus influenzae*. This group applied for patents on the nucleotide sequence for *H. influenzae* and on the bioinformatics technology used to analyze this genome.

Previously, Venter and his colleagues described a set of experiments in which they randomly cloned short pieces of cDNAs from human brain cells. These short sequences—called **expressed sequence tags (ESTs)**—could in theory be used as probes to identify full-length cDNAs. Some of Venter's ESTs were found to be identical to genes that had already been cloned or to a portion of a gene; others appeared to be novel gene sequences or junk DNA. Hoping to gain proprietary rights to full-length genes that could be identified from Venter's ESTs, TIGR applied for a patent. This request generated a great deal of controversy. For instance:

- Should scientists be allowed to patent DNA sequences from naturally living organisms?
- What if a patent is awarded for only small *pieces of* a gene—even if no one knows what a DNA sequence does—just because some individuals or a company wants a patent to stake their claim to having cloned a piece of DNA first?
- What if there are no clear uses for the DNA sequences cloned?
- Can or should investigators who use a gene microarray or create a DNA library be allowed to patent the entire genome of any organism they have studied?
- When they are granted a patent, scientists essentially have a monopoly on patented information for two decades from the patent filing date. Many

believe that the hoarding of genome information is against the tradition of sharing information to advance science. Would awarding patents slow progress in cloning genes if groups hoarded data and did not share information? Could or should a group stake a claim to a gene, thereby preventing others from working on it or developing a product from it?

- What about individuals who figure out *what to do* with the gene?
- Should a genetically engineered living organism be patented? Engineered bacteria (for example, those used to clean up environmental pollution) and transgenic animals have been patented, as have clinically important genes such as beta interferon.
- Can a group claim rights to anticipated future uses of the gene even if there are no data to substantiate such claims?
- What if a gene sequence is involved in a disease for which a genetic therapy may be developed?

Since 1980, the U.S. Patent and Trademark Office has granted patents on more than 35,000 genes or gene sequences, and an estimated 20% of human genes have been patented. Some scientists are concerned that patents awarded for simply cloning a piece of DNA is awarding a patent for too little work. Many scientists believe it is more appropriate to patent the novel technology used to discover and study genes and the applications of genetic technology such as gene therapy approaches than to patent the gene sequences themselves. Generally most gene patents cover technology uses for a sequence, such as a genetic test, and not the DNA sequence itself. From a commercial standpoint, one advantage of patenting is that it provides private companies with a financial incentive to get a medicine or technology to the marketplace and enables them to make a profit after many millions (or billion in some cases) of dollars are spent on R&D to make a product. To patent or not? You decide.

High-throughput computer-automated sequencing replaces the original Sanger sequencing technique

Because of limitations in running a sequencing gel, the original Sanger method could be used only to sequence approximately 200 to 400 nucleotides in a single reaction; therefore, in sequencing a longer piece of DNA—for example, 1,000 base pairs—it was necessary to run multiple reactions to create overlapping sequences

that could be pieced together to determine the entire, continuous sequence of 1,000 base pairs.

Because of this limitation, the Sanger sequencing approach was a cumbersome method for the large-scale sequencing efforts such as those used for the Human Genome Project. The project was completed ahead of schedule in part because of the development of **computer-automated DNA sequencing methods** capable of sequencing long stretches of DNA (more than 500 bp in a single reaction). Initially computer-automated sequencing reactions used either ddNTPs, each labeled with a different colored nonradioactive fluorescent dye, or a sequencing primer labeled at its 5′ end with a dye (Figure 3.12). A single reaction tube is used, and the manual-style approach of using polyacrylamide gels followed by autoradiography has been replaced by separating sequencing reactions on a single lane of an ultrathin diameter tube gel called a *capillary gel*. As DNA fragments move through the gel, they are scanned with a laser beam. The laser stimulates the fluorescent dye on each DNA fragment, which emits different wavelengths of light for each ddNTP. The emitted light is collected by a detector that amplifies and then feeds this information to a computer to process and convert the light patterns and reveal the DNA sequence (**Figure 3.13**).

Automated DNA sequencers often contain multiple capillary gels several feet long, allowing many bases to be separated. Therefore some instruments run as many as 96 capillary gels, each producing around 900 bases of sequence. With these instruments it became possible to generate approximately 2 million bases of sequence in a day. As a result, these sequencers became known as *high-throughput* sequencers because of their ability to process and generate large amounts of sequence data in relatively short periods of time.

Next-Generation Sequencing (NGS)

Genomics has spurred a demand for sequencers that are faster and capable of generating millions of bases of DNA sequence in a relatively short time, leading to the development of **next-generation sequencing (NGS)** technologies designed to produce highly accurate and long stretches of DNA sequence, greater than 1 gigabase (billion bases) of DNA per reaction, at a low cost. Next-generation sequencing approaches dispense with the Sanger technique and capillary electrophoresis methods in favor of sophisticated, parallel formats (simultaneous-reaction formats) that use state-of-the art fluorescence imaging techniques. NGS technologies are providing an unprecedented capacity for generating massive amounts of DNA sequence data rapidly (up to 200 times faster than Sanger approaches in some cases!) and at dramatically reduced costs per base.

The desire for next-generation sequencing among the research community and challenges such as the $1,000 genome (discussed later in this chapter) has led to an intense technology race among many companies eager to produce NGS methods. In 2005, Roche 454 Life Systems was the first company to commercialize a next-generation sequencing technology (Figure 3.13). This approach sequences genomes using a so-called solid-phase method in which beads are attached to fragmented genomic DNA, which is then PCR-amplified in separate water droplets in oil for each bead, loaded into multiwell plates, and mixed with DNA polymerase. Multiwell plates often contain more than a million wells with one bead per well, each serving as a reaction tube for sequencing (see Figure 3.13). A reaction called **pyrosequencing** is then used to sequence DNA on the beads in each well.

In pyrosequencing, a single labeled nucleotide (e.g., dGTP) is flowed over the wells. Each well contains a single bead along with primers annealed to the DNA on the bead. When a complementary nucleotide crosses a template strand adjacent to the primer, it is added to the 3′ end of the primer by DNA polymerase. Incorporation of a nucleotide results in the release of pyrophosphate, which initiates a series of chemiluminescent (light-releasing) reactions that ultimately produce light by using the firefly enzyme luciferase. Emitted light from the reaction is captured and recorded to determine when a single nucleotide has been incorporated into a strand. By rapidly repeating the nucleotide-flow step with each of the four nucleotides to determine which base is next in the sequence, this approach can generate read lengths of about 400 bases and on the order of 400 million bases (Mb) of data per 10-hour run. NGS sequencing technologies are also creating major data-management challenges for saving and storing such large image data files.

The company Applied Biosystems (ABI) has developed an approach called **SOLiD** (supported oligonucleotide ligation and detection) that can produce 6 *gigabases* of sequence data per run! The SOLiD method combines a variety of approaches to sequence DNA fragments that are linked to beads and amplified, similar to the 454 approach, but different sequencing technologies are used, which can provide greater output of sequencing data per instrument run. The instrumentation to run these platforms is expensive, but given the massive amounts of sequence that the NGS methods can generate, the average cost per base is much lower than that of Sanger sequencing. Through 2006, new sequencing technologies were cutting sequencing costs in half about every 2 years.

Work on third-generation sequencers is actively under way and there is every reason to believe that these instruments will be available by the middle part of this decade. For example, one promising approach on the immediate horizon involves the use of nanotechnology by pushing single-stranded DNA

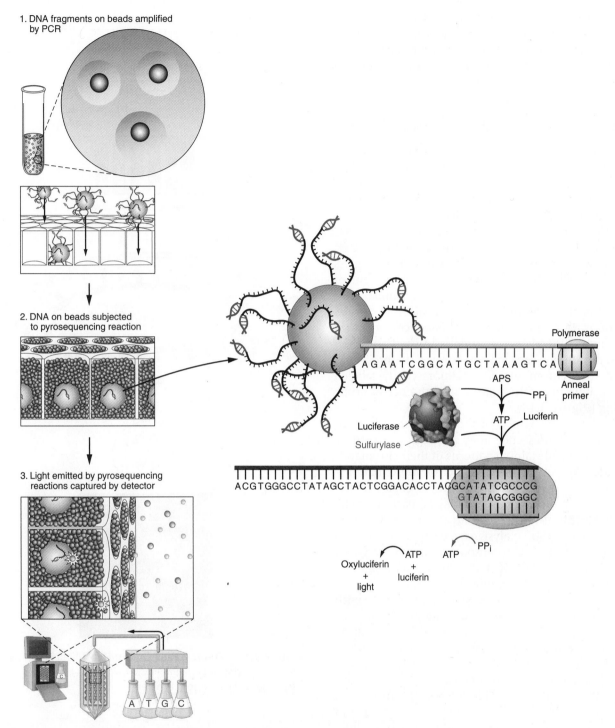

1. DNA fragments on beads amplified by PCR

2. DNA on beads subjected to pyrosequencing reaction

3. Light emitted by pyrosequencing reactions captured by detector

Polymerase

Anneal primer

A G A A T C G G C A T G C T A A A G T C A

APS

PP$_i$

ATP

Luciferin

Luciferase

Sulfurylase

ACGTGGGCCTATAGCTACTCGGACACCTACGCATATCGCCCG
GTATAGCGGGC

Oxyluciferin
+
light

ATP
+
luciferin

ATP

PP$_i$

A T G C

FIGURE 3.13 Roche 454 Next-Generation Sequencing Technology Roche 454 sequencing technology binds DNA fragments to beads. DNA fragments are amplified by PCR and then added to wells (Step 1) and subjected to pyrosequencing, as described in the text. (Step 2) During pyrosequencing, an individual fluorescent nucleotide, the blue G shown in this case, is flowed over the well. As a nucleotide is added to a primer by DNA polymerase, inorganic pyrophosphate (PPi) is released which reacts with adenosine 5'-phosphosulfate (APS) and the enzyme sulfurylase to produce ATP (blue arrow). The firefly enzyme luciferase uses ATP to generate light, which can be detected and quantified. (Step 3) Light captured by a detection system is analyzed to trace the pattern of nucleotides added to each well. The flow cycle is subsequently repeated with each of the other three nucleotides. Continual cycling of this process generates sequencing reads of approximately 400 bases.

fragments into nanopores and then cleaving off individual bases to produce a signal that can be captured. This method does not involve DNA amplification or fluorescent tags and thus provides direct sequencing of the DNA in a single strand.

Fluorescence In Situ Hybridization

A technique called **fluorescence in situ hybridization** (FISH; *in situ* means "in place") can be used to identify which chromosome contains a gene of interest. For example, if you just cloned a human gene believed to be involved in intelligence, you could use FISH to determine on which chromosome this gene resides. In FISH, chromosomes are isolated from cells such as white blood cells and spread out on a glass microscope slide. A DNA or RNA probe for the gene of interest is labeled with fluorescent nucleotides and then incubated in solution with the slide. The probe hybridizes with complementary sequences on chromosomes on the slide. The slide is washed and then exposed to fluorescent light. Wherever the probe has bound to a chromosome, the fluorescently labeled probe is illuminated to indicate the presence of probe binding (**Figure 3.14**).

To determine which of the 23 human chromosomes show fluorescence, they are aligned according to the length and staining patterns of their chromatids to create a karyotype. Fluorescence on more than one chromosome indicates either multiple copies of the gene or related sequences that may be part of a gene family. FISH is also used to analyze genetic disorders. For example, FISH analysis can be performed on a karyotype of fetal chromosomes from a pregnant woman to determine if a developing fetus has an abnormal number of chromosomes.

When a gene is expressed in an organ with many different cell types—for example, kidney or brain tissue—FISH can also be used to determine the cell type that is expressing a particular mRNA. In this approach, the tissue of interest is preserved in a fixative solution and then embedded in a wax-like material or resin. This allows researchers to slice the tissue into thin sections about 1 to 5 mm thick and to attach them to a microscope slide. Sometimes frozen sections of tissue are used for these experiments. The slide is incubated with a fluorescent dye-tagged RNA or DNA probe for the gene of interest. The probe hybridizes to mRNA within cells in their native place and probe binding is determined by detecting fluorescence. For some studies, PCR can even be performed directly on tissue sections in situ as a way of determining cell-type expression for a given gene.

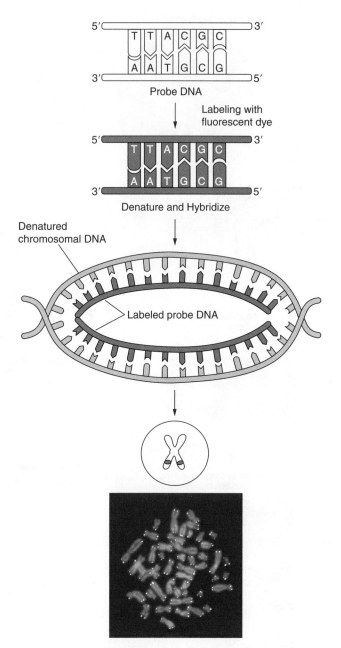

FIGURE 3.14 **Fluorescence in Situ Hybridization** White spots at the tips of each chromosome indicate fluorescence from a probe binding to telomeres.

Southern Blotting

Another technique, called **Southern blot analysis** (Southern blotting or hybridization), is frequently used to determine gene copy number among other applications. Developed by Ed Southern in 1975, Southern blotting begins by digesting chromosomal DNA into small fragments with restriction enzymes. DNA fragments are separated by agarose gel electro-

(a)

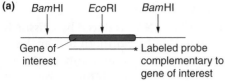

BamHI EcoRI BamHI

Gene of interest ———— * Labeled probe complementary to gene of interest

(b)

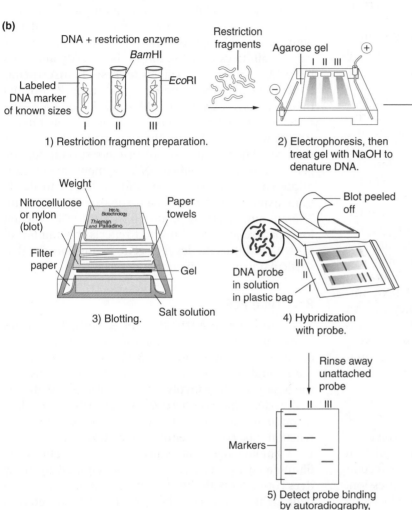

DNA + restriction enzyme
BamHI
EcoRI
Labeled DNA marker of known sizes
I II III
1) Restriction fragment preparation.

Restriction fragments
Agarose gel
I II III
(+)
(−)
2) Electrophoresis, then treat gel with NaOH to denature DNA.

Weight
Nitrocellulose or nylon (blot)
Paper towels
Intro to Biotechnology
Thieman and Palladino
Filter paper
Gel
Salt solution
3) Blotting.

Blot peeled off
III
II
I
DNA probe in solution in plastic bag
4) Hybridization with probe.

Rinse away unattached probe
I II III
Markers
5) Detect probe binding by autoradiography, chemiluminescence or other imaging techniques.

FIGURE 3.15 Southern Blot Analysis of DNA Fragments (a) Region of DNA for a gene of interest to be studied by Southern blot analysis (b). Steps involved in Southern blotting.

phoresis (**Figure 3.15**). However, the number of restriction fragments generated by digesting chromosomal DNA is often so great that simply running a gel and staining the DNA does not resolve discrete fragments. Rather, digested DNA appears as a continuous smear of fragments in the gel. Southern blotting is used to visualize only *specific* fragments of interest. Following electrophoresis, the gel is treated with an alkaline solution to denature the DNA; then the fragments are transferred onto a nylon or nitrocellulose membrane using a technique called *blotting*.

Blotting can be achieved by setting up a gel sandwich in which the gel is placed under the nylon mem-

brane, filter paper, paper towels, and a weight to allow for wicking of a salt solution through the gel, which will transfer DNA onto the nylon by capillary action (Figure 3.15). The single-strand of DNA stick to the blot, positioned in bands exactly as on the gel. Alternatively, pressure or vacuum blotters can be used to transfer DNA onto nylon. The nylon blot is then baked or briefly exposed to UV light to attach the DNA permanently. Next the blot is incubated with a labeled probe in much the same way that colony hybridizations are carried out. Increasingly, nonradioactive probes are being used for Southern blotting and many other hybridization techniques.

The blot is washed to remove extraneous probes and then exposed to film by autoradiography if a radioactive or chemiluminescent probe is used. Wherever the probe has bound to the blot, radioactivity or light released by the probe develops silver grains on the film to expose bands on the blot, creating an autoradiogram (see Figure 3.15). Chemiluminescent probes can also be detected by digital camera imaging systems. By interpreting the number of bands revealed, it can be possible to determine gene copy number.

The development of Southern blot analysis was an important technique, which formed the basic principles for **Northern blotting** (the separation and blotting of RNA molecules, as discussed in the next section) and **Western blotting** (the separation and blotting of proteins). Northern and Western blotting techniques were not named after scientists named "Northern" and "Western"; rather, they were named as tongue-in-cheek references to Ed Southern, the founder of Southern blots. Southern blotting has many other applications, including gene mapping, related gene-family identification, genetic mutation detection, PCR product confirmation, and DNA fingerprinting (a topic we discuss in Chapter 8). An excellent animation of how Southern blot analysis is used in DNA fingerprinting can be found at the website for the Cold Spring Harbor DNA Learning Center, which can be found on the Companion Website.

Studying Gene Expression

Molecular biologists around the world are involved in research studying gene expression and the regulation of gene expression. Numerous different molecular techniques are available for studying gene expression. Most methods involve analyzing mRNA produced by a tissue. This is often a good measure of gene expression because the amount of mRNA produced by a tissue is often equivalent to the amount of protein the tissue makes.

Northern Blot Analysis

The basic methodology of a Northern blot is similar to Southern blot analysis. In Northern blotting, RNA is isolated from a tissue of interest and separated by gel electrophoresis (the RNA is not digested with enzymes). RNA is blotted onto a nylon membrane and then hybridized to a probe, as described for Southern blots. Exposed bands on the autoradiogram show the presence of mRNA for the gene of interest and the size of the mRNA (**Figure 3.16a**). In addition, the amounts of mRNA produced by different tissues can be compared and quantified.

Reverse transcription PCR

Sometimes the amount of RNA produced by a tissue is below the level of detection by Northern blot analysis. PCR allows for detecting minute amounts of mRNA from even very small amounts of starting tissue. For instance, PCR has been a great tool for molecular biologists studying gene expression in embryos and developing tissues where the amount of tissue for analysis is very small. Because RNA cannot be directly amplified by PCR, a technique called **reverse transcription PCR (RT-PCR)** is carried out. In RT-PCR, isolated mRNA is converted into double-stranded cDNA by the enzyme reverse transcriptase in a process similar to the way in which cDNA for a library is made. The cDNA is then amplified with a set of primers specific for the gene of interest. Amplified DNA fragments are electrophoresed on an agarose gel and evaluated to determine expression patterns in a tissue (see Figure 3.16b). The amount of cDNA produced in a RT-PCR reaction for a particular gene of interest reflects the amount of mRNA, and thus the level of gene expression, for that particular gene in a given tissue.

Real-time PCR

Real-time or quantitative PCR (qPCR), enables researchers to quantify amplification reactions as they occur in "real time" (**Figure 3.17** on page 86). There are several ways to run real-time PCR reactions, but the basic procedure involves the use of specialized thermal cyclers that use a laser to scan a beam of light through the top or bottom of each PCR tube. Each reaction tube contains either a dye-containing probe or DNA-binding dye that emits fluorescent light when illuminated by the laser. The light emitted by these dyes correlates with the amount of PCR product amplified. Light from each tube is captured by a detector, which relays information to a computer to provide a readout on the amount of fluorescence after each cycle; this readout can be plotted and analyzed to quantitate the number of PCR products produced after each cycle.

Two commonly used approaches for real-time PCR involve the use of TaqMan probes and a dye called SYBR Green. SYBR Green binds double-stranded DNA. As more double-stranded DNA is copied with each round of real-time PCR, there are more DNA copies to bind SYBR Green, which increases the amount of fluorescent light emitted.

TaqMan probes are complementary to specific regions of the target DNA between where the forward and reverse primers for PCR bind (Figure 3.17). TaqMan probes contain two dyes. One dye, the reporter, is located at the 5' end of the probe and can release fluorescent light when excited by laser light from the thermal cycler. The other dye, called a quencher, is

(a)

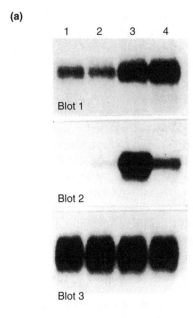

(b)

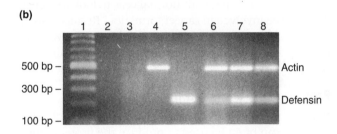

FIGURE 3.16 **Analyzing Gene Expression by Northern Blot Analysis and RT-PCR** Northern blot analysis and RT-PCR are two common techniques for analyzing the amount of mRNA produced by a tissue (gene expression). (a) Blot 1 is a portion of an autoradiogram from a Northern blotting experiment in which RNA from four rat tissues—1, seminal vesicles; 2, kidney; 3 and 4, different segments of the epididymis (a male reproductive organ)—was blotted onto nylon and then probed with a radioactive cDNA probe for a gene involved in protecting tissues from damage by harmful free radicals, atoms, or molecules with unpaired numbers of electrons. Notice how the amount of mRNA detected in lanes 3 and 4 (as indicated by the size and darkness of each band) is greater than the amount of mRNA in lanes 1 and 2. Blot 2 shows an autoradiogram from blot 1 that was stripped of bound probe and reprobed with a probe for a different gene. Blot 3, the same blot shown in the other two panels, was stripped of bound probe and reprobed with a radioactive cDNA for a gene (cyclophilin) that is expressed at nearly the same levels in virtually all tissues. (b) Agarose gel from an RT-PCR experiment in which RNA from rat tissues was reverse-transcribed and amplified with primers for β-actin (an important component of the cytoplasm of cells) and/or β-defensin-1 (a gene that encodes a peptide that provides protection against bacterial infections in many tissues). Lane 1 contains DNA size standards of known size (often called a ladder) increasing in 100-bp increments. Lane 2 is a negative control sample in which primers were added to a PCR experiment without cDNA. Lane 3 is a negative control sample in which cDNA was added to a PCR experiment without primers. Notice that lanes 2 and 3 do not show any amplified PCR product because amplification will not occur without cDNA as target DNA (lane 2) or without primers (lane 3). Lane 4 shows kidney cDNA amplified with actin primers. Lane 5 shows kidney cDNA amplified with defensin primers. Lanes 6, 7, and 8 show PCR products from cDNA of three different rat reproductive tissues that were amplified with both actin and defensin primers. Notice how lanes 6 and 8 show relatively even amounts of actin and defensin PCR products; greater amounts of defensin PCR product are shown in lane 7. These differences reflect the different amounts of defensin mRNA made by these tissues.

attached to the 3′ end of the probe. When these two dyes are close to each other, the quencher dye interferes with the fluorescent light released from the reporter dye. However, as *Taq* DNA polymerase extends each primer, it removes the reporter dye from the end of the probe (and eventually removes the entire probe). Now that the reporter dye is separated from the quencher, the fluorescent light released by the reporter can be detected by the thermal cycler. Detected light is analyzed by a computer to produce a plot displaying the amount of fluorescence emitted with each cycle. Because real-time PCR does not involve running gels, it is a powerful and rapid technique for measuring and quantitating changes in gene expression, particularly when multiple samples and different genes are being analyzed simultaneously.

Gene microarrays

DNA microarray analysis is another technique for studying gene expression that rapidly gained popularity over the past 15 years because it enables researchers to analyze all the genes expressed in a tissue very quickly (**Figure 3.18** on page 87). A microarray, also known as a **gene chip,** is created with the use of a small glass microscope slide. Single-stranded DNA molecules are attached or "spotted" onto the slide using a computer-controlled high-speed robotic arm called an *arrayer,* which is fitted with a number of tiny pins. Each pin is immersed in a small amount of solution containing millions of copies of different DNA molecules, such as cDNAs for different genes. The arrayer fixes this DNA onto the slide at specific locations (points or spots) recorded by a computer. A single microarray can have

(a)

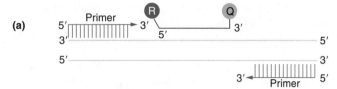

1. *Hybridization.* Forward and reverse PCR primers bind to denatured target DNA. TaqMan probe with reporter (R) and quencher (Q) dye binds to target DNA between the primers. When probe is intact, emission by the reporter dye is quenched.

2. *Extension.* As DNA polymerase extends the forward primer, it reaches the TaqMan probe and cleaves the reporter dye from the probe. Released from the quencher the reporter can now emit light when excited by a laser.

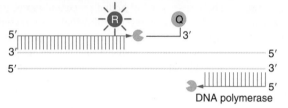

(b)

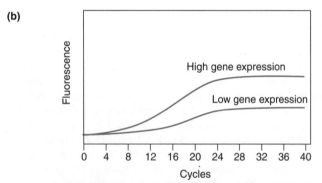

3. *Detection.* Emitted light from the reporter is detected and interpreted to produce a plot that quantitates the amount of PCR product produced with each cycle.

FIGURE 3.17   Real-Time PCR (a) The TaqMan method of real-time PCR involves a pair of PCR primers along with a probe sequence complementary to the target gene. The probe contains a reporter dye (R) at one end and a quencher dye (Q) at the other. When the quencher dye is close to the reporter dye, it interferes with fluorescence released by the reporter dye. When *Taq* DNA polymerase extends a primer to synthesize a strand of DNA, it cleaves the reporter dye off the probe, allowing the reporter to give off energy. (b) Each subsequent PCR cycle removes more reporter dyes, so that increased light emitted from the dye can be captured by a computer to produce a read-out of fluorescence intensity with each cycle.

over 10,000 spots of DNA, each containing unique sequences of DNA for a different gene.

There are several different kinds of microarrays. Figure 3.18 shows a representative example of how a microarray can be used to study gene expression. Scientists begin by extracting extract mRNA from a tissue

of interest. Then cDNA synthesized from this mRNA is labeled with a fluorescent dye. Labeled cDNA is incubated overnight with the microarray where it hybridizes to spots on the microarray that contain complementary DNA sequences. The microarray is washed and scanned by a laser that causes the cDNA hybridized to the microarray to fluoresce. Fluorescent spots reveal which genes are expressed in the tissue of interest, and the intensity of fluorescence indicates the relative amount of expression. The brighter the spot, the more mRNA is expressed in that tissue. Microarrays can also be run by labeling cDNAs from two or more tissues with different-colored fluorescent dyes. Gene expression patterns from the different tissues are compared based on the color of spots that appear following hybridization and detection.

For many species, including humans, entire genomes are available on microarrays. Researchers are also using microarrays to compare patterns of expressed genes in tissues under different conditions. For example, cancer cells can be compared with normal cells to look for genes that may be involved in cancer formation. Results of such microarray studies can be used to develop new drug therapy strategies to combat cancer and other diseases (Chapter 11).

Protein expression and purification

Bacteria that have been transformed with recombinant plasmids can often be used to produce the protein product of the isolated gene. Large quantities of bacteria can be grown in a fermenter, and the protein can be isolated (Chapter 4).

Gene mutagenesis studies

Of the many different ways that scientists work with and study cloned genes, there are a variety of techniques that can be used to study the structure and function of protein produced by a specific gene. One approach is called **site-directed mutagenesis.** In this technique, mutations can be created in specific nucleotides of a cloned gene contained in a vector. The gene can then be expressed in cells, which results in the translation of a mutated protein. This allows researchers to study the effects of particular mutations on protein structure and serves as a way of determining which nucleotides are important for specific functions of the protein. Site-directed mutagenesis can be a very valuable way to help scientists identify critical sequences in genes that produce proteins involved in human diseases.

RNA interference

In 1998, researchers Craig C. Mello of the University of Massachusetts Medical School and Andrew Z. Fire of Stanford University published groundbreaking work

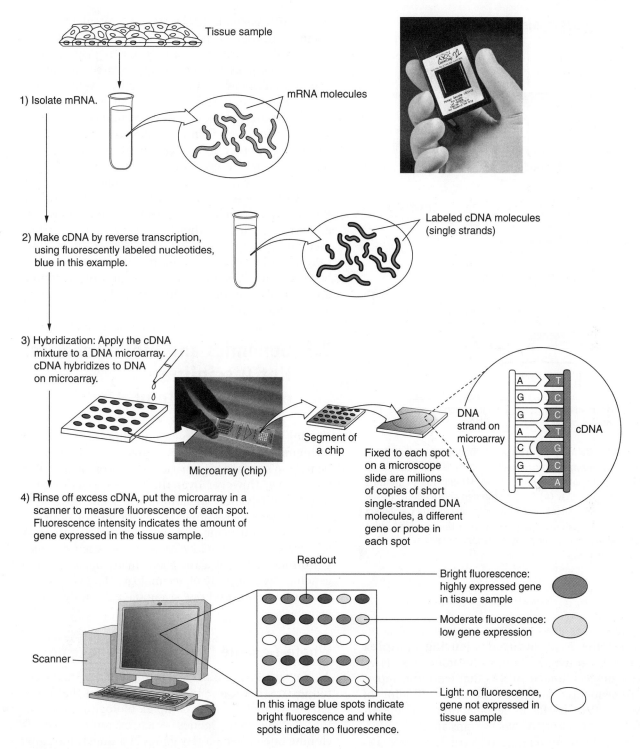

FIGURE 3.18 Gene Microarray Analysis

in which they used double-stranded pieces of RNA (dsRNA) to inhibit or silence expression of genes in the nematode roundworm *Caenorhabditis elegans*. This naturally occurring mechanism for inhibiting gene expression is known as **RNA interference (RNAi).** As Mello,

Fire, and other researchers discovered, dsRNA can be bound by an RNA-digesting enzyme called dicer, which cuts dsRNA into 21- to 25-nucleotide-long snippets of RNA molecules called **small interfering RNAs (siRNAs).**

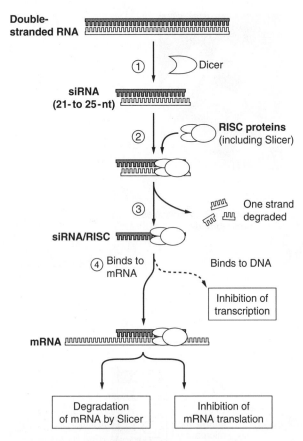

FIGURE 3.19 **RNA Interference** (1) Double-stranded RNA is cleaved by the enzyme dicer into siRNAs. (2) RISC proteins bind siRNAs and degrade one of the two strands (3) to produce single-stranded siRNAs. (4) Single-stranded siRNAs bind to complementary sequences on mRNA molecules in the cytoplasm and interfere with (silence) gene expression through triggering mRNA degradation by slicer or by inhibiting translation of the mRNA by ribosomes.

These siRNAs are bound by a protein-RNA complex called the **RNA-induced silencing complex (RISC).** RISC unwinds the double-stranded siRNAs, releasing single-stranded siRNAs that bind to complementary sequences in mRNA molecules. Binding of siRNAs to mRNA results in degradation of the mRNA (by the enzyme slicer) or blocks translation by interfering with ribosome binding (**Figure 3.19**). RNAi is similar in mechanism to the gene-silencing actions of miRNAs, but a primary difference is that siRNAs are generated from dsRNA (see Chapter 2).

Initially it appeared that the discovery of siRNAs and miRNAs was perhaps a relatively esoteric finding. Predictions now indicate that mammalian cells express thousands of siRNAs or miRNAs, and these may in turn regulate hundreds of genes. Biotechnology and pharmaceutical companies are looking for ways in which siRNAs and miRNAs may be exploited for therapeutic purposes. Techniques incorporating RNAi have developed rapidly as methods for regulating gene expression and as a potential way to target and inactivate specific genes with high efficiency. We'll consider examples of how biotechnology and pharmaceutical companies are working on RNAi techniques for silencing genes involved in human diseases later (Chapter 11). RNAi has become such a rapidly developing technology that recent estimates indicate the use of RNAi reagents will grow from about $400 million in 2005 to over $900 million by 2012. The 2006 Nobel Prize in Physiology or Medicine was awarded to Andrew Fire and Craig Mello for their discovery of this natural method for switching genes off. Nobel Prizes are typically awarded decades after the honored work has been completed, so this unusually quick recognition is a clear indication of the value of RNAi as a powerful and promising research tool.

3.5 Genomics and Bioinformatics: Hot Disciplines of Biotechnology

Cloning individual genes using libraries and other techniques described in this chapter will continue to be used in specialized approaches in recombinant DNA technology. However over the last 15 years or so, a number of advances in cloning and sequencing technologies have increasingly led to the use of strategies for cloning, sequencing, and analyzing entire genomes—an exciting and rapidly developing science called **genomics.** You will learn about many applications of genomics as you study biotechnology. Here we provide an introduction to how scientists can study whole genomes.

Whole-Genome "Shotgun" Sequencing

As powerful as traditional recombinant DNA techniques are, it became increasing apparent that if scientists wanted to study complex biological processes that involve many genes—such as most cancers—cloning one or even a few genes at a time is too slow, yielding only incremental information about genes. As a result, scientists started working on strategies for cloning and sequencing entire genes—a strategy commonly called **whole-genome "shotgun" sequencing.** The analogy is that cloning individual genes using libraries is equivalent to using a rifle to hit a specific spot on a target (e.g., cloning a specific gene), whereas the pellets from a shotgun would randomly hit many spots on a target with little preci-

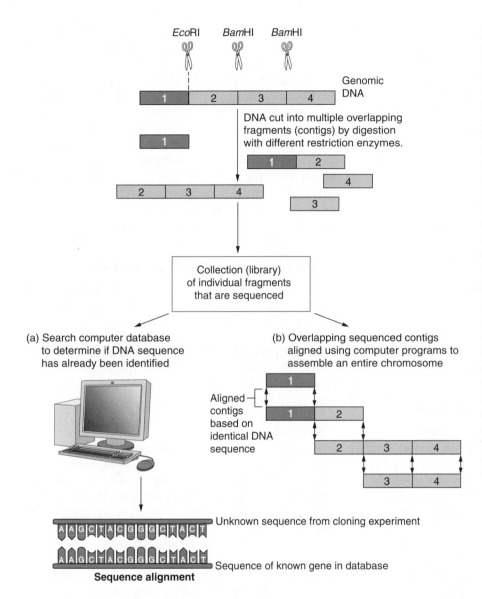

EcoRI   *Bam*HI   *Bam*HI

Genomic DNA

DNA cut into multiple overlapping fragments (contigs) by digestion with different restriction enzymes.

Collection (library) of individual fragments that are sequenced

(a) Search computer database to determine if DNA sequence has already been identified

(b) Overlapping sequenced contigs aligned using computer programs to assemble an entire chromosome

Aligned contigs based on identical DNA sequence

Unknown sequence from cloning experiment

Sequence of known gene in database

Sequence alignment

FIGURE 3.20 Whole-Genome Shotgun Sequencing and Two Examples of Bioinformatics Applications Shown is a simplified example of a piece of genomic DNA cut into smaller pieces (labeled 1, 2, 3, and 4) by *EcoRI* and *Bam*HI. Regardless of how a piece of DNA is cloned, DNA sequence analysis is an essential part of learning more about the cloned DNA. (a) Typically such analysis begins by searching the unknown sequence against a database of known sequences. In this example, sequence alignment reveals that the unknown sequence cloned is an exact match with a sequence of a "known" gene already in the database. Note: For simplicity, only one strand of sequenced DNA is being compared with the database. (b) Another application of bioinformatics involves the use of computer programs to align DNA fragments (contigs) based on nucleotide sequence overlaps. The scheme shown here is similar to one approach used in the Human Genome Project to assemble completed sequences of entire chromosomes. Although when contigs are compared, typically only relatively short sequences at the end of each contig are aligned to assemble an entire chromosome.

sion. In shotgun sequencing, the entire genome, introns and exons, is sequenced. The complete genome sequence is assembled with the help of software programs, and then individual genes are identified through bioinformatics.

One common shotgun strategy for constructing the sequence of whole chromosomes involves using restriction enzymes to digest pieces of entire chromosomes (**Figure 3.20**). This process can produce thousands of overlapping fragments called **contigs (contiguous sequences),** which have to be sequenced. Each contig is sequenced, and then computer programs are used to align the fragments based on overlapping sequence pieces (Figure 3.20b). This approach enables scientists to reconstruct the entire sequence of a whole chromosome and, as you can see in the figure, the use of computers to organize and compare the DNA sequences of these fragments is essential. This is **bioinformatics** in action—an interdisciplinary field that applies computer science and information technology to promote an understanding of biological processes.

In 1995, scientists at The Institute for Genomic Research were the first to use a shotgun approach to successfully sequence the genome of any organism when they sequenced the 1.8 million base pair genome of the bacterium *Haemophilus influenzae*. Many doubted that shotgun-sequencing strategies could be used effectively to sequence larger genomes, but development of this technique and novel sequencing strategies rapidly accelerated the progress of the Human Genome Project.

Bioinformatics: Merging Molecular Biology with Computing Technology

When scientists sequence a newly identified gene or DNA sequence, they report their findings in scientific publications and submit the sequence data to databases so that other scientists who may be interested in this sequence information can have access to it. Database manipulations of DNA sequence data were among the first applications of bioinformatics, which involves the use of computer hardware and software to study, organize, share, and analyze data related to gene structure, gene sequence and expression, and protein structure and function.

As genome data have rapidly accumulated, resulting in an enormous amount of information being stored in public and private databases, bioinformatics has become an essential tool that allows scientists to share and compare data, especially DNA sequence data. We discuss basic applications of bioinformatics throughout this book; however, we introduce the field here with a few very common applications.

Examples of Bioinformatics in Action

Even before whole-genome sequencing projects began, scientists were accumulating sequence information from a variety of different organisms, necessitating the development of sophisticated databases that could be used by researchers around the world to store, share, and obtain the maximum amount of information from protein and DNA sequences. Databases are essential tools for archiving and sharing data with researchers and the public. Because high-speed computer-automated DNA sequencing techniques were developed nearly simultaneously with expansion of the Internet as an information tool, many DNA sequence databases became available through the Internet.

When scientists who have cloned a gene enter their sequence data into a database, these databases search the new sequence against all other sequences in the database and create an *alignment* of similar nucleotide sequences if a match is found (Figure 3.20a). This type of search is often one of the first steps taken after a gene is cloned using recombinant DNA techniques because it is important to determine if the sequence has already been cloned and studied or if the gene sequence is a novel one. Among other applications, these databases can also be used to predict the sequence of amino acids encoded by a nucleotide sequence and to provide information on the function of the cloned gene.

One of the most widely used DNA sequence databases worldwide is called **GenBank.** It is the largest publicly available database of DNA sequences and contains the National Institutes of Health (NIH) collection of DNA sequences. GenBank shares and acquires data from databases in Japan and Europe. Maintained by the National Center for Biotechnology Information (NCBI) in Washington, D.C., GenBank contains more than 100 billion bases of sequence data from over 100,000 species, and it is growing rapidly every year. The NCBI is a gold mine of bioinformatics resources that creates public access databases and develops computing tools for analyzing and sharing genome data. The NCBI has also designed user-friendly ways to access and analyze an incredible amount of data on nucleotide sequences, protein sequences, molecular structures, genome data, and even scientific literature.

DNA database searching: Try it yourself

For example, an NCBI program called **Basic Local Alignment Search Tool** (**BLAST;** www.ncbi.nlm.nih.gov/BLAST) can be used to search GenBank for sequence matches between cloned genes and to create DNA sequence alignments. Go to the BLAST website and click "standard nucleotide-nucleotide BLAST [blastn]." In the search box, type in the following sequence: AATAAAGAAC CAGGAGTGGA. Imagine that this sequence is from a piece of a gene that you just cloned and sequenced and you want to know if anyone cloned this gene before you. Click the "Blast!" button. Your results will be available in a minute or two. Click the "Format!" button to see the results of your search. A page will appear with the results of your search (you may need to scroll down the page to find the sequence alignment). What did you find?

Figure 3.21 shows a BLAST search alignment comparing the human and mouse gene sequences for an obesity gene (*ob*) that produces a hormone called *leptin*. Leptin plays a role in fat metabolism, and mutations in the leptin gene can contribute to obesity (see also Figure 11.1). Notice that the nucleotide sequence for these two genes is very similar, as indicated by the vertical lines between identical nucleotides.

The Human Genome Nomenclature Committee, supported by the NIH, establishes rules for assigning names and symbols to newly cloned human genes. Each entry into GenBank is provided with an **accession number** that scientists can use to refer back to that cloned sequence. For example, go to the GenBank website listed at the Companion Website and then type in the accession number U14680 and click "GO." What gene is identified by this accession number? Click on the accession number link. Notice that GenBank provides the original journal reference that reported this sequence, the single-letter amino acid code of the protein encoded by this gene, and the nucleotide sequence (cDNA) of this gene. Alternatively, you can type in the name of a gene or a potential gene that you are

FIGURE 3.21 Comparison of the Human and Mouse *ob* Genes Partial sequences for these two genes are shown, with the human gene sequence on top and the mouse gene sequence below it.

interested in to see if it has already been cloned and submitted to GenBank. (The BLAST search will identify a match with a human gene for early-onset breast cancer, *BRCA1*. The GenBank search identifies the same gene by its accession number).

GenBank is only one example of an invaluable database that is an essential for bioinformatics. Many other specialized databases exist with information such as single-nucleotide polymorphism data, BAC and YAC library databases, and protein databases that catalog amino acid sequences and three-dimensional protein structures.

A Genome Cloning Effort of Epic Proportion: The Human Genome Project

It is a very exciting time to be studying biotechnology. We are fortunate to witness the completion of an unprecedented project that involved many of the techniques described in this chapter, the **Human Genome Project (HGP).** Initiated in 1990 by the U.S. Department of Energy (DOE), the HGP was an international collaborative effort with a 15-year plan to identify all human genes, originally estimated at 80,000 to 100,000 in number, and to sequence the approximately 3 billion base pairs thought to make up the 24 different human chromosomes (chromosomes 1 to 22, X, Y). The HGP was also designed to accomplish the following:

- Analyze genetic variations among humans. This included the identification of single-nucleotide polymorphisms (SNPs; see Figure 1.11).

- Map and sequence the genomes of **model organisms,** including bacteria, yeast, roundworms, fruit flies, mice, and others.

- Develop new laboratory technologies such as high-powered automated sequencers and computing

technologies, as well as widely available databases of genome information, which can be used to advance our analysis and understanding of gene structure and function.

- Disseminate genome information among scientists and the general public.

- Consider the ethical, legal, and social issues that accompany the HGP and genetic research.

In the United States, public research on the HGP was coordinated by the National Center of Human Genome Research, a division of the NIH and the DOE. The project funded seven major sequencing centers in the United States. Over time, the project grew to become an international effort with contributions from scientists in 18 countries. The work was primarily carried out by the International Human Genome Sequence Consortium, which involved nearly 3,000 scientists working at 20 centers in six countries: China, France, Germany, Great Britain, Japan, and the United States. The estimated budget for completing the genome was $3 billion, a cost of $1 per nucleotide. Driven in part by competition from private companies and the development of computer-automated DNA sequences (such as the one described in Figure 3.12) and bioinformatics, the HGP turned out to be a rare government project that completed all of its initial goals and several additional goals more than 2 years ahead of schedule—and under budget!

The most aggressive competitor on the project was a private company called Celera Genomics (aptly named from a Latin word meaning "swiftness") directed by Dr. J. Craig Venter, who was the scientist directing The Institute for Genomic Research when it used shotgun cloning to sequence the genome for *H. influenza*, as discussed earlier in this section. Celera announced its intention to use its novel shotgun sequencing approach, similar to the methods they had used to sequence the genome for *H. influenza*, as well as newly developed, high-powered computer-automated DNA sequencers, to sequence the entire human genome in 3 years. Fearful of how private corporations might control the release of genome information, U.S. government groups involved in the HGP were effectively forced to keep pace with private groups to stay competitive in the race to complete the genome.

In 1998, as a result of accelerated progress, a revised target date of 2003 was set for completion of the project. On June 26, 2000, leaders of the HGP and Celera Genomics participated in a press conference with President Bill Clinton to announce that a rough "working draft" of approximately 95% of the human genome had been assembled (nearly 4 years ahead of the initially projected timetable). At a joint press conference on February 12, 2001, Dr. Francis Collins, director of the NIH National Human Genome Research Institute and of the HGP, and Dr. J. Craig Venter of Celera announced that a series of papers describing the initial analysis of the genome working draft sequence were published by their research groups in the prestigious journals *Nature* and *Science*, respectively. Scientists spent the next 2 years working to fill in thousands of gaps in the genome by completing the sequencing of pieces not yet finished, correcting misaligned pieces, and comparing sequences to ensure the accuracy of the genome. On April 14, 2003, the International Human Genome Sequencing Consortium announced that its work was done. A "map" of the human genome was essentially complete, with virtually all bases identified and placed in their proper order and potential genes assigned to a chromosome (see **Figure 3.22**).

What Have We Learned from the Human Genome?

One of the most surprising findings of the project was that the human genome consists of approximately 20,000 protein-coding genes, not 100,000 genes as predicted. The prediction was based primarily on estimates that human cells make approximately 100,000 to 150,000 proteins. One reason why the actual number of genes is so much lower than the predicted number is the discovery of large numbers of gene families with related functions. In addition, many genes that code for multiple proteins through alternative splicing have been found. It has been estimated that 92% to 95% of human genes produce multiple proteins through alternative splicing.

Analysis of human genes by functional categories has provided genome scientists with a snapshot of the numbers of genes involved in different molecular functions. **Figure 3.23** shows one proposed interpretation of assigned functions to human genes based on gene sequence similarity to genes of known function. Not surprisingly, many genes encode enzymes, while other large categories of genes encode proteins involved in signaling and communication within and between cells and DNA- and RNA-binding proteins. Notice that this estimate also shows that approximately 42% of human genes have no known function, although recent evidence suggests that functions for over half of our genes remain unknown. Keep this in mind if you are interested in a career in genetics research, because efforts to discover what these genes do will provide exciting career opportunities for many years into the future.

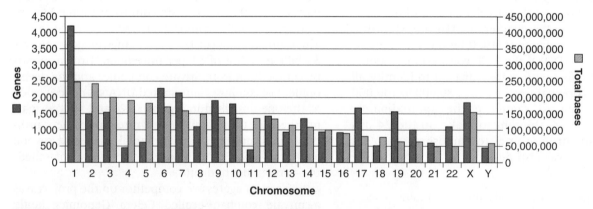

FIGURE 3.22 Estimated number of genes on each chromosomes and approximate size in base pairs (bp) of each human chromosome Figure-compiling statistics for human chromosomes are based on the Sanger Institute's human genome information in the Vertebrate Genome Annotation (VEGA) database. Number of genes is an estimate, as it is in part based on gene predictions, which is why this value is higher than the estimated 20,000 genes generally accepted as the number of human genes. Total chromosome length is an estimate as well, based on the estimated size of some noncoding sections of the genome that remain unsequenced.

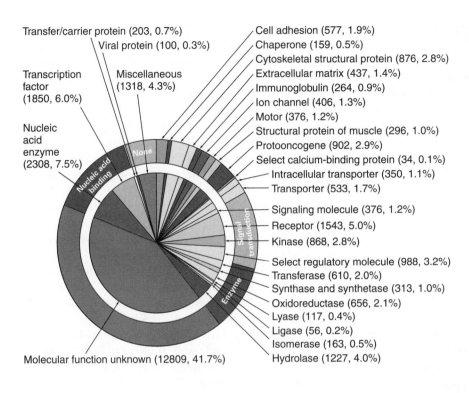

Transfer/carrier protein (203, 0.7%)
Viral protein (100, 0.3%)

Transcription factor (1850, 6.0%)

Miscellaneous (1318, 4.3%)

Nucleic acid enzyme (2308, 7.5%)

Cell adhesion (577, 1.9%)
Chaperone (159, 0.5%)
Cytoskeletal structural protein (876, 2.8%)
Extracellular matrix (437, 1.4%)
Immunoglobulin (264, 0.9%)
Ion channel (406, 1.3%)
Motor (376, 1.2%)
Structural protein of muscle (296, 1.0%)
Protooncogene (902, 2.9%)
Select calcium-binding protein (34, 0.1%)
Intracellular transporter (350, 1.1%)
Transporter (533, 1.7%)

Signaling molecule (376, 1.2%)
Receptor (1543, 5.0%)
Kinase (868, 2.8%)

Select regulatory molecule (988, 3.2%)
Transferase (610, 2.0%)
Synthase and synthetase (313, 1.0%)
Oxidoreductase (656, 2.1%)
Lyase (117, 0.4%)
Ligase (56, 0.2%)
Isomerase (163, 0.5%)
Hydrolase (1227, 4.0%)

Molecular function unknown (12809, 41.7%)

FIGURE 3.23   Proposed Functions for the Numbers of Human Genes Assigned to Different Functional Categories Parentheses show percentages based on the 2001 published data by Venter et al. for 26,383 genes.

Here is a summary of important highlights of some of the basic concepts scientists have learned from the HGP:

- The human genome consists of approximately 3.1 billion base pairs.

- The genome is approximately 99.9% the same between individuals of all nationalities.

- Single-nucleotide polymorphisms (SNPs) and **copy number variations (CNVs)**—such as long deletions, insertions and duplications in the genome—account for much of the genome diversity identified between humans.

- Less than 2% of the genome codes for genes.

- The vast majority of our DNA is non–protein-coding, and repetitive DNA sequences account for at least 50% of the noncoding DNA.

- The genome contains approximately 20,000 protein-coding genes.

- Many human genes are capable of making more than one protein, allowing human cells to make at least 100,000 proteins from only about 20,000 genes.

- The functions of over half of all human genes are unknown.

- Chromosome 1 contains the highest number of genes. The Y chromosome contains the fewest genes.

- Many of the genes in the human genome show a high degree of sequence similarity to genes in other organisms.

- Thousands of human disease genes have been identified and mapped to their chromosomal locations.

How will we benefit from the HGP? The complete sequence of the human genome has been described as the genetic "blueprint" of humanity, containing the scientific keys to understanding our biology and behaviors. Identifying all human genes is not as important as understanding *what* these genes do and *how* they function. We will not fully understand the function of all human genes for many years, if ever. One immediate impact of the HGP will be the identification of genes associated with human genetic diseases. We will discuss some of the ways the HGP is changing diagnosis and treatment of genetic diseases (Chapter 11). A better understanding of the genetic basis of many diseases will lead to the development of new strategies for disease detection and innovative therapies and cures. For updated information on the HGP and for an outstanding overview of goals, sequencing and mapping technologies, and the ethical, legal, and social issues of the HGP, visit the human genome sites listed at the Companion Website.

The Human Genome Project Started an "Omics" Revolution

The HGP and genomics are largely responsible for ushering in a new area of biological research—the "omics." It seems that every year, new or existing areas of biological research are being described as having an omics connection. For example:

- Proteomics—studying all of the proteins in a cell
- Metabolomics—studying proteins and enzymatic pathways involved in cell metabolism
- Metabonomics–measuring metabolic products produced by cells in response to stimuli (such as drug treatment) and genetic manipulation
- Glycomics—studying the carbohydrates of a cell
- Transcriptomics—studying all genes expressed (transcription) in a cell
- Metagenomics—the analysis of genomes of organisms collected from the environment
- Pharmacogenomics—customized medicine based on a person's genetic profile for a particular condition

If any of the "omics" areas defined here sound of interest to you, do some Internet research on these topics and you will find vast amounts of information about each area. We will touch on many of these omics areas throughout the book. As further evidence of the impact of genomics, a new field of nutritional science, called nutritional genomics or **nutrigenomics,** has emerged. Nutrigenomics is focused on understanding interactions between diet and genes. Several companies provide nutrigenomics tests, using microarrays or other genetic tests to analyze your genotype for genes thought to be associated with different medical conditions or aspects of nutrient metabolism. These companies then provide a customized report on nutrition and claim that they can suggest dietary changes you should make to improve your health and prevent illness based on your genes. Whether nutrigenomics is actually a valid scientific approach is a subject of debate among scientists.

Comparative genomics

It might surprise you that in addition to studying the human genome, the HGP involved mapping and sequencing genomes from a number of model organisms, including *E. coli*, a model plant called *Arabidopsis thaliana*, the yeast *Saccharomyces cerevisiae*, the fruit fly *Drosophila melanogaster*, the nematode roundworm *Caenorhabditis elegans*, and the mouse *Mus musculus*, among other species. Complete genome sequences of these model organisms have been incredibly useful for **comparative genomics** studies, which allow researchers to study gene structure and function in

these organisms in ways designed to understand gene structure and function in other species, including humans. Because we share many of the same genes as flies, roundworms, and mice, such studies will also lead to a greater understanding of human evolution. The number of genes we share with other species is very high, ranging from about 30% of the genes in yeast to about 80% of the genes in mice and about 95% of the genes in chimpanzees (Table 3.3).

Comparative genomic analysis of the genome for "man's best friend" has revealed that we share about 75% of our genes with dogs. Human DNA even contains around 100 genes that are also present in many bacteria. When researchers completed the 814 million-bp genome of the sea urchin (*Strongylocentrus purpuratus*), an invertebrate that has served as an important model organism particularly for developmental biologists, we learned that of the 23,500 genes in the urchin genome, many are genes with important functions in humans.

The genomes of many model organisms have been completely sequenced, and there are literally hundreds of genomics projects under way worldwide. Genome scientists around the world have proposed to sequence 10,000 vertebrate genomes, the **Genome 10K plan.** This ambitious plan proposes to assemble 10,000 genomes in 5 years—about one genome a day. The first genome for a tree, the black cottonwood (a type of poplar), was sequenced several years ago. To date, the poplar's 45,555 genes are the highest number found in a genome. Scientists anticipate using data from this project to help the forestry industry make better products, including biofuels or even genetically engineered poplars to capture high levels of carbon dioxide from the atmosphere. The honeybee genome has also been completed, and information from this project will be used to understand bee genetics and behavior to help the honey-producing industry as well as advance an understanding of how bee toxins produce allergic responses.

We will discuss genomics projects in other chapters of this book, and throughout it you will learn about many applications of genomics. For example, in Chapter 5 we discuss microbial genome sequencing projects that are sequencing genomes for literally hundreds of newly identified microbes, including microbes that live and on humans (the **Human Microbiome Project**) and microbes present in water, soil, and air samples from around the world **(metagenomics).** We will also discuss **synthetic genome** approaches designed to create novel life forms from artificially constructed genomes.

Stone Age Genomics

As yet another intriguing example of genomics, a number of labs around the world are involved in analyzing "ancient" DNA. These studies are generating fascinat-

TABLE 3.3   **COMPARISON OF SELECTED GENOMES**

Organism (scientific name)	Approximate Size of Genome (date completed)	Number of Genes	Approximate Percentage of Genes Shared with Humans	Web Access to Genome Databases
Bacterium (*Escherichia coli*)	4.1 million bp (1997)	4,403	Not determined	www.genome.wisc.edu/
Chicken (*Gallus gallus*)	1 billion bp (2004)	~ 20,000–23,000	60%	http://genomeold.wustl.edu/projects/chicken
Dog (*Canis familiaris*)	6.2 million bp (2003)	~ 18,400	75%	http://www.ncbi.gov/genome/guide/dog
Chimpanzee (*Pan troglodytes*)	~ 3 billion bp (initial draft, 2005)	~ 20,000–24,000	96%	http://www.nature.com/nature/focus/chimpgenome/index.html
Fruit fly (*Drosophila melanogaster*)	165 million bp (2000)	~ 13,600	50%	www.fruitfly.org
Humans (*Homo sapiens*)	~ 2.9 billion bp (2004)	~ 20,000–25,000	100%	www.doegenomes.org
Mouse (*Mus musculus*)	~ 2.5 billion bp (2002)	~ 30,000	~ 80%	www.informatics.jax.org
Plant (*Arabidopsis thaliana*)	119 million bp (2000)	~ 26,000	Not determined	www.arabidopsis.org
Rat (*Rattus norvegicus*)	~ 2.75 billion bp (2004)	~22,000	80%	www.hgsc.bcm.tmc.edu/projects/rat
Roundworm (*Caenorhabditis elegans*)	97 million bp (1998)	19,099	40%	genomeold.wustl.edu/projects/celegans
Yeast (*Saccharomyces cerevisiae*)	12 million bp (1996)	~5,700	30%	genomeold.wustl.edu/projects/yeast.index.php

ing data from minuscule amounts of ancient DNA from bone and other tissues and fossil samples that are tens of thousands of years old. Analysis of DNA from a 2,400-year-old Egyptian mummy, mammoths, platypuses, Pleistocene-age cave bears, and Neanderthals are some of the most prominent examples of **Stone Age genomics,** also called **paleogenomics.**

Researchers from McMaster University in Canada and Pennsylvania State University published partial sequence data (about 13 million bp) from a 27,000-year-old woolly mammoth. This study showed that there is approximately 98.5% sequence identity between mammoths and African elephants. Subsequent work by other scientists has used whole-genome shotgun sequencing of mitochondrial and nuclear DNA from Siberian mammoths to provide data on the mammoth genome. These studies suggest that the mammoth genome differs from that of the African elephant by as little as 0.6%. These studies are a great demonstration of how stable DNA can be under the right conditions, particularly when frozen. Also intriguing are similarities that have been revealed between the mammoth and human genomes! When the gene sequences from human chromosomes were aligned with sequences from the mammoth genome, approximately 50% of mammoth genes show sequence alignment with human genes on autosomes.

In 2009, a team of scientists led by Svante Pääbo at the Max Planck Institute for Evolutionary Anthropology in Germany and 454 Life Sciences reported completion of a rough draft of the Neanderthal (*Homo neanderthalensis*) genome encompassing more than 3 billion bp of Neanderthal DNA and about two-thirds of the genome. In 2006, Pääbo's group, along with a number of scientists in the United States, reported the first sequence of about 65,000 bp of nuclear DNA isolated from bone of a 38,000-year-old Neanderthal sample from Croatia. Because Neanderthals are close relatives of humans, the sequencing of the Neanderthal genome provides a tremendous opportunity to use comparative genomics to advance our understanding of evolutionary relationships between humans.

The human and Neanderthal genomes are 99% identical. Some of these sequences are involved in cognitive development and sperm motility. Of the many genes shared by these species, *FOXP2* is a gene which has been linked to speech and language ability. There are many genes that influence speech, so this finding does not mean that Neanderthals spoke as we do. But because Neanderthal had the same modern human *FOXP2* gene, scientists have speculated that Neanderthals possessed linguistic abilities. The realization that modern human and Neanderthals lived in overlapping ranges as recently as 30,000 years ago has led to speculation about the interactions between modern humans and Neanderthals. Genome studies suggest that there was interbreeding between Neanderthals and modern humans an estimated 45,000 to 80,000 years ago in the eastern Mediterranean. These exciting studies, previously thought to be impossible, are having ramifications in many areas of human evolution, and it will be interesting indeed to follow the progress of this work.

10 Years After the HGP: What Is Next?

In the 10 years since completion of a draft sequence of the human genome, studies on the human genome continue at a very rapid pace. As a result of the HGP, many other major theme areas for human genome research have emerged, including cancer genome projects, analysis of the epigenome (including the **Human Epigenome Project,** which is creating hundreds of maps of epigenetic changes in different cell and tissue types and evaluating potential roles of epigenetics in complex diseases), characterization of SNPs (the **International HapMap Project**) and CNVs for their role in genome variation, disease, and pharmacogenomics applications. As the HGP was being completed, a group of about three dozen research teams around the world began the **Encyclopedia of DNA Elements (ENCODE) Project.** The main goal of ENCODE is to use both experimental approaches and bioinformatics to identify and analyze functional elements (such as transcriptional start sites, promoters, and enhancers) that regulate expression of human genes. Here we discuss two major human genome-related research project areas: personalized genome projects and cancer genome projects.

Personalized Genome Projects

Because of rapid advances in sequencing technologies, several companies have even proposed to sequence genomes for individual people, or **personalized genomics.** In 2006, the X Prize Foundation announced the Archon X Prize for Genomics, a project to award $10 million to the first group to develop technology capable of sequencing 100 human genomes with a high degree of accuracy in 10 days for under $10,000 per genome. Other groups are working on sequencing a personalized genome for a mere $1,000. Pursuit of the $1,000 genome is evidence that DNA sequencing may eventually be affordable enough for individuals to consider acquiring a readout of their own genetic blueprint.

In 2007, the company 454 Life Sciences together with researchers at Baylor University sequenced James Watson's genome for approximately $1 million. 454 Life Sciences decided that "Project Jim," sequencing the genome of the codiscoverer of the DNA structure, was a good high-profile way to develop and promote their new next-generation sequencing technology. James Watson provided 454 scientists with a blood sample in 2005 and, in mid-2007, they presented Dr. Watson with two DVDs containing his genome sequence, which was completed at a rough cost of just under $1 million. Watson has allowed his sequence to be available to researchers except for the sequence of his apolipoprotein E gene (*ApoE*). *ApoE* gene mutations can indicate a disposition for Alzheimer's disease. Human genome pioneer Craig Venter also had his genome sequenced by scientists at The Craig J. Venter Institute.

George Church and colleagues at Harvard University have started the **Personal Genome Project (PGP)** and have recruited volunteers to provide DNA for individual genome sequencing with the

YOU DECIDE

To Sequence or Not to Sequence?

As of early 2010, thirteen individual human genome sequences have been reported, including sequences for a Yoruba African person, two individuals of northwest Europe origin, a Han Chinese individual, two persons from Korea, African Archbishop Desmond Tutu, a family of four, and several others. Would you want your genome sequenced? What would you pay for your genome sequence? What would you do with this sequence information? Who would have access to your sequence information? You Decide.

Career Options in the Biopharmaceutical Industry: Perspectives from a Recent Graduate

Biopharmaceutical companies discover new drug treatments and develop them from the research and development stages through to commercial manufacturing, marketing, and finally sales. What makes the biopharmaceutical industry or "biopharma," different from the traditional pharmaceutical industry? Pharmaceutical companies in general develop chemically based drug compounds that treat illness nonspecifically, whereas biopharma uses knowledge of biological systems to develop biologically derived drug compounds that treat illness in a highly specific manner. Although biopharma is currently much smaller than the pharmaceutical industry, its growth is triggered by the increasing awareness of the potential to treat illnesses effectively and possibly to find cures for illnesses that were once thought incurable. The lines between traditional pharma companies and biopharma are blurred. Many large pharma companies have biotechnology divisions that work on biopharma. For example, the pharma company Roche recently purchased Genentech.

Career paths within the biopharmaceutical industry are similar to those in the pharmaceutical industry, owing to the similarity in the path of drug development. The career possibilities are wide-ranging; however, gaining employment with a biopharmaceutical company is not easy. Competition is fierce, and the truth is that no matter how well a person may fit a position, there is always someone else who will be just as qualified. For that reason, start thinking about career options shortly after you have selected your major. The best way to begin is by becoming familiar with the industry. Many of the websites cited on the Companion Website, are excellent resources for learning about different job opportunities. This knowledge will enable you to customize your education to better serve a particular job function—a sure way to get ahead of the competition.

Many entry-level positions are available to those with associate's and bachelor's degrees; however, having a bachelor's degree will guarantee a slightly higher starting salary. Many higher-paying entry-level positions require an advanced degree, such as a master's degree or even a doctorate (PhD.). Knowing this may greatly influence your decision to continue your education before entering the workforce. However, as important as educational training is, employers are constantly looking for experienced individuals. Gaining real-life experience through volunteer work, independent study research, or an internship may offer you the best opportunity to explore your options while you are completing your studies. Industry internships are an excellent way to learn about different roles in a company and to see what you like, and they may also lead to employment after graduation.

If you cannot find the ideal job in the industry after graduation, temporary employment is another good way to begin with a biopharmaceutical company. Many placement services work with job seekers and biotechnology companies to find people who can fill a need for a short period of time; however, if you do a good job in a temporary position, the company may try to find a permanent position for you.

In considering a position, it is important to understand the job description as well as what is required of that position. Numerous companies hire people as lab technicians, but the job descriptions for this position are as numerous as the jobs. For example, quality-control lab technicians follow very controlled and highly precise procedures to test the quality of a product; their main function is to ensure that a good product is being produced. Technicians working in a diagnostic capacity are responsible for testing a new product or procedure to make sure that it works correctly and then for recommending improvements. Research and development or discovery lab technicians are responsible for just that: their main function is to develop new products as well as to find new uses for old products. It is not required that you have prior experience in every experimental assay or lab procedure to become a lab technician. A good company will always teach you what you need to know to do the specific job; however, all lab technicians must have good lab skills and be organized, detail-oriented, motivated, and, most importantly, enthusiastic.

A potential down side of working in the biopharmaceutical industry is that all work is tailored around developing, producing, and selling a product. Creativity and originality are not always incorporated into the daily routine of life in the laboratory. The work environment is usually very controlled because all operations are regulated by the FDA and must be documented rigorously. Also, since products are produced through biological processes, work schedules are linked directly to these processes, such as waiting for cells to grow to the right density. In some cases, this can mean a shorter work week or less structured work hours.

In the biopharmaceutical industry, the fast-paced work environment is always changing and always demanding that its employees work hard and efficiently. Each day brings new obstacles and challenges, which makes working in this industry so exciting.

Contributed by Robert Sexton (B.S., Biology, M.B.A., Monmouth University), Oncology Development Operations, Novartis Pharmaceuticals Corporation.

understanding that the resulting genome data will be made publicly available. Church's genome has been completed and been made available online. Since the Watson and Venter genomes were completed the first complete genome sequence was provided for an individual "ancient" human, a Palaeo-Eskimo, obtained from a roughly 4,000-year-old permafrost-preserved hair. This work recovered about 78% of the diploid genome and revealed many interesting SNPs (of which about 7% have not been previously reported).

Cancer genome projects

Cancer as a disease has many genetic components. The NIH began work on a cancer genome project called the **Cancer Genome Atlas Project (TCGA)** to map important genes and genetic changes involved in cancer. The TCGA project has sequenced over 100 partial genomes for various cancers to date. Together with a group of researchers in 11 countries, called the International Cancer Genome Consortium (JCGC), there are plans to sequence genomes from over 500 tumor samples representing more than 20 different cancers. Not surprisingly, most common cancers—such as breast, lung, colon, pancreatic, brain, and ovarian cancer—are among the most actively studied cancer types and are being analyzed first. Ultimately it is expected that the identification of key genes involved in tumor formation and metastasis (spreading) will lead to improved diagnostic techniques for the detection of cancer and more effective treatments for curing cancer (this is already happening in some cases, as we will discuss in Chapter 11).

MAKING A DIFFERENCE

One of the most successful examples of how recombinant DNA technology is saving lives is also one of its first applications for treating human disease. Earlier in the chapter we discussed how the recombinant form of insulin, called Humulin, became the first product of recombinant DNA technology to be approved by the FDA for use in humans. Prior to recombinant insulin, insulin was isolated from animals. In many diabetics, insulin from animal sources was ineffective and caused a range of complications. Producing recombinant insulin was not easy (see Figure 5.9). Insulin is composed of two different polypeptides chains. Both polypeptides had to be expressed in recombinant form and assembled to form the active hormone. Needless to say, it was quite a feat to successfully engineer insulin, such a challenging recombinant protein, as the first product of this technology. This demonstration that a recombinant approach for producing a complex protein was possible made scientists highly optimistic about the possibilities of recombinant DNA approaches to making other proteins of therapeutic value. Humulin was originally developed

by Genentech and then widely distributed by Eli Lilly and Company, which continues to produce Humulin today. Since it first became available in 1982, Humulin has been used successfully and safely to treat millions of patients. The vast majority of insulin given to insulin-dependent diabetes mellitus patients worldwide is recombinant insulin. There are now many variations of Humulin available, most of which are different mixtures that determine whether the insulin is fast- or slow-acting once injected. Recently researchers have been working on techniques to produce recombinant insulin in plants, which is expected to dramatically reduce biomanufacturing costs. The insulin story is an outstanding example of how recombinant DNA techniques are being used to make a difference in alleviating human pain and suffering.

QUESTIONS & ACTIVITIES

Answers can be found in Appendix 1.

1. Distinguish among gene cloning, recombinant DNA technology, and genetic engineering by describing each process and discussing how they are interrelated. Provide examples of each approach as described in this chapter.

2. Describe the importance of DNA ligase in a recombinant DNA experiment. What does this enzyme do and how does its action differ from the function of restriction enzymes?

3. Your lab just determined the sequence of a rat gene thought to be involved in controlling the fertilizing ability of rat sperm. You believe a similar gene may control fertility in human males. Briefly describe how you could use what you know about this rat gene combined with PCR to clone the complementary human gene. Be sure to explain your experimental approach and the necessary lab materials. Also explain in detail any procedures necessary to confirm that you have a human gene that corresponds to your rat gene.

4. What features of plasmid cloning vectors make them useful for cloning DNA? Provide examples of different types of cloning vectors and discuss their applications in biotechnology.

5. If you performed a PCR experiment starting with only one copy of double-stranded DNA, approximately how many molecules would be produced after 15 cycles of amplification?

6. Compare and contrast genomic libraries with cDNA libraries. Which type of library would be your first choice to use if you were attempting to clone a gene in adipocyte (fat) cells that encodes a protein thought to be involved in obesity? Explain

your answer. What type of library would you choose if you were interested in cloning gene regulatory elements such as promoter and enhancer sequences?

7. Visit the Online Mendelian Inheritance in Man Site (OMIM) at the Companion Website and then click on "Search the OMIM database." Type *diabetes* in the search box and then click "Submit Search." What did you find? Try typing *114480* in the search box. What happened this time? Alternatively, search for a gene you might be interested in and see what you can find in OMIM. If the results of your OMIM search are too technical, visit the "Genes & Disease" section of the NCBI site from the Companion Website then search for *diabetes.*

8. Software analysis of DNA sequences makes it much easier for molecular biologists to study gene structure. This activity is designed to help you experience applications of DNA analysis software. Imagine that the following very short sequence of nucleotides, GGATCCGGCCGGAATT CGTA, represents one strand of an important gene that was just mailed to you for your research project. Before you can continue your research, you must find out which restriction enzymes, if any, cut this piece of DNA.

Go to the Webcutter site from the Companion Website. Scroll down the page until you see a text box with the title, "Paste the DNA Sequence into the Box Below." Type the sequence of your DNA piece into this box. Scroll down the page, leaving all parameters at their default settings until you see "Please Indicate Which Enzymes to Include in the Analysis." Click on "Only the following enzymes:" and then use the drop-down menu

and select *Bam*HI. Scroll to the bottom of the page and click the "Analyze sequence" button. What did you find? Is your sequence cut by *Bam*HI? Analyze this sequence for other cutting sites to answer the following questions. Is this sequence cut by *Eco*RI? How about *Sma*I? What happens if you do a search and scan for cutting sites with all enzymes in the database?

9. Go to the molecular biology section of the Biology Project website from the University of Arizona (you can find the address on the Companion Website). Link to "Recombinant DNA Technology" and test your knowledge of recombinant DNA technology by working on the questions at this site.

10. Search the Web for the company Sciona, which markets the MyCellf DNA evaluation kit for nutrigenomics. Do you think kits such as this should be used, even though they are largely based on unproven information?

11. What is the structural difference between a deoxyribonucleotide (dNTP) and a dideoxyribonucleotide (ddNTP) used for DNA sequencing?

12. Describe several findings of the Human Genome Project.

13. Modern biology is experiencing an "omics" revolution. What does this mean?

Visit www.pearsonhighered.com/biotechnology

To download learning objectives, chapter summary, "Keeping Current" web links, glossary, flashcards, and jpegs of figures from this chapter.

Proteins as Products

After completing this chapter, you should be able to:

- Explain the uses of some biotechnologically produced enzymes in industry.

- Provide three examples of medical applications of proteins.

- List common household products that may include manufactured proteins as ingredients.

- Discuss the advantages and disadvantages of bacterial, fungal, plant, and animal sources of protein expression.

- Explain why *Escherichia coli* is frequently used for protein production, and describe *E. coli*'s limitations.

- Explain why protein glycosylation may determine the choice of a protein expression system.

- Describe a general scheme for protein purification of a protein like hemoglobin.

- Explain how a target protein is separated from other cell proteins given a specific purification sequence.

- Discuss the benefits of being able to predict protein structure from the DNA sequence (proteomics).

Lab Supervisor Preparing a Purification Column to Separate a Commercial Biological Product.

FORECASTING THE FUTURE

Biotechnologically produced medications (or drugs) are all proteins. Until recently, the chemistry of the purified protein molecule dictated the method of drug delivery (injection, absorption, ingestion, or other means). Unfortunately, these delivery methods can lead to serious systemic and autoimmune side effects. Interferon alpha-2b, traditionally given by injection and approved for treatment of melanoma, hepatitis C, and conditions caused by the human papillomavirus (HPV), is a good example of a medicine with a traditionally problematic delivery method. Recently, to address this problem, the company Helix BioPharma has developed a multilamellar nanovesicle (a multilayered submicroscopic particle that can be easily assimilated by the body) that enables a timed, controlled release of the protein applied in a skin cream. This topical delivery is adapted to large hydrophilic molecules such as interferon alpha-2b, which helps avoid the serious side effects caused by more traditional delivery methods. In the future, the design of delivery methods that are adapted to the chemistry of the protein biomolecule may be as important as the discovery of the biomolecule itself.

Tropical rainforests, the deepest reaches of the ocean, boiling geysers in Yellowstone National Park, and whale skeletons—these are on the frontier of the scientific quest for proteins. **Proteins** are large molecules required for the structure, function, and regulation of living cells. Each protein molecule has a unique function in the biochemical reactions that sustain life. As researchers explore the proteins that occur in nature, they unlock secrets that govern growth, speed of chemical decomposition, and protection from disease.

The applications of proteins are as numerous as proteins themselves. Consider whale skeletons: during the natural decomposition process, bones are often colonized by bacteria, some of which have evolved especially to digest the fatty residue remaining on them. The proteins that the bacteria produce to break down the fats are adapted to the frigid waters of the deep sea. Researchers recognized that a substance with the ability to dissolve fats at cold temperatures would make a great additive for commercial laundry detergents.

Even after a protein is discovered in nature and an application is matched to its characteristics, a great deal of ingenuity is required to produce the protein of the necessary quality and quantity required. For example, if we plan to mass-produce a new cold-water detergent, we cannot rely on whale skeletons; rather, we must find another source of the proteins needed. Fortunately, biotechnology facilitates the production of virtually all proteins. We focus on those production processes in this chapter.

In 2000, the National Institutes of Health launched the Protein Structure Initiative (PSI), a 10-year, $600 million effort to identify the structure of human proteins. The PSI is a federal university and industry effort aimed at reducing costs and lessening the time it takes to determine three-dimensional protein structures. The initial goal of the PSI was to make the structures of most proteins easily obtainable from knowledge of their corresponding DNA sequences. In the current phase, which began in 2010, investigators are using high-throughput structure determination, which was successfully developed during the earlier phases of the PSI, to study a broad range of important biological and biomedical problems.

This massive effort to further understand the structure of proteins helped researchers understand their function. More than 1,200 superfamilies (groups of related protein structures) of proteins have been identified so far, and the relationship between a protein's amino acid sequence to its structure is now well understood. The public database that is part of the initiative currently holds more than 33,000 protein sequences. The goal of the initiative is to be able to model unknown protein structures based on structural comparisons to those stored in the database. For example, the Human Genome Project produced the 3 billion base-pair sequences of human DNA, but the predicted 100,000 proteins resulting from this DNA will be hard to determine without the effort of scientists supported by major governmental initiatives like this.

We begin with a quick survey of the many applications of proteins in a variety of industries. Then we look at the nature of protein structures, paying special attention to the process of protein folding. With that as a foundation, we delve into some of the details of protein processing, beginning with the methods of expressing proteins. We then learn how expressed proteins are purified and examine the processes used to analyze and verify the final product. Although there is no one best method for processing proteins, several generally useful techniques are available. In this chapter, we look at those techniques, keeping in mind that the specifics of protein processing vary from case to case.

4.1 Proteins as Biotechnology Products

The use of proteins in manufacturing processes is a time-tested technology. For example, two of the oldest food-processing endeavors depend on proteins: beer brewing and winemaking. Fermentation depends not only on yeast-produced enzymes but also on others added directly to the batch. Cheese making is another food-processing industry that has always used proteins and, thanks to bioengineering, the protein source now used is from engineered bacteria rather than the calves' stomachs that originally provided the cheese-making enzyme (**Figure 4.1**).

Even though the value of proteins in manufacturing had long been evident, we were not able to further

this knowledge until the 1970s, when recombinant DNA technology was first developed and it became possible to produce specific proteins on demand. Since that time, the production of proteins has been the driving force behind the development of new products in a wide variety of industries.

Other industries are also benefiting from the ready availability of bioengineered proteins. Many of these applications depend on the power of a group of proteins called **enzymes** to speed up chemical reactions. Enzymes serve myriad purposes, such as making detergents work better, increasing the flow of oil in drilling operations, and cleaning contact lenses. Enzymes are used in manufacturing to break down large molecules—a process called **depolymerization.** These enzymes include glycosidases (carbohydrases) like **amylase,** which breaks down starch; **proteases,** which break down other proteins; and **lipases,** which break down fats (Table 4.1). Such enzymes (like chymosin, for cheese) are used in food and beverage production and for bulk processing in industries (see Figure 4.1).

Hormones, which carry chemical messages, and **antibodies**, which protect the organism from disease, are two other groups of proteins produced commercially, primarily for the medical industry. Hormones also have agricultural uses. For instance, hormones can stimulate the rooting of plant cuttings and encourage more rapid growth of meat animals. We discuss hormones used in agriculture in more detail later (Chapters 6 and 7) .

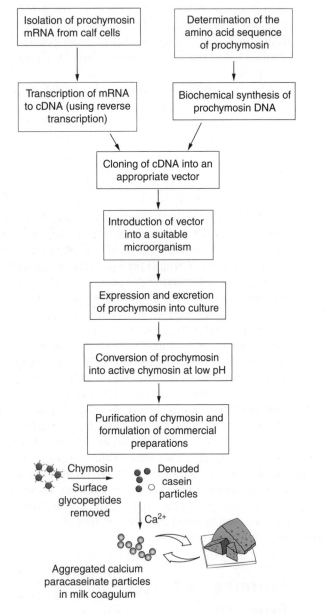

FIGURE 4.1   Cheese Production Casein, the primary ingredient in cheese, is the result of a chemical conversion that depends on chymosin. This enzyme has been obtained from the stomach walls of unweaned calves for centuries. Today 80 percent of the chymosin manufactured is produced (and purified) from genetically modified cells.

TABLE 4.1   SOME ENZYMES AND THEIR INDUSTRIAL APPLICATIONS

Enzyme	Application
Amylases	Digest starch in fermentation and processing
Proteases	Digest proteins for detergents, meat/leather, cheese, brewing/baking, animal/human digestive aids
Lipases	Digest lipids (fats) in dairy and vegetable oil products
Pectinases	Digest enzymes in fruit juice/pulp
Lactases	Digest milk sugar
Glucose isomerase	Produce high-fructose syrups
Cellulases/hemicellulases	Produce animal feeds, fruit juices, brewing converters
Penicillin acylase	Produces penicillin

Biotech Drugs and Other Medical Applications

Biotechnology proteins have revolutionized the health care and pharmaceutical industries in the past several decades. Many illnesses, from common conditions like diabetes to rare diseases like Gaucher's disease, can be treated by replacing missing proteins. In the case of diabetes, the missing protein is the hormone insulin. Insulin once had to be harvested from pigs and cows. This was less than ideal because human bodies often rejected this foreign protein. Researchers overcame the problem by turning to an unlikely source: the bacterium *E. coli*. By inserting human genes into *E. coli*, they created microscopic insulin factories. (We look at the remarkable use of genetically engineered organisms as a source of proteins more closely later in this chapter.) The U.S. Food and Drug Administration (FDA) approved this new insulin in 1982, making it the first recombinant DNA drug. The ability to produce an abundant supply of human insulin has improved the health and lives of millions of people. (Table 4.2 lists some other protein-based pharmaceutical products.)

Therapeutic proteins—such as monoclonal antibodies, blood proteins, and enzymes produced by living organisms to fight disease—can also be thought of as biotech drugs. Unlike other medicines, these drugs are not synthetically produced (i.e., chemically synthesized by adding one compound at a time) but are usually produced through microbial fermentation or by mammalian cell culture. Today, there are nearly 400 biotechnology medicines in the pipeline, and the majority are proteins. If even a small percentage of these drugs succeed, it will significantly add to the approximately 40 biotech drugs currently in use.

Producing biotech drugs is a complicated and time-consuming process. Researchers can spend many years just identifying the relevant therapeutic protein, determining its gene sequence, and working out a process to make sufficient quantities of the protein molecules using biotechnology. Once this method is determined, technicians can produce large batches of the protein products in bioreactors, under carefully controlled conditions, by growing host cells that have been transformed to contain the therapeutic gene (a bioreactor is a sterile production container designed to produce biological products). The cells are stimulated to produce the target proteins through precise culture conditions that include a balance of temperature, oxygen, acidity, and other variables. At the appropriate time, the proteins are isolated from the cultures, stringently tested at every step of purification (which we discuss later in this chapter), and formulated into pharmaceutically active products. Manufacturing technicians monitoring bioreactors must strictly comply with FDA regulations at all stages of the procedure (see Chapter 12 for examples of these regulations).

Another dramatic example of the potential use of proteins in health care is in the treatment of damaged

---

TABLE 4.2 SOME PROTEIN-BASED PHARMACEUTICAL PRODUCTS (MOST PRODUCED AS RECOMBINANT PROTEINS)

Protein	Application
Erythropoietins	Treatment of anemia
Interleukins 1, 2, 3, 4	Treatment of cancer, AIDS; radiation- or drug-induced bone marrow suppression
Monoclonal antibodies	Treatment of cancer, rheumatoid arthritis; used for diagnostic purposes
Interferons (α, β, γ, including consensus)	Treatment of cancer, allergies, asthma, arthritis, and infectious disease
Colony-stimulating factors	Treatment of cancer, low blood cell count; adjuvant chemotherapy; AIDS therapy
Blood clotting factors	Treatment of hemophilia and related clotting disorders
Human growth factor	Treatment of growth deficiency in children
Epidermal growth factor	Treatment of wounds, skin ulcers, cancer
Insulin	Treatment of types 1 and 2 diabetes mellitus
Insulin-like growth factor	Treatment of type 1 diabetes mellitus
Tissue plasminogen factor	Treatment after heart attack, stroke
Tumor necrosis factor	Cancer treatment
Vaccines	Vaccination against hepatitis B, malaria, herpes

corneas. A study from researchers in Canada and Sweden has shown that biosynthetic corneas can help regenerate and repair damaged eye tissue and improve vision. The discovery is significant because this approach could help restore sight to millions of people who are waiting for donated human corneas for transplantation. In the study, each patient underwent surgery on one eye to remove damaged corneal tissue and replace it with the biosynthetic cornea, made from synthetically cross-linked recombinant human collagen. The protein was produced in yeast cells and chemically cross-linked for the experiments. Over 2 years of follow-up, the researchers observed that cells and nerves from the patients' own corneas had grown into the implant, resulting in a "regenerated" cornea that resembled normal, healthy tissue. The biosynthetic corneas also became sensitive to touch. Vision improved in many of the patients, and after contact lens fitting, vision was comparable to conventional corneal transplantation.

Using a new method for rapidly screening molecules associated with disease, Joshua LaBaer and colleagues from the Biodesign Institute at Arizona State University have identified a broad panel of 28 early predictors, or **biomarker proteins,** which may one day aid in the early diagnosis of breast cancer. Studies have shown that proteins produced by cancers can trigger the body to produce antibodies not found in healthy individuals. These "autoantibodies" can be measured in the blood and used to reveal the presence of cancer. In the breast cancer study, protein microarrays were used to display thousands of different potential biomarker proteins on a single microscope slide. A tiny drop of blood was added to the microarray to look for those proteins that are recognized by antibodies from cancer patients but not from the healthy women (see section 4.10). The initial results narrowed the number of potential biomarkers candidates from 5,000 to 761. Finally, these 761 proteins were tested in a blind study to find the final 28 protein biomarkers. Autoantibody biomarkers in patients can be readily used for the detection of many other cancers: ovarian, prostate, and lung, among others.

Food Processing

You can find plenty of examples of industrial uses of proteins in your kitchen. For decades the food processing industry has relied on enzymes to improve baby food, canned fruit, cheeses, baked goods, beer, desserts, and dietetic foods. The enhancements are quite varied.

In bread, for example, enzymes may be used to make starches easier for yeast to act on, allowing the dough to rise more quickly. As a result, the bread dough is easier to handle and the final texture is more consistent.

Textiles and Clothing

Other examples of industrial proteins are right in your closet. For more than a century, enzymes have been used in the textile industry to break down the starches used to "size" (stiffen) fabrics during the manufacturing process. Enzymes are also replacing harsh chemicals and processes used to lighten and soften fabrics.

One exciting example of the textile uses of proteins is illustrated by recent research into spider silk. Spider-silk fibers have astonishing mechanical properties: they have strength comparable to steel, toughness greater than Kevlar and are less dense than cotton or nylon. However, spiders don't spin nearly enough silk to be harvested for use in industrial products. Researchers in Germany have recently discovered that protein subunits are responsible for silk's amazing properties. These findings provide a clearer understanding of the mechanical nature of spider-silk fibers and may be useful for design of silk-like products. The transfer of spider-silk genes into host organisms that can produce the quantities of protein needed has already been completed and the new structural information should lead to new products soon.

YOU DECIDE

Testing for the Best Product: Who Should Pay?

It takes about $800 million to bring a drug to market. Included in this cost is the price of researching drugs that are not approved owing to an adverse reaction or ineffectiveness. Usually, the purification process has already been developed and the product is in human trials when this happens Although drug prices are often high, many people do not realize that these prices reflect, in part, the cost of research on products that were not approved. If we also add the ability to determine which drug is best for each patient (pharmacogenetics) to the cost of bringing a drug to market, the cost climbs higher (see Chapter 1). Because biotechnology companies are supposed to make a profit for their stockholders, it is more profitable for the company to have everyone buy their drug, even if the drug is not entirely suited for each individual. What do you think is the best solution: higher prices and better drugs, or lower prices and drugs that do not always work and may even have negative side effects? You decide.

Detergents

As noted earlier, when enzymatic ingredients are added to detergents, they do a better job of cleaning and are more biodegradable. Laundry detergents take advantage of the specific roles of proteases, lipases, and amylases to dissolve stains in cooler water. If you look at the labels of many laundry stain removers, you will see enzymes listed as the first and sometimes only active ingredients.

Bioremediation: Treating Pollution with Proteins

In addition to reducing the quantity of pollutants produced during industrial processes, proteins can be used to clean up harmful wastes. Organic wastes from feedlots, homes, and businesses are a growing threat to the environment, especially aquatic ecosystems. Enzymes can be used to digest such organic wastes before they cause trouble.

Another promising new application for proteins is in neutralizing heavy metal pollutants such as mercury and cadmium. These dangerous elements can persist in the environment, causing harm to organisms throughout the food chain. These metals resist enzymatic breakdown, but this does not mean that proteins cannot be used to neutralize them (see Chapter 9). Some microorganisms have a sticky coat of **metallothioneins,** proteins that actually capture heavy metals. In this case, the pollutant is not dismantled or digested but simply made less dangerous. When toxic metals are bound to bacteria, they are less likely to be absorbed by plants and animals.

Researchers are currently using the power of genetic engineering to create new, better biological tools to attack and destroy toxic substances. Because the research process sometimes depends on randomly shuffling genes in the bacteria, the enzymes produced by the rearranged genes can be more or less reactive than those naturally produced. In a sense, scientists are accelerating the process of random mutation and evolution in the hope of discovering new, more efficient pollution-eating proteins (described in greater detail in Chapter 9).

4.2 Protein Structures

Ribosomes are factories that produce proteins (Chapter 2). To understand the processes of expressing and harvesting proteins, we must look at the molecular structure of proteins more closely.

Proteins are complex molecules built of chains of amino acids. Like all molecules, proteins have specific molecular weights. They also have an electrical charge that cause them to interact with other molecules. This ability to interact is the key to the biological activity of proteins. Consider, for example, the way the chemical structure and electrical charge of an amino acid can influence its interactions with water: The molecules will be either **hydrophilic** (water loving, attracted to water molecules) or **hydrophobic** (water hating, as if the water molecules and amino acids repelled one another).

Structural Arrangement

Proteins are capable of four levels of structural arrangement. These are the primary, secondary, tertiary, and quaternary structures of proteins (see **Figure 4.2**). The exact arrangement that a protein has depends on the specific chemical sequence of its amino acids and the types of side groups that are present. (See Appendix 2, side groups of amino acids for clarity.)

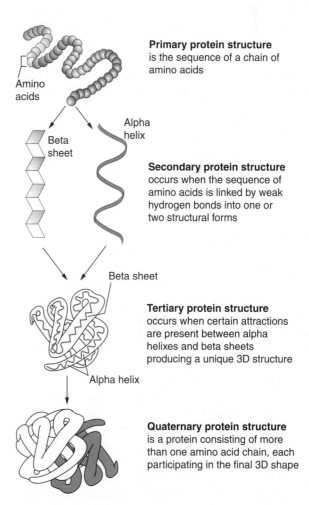

Primary protein structure is the sequence of a chain of amino acids

Amino acids

Beta sheet

Alpha helix

Secondary protein structure occurs when the sequence of amino acids is linked by weak hydrogen bonds into one or two structural forms

Beta sheet

Tertiary protein structure occurs when certain attractions are present between alpha helixes and beta sheets producing a unique 3D structure

Alpha helix

Quaternary protein structure is a protein consisting of more than one amino acid chain, each participating in the final 3D shape

FIGURE 4.2 The Four Levels of Protein Structure The proper folding of proteins is necessary for full functional capability. Purification methods must guarantee that proper folding is retained.

Primary structures

The 20 commonly occurring amino acids are the building blocks that make up proteins. Ten to 10,000 amino acids can be linked together in a head-to-tail fashion to form the sequence of a protein. The sequence in which amino acids are linked together at the ribosome is known as the *primary structure*. Altering a single amino acid in the sequence can mean that the protein loses all function. Genetic diseases are often the result of these protein mutations.

Secondary structures

Secondary protein structures occur when chains of amino acids fold or twist at specific points, forming new shapes due to the formation of hydrogen bonds between hydrophobic amino acids. The most common shapes, alpha helices and beta sheets, are described in the section on protein folding. In the alpha-helix arrangement, the amino acids form a right-handed spiral. The hydrogen bonds stabilize the structure, linking an amino acid's nitrogen atom to the oxygen atom of another amino acid. Because the links occur at regular intervals, a spiraling chain is formed. In the beta-sheet structure, the hydrogen bonds also link the nitrogen and oxygen atoms; however, because the atoms belong to amino acid chains that run side by side, an essentially flat sheet is formed. The sheets can either be "parallel" (if the chains all run in the same direction) or "antiparallel" (in which case the chains alternate in direction). One of the fundamental elements of protein structure, the beta turn, occurs when a single chain loops back on itself to form an antiparallel beta sheet. Both the alpha-helix and beta-sheet structures exist because they are the most stable structures the protein can assume.

Tertiary structures

Tertiary protein structures are three-dimensional polypeptides (large molecules made up of many similar, smaller molecules) that are formed when secondary structures are cross-linked. The bonds that hold together tertiary structures occur between amino acids capable of forming secondary covalent bonds (like cysteine, which can form disulfide bonds with another adjacent cysteine, cross-linking the protein into a unique shape). (A quick look at Appendix 2 will show the SH side group of cysteines that can cross-link to form disulfide bridges.) The tertiary structure of a protein determines its function, such as binding a cellular receptor or catalyzing a chemical reaction.

No matter what secondary and tertiary structure a protein takes, it is important to remember that structures are fragile. Hydrogen bonds can be broken easily, damaging a valuable protein. Anyone who has heated an egg has realized that the fluid egg proteins (albumin) quickly change shape (egg white) as the hydrogen and cross-linking tertiary bonds are broken by heat. Most proteins used in the lab are kept on ice to maintain their fragile structure and prevent them from being destroyed by the heat in the lab.

Quaternary structures

Quaternary protein structures are unique, globular, three-dimensional complexes built of several polypeptides. Hemoglobin, which carries oxygen in the blood, is an example of a protein with quaternary structure; it is composed of two tertiary proteins linked together.

Protein Folding

Everything that is important about a protein—its structure, its function—depends on folding. Folding describes how different strands of amino acids take their shape; for example, sickle cell anemia results from a misfolding due to a single amino acid replacement at a strategic location in the primary structure. If a protein folds incorrectly, not only will the desired function of the protein be lost but the resulting misfolded protein can be detrimental. For example, the plaque that forms in Alzheimer's disease accumulates because a misfolded protein cannot be broken down by the enzymes present in brain cells.

The first breakthrough in understanding the fundamental forms of proteins came in 1951, when Pauling and Corey described the alpha helix and beta sheet, which are the most common components of protein structure. Both structures result from hydrogen bonding that ties together the chain of amino acids. Not only Alzheimer's disease but cystic fibrosis, "mad cow" disease (bovine spongiform encephalitis, or BSE), many forms of cancer, and some heart attacks have all been linked to clumps of incorrectly folded proteins. Because protein folding occurs naturally, it is easy to see why one of the biggest challenges biotechnology faces is understanding and controlling the protein folding in the manufacturing process.

Glycosylation

More than 100 **posttranslational modifications** (like glycosylation) occur within a eukaryotic cell. In **glycosylation,** carbohydrate units (sugar molecules) are added to specific locations on proteins (see **Figure 4.3**). This change can have a significant effect on a protein's activity: it can increase solubility and orient proteins into membranes and extend the active life of a molecule in an organism.

Glycoproteins can be used in the treatment of disease, as scientists with the Scripps Research Institute (reported in the June 10, 2010, edition of the journal

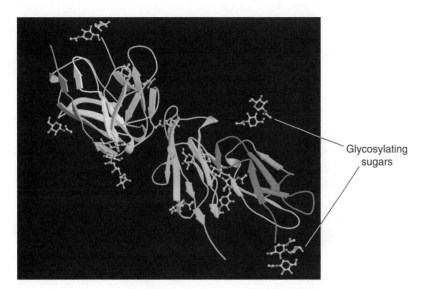

FIGURE 4.3 Three-Dimensional Protein Structure with Glycosylated Side Chains Glycosylation occurs within eukaryotic cells and probably extends the life of the protein by protecting it from natural cellular mechanisms that destroy proteins.

Glycosylating sugars

Blood) have discovered. This discovery represents a new way to target and destroy a type of cancerous cell. A glycoprotein can be combined with a nanoparticle (a small synthetic molecule) loaded with a chemotherapy drug, resulting in a new way to target and destroy cancer cells. By targeting B lymphoma cancers with chemotherapeutic-loaded nanoparticles, the effective dose of the drug is increased while simultaneously protecting normal tissues (as described in *Forecasting the Future*, on page 101). It's clear that the findings of the Scripps scientists could lead to the development of other novel drug therapies based on glycosylation to treat leukemia, lymphomas, and related cancers.

Protein Engineering by Directed Molecular Evolution

Biotechnology often relies on protein engineering At times it is necessary to introduce specific, predefined alterations in the amino acid sequence. This can be done with directed molecular evolution technology. A major biotech company, for example, induces mutations randomly into genes and then selects the bacterium with the protein product (enzyme) that has the highest activity. In this way, the company has been able to produce organisms (and industrial enzymes) that tolerate a cyanide concentration of more than 1.0 mole (M) per liter, which is too high for most bacteria to survive. No "natural" environment has experienced cyanide at this level. The resulting selected organisms can be used to remediate cyanide contamination resulting from mining and other industrial waste accumulation. A mutation resulting in this type of organism is not possible in a natural selection event because the natural environment has not changed

enough to select for these types of bacterial survivors. Directed molecular evolution requires introducing specific changes in the nucleotide sequences of a particular gene, resulting in new arsenic-metabolizing enzymes, as seen in **Figure 4.4** on the next page.

Such newly modified genes can then be introduced into a host cell, where the required amino acid sequence is produced by the host system. This technique allows researchers to create proteins with specific enhancements. Unlike naturally occurring mutation, directed molecular evolution focuses only on mutations that occur in a specific gene and selects the best proteins from that gene, irrespective of the potential benefits it may have for the original organism. For example, when *E. coli* manufactures human insulin, there is no benefit to the bacteria. For more information about directed molecular evolution, visit Maxygen Corporation's website at www.maxygen.com.

Enzymes have been modified over millions of years of evolution to catalyze specific chemical reactions in cells. Within the last decade, scientists have used directed evolution and rational design (designing a protein to fit a surface) with varying degrees of success to improve activity, stability, and selectivity of native enzymes. Much progress has been made in David Baker's lab at the University of Washington using a rational design program called Rosetta. The team designed an enzyme not found in nature by modeling the active site part of the enzyme and then finding a protein scaffold structure to attach to the part. After testing 84 constructs for activity, they selected 50 and inserted their DNA sequence into *E. coli* for expression. The final chosen enzyme functioned, but not as well as the native enzymes. This required them to mutate the active site many times to

a.

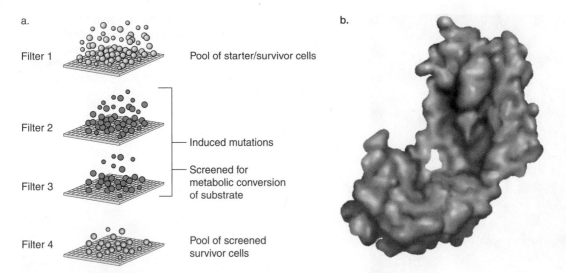

Filter 1   Pool of starter/survivor cells

Filter 2   — Induced mutations

                                           — Screened for
Filter 3   metabolic conversion
                                              of substrate

Filter 4   Pool of screened
                                              survivor cells

b.

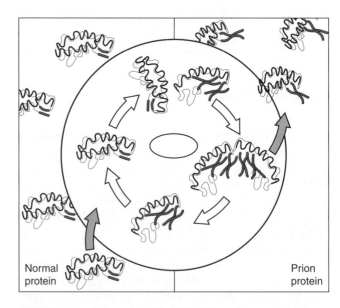

FIGURE 4.4   Directed Molecular Evolution Technology a. Genes that express proteins can be subjected to mutational events, generating a diverse group of novel gene sequences. Organisms with mutated genes are screened for unique properties. After an improvement in protein function, the process can be repeated until the maximum function is obtained. Unlike natural selection, this process focuses on the properties of the protein, not the organism, and can achieve changes that may never occur in nature. It may not have an evolutionary advantage, but the proteins it produces will have more commercial value. b. Scientists have discovered a protein used by bacteria to improve protein folding. The "spy" protein (shown) reduces the misfolding that commonly occurs when bacteria produce recombinant proteins for drug or industrial use. The research team used "directed evolution" to successively select bacteria for their abilities to refold proteins that protect them against antibiotics.

achieve better activity. This research shows that the day when proteins can be designed rationally is yet to come, but progress is being made.

In addition to naturally occurring and mutation-produced proteins, biotechnology is also creating entirely new protein molecules. These molecules, designed and built in the laboratory, indicate that it might be possible to invent proteins that are tailored for specific applications. Faulty protein folding is often expressed as a disease caused by infectious protein particles, called **prions** (**Figure 4.5**), which attract normal cell proteins and induce changes in their structure, leading to the accumulation of useless proteins that damage cells. Prion can occur in sheep and goats (scrapie) and cows (bovine spongiform encephalitis, or "mad cow" disease). Human forms of these brain-destroying diseases include kuru and **transformable spongiform encephalitis (TSE).** All these diseases involve changes in the conformation of the prion precursor protein, a protein normally found in mammalian neurons as a membrane glycoprotein. The inability to detect the disease until the infected animal is either sick or dead has complicated control of these diseases. Biotechnology research has sought to form synthetic infectious particles that can be studied, which will aid in developing detection and control.

FIGURE 4.5   Prions Are Misfolded Proteins A misfolding of proteins that occurs in prion disorders can be duplicated in the lab to produce large quantities of prion proteins, which can then be studied to create diagnostic kits. Follow the arrows from the left as a prion protein attracts a normal protein and changes it into a prion, resulting in an accumulation of prions within the cell.

Normal
protein

Prion
protein

4.3 Protein Production

By now, two things should be evident: (1) proteins are valuable, and (2) they are complex, fragile products. With these points in mind, we now examine the work of biotechnology in the production of proteins.

Producing a protein in the lab is a long, painstaking process, and at every stage there are many methods of production from which to choose. We refer to the two major phases used in producing proteins as **upstream processing** and **downstream processing.** Upstream processing includes the actual expression of the protein in the cell. During downstream processing, the protein is first separated from other parts of the cell and isolated from other proteins. Purity and functional abilities are then verified. Finally, a stable means of preserving the protein is developed. The choices made during upstream processing can simplify downstream processing.

Protein Expression: Upstream Processing

We begin a detailed discussion of protein processing by looking at the first decision made in upstream processing: selecting the cell to be used as a protein source. Microorganisms, fungi, plant cells, and animal cells all have unique qualities that make them good choices in the right circumstances.

Bacteria

Bacteria are an attractive protein source for several reasons. First, the fermentation processes of bacteria are well understood. Also, they can be cultured in large quantities in a short time. In industrial applications, this ability to generate the product on a large scale is often essential. Bacteria are also relatively easy to alter genetically.

Several methods of recombinant DNA technology can be used to increase the level of production of a bacterial protein. One is the introduction of additional copies of the relevant gene to the host cell. In most cases the relevant gene introduced into the organism is under the control of expression by a more powerful transcriptional promoter (see Chapter 3).

The bacterial species most commonly used to produce genetically engineered proteins is *E. coli*. Because early research into bacterial genetics focused on *E. coli* as a model system, we now understand the genetic characteristics of *E. coli* reasonably well.

In some instances, the foreign gene (in the form of **cDNA or complementary DNA**) for the desired protein product is attached directly to a complete or partial *E. coli* gene. In these cases, the genetically engineered *E. coli* produce the desired protein, but it is in the form of a **fusion protein.** In fusion proteins, a tar-

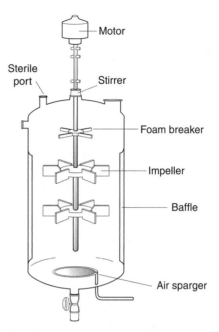

FIGURE 4.6 Bioreactor Large-volume cell culture is commonly performed in a self-contained, closed (sterile) system until the products are harvested. Through sterile ports, workers can adjust pH, gas concentrations, and other variables based on input from internal sensors. Sterile bags as large as 2,000 liters can be used on rocking tables as alternatives to stainless steel bioreactors.

get protein is fused to a bacterial protein; therefore, an additional step is required to break the two apart. The fused bacterial protein is usually an enzyme that will bind to its substrate and can be attached to a purification column (see description of affinity columns, below). The majority of proteins synthesized naturally by *E. coli* are intracellular (within the cell). In most cases, the resultant foreign protein accumulates in the cell's cytoplasm in the form of insoluble clumps called **inclusion bodies,** which must be purified from the other cell proteins before they can be used.

There are some limitations to the use of microorganisms in protein production. All bacteria, including *E. coli,* are prokaryotic. Prokaryotes are unable to carry out certain processes, such as glycosylation. For this reason, some proteins can be produced only by eukaryotic cells, as seen in Table 4.3.

Although it is possible to conduct the entire protein production process in a small flask in the laboratory, genetically engineered microorganisms can also be grown in large-scale fermenters (anaerobic) or **bioreactors** (aerobic) (**Figure 4.6**).

Computers monitor the environment in bioreactors, keeping oxygen levels and temperature ideal for cell growth. Cell growth is monitored carefully, because when the phase of growth is highest, the pro-

TABLE 4.3   ADVANTAGES AND DISADVANTAGES OF RECOMBINANT PROTEIN PRODUCTION IN E. COLI

Advantages	Disadvantages
E. coli genetics are well understood	Foreign proteins produced as inclusion bodies must be refolded
Almost unlimited quantities of proteins can be generated	Proteins cannot be folded in ways needed for many proteins active in mammalian systems
Fermentation technology is well understood	Some proteins are inactive in humans

moter must be activated to stimulate foreign gene expression. Activating a gene in a recombinant organism requires correct timing. It must be done after the organism has completed synthesizing important natural proteins needed for its metabolism.

Fungi

Fungi are the source of a wide range of proteins used in products as diverse as animal feed and beer. Naturally existing proteins found in some fungi are nutritious and used as foods In addition to naturally occurring proteins, many species of fungi are good hosts for engineered proteins. Unlike bacteria, fungi are eukaryotic and capable of some posttranslational modification (like folding human proteins correctly) and are used for that synthesis, as illustrated in Table 4.4.

Plants

Plant cells can also be used for protein expression. In fact, plants are an abundant source of naturally occurring, biologically active molecules, and 85 percent of all current drugs originated in plants. One example of a plant-derived protein produced on an industrial scale is the **proteolytic** (protein-degrading) enzyme **papain.**

TABLE 4.4   SOME RECOMBINANT PROTEINS FROM FUNGI

Protein	Fungi
Human interferon	*A. niger, A. nidulans*
Human lactoferrin	*A. oryzae, A. niger*
Bovine chymosin	*A. niger, A. nidulans*
Aspartic proteinase	*A. oryzae*
Triglyceride lipase	*A. oryzae*

Papain, or vegetable pepsin, is a protease used as a meat-tenderizing agent. It digests the collagen present in connective tissue and blood vessels that makes meat tough. Enzymes produced by plants will undoubtedly also be used in the near future for increased drug production.

Plants can be genetically modified to produce specific proteins that do not occur naturally. This process encourages rapid growth and reproductive rates in plants, which can be a distinct advantage. For example, tobacco, the first plant to be genetically engineered, can produce a million seeds from a single plant. As a nonfood plant, it makes a good choice for biotech protein production. Once the genetic material is integrated, a million new "plant protein factories" can fill the fields (as described in Chapter 6).

There are also disadvantages to using plants as protein producers. Not all proteins can be expressed in plants, and, because they have tough cell walls, the process of extracting proteins from them can be time-consuming and difficult. Finally, although plant cells can often properly glycosylate proteins, the process is slightly different from that of animal cells. This may rule out using plants as biofactories for the expression of some proteins. We will discuss transgenic plants in greater detail later (Chapter 6).

Mammalian cell culture systems

Sometimes it is possible to culture animal cells, growing them in a medium until it is time to harvest the proteins. This process is challenging because the nutritional requirements of mammalian animal cells are complex. Mammalian cells also grow relatively slowly, and the opportunity for mammalian cell cultures to become contaminated is greater than that of other culture systems. Despite these issues, mammalian cells are still the best if not the only choice for proteins destined to be used in humans.

Animal bioreactor production systems

Cells in culture are not the only option in using animal cells; sometimes living animals are protein producers. Consider, for example, the technique used to harvest monoclonal antibodies. Monoclonal antibodies react against only one target, making them valuable in diagnostic and therapeutic applications (see Chapter 7). Antibodies are proteins produced in reaction to **antigens** (usually an invading viruses or bacteria). Antibodies can combine with and neutralize an antigen, protecting the organism. The production of antibodies is part of the immune response that helps living things resist infectious disease. When the production of a monoclonal antibody is the goal, mice are injected with an antigen. The mouse either secretes the desired antibody or the mouse's antibody-producing tissue is fused with cancer cells (to make them immortal). When fluid from the tumor that results is collected, the monoclonal antibodies can be purified from it.

Another method of animal bioreactor protein production uses the milk or eggs from transgenic animals (animals that contain genes from other organisms). These animal products contain the proteins from the recombinant gene that was introduced and can be purified from the milk or egg proteins. In 2009, the FDA approved the first human drug produced from a goat: ATryn, which treats a rare bleeding disorder in humans (see Chapter 7).

Insect systems

Insect systems are another avenue of protein production from animal cells. **Baculoviruses** (viruses that infect insects) are used as vehicles to insert DNA, causing the desired proteins to be produced by the insect cells. However, there are instances in which the posttranslational modification of proteins is slightly different in insects than it is in mammals; therefore the use of insect expression systems may be of limited value. For the time being, insect expression systems are primarily used when small quantities of proteins are needed in research.

Protein Purification Methods: Downstream Processing

Once a protein is produced, downstream processing begins (**Figure 4.7**). First, the protein must be harvested. If the protein is intracellular, the entire cell is harvested; if it is extracellular, the protein is excreted into the culture medium that is collected. Harvesting, though, is just the beginning of downstream processing. Next, the real work begins: the protein must be purified. This is the process of separating the target protein from the complex mixture of biological molecules in which it was produced.

Purity in this context is a relative term. Generally, the FDA requires that a sample be composed of 99.99 percent of the target protein. Separating proteins from all other cellular contents is not easy, and isolating the target protein from the other proteins in the sample can be even more difficult. To understand the process of purifying proteins, we look at some steps commonly followed in it.

Step one: Preparing an extract for purification

It's helpful to note the relative volume of the extract. Often, the medium, or culture filtrate, harvested from a large fermenter or bioreactor is enough to fill a swimming pool, and the target protein represents less than 1 percent of that pool. Even if the protein is being expressed on a much smaller scale, finding the essential protein can seem like finding a needle in a haystack.

If the protein is intracellular, the first task is **cell lysis,** disrupting the cell wall to release the protein. There are many methods for doing this: freezing and thawing (which disrupts cell membranes and releases cell contents); detergents (used to dissolve cell walls), and mechanical methods (ultrasonics or grinding with tiny glass beads). Given the fragility of proteins, freeing them from the cell without degrading them entirely is challenging. The disruption process releases the protein of interest as well as the entire intracellular content of the cell.

After the cells have been ruptured, organic alcohols and salts may be added to the mixture. Both of these take advantage of the hydrophobic orientation of the proteins by attracting water from the proteins, causing them to coalesce. These agents increase the interactions between the protein molecules to separate them from the mixture.

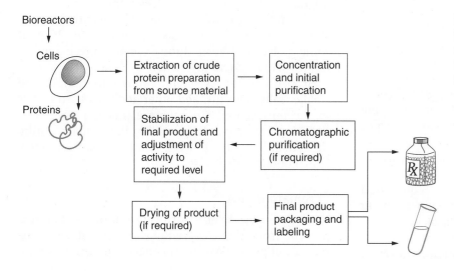

FIGURE 4.7 Basic Steps in Bioprocessing Purifications can be accomplished from raw materials or from bioreactors. The steps in the process must be devised and are often unique (and patentable).

Step two: Stabilizing proteins in solution

Next, the proteins must be stabilized. Recall that it is important to maintain the bioactivity of the protein and that proteins are relatively fragile molecules. As a consequence, precautions must be taken to protect the protein during the purification process.

Maintaining a low temperature is crucial to protecting proteins, so most purifications must occur at low temperatures. Heat as moderate as room temperature limits the activity of proteins. Maintaining the proper pH for the activity of a protein is also important, and most active proteins are suspended in buffering agents to preserve maximal function.

Natural proteases that can digest the target proteins in a preparation are another threat. Protease inhibitors and antimicrobials can be added to prevent the protein molecules from being dismantled but must be removed later, as must any additive used in the purification process. Still another potential problem is mechanical destruction by foaming or shearing of the proteins into useless fragments. Once again, additives can help prevent foaming and shearing from destroying the protein, but the additives must be removed later.

As we've seen, some purification methods are powerful enough to damage the target protein. It takes a balancing act to extract and purify proteins successfully. Although it is essential that the protein be purified, it is equally important that the protein maintain its biological activity.

Step three: Separating the components in the extract

The last step in the purification process can be the most important. Similarities between proteins permit us to separate them from material such as lipids (fats), carbohydrates, and nucleic acids, which are also released when a cell is disrupted. Differences between individual proteins are then used to separate the target proteins from others. Following are several methods used for protein separation.

Protein precipitation

Proteins often have hydrophilic amino acids on their surfaces that attract and interact with water molecules. That characteristic is used as the basis for separating proteins from other substances in the extract. Salts, most commonly ammonium sulfate, can be added to the protein mixture to **precipitate** the proteins (to cause them to settle out of solution).

Ammonium sulfate precipitation is frequently a first step in protein purification, resulting in a protein precipitate that is relatively stable. Problems associated with ammonium sulfate precipitation make it a poor choice in some industrial situations, however.

Ammonium sulfate is highly reactive when it contacts stainless steel, for example, and many industrial purification facilities are made of stainless steel. Other solvents frequently used to promote protein precipitation include ethanol, isopropanol, acetone, and diethyl ether. Just like ammonium sulfate, these solvents cause protein precipitation by removing water from between the protein molecules.

Filtration (size-based) separation methods

There are a variety of ways to separate molecules based on size and density. **Centrifugation** separates samples by spinning them at high speed. With this process, proteins can often be isolated in a single layer or separated from heavier cell components. Small-volume centrifuges are capable of processing only a few liters at each run. Large reactors can process hundreds or thousands of liters (see **Figure 4.8**). Industrial-scale centrifugation is normally achieved using continuous-flow centrifuges that allow continuous processing of the contents of a bioreactor.

Filters of various sizes and types can also be used to separate protein from other molecules in the mixture. In this process, known as **membrane filtration,** thin membranes of nylon or other engineered substances with varying pore sizes are used to filter out all of the cellular debris from a solution. First, **microfiltration** removes the precipitates and bacteria. **Ultrafiltration** then uses filters that can catch molecules such as proteins and nucleic acids. Some ultrafiltration processes can even separate large proteins from smaller ones. (Refer to www.amicon.com to view some of these devices.) One of the main shortcomings of membrane filtration systems is their tendency to clog easily. On the plus side, the use of these filtration systems takes less time than centrifugation.

Diafiltration and **dialysis** are filtration methods that rely on the chemical concept of equilibrium, the migration of dissolved substances from areas of higher concentration to areas of lower concentration. As shown in **Figure 4.9**, dialysis depends on the ability of some molecules to pass through semipermeable membranes while others are halted or slowed because of their size. Dialysis is often required to remove the smaller salts, solvents, and other additives used earlier in the purification. The salts are then replaced with buffering agents that help stabilize the proteins during the remainder of the process. Diafiltration adds a filtering component to dialysis.

Chromatography

The initial steps in any purification process liberate a protein from the cell, remove undesired contaminants and particulates, and concentrate the proteins.

(a) Small - volume fixed angle centrifuge

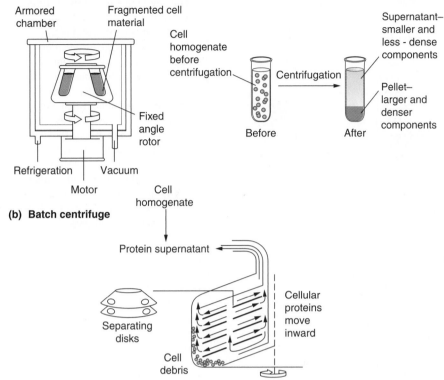

Armored chamber

Fragmented cell material

Cell homogenate before centrifugation

Fixed angle rotor

Refrigeration

Vacuum

Motor

Before

Centrifugation

After

Supernatant–smaller and less - dense components

Pellet–larger and denser components

(b) Batch centrifuge

Cell homogenate

Protein supernatant

Separating disks

Cellular proteins move inward

Cell debris

FIGURE 4.8 Fixed-Angle and Batch Centrifugation Fixed-angle centrifuges (a) can develop extremely high gravitational forces (*g* forces) but are limited to smaller quantities and must be run separately for each batch. Batch centrifuges (b) were developed to allow continuous flow of materials and separation of cell debris from cell proteins. The continuous pressure of the inflow and the centrifugal force can be adjusted to maximize the separation an outflow of the protein supernatant.

Chromatography methods allow us to sort proteins by size or by how they cling to or separate from other substances. In chromatography, long glass tubes are filled with microscopic resin beads and a buffered solution. The protein extract is then added and flows through the resin beads in a glass column. Depending on the resin used, the protein either sticks to the beads or passes through the column while the beads act as a filtration system.

Size-exclusion chromatography (SEC) uses gel beads as a filtering system. Larger protein molecules quickly work their way around the gel beads while smaller molecules pass through more slowly because they are able to slip through tiny holes in the

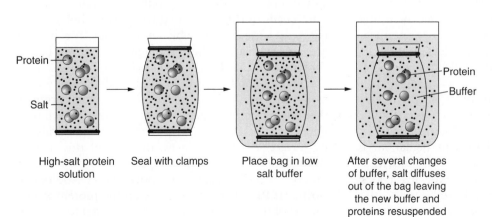

Protein

Salt

High-salt protein solution

Seal with clamps

Place bag in low salt buffer

After several changes of buffer, salt diffuses out of the bag leaving the new buffer and proteins resuspended

Protein

Buffer

FIGURE 4.9 Dialysis Dialysis can be used to remove salt (large dots in figure), by the process of diffusion, and replace it with a buffer that is better suited for protein stability.

(a)

Sample of low-molecular-weight and high-molecular-weight proteins applied to top of column

As buffer flows, the low-molecular-weight proteins enter the gel beads, retarding their flow

The higher molecular weight proteins are less able to enter the gel beads, so they elute first

Low molecular weight

High molecular weight

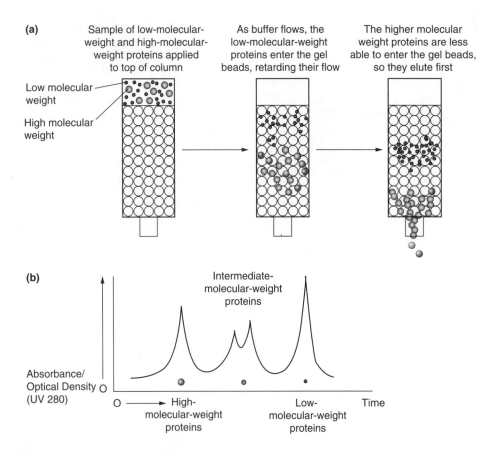

(b)

Intermediate-molecular-weight proteins

Absorbance/ Optical Density (UV 280)

High-molecular-weight proteins

Low-molecular-weight proteins

Time

FIGURE 4.10 Protein Purification by Size Exclusion Chromatography (a) Low- and high-molecular-weight proteins travel through a size-exclusion chromatography column. (b) The diagram shows how the low and high molecular weight proteins appear as they come off the column when monitored for proteins (ultraviolet [UV] 280 absorption). Notice that the high-molecular-weight proteins move quickly through the buffer, whereas the low-molecular-weight proteins are slowed by the matrix of the column resin. Resins may be purchased with many different pore sizes.

beads, as shown in **Figure 4.10**. The gels are available in a variety of pore sizes, and the necessary gel for proper separation depends on the molecular weight of the contaminants or proteins being separated. This method can make only preliminary separations, however, and can pose problems in industrial settings because it requires very large columns.

Ion-exchange (IonX) chromatography, relies on an electrostatic charge (like static cling) to bind proteins to resin beads in a column. While the charged proteins cling to the resin, other contaminants pass through and out of the column, as shown in **Figure 4.11**. The proteins can then be eluted (released from the resin) by changing the electrostatic charge; this is done by rinsing the column with salt solutions of increasing concentrations. The bound protein is then released from its attachment (detected by viewing under UV 280) and collected.

Affinity chromatography relies on the ability of most proteins to bind specifically and reversibly to uniquely shaped compounds called **ligands.** Ligands are small molecules that bind to a particular large molecule in a protein. Think of ligands fitting with a unique protein molecule the way a key fits a lock (**Figure 4.12**). After the proteins have bound to the resin beads, a buffer solution is used to wash out the

unbound molecules. Finally, special buffer solutions are used to cause desorption (to break the ligand bonds) of the retained proteins. Fusion proteins, as mentioned earlier, can be used in affinity chromatography because the substrate (ligand) of the bacterial protein can be part of the affinity column, attracting the fused protein (bacterial enzyme protein) to the column. Affinity chromatography may shorten the purification process by reducing the number of steps.

As we have seen, amino acids are either attracted to or repelled by water molecules. In **hydrophobic interaction chromatography (HIC),** proteins are sorted on the basis of their repulsion of water. The column beads in HIC are coated with hydrophobic molecules, and the hydrophobic amino acids in a protein are attracted to the similar chemicals in the beads, shown in **Figure 4.13** on page 116.

Isoelectric focusing is often used in quality control during purification to identify two similar proteins that are difficult to separate by other means. Each protein has a specific number of charged amino acids on its surface in specific places. Because of this unique combination of charged groups, each protein has a unique electronic signature known as its **isoelectric point (IEP),** where the charges on the protein match the pH of the solution. The IEP can be used to separate

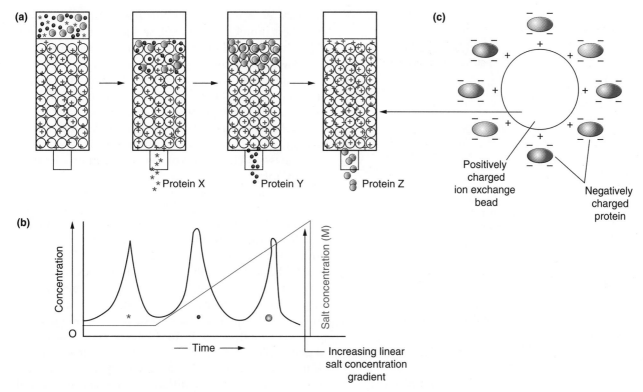

FIGURE 4.11   **Ion-Exchange Chromatography** (a) Charged amino acids bind to ionic resin beads. Increasing the ionic strength of the buffer displaces proteins (based on their binding strength) after they have bound to the column resin. (b) Protein X is released at the lowest concentration of the displacing salt gradient; protein Y has a higher binding strength and is released second; protein Z has the highest binding strength and requires a high salt concentration to displace it from the ion exchange beads of the column. (c) Anion-exchange resin is positive; cation-exchange resins are negatively charged.

similar proteins from one another. Isoelectric focusing is the first dimension of **two-dimensional electrophoresis.**

Two-dimensional electrophoresis separates proteins based on their electrical charge and size. It is essentially the combination of two methods, IEP and gel electrophoresis. In this technique, researchers introduce a solution of cell proteins onto a specially prepared strip of polymer. When the strip is exposed to an electrical current, each protein in the mixture settles into a layer according to its charge. Next, the strip is aligned with a gel and again exposed to electrical current. As the proteins migrate through the gel, they separate according to their molecular weight.

Analytic methods

High-performance liquid chromatography (HPLC)
adds a twist to the previously described chromatography

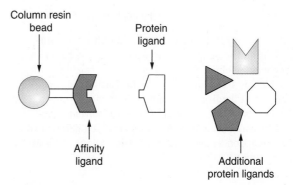

FIGURE 4.12   **Affinity Chromatography** Affinity ligands are designed to bind specifically to unique three-dimensional chemical components of the protein being purified. The nonbound proteins wash through the column. Increasing the ionic strength of the buffer can displace the bound protein (after preferential binding) and regenerate the affinity column. The displaced (pure) protein can be collected and concentrated.

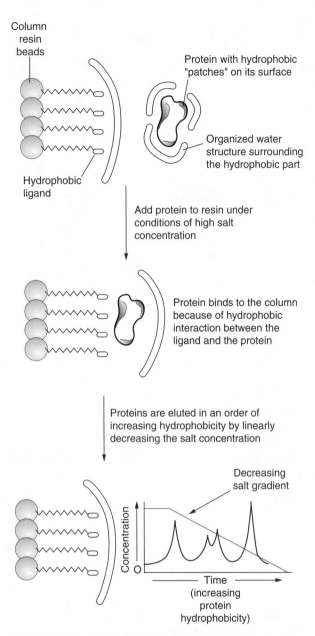

FIGURE 4.13 Hydrophobic Interaction Chromatography Increasing the nonpolar concentration of the buffer can cause the hydrophobic portions of proteins to combine with a hydrophobic ion-exchange-column resin. Further reducing the ionic concentration displaces the protein from the column resin and replaces its attachment with a nonpolar solvent. Fractions can be collected and protein concentration determined based on detection by spectrophotometric analysis at UV 280 nanometers (nm).

methods, which depend on gravity flow or very low pressure pumps to move the extract through the columns. Such low-flow methods can take several hours to process a single sample. In contrast, HPLC systems use greater pressure to force the extract through the column

in a shorter time. HPLC systems have limitations, however. Less protein is separated, so the technique is more useful in analytical situations than in mass production.

Mass spectrometry (mass spec) is a highly sensitive method used to identify small differences between proteins. In fact, it is used frequently on the outflow of HPLC systems. All mass spectrometers do three things: suspend the sample molecules into a charged gas phase, separate the molecules based on their mass-to-charge ratios by acceleration down a narrow tube, and finally detect the separated ions. A sample as small as one picogram (one billionth of a gram) can be analyzed by this process, which is illustrated in **Figure 4.14**. A definitive readout is produced, which indicates the identity and size of most of the proteins or fragments analyzed.

An important application of this process is in protein sequencing. A larger protein can be digested into fragments (peptides) and analyzed to determine the amino acid sequence. Mass spectrometry is now the preferred method and, in biotech companies, has largely replaced the slower Edmond end-group analysis method used for amino acid sequencing. Mass spectrometry can detect the difference between isomers of the same protein, and its capabilities are improving faster than most lab instruments in terms of critical analysis.

Verification

At each step of the purification process, it is important to verify that the target protein has not been lost and that concentration efforts have been successful. **SDS-PAGE** (polyacrylamide gel electrophoresis) is often used for verification. In this method, a detergent called sodium dodecyl sulfate (SDS) is added to a sample of the protein mixture and the mixture is heated. The sulfate charges are distributed evenly along the denatured (heated) protein, making separation dependent on the size of the protein. After this treatment, the protein sample is loaded onto a special gel matrix (PAGE) where it forms a single band at a specific location depending on its molecular size and mass, as seen in **Figure 4.15**.

By adding a dye that combines with proteins, such as **Coomassie stain,** a colored band results, and a known size marker can be compared with the stained sample. When the sample and the known marker are equivalent, we have evidence that the protein of interest is indeed being concentrated. Since this test is run at each step in the purification process, the colored band should become increasingly intense, proving that the protein has not been lost.

A specific detection method for proteins separated by SDS-PAGE is Western blotting (similar to Southern blotting, Chapter 3). In Western blotting, protein is

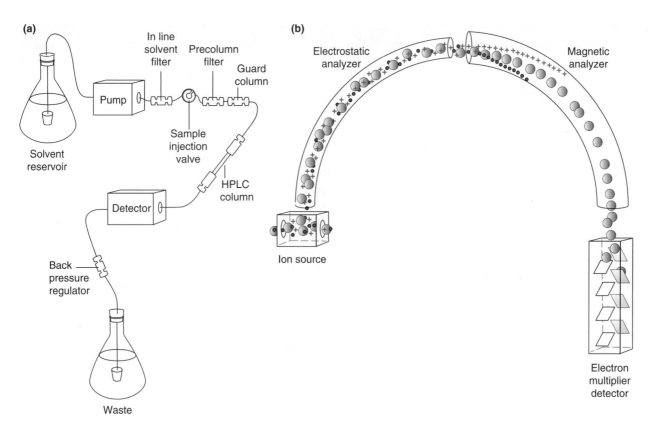

FIGURE 4.14 **High-Performance Liquid Chromatography and Mass Spectrometry**
High-performance liquid chromatography is often coupled with mass spectrometry to
analyze proteins. (a) HPLC uses noncompressible resin beads under very high pressures to
separate proteins that are often similar in size. (b) Mass spectrometry follows this initial
separation to detect subtle differences in ionized and accelerated proteins that will be
analyzed (by distance or time) as they travel down a vacuum-filled tube.

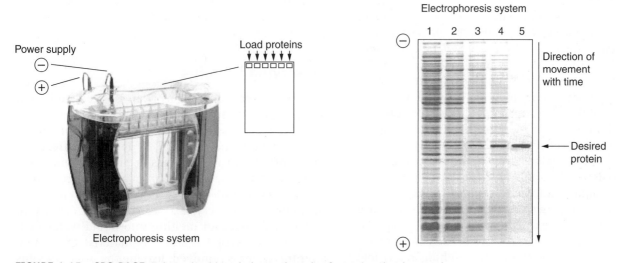

FIGURE 4.15 **SDS-PAGE** Polyacrylamide gel electrophoresis of proteins that have uniform charges (produced by heating in sodium dodecyl sulfate) can be separated by size,
based on their migration in a vertical electrical field. This procedure commonly follows
each major step in purification to verify that the band of desired protein is becoming
more concentrated (and has not been lost in the separation). Notice that the 65-kilodal-
ton (kD) protein has the highest concentration in the final step (lane 2) of this procedure (lane 1) when proteins of known size are used for comparison. The smallest proteins
will reach the + pole first.

transferred from a gel to a nitrocellulose membrane and detected with a specific antibody that recognizes that protein by its unique structure. An enzyme attached to the antibody can be used to convert a fluorescent substrate so as to produce a permanent record of the detection. In this procedure, an electric current is applied to the gel. The separated proteins then migrate through the gel and onto the membrane in the same pattern as they separate on the SDS-PAGE. All sites on the membrane can then be blocked in such a way that antibody (serum) will not bind to them nonspecifically to detect the antigen blotted on the membrane; a primary antibody (serum) is then added at an appropriate dilution and incubated with the membrane. If there are any antibodies present that are directed against one or more of the blotted antigens, those antibodies will bind to the protein(s) while other antibodies will be washed away at the end of the incubation. To detect the antibodies that have bound, anti-immunoglobulin antibodies coupled to a reporter group, such as the enzyme alkaline phosphatase, are added. Finally, after excess second antibody is washed free of the blot, a substrate is added that precipitates upon reaction with the conjugate, resulting in a visible band where the primary antibody bound to the protein (see Figure 4.15).

Antibody detection of a specific protein can also be carried out using an enzyme-linked immunosorbent assay (ELISA). The most commonly used ELISA requires two antibodies: one to capture the unique protein and one attached to an enzyme to produce a color reaction (see Figure 4.12). ELISA occurs on a multiwell plate using affinity chromatography. The first antibody to the unique protein is plated on to a multiwell ELISA plate, the protein is added, and—after a series of wash and blocking steps—the second antibody (enzyme attached) is added. The addition of a substrate permits a color reaction to occur if the antibody has bound. ELISA plates are designed to capture many important proteins currently under research.

Preserving Proteins

After the target protein has been isolated, collected, and purified, it must be saved in a manner that will preserve its activity until it can be used. One means of preserving the protein is **lyophilization,** or freeze-drying. In this process, a purified liquid protein solution is frozen. A vacuum is then used to hasten the evaporation of water from the fluid. In lyophilization, ice crystals are converted directly to water vapor without melting into liquid water first. The containers of freeze-dried material are sealed after the water is removed, leaving the dried proteins behind. Lyophilization has been demonstrated to maintain protein structure, making it a commonly chosen method for the preservation of biotechnologically derived proteins. Many freeze-dried proteins may be stored at room temperature for prolonged periods of time.

Scaling Up Protein Purification

Protein purification protocols are usually designed in the laboratory on a small scale. These techniques at this level of productivity are feasible if the product is needed in only small amounts. Orders of monoclonal antibodies, for example, are in the range of grams per year; a relatively small demand. A single laboratory-scale bioreactor can usually produce adequate quantities. However, other proteins are required in much larger quantities, which means that the production methods must be scaled up.

Scaling up is not always easy to accomplish. Laboratory methods that may work on a small scale may not always be adaptable to large-scale production. Furthermore, changes in the purification process can invalidate earlier laboratory-scale clinical studies. For example, when the FDA approves a bioengineered protein, it also approves the process for producing it. To change the process, it may be necessary to seek FDA approval once again. For this reason, bioprocess engineers are involved in the early stages of purification and work to ensure that it will be possible to scale up the process later on.

Postpurification Analysis Methods

During research, it is often helpful to look closely at a purified protein. The goal may be to better understand the molecular structure of a specific protein or to alter its structure to change the protein's function. The following two methods are used in the postpurification analysis of proteins.

TOOLS OF THE TRADE

Piecing Together the Human Proteome with Protein

Proteomes, the collection of proteins associated with a specific life function, have become more important since the discovery of the human genome. As mentioned earlier, the cost of bringing a drug to market is about $500 million and takes between 5 and 8 years. Because most of the drugs produced by the biotechnology industry are proteins (such as growth factors, antibodies, and synthetic hormones) that replace missing or nonfunctional proteins in humans, companies are particularly interested in developing less expensive and faster methods of detecting how proteins function in disease detection and treatment—specifically, new protein microarrays. Like DNA microarrays, these miniature devices detect proteins associated with disease and those present in abnormal concentrations. As new protein structures are identified, new microarrays are constructed.

Protein microarrays have many uses. Most current biotech drugs function at the protein level, interacting with receptors, triggering events, and targeting other proteins in cells of the body. The detection of biomarker proteins using microarrays has improved the monitoring of drug action in many diseases. For example, we have all experienced the benefit of vaccines: proteins that stimulate our immune systems to recognize disease organisms without contracting the disease. Antibody production in response to vaccination is an example of the effect of a protein biomarker and can indicate positive immunity.

Protein microarrays are usually constructed of glass slides coated with a material that binds proteins. Attached to each slide is an antibody that is specific to the protein to be detected as well as a signaling mechanism that indicates capture of the protein. Cambridge Antibody Technology (in the United Kingdom) has a library of more than 100 billion antibodies collected from the blood of healthy individuals. Packard BioScience (in Connecticut) has developed a protein microarray that allow the attached protein to maintain its three-dimensional shape while embedded in the coating material. Ciphergen Biosystems (in California) has adapted mass spectroscopy to perform rapid on-chip detection and analysis of proteins directly from biological samples. So, what is left to do?

There are countless proteins from numerous diseases yet to be discovered (Chapter 11). The ability to diagnose a disease and determine the most effective treatment will depend on the ability to purify proteins and develop antibodies that are related to the disease. The opportunity to change human disease conditions that were previously incurable depends on discovering and purifying the important proteins of life's proteome. Take the time to access the websites of some of the companies that develop these devices to keep up with this important technology (e.g., www.ciphergen.com, www.packardbioscience.com, or www.biacore.com).

Protein sequencing

Recall from our discussion of protein structures that each protein has a specific sequence of amino acids, known as the primary sequence. To understand a protein completely, it is important to determine its primary sequence. Automated protein sequencers make this task possible (see Figure 4.14). In the mass spectrometry method, peptide masses are identified by their unique signatures (retention times). By converting peptides into ions and subjecting them to acceleration in a vacuum, it is possible to identify many unique proteins in a matter of minutes.

X-ray crystallography

X-ray crystallography is used to determine the complex tertiary and quaternary structures of proteins. The method requires pure crystals of proteins that have been carefully dehydrated from solutions. Bombarded with x-rays, a pure protein creates specific shadow patterns based on its configuration (see Chapter 2). Computer analysis of these patterns allows the generation of "ribbon" diagrams, which not only describe the structure of the proteins but also make it possible to plan potential modifications of the protein molecule to improve function. Protein crystallography is usually a requirement for approval by the FDA because it verifies that the process produces the product being approved.

4.4 Proteomics

Many diseases are the result of flaws in protein expression, but not all of those diseases can be understood simply in terms of genetic mutation. Because proteins undergo posttranslational modification, the puzzle can be far more complicated. A new scientific discipline, **proteomics,** is dedicated to understanding the complex relationship of disease and protein expression.

In proteomics, **proteomes** (the PROTEin complement to a genOME) are compared between healthy and diseased states. The variations of protein expression are then correlated to the onset or progression of a specific disease. The goal of proteomic research is the discovery of protein markers that can be used in new diagnostic methods and the development of targeted drugs for treating disease. For example, scientists at the University of California, Davis, purified the protein produced by BRCA2, an oncogene linked to breast cancer. The researchers were then able to synthesize the protein to study BRCA2's role in DNA repair. BRAC2 is susceptible to the drug trastuzumab (Herceptin; Genentech, San Francisco), which has increased the breast cancer survival rate to nearly 70 percent. In this way, only patients with BRAC2 biomarkers receive trastuzumab, saving discomfort and unnecessary treatment of all breast cancer patients.

Although the BRCA2 gene was discovered in 1994, purifying the protein made by the gene proved difficult, since it is very large, does not express well, and degrades easily. The U.C. Davis researchers tested many different cell lines and succeeded in introducing a BRCA2 gene into a human cell line and expressing it as a whole protein. Other researchers used genetic engineering techniques to manufacture the human protein in yeast. They then tested the purified protein for its function in repairing damaged DNA. Experiments with the BRCA2 protein confirmed that it plays a role in repairing damaged DNA, and when it is damaged in breast cancer, the DNA fails to function. Research continues on the function of BRCA2 and other proteins involved in breast cancer, but this progress would not have been possible without proteomics.

Proteomics applies several techniques described earlier: two-dimensional electrophoresis is used to separate proteins, mass spectrometry verifies the protein's identity, and the protein is characterized using amino acid sequencing. Some of this labor-intensive work may be assisted by automation in the future. A **protein microarray** is a set of proteins immobilized on a surface—usually a glass slide—which has been coated with a reagent that will indicate binding by color change (**Figure 4.16**). The capacity of these arrays has increased but does not rival that of DNA arrays because of the difficulties of production. A single

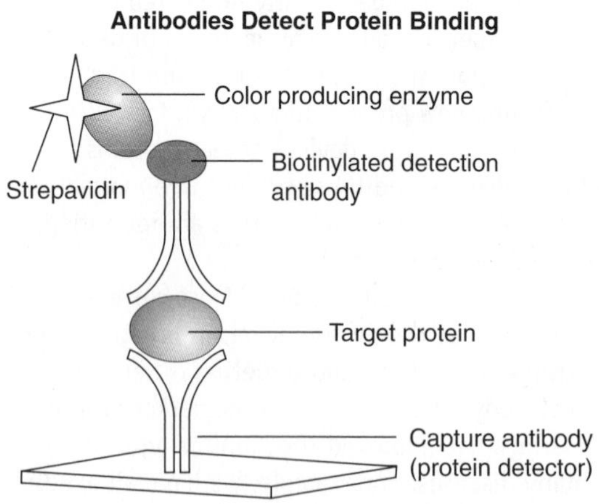

Antibodies Detect Protein Binding

- Color producing enzyme
- Strepavidin
- Biotinylated detection antibody
- Target protein
- Capture antibody (protein detector)

FIGURE 4.16   Protein Microarrays protein microarrays that uniquely bind to single proteins are being perfected. When a protein binds, it releases a fluorescent signal. The ability to detect the presence of unique proteins (proteomics) will permit proper decisions in disease diagnosis and therapy to be made.

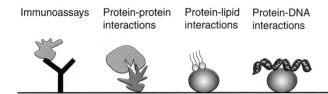

Immunoassays   Protein-protein   Protein-lipid   Protein-DNA
                interactions   interactions   interactions

Protein-small   Protein-peptide   Kinase   Profiling immune
molecule   interactions   responses
interactions

ATP → ADP

FIGURE 4.17   Protein Microarrays Can Detect More Than Just Proteins The capacity of protein microarrays has increased. Functional protein microarrays have recently been applied to the discovery of protein interactions (i.e., protein-protein, protein-lipid, protein-DNA, protein-drug, and protein-protein interactions), expanding knowledge of the significance of protein interactions to normal cell function. Binding antibodies or other proteins to the matrix allows detection when the binding occurs.

DNA array can monitor the expression of an entire genome. Functional protein microarrays have recently been applied to the discovery of protein interactions: protein-protein, protein-lipid, protein-DNA, protein-drug, and protein-protein interactions (**Figure 4.17**).

MAKING A DIFFERENCE

As we have seen, disease diagnosis and treatment have been vastly improved by protein purification technology. Despite this, there are still problems to be solved: biological fluids (plasma, sera, urine, and saliva) have a wide range of different protein concentrations, and high concentrations of some proteins mask the presence of others—like disease-specific proteins called biomarkers. These low-concentration biomarker proteins are often degraded by natural enzymes or the purification process itself. Ceres Nanosciences of Virginia is applying nanotechnology to production purification to address this problem. The Nanotrap uses porous nanoparticles to attract specific biomark-

ers The nanoparticles exclude other proteins owing to the small size of the pores on the surface that they must penetrate. Nanotrap technology adds another dimension to the protein purification process, permitting the detection of proteins that "mark" the presence of a disease. Nanotechnology is making a difference in disease treatment.

QUESTIONS & ACTIVITIES

Answers can be found in Appendix 1.

1. What can the public database on proteins tell us?

2. How does directed molecular evolution technology differ from mutations that occur naturally?

3. How has an understanding of protein structure benefited from the results of the Protein Structure Initiative?

4. Why are proteins being primarily searched through cDNA products?

5. If organisms produce proteins on their own, why should companies be allowed to patent proteins?

6. If you are purifying a small protein from an SEC column, which fractions do you want to collect?

7. Do proteins strongly bound to an IonX column elute first or last from the column?

8. Is affinity chromatography more or less selective at separating proteins than IonX chromatography?

9. List the protein separation methods primarily used in analytic (rather than production) methods.

10. If you carry out an SDSPAGE analysis after each step in a protein purification sequence and found that the last step resulted in the lack of a band at the location you expected, what would this mean?

Visit www.pearsonhighered.com/biotechnology
To download learning objectives, chapter summary, "Keeping Current" web links, glossary, flashcards, and jpegs of figures from this chapter.

Microbial Biotechnology

After completing this chapter you should be able to:

- Provide examples of how prokaryotic and eukaryotic cell structures differ.

- Describe features of bacteria that make them useful for applications in biotechnology.

- Provide examples of how yeast are used in biotechnology.

- List examples of medically important proteins that are produced in bacteria using recombinant DNA technology.

- Explain how alcohol and lactic acid fermentation processes can be used to produce common foods and beverages.

- Describe the role microorganisms play in the development and production of vaccines; provide examples of different vaccines.

- Discuss why studying microbial genomes is valuable.

- Understand why scientists are interested in using synthetic genomes and synthetic biology for biotechnology applications.

- Explain the value of metagenomics; describe the goals of the Human Microbiome Project.

- Define bioterrorism, identify microorganisms that may be used as bioweapons, and discuss strategies to combat them.

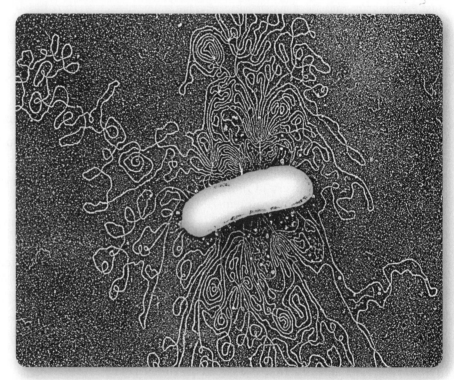

Microbes, such as the *Escherichia coli* (*E. coli*) cell shown in the center of this photo, have a long history of applications in biotechnology. The cell wall and plasma membrane of this cell have been burst open to reveal the bacterial chromosome.

Just when Italian art restorers were at a loss to save one of the world's largest collections of four-teenth- and fifteenth-century medieval frescoes (paintings done on plaster) decorating a cemetery in Pisa near the Leaning Tower, soil bacteria called *Pseudomonas stutzeri* came to the rescue. Restorers were trying to clean the frescoes of paint-clouding grime contained within glue that had been used to repair them after damage in World War II. Traditional chemical reagents only further damaged the paintings. Knowing that *P. stutzeri* could degrade nitrogens, plastic resins, and pollutants such as tetrachloroethylene, scientists cultured a strain of these microbes in the lab and applied them to the frescoes. Within 6 to 12 hours, the bacteria had cleaned approximately 80% of the glue residue, and the remaining residue was removed with a light washing. The next proposed project for these **microorganisms** is removing black crust from limestone and marble monuments in Greece.

Microorganisms, or **microbes,** are tiny organisms too small to be seen individually by the naked eye; they must be viewed with the help of a microscope. Although the most abundant microorganisms are bacteria, microbes also include viruses; fungi such as yeast and mold; algae; and single-celled organisms called protozoa. Bacteria have existed on the earth for over 3.5 billion years, and they greatly outnumber humans. It has been estimated that microbes comprise over 50% of the earth's living matter. Yet less than 1% of all bacteria have been identified, cultured, and studied in the laboratory. We are literally surrounded by bacteria. They live on our skin, in our mouths, and in our intestines; they are in the air and on virtually every surface we touch. Bacteria have also adapted to live in some of the harshest environments on the planet: polar ice caps, deserts, boiling hot springs, and under extraordinarily high pressure in deep-sea vents miles under the ocean's surface.

The rich abundance of bacteria and other microbes provides a wealth of potential biotechnology applications. Well before the development of gene-cloning techniques, humans used microbes in biotechnology. In this chapter, we primarily discuss the important roles bacteria have played in both old and new practices of biotechnology. We conclude by discussing the dangers of microbes as agents of bioterrorism.

FORECASTING THE FUTURE

Speculating on the future directions of microbial biotechnology is not particularly easy because there are so many active areas of research with great potential. Two areas of great intrigue are synthetic genomes and biofuels. As you will learn later in the chapter, a synthetic genome has been assembled for a bacterial strain and transplanted into another strain. What will be some of the future applications of this technology and of the field of synthetic biology? Will it eventually be possible for a synthetic eukaryotic genome to be produced? For biofuels, there is a great need to reduce our global dependence on fossil fuels and generate alternative energy sources. Can microbes be used in metabolic processes to create fuels or to break down materials to release components that can become fuels in a way that is effective and cost-efficient for widespread use? Microbial biotechnology is a dynamic field that is producing many new applications. We could forecast several other topics as hot areas of microbial biotechnology in the future, but we think it will be particularly interesting to keep an eye on developments in biofuels and synthetic biology.

5.1 The Structure of Microbes

Before you can consider the many applications of microbial biotechnology, you must be able to distinguish microorganisms from plant and animal cells. Recall that cells can be broadly classified based on the presence (eukaryotes) or absence (prokaryotes) of the nucleus that contains a cell's DNA (Chapter 2). *Eukaryotic* cells include plant and animal cells; fungi such as yeast; algae; and single-celled organisms called protozoans, among which are amoebas like those you may have studied in high school biology. Unlike eukaryotic cells, *prokaryotic cells* lack most membrane-bound organelles, such as a nucleus. Prokaryotes include the **domains** (taxonomic categories above the kingdom level) **Bacteria** and **Archaea**—organisms that share properties of both eukaryotes and prokaryotes. The cellular structure of microorganisms is important in determining both where they will thrive and how they can be used in biotechnology. For example, Archaea that live in extreme environments such as very salty conditions are called halophiles or hot environments are called thermophiles; as a result, they have very unique metabolic properties. Moreover, structural features of bacteria, in particular, make them excellent microorganisms to use for biotechnology research.

Bacterial cells are much smaller (1–5 micrometers, or μm; 1 μm = 0.001 millimeter) than eukaryotic cells (10–100 μm) and have a much simpler structure (refer to Figure 2.1 for a diagram of prokaryotic cell structure). Bacterial cells also exhibit the following structural features:

- DNA is not contained within a nucleus and typically consists of a single circular chromosome that lacks histone proteins.

- Bacteria may contain plasmid DNA.
- Bacteria lack membrane-bound organelles.
- The cell wall that surrounds the cell (plasma) membrane is structurally different from the plant cell wall. Composed of a complex polysaccharide and protein substance called **peptidoglycan,** the cell wall forms a rigid outer barrier that protects the cells and determines their shape. In Archaea, this structure does not contain peptidoglycans.
- Some bacteria contain an outer layer of carbohydrates, which form a structure called the *capsule.*

Most bacteria are classified by the **Gram stain,** a technique in which dyes are used to stain the cell wall of bacteria. Gram-positive bacteria stain purple and have simple cell wall structures rich in peptidoglycans, whereas gram-negative bacteria stain pink and have complex cell wall structures with less peptidoglycan. Bacteria do not form multicellular tissues like animal and plant cells, although some bacteria can associate with each other to form chains or filaments of many connected cells.

Bacteria vary in their size and shape. The most common shapes include spherical cells called **cocci** (singular, coccus), rod-shaped cells called **bacilli** (singular, bacillus), and corkscrew-shaped or spirillar bacteria (**Figure 5.1**). As you study microbes, you will learn that the names of bacteria frequently give you with a tip about the shapes of those cells. For instance, *Staphylococci* are spherical bacteria that live on the surface of our skin. These grow in bunches likes grapes, which reflects their naming from a Greek term referring to bunches of grapes.

The single circular chromosome that comprises the genome of most bacteria is relatively small. Bacterial chromosomes average in the range of 2 million to 4 million base pairs in size, compared with 200 million base pairs for a typical human chromosome. As you learned in Chapter 3, some bacteria also contain plasmids in addition to their chromosomal DNA. Plasmid DNA often contains genes that are not essential for life—for example, genes for antibiotic resistance and genes encoding proteins that form connecting tubes called *pili,* which allow bacteria to exchange DNA between cells (see Figure 2.1). But plasmid DNA is an essential tool for molecular biologists because it can be used to clone pieces of DNA in recombinant DNA experiments.

Bacteria grow and divide rapidly. Under ideal growth conditions, many bacterial cells divide every 20 minutes or so, whereas eukaryotic cells often grow for 24 hours or much longer before they divide. Therefore, under favorable growth conditions in the laboratory, a small population of bacteria can divide rapidly to produce millions of identical cells. Because bacteria are so small, millions of cells can easily be grown on small dishes of agar or in liquid culture media. When grown on culture plates, each bacterial cell typically divides to form circular colonies that contain thousands or millions of cells (**Figure 5.2a**). For many applications in biotechnology, bacteria are often grown in fermenters that can hold several hundred or thousand liters of liquid culture medium (Figure 5.2b).

It is also relatively easy to make mutant strains of bacteria that can be used for molecular and genetic studies. Mutants can be created by exposing bacteria to x-rays, ultraviolet light, and a variety of chemicals that mutate DNA (mutagens). Literally thousands of mutant strains are available. For these reasons and many more, bacteria are not only the favorite organisms of many microbiologists but also ideal model organisms for studies in molecular biology, genetics, biochemistry, and biotechnology.

Yeasts Are Important Microbes Too

Although the primary focus of this chapter is applications of bacteria in biotechnology, **yeasts** have served many important roles in biotechnology. In fact, archeologists have uncovered recipes on ancient Babylonian tablets from 4300 B.C. for brewing beer using yeast, which is one of the oldest documented applications of

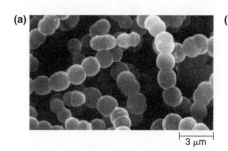
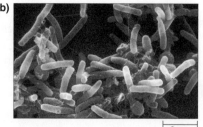
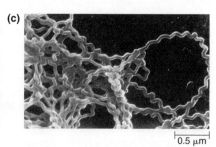

(a)   3 μm   (b)   2 μm   (c)   0.5 μm

FIGURE 5.1 **Shapes of Bacteria** The most common shapes of bacteria are (a) spherical, (b) rod-shaped, and (c) corkscrew-shaped.

(a)

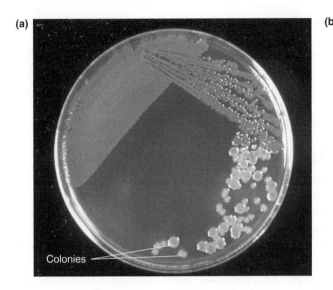

Colonies

(b)

FIGURE 5.2   **Bacteria in Culture** (a) Most bacteria can be grown under a variety of conditions in liquid culture media or on a solid medium such as in the petri dish shown here. On solid media, many bacteria grow in circular clusters called colonies, which typically begin with a single cell that divides rapidly to produce a spot visible to the naked eye. A single colony may contain millions of individual bacterial cells. (b) Bacteria can also be grown in large quantities. The fermenters shown here contain several hundred liters of bacteria growing in liquid culture broth. These fermenters function as bioreactors that serve many purposes, such as growing bacteria for isolating recombinant proteins and culturing yeast to produce wine.

biotechnology. Much more recently, yeast-producing recombinant human antibodies have been generated. Yeasts are single-celled eukaryotic microbes that belong to the **Kingdom Fungi.**

There are well over 1.5 million species of fungi, yet only around 10% of these have been identified and classified, so there is significant potential for identifying more valuable products from fungi. For instance, fungi are important sources of antibiotics and drugs that lower blood cholesterol. In addition to having many structures that are similar those of to other eukaryotic cells, such as plant and animal cells, yeasts also contain a number of membrane-enclosed organelles in the cytoplasm, a cytoskeleton, and chromosome structures similar to human chromosomes. Yeast cells also have larger genomes than most bacteria. These features make yeasts very valuable model organisms for studying chromosome structure, gene regulation, cell division, and cell-cycle control. Recall that yeast artificial chromosomes (YACs) played important roles in the Human Genome Project (Chapter 3).

Different types of yeast vary greatly in size, but a majority are larger than bacteria and spherical, elliptical, or cylindrical in shape. Many can grow in the presence of oxygen **(aerobic conditions)** or in the absence of oxygen **(anaerobic conditions)** and under a variety of nutritional growth conditions. A wide number of different types of yeast mutants are also available. *Saccharomyces cerevisiae,* a commonly studied strain of yeast, was the first eukaryotic organism to have its complete genome sequenced. It has 16 linear chromosomes that contain over 12 million base pairs of DNA and approximately 6,300 genes. Mechanisms of gene expression in yeast resemble those in human cells. Several human disease genes have also been discovered in yeast and, by studying them, scientists can learn a great deal about how these genes function in human and our evolutionary relatedness to yeast.

A strain of yeast called *Pichia pastoris* has proven to be a particularly useful organism. *Pichia* grows to a higher density (biomass) in liquid culture than most laboratory strains of yeast, has a number of strong promoters that can be used for high production of proteins, and can be used in **batch processes** to produce large numbers of cells. In the next section we consider how microorganisms can be used by scientists as valuable tools for biotechnology research.

5.2 Microorganisms as Tools

Microorganisms in either their natural state or genetically modified forms have served as useful tools in a variety of fascinating ways.

Microbial Enzymes

Microbial enzymes have been used in applications from food production to molecular biology research. Because microbes are an excellent and convenient source of enzymes, some of the first commercially available enzymes isolated for use in molecular biology were DNA polymerases and restriction enzymes from bacteria. Initially isolated primarily from *E. coli*, DNA polymerases became available for a range of recombinant DNA techniques such as labeling DNA sequences to make probes and using the polymerase chain reaction (PCR) to amplify DNA.

Taq is a heat-stable enzyme essential for PCR that was isolated from the hot-spring Archaean *Thermus aquaticus* (in Chapter 3, you were introduced to Taq DNA polymerase as a **thermostable enzyme**). Because of their ability to grow and thrive under extreme heat, these microbes are called **thermophiles** (from the Greek words meaning "heat-loving"). Many similar thermostable enzymes have been identified in other thermophiles. For example, Pfu DNA polymerase, a popular thermostable enzyme widely used for PCR, is derived from the thermophile *Pyrococcus furiosus,* a species of Archaea originally present in geothermally heated marine sediments. Several companies have permission from the U.S. government to prospect geysers in Yellowstone National Park to identify other potentially valuable microorganisms that might contain novel and valuable enzymes. Such so-called **bioprospecting** projects are occurring all around the world. Recently a strain of yet-to-be named bacteria was isolated from a hydrothermal vent on the floor of the northeast Pacific Ocean. This strain has been shown to survive at 121°C, which is believed to be the record so far for the upper temperature limit at which life can exist.

The enzyme **cellulase,** produced by *E. coli,* degrades cellulose, a polysaccharide that forms the plant cell wall. Cellulase is widely used to make animal food more easily digestible. Have you ever owned a pair of stone-washed denim jeans? These soft and faded jeans are not produced by a literal washing with stones. Instead, the denim is treated with a mixture of cellulases from fungi such as *Trichoderma reesei* and *Aspergillus niger.* These cellulases mildly digest cellulose fibers in the cotton used to make the pants, resulting in a softer fabric. The protease **subtilisin,** derived from *Bacillus subtilis,* is a valuable component of many laundry detergents, where it functions to degrade and remove protein stains from clothing. Several bacterial enzymes are also used to manufacture foods, such as carbohydrate-digesting enzymes called amylases that are used to degrade starches for making corn syrup.

Bacterial Transformation

Recall that *transformation*—the ability of bacteria to take in DNA from their surrounding environment—is an essential step in the recombinant DNA cloning process (Chapter 3). In DNA cloning, recombinant plasmids can be introduced into bacterial cells through transformation so that the bacteria can replicate the recombinant plasmids. Most bacteria do not take up DNA easily unless they are treated to make them more receptive, so-called **competent cells.** One technique for preparing competent cells involves treating cells with an ice-cold solution of calcium chloride. Positively charged atoms (cations) in the calcium chloride disrupt the bacterial cell wall and membrane to create small holes through which DNA can enter. These cells can then be frozen at ultralow temperatures (–80°C to –60°C) to maintain their competent state and then be used in the laboratory as needed.

Once competent cells are prepared, they can be transformed with DNA relatively easily, as shown in **Figure 5.3a.** Typically, the DNA to be introduced into bacteria is inserted in a plasmid containing one or more antibiotic-resistance genes. The recombinant plasmid vector is mixed in a tube with the competent cells and the mixture placed on ice for a few minutes. The exact mechanism of transformation is not fully understood, but we do know that the cells must be kept cold, during which time DNA sticks to the outer surface of the bacteria, and the cold conditions probably also serve to create gaps in the lipid structure of the cell membrane that allow for the entry of DNA. Cations in the calcium chloride are thought to play a significant role in neutralizing the negative charges of phosphates in the cell membrane and the DNA that would otherwise cause them to repel each other. The cells are then heated briefly (for about a minute) at temperatures between 37°C and 42°C. During this brief heat "shock," DNA enters the bacterial cells.

After these cells have been allowed to grow in culture broth, they can be plated onto an agar medium containing antibiotics. Only those cells that were transformed with plasmid DNA containing the appropriate antibiotic-resistant genes will grow to produce colonies. This technique is called *antibiotic selection* (see Chapter 3). Plasmid DNA is replicated (cloned) by

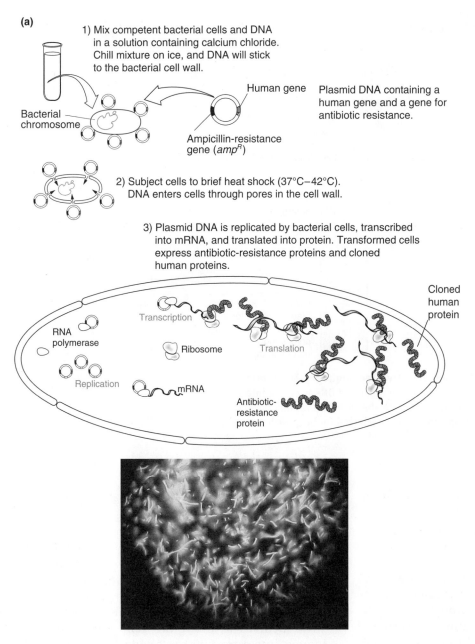

(a)

1) Mix competent bacterial cells and DNA in a solution containing calcium chloride. Chill mixture on ice, and DNA will stick to the bacterial cell wall.

Bacterial chromosome

Human gene

Ampicillin-resistance gene (*amp^R*)

Plasmid DNA containing a human gene and a gene for antibiotic resistance.

2) Subject cells to brief heat shock (37°C–42°C). DNA enters cells through pores in the cell wall.

3) Plasmid DNA is replicated by bacterial cells, transcribed into mRNA, and translated into protein. Transformed cells express antibiotic-resistance proteins and cloned human proteins.

RNA polymerase

Transcription

Ribosome

Translation

Replication

mRNA

Antibiotic-resistance protein

Cloned human protein

FIGURE 5.3 Transformation of Bacterial Cells (a) In a laboratory environment, most bacterial cells can be induced to take up foreign DNA by calcium chloride treatment. (b) *E. coli* transformed with a plasmid containing the jellyfish gene for green fluorescent protein (GFP). These bacterial cells glow bright green—a dramatic example of transformation—indicating that they have taken up plasmids containing the GFP gene and expressed GFP mRNA and protein.

transformed bacteria, and genes in the plasmid are transcribed and translated into protein. Thus, the transformed bacterial cells now express recombinant proteins.

This process is called transformation because one can "transform" the properties of bacterial cells by introducing foreign genes. Transformed cells have been genetically altered with new properties encoded by the DNA, enabling them to produce substances they would not normally produce. For example, *E. coli* can be transformed with a gene called green fluorescent protein (GFP), which comes from jellyfish (Figure 5.3b; we look more closely at useful applica-

tions of GFP in Chapter 10). As we will discuss in Section 5.3, bacterial cells have been transformed to express a large number of valuable genes including many human genes.

Electroporation

Another common technique for transforming cells is called **electroporation** (Figure 5.4, on page 128). In this approach, an instrument called an electroporator produces a brief electrical shock that introduces DNA into bacterial cells without killing most of them. Electroporation offers several advantages over the

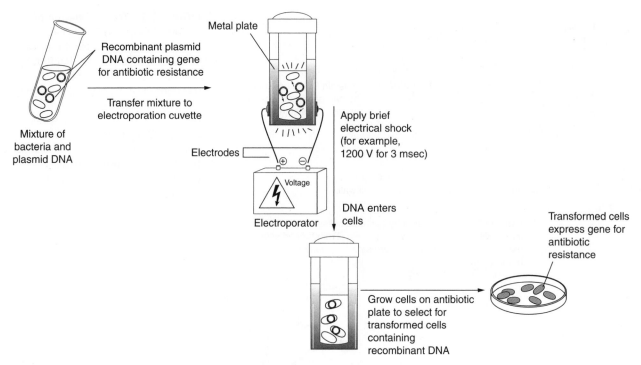

FIGURE 5.4 **Electroporation Is a Rapid and Effective Technique for Transforming Bacteria** In electroporation, a mixture of bacteria and plasmid DNA is placed into a cuvette. Upon the application of a brief electrical shock to the cuvette, DNA quickly enters cells. Cells containing recombinant DNA can then be selected for by growth on agar containing an antibiotic or another selection component.

calcium chloride treatment although competent cells are still required for electroporation. Electroporation is rapid, requires fewer cells, and can also be used to introduce DNA into many other cell types including yeast, fungi, animal, and plant cells. In addition, electroporation is a much more efficient process than calcium chloride transformation. A greater majority of cells will receive foreign DNA through electroporation than by calcium chloride treatment. Because of this, much less DNA can be used to transform cells (picogram amounts of DNA are sufficient). One drawback to this technique is that it is more costly than calcium chloride transformation because of the cost of an electroporator and competent cells that can tolerate electrical shock. Regardless of how bacterial cells are transformed, once the DNA of interest is introduced into bacteria, a variety of useful techniques can be carried out.

Cloning and Expression Techniques

In addition to replicating recombinant DNA, transformed bacteria are valuable because they can frequently be used to mass-produce proteins for a variety of purposes.

Creating bacterial fusion proteins to synthesize and isolate recombinant proteins

One popular technique for using bacteria for the synthesis and isolation of recombinant proteins involves making a **fusion protein.** There are a variety of ways to produce a fusion protein, but the basic concept of this technique is to use recombinant DNA methods to insert the gene for a protein of interest into a plasmid containing a gene for a well-known protein that serves as a "tag" (**Figure 5.5**). The tag protein then allows for the isolation and purification of the recombinant protein as a fusion protein. Plasmid vectors for making fusion proteins are often called **expression vectors** because they enable bacterial cells to produce or express large amounts of protein. Commonly used expression vectors include those that synthesize proteins such as maltose-binding protein (shown in Figure 5.5), glutathione S-transferase, luciferase, green fluorescent protein, and β-galactosidase.

Expression vectors incorporate prokaryotic promoter sequences so that, once the recombinant expression vector containing the gene of interest is introduced into bacteria by transformation, bacteria synthesize

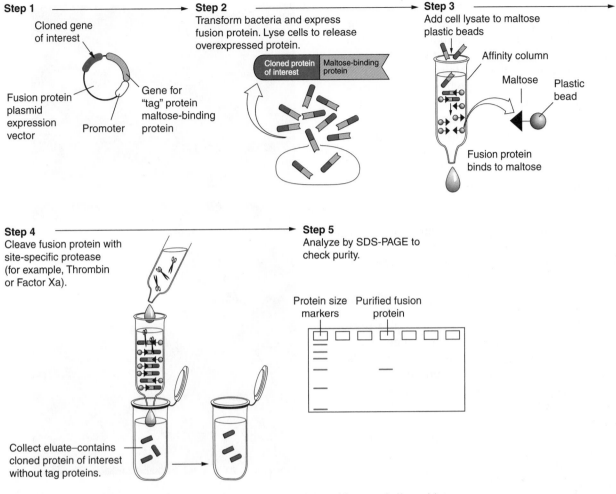

Step 1
Cloned gene of interest

Fusion protein plasmid expression vector

Promoter

Gene for "tag" protein maltose-binding protein

Step 2
Transform bacteria and express fusion protein. Lyse cells to release overexpressed protein.

Cloned protein of interest | Maltose-binding protein

Step 3
Add cell lysate to maltose plastic beads

Affinity column

Maltose | Plastic bead

Plastic bead

Fusion protein binds to maltose

Step 4
Cleave fusion protein with site-specific protease (for example, Thrombin or Factor Xa).

Collect eluate–contains cloned protein of interest without tag proteins.

Step 5
Analyze by SDS-PAGE to check purity.

Protein size markers | Purified fusion protein

FIGURE 5.5 **Fusion Proteins** To make a fusion protein, a gene of interest is ligated into an expression vector. Bacterial cells are transformed with recombinant DNA. Transformed cells express the fusion protein, which is isolated by binding it to an affinity column.

mRNA and protein from this plasmid. The mRNA strands transcribed are hybrid molecules that contain sequence coding for the gene of interest and the tag protein. As a result, a fusion protein is synthesized from this mRNA consisting of the protein of interest joined to (fused) to the tag protein, in this case maltose-binding protein (Figure 5.5, step 2).

To isolate the fusion protein and separate it from other proteins normally made by the bacteria, cells are broken open (lysed) and homogenized to create a bacterial milkshake of sorts known as an *extract*. The extract is then passed through a tube called a *column*. One common approach is to fill a column with plastic beads coated with molecules that will bind to the tag protein portion of the fusion protein. This technique is called **affinity chromatography** because the beads in the column have an attraction, or "affinity," for binding to the tag protein. For

example, in Figure 5.5, plastic beads are attached to the sugar maltose, which will be bound by maltose-binding protein. Next, an enzyme treatment that uses protein-cutting enzymes called *proteases* cuts off and releases the protein of interest from the tag protein. Some techniques for making fusion proteins incorporate short peptide tags of just a few amino acids. For example, poly-His tags are a short string of the amino acid histidine. One benefit of this approach is that, unlike maltose-binding protein and other tags that are large proteins, the small tags typically don't affect the structure and function of the protein they are fused to, so they usually don't need to be removed. Fusion protein techniques are used to provide purified proteins to study protein structure and function and used to isolate insulin and other medically important recombinant proteins (see Figure 5.8 later in the chapter).

E. coli and the Gram-negative rod-shaped bacterium *Bacillus subtilis* are commonly used microbes for producing fusion proteins. In particular, *B. subtilis* is a favored microbe for many applications when producing fusion proteins for human proteins because it will secrete them into growth media where they can easily be harvested and purified. And unlike some bacteria, *B. subtilis* often processes proteins in such a way as to maintain their three-dimensional folding and function.

Microbial proteins as reporters

According to recent estimates, close to three-fourths of all marine organisms can release light through a process known as **bioluminescence.** For marine fish, bioluminescence in lines of cells along the side of a fish and in its fins can be used to attract mates in dark ocean environments. Bioluminescence in many marine species is created by bacteria such as *Vibrio fisheri* that use the marine organism as a host (**Figure 5.6**). Bacteria such as *Vibrio* have been used as biosensors to detect cancer-causing chemicals called carcinogens, environmental pollutants, and chemical and bacterial contaminants in foods. *Vibrio fisheri* and another marine bioluminescent strain called *Vibrio harveyi* create light through the action of genes called *lux* genes. Several *lux* genes encode protein subunits that form an enzyme called **luciferase** (derived from the Latin *lux ferre,* meaning "light bearer").

Luciferase is the same enzyme that allows fireflies to produce light. The *lux* genes have been cloned and used to study gene expression in a number of unique ways. For instance, by cloning *lux* genes into a plasmid, the *lux* plasmid can be used to produce fusion proteins. Also, *lux* genes can serve as valuable **reporter genes.** If inserted into animal or plant cells, the luciferase encoded by the *lux* plasmid cause these cells to fluoresce (Figure 5.6). In this manner, the *lux* plasmid is acting as a "reporter" to provide a visual indicator of gene expression.

Lux genes have been used to develop a fluorescent bioassay to test for tuberculosis (TB). TB is caused by the bacterium *Mycobacterium tuberculosis,* which grows slowly and can exist in a human for several years before TB develops (TB symptoms are discussed in Section 5.4). For the TB bioassay, scientists introduced *lux* genes into a virus that infects *M. tuberculosis.* Saliva from a patient who may be infected with *M. tuberculosis* is mixed together with the *lux*-containing virus. If *M. tuberculosis* is in the saliva sample, the virus infects these bacterial cells, which can be detected by their glowing. Similarly engineered strains of *E. coli* have been used to detect arsenic contamination in

(a)

(b)

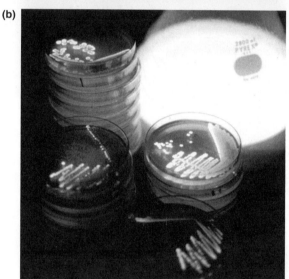

(c) The light-releasing chemical reaction catalyzed by luciferase

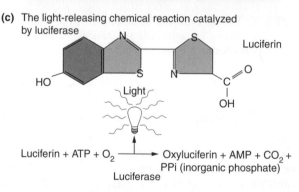

FIGURE 5.6   Bioluminescent Marine Bacteria   Bioluminescent marine bacteria, such as *Vibrio fisheri* shown (a) glowing in the light-releasing organs of a deep-sea fish and (b) growing in the laboratory, generate light. (c) *Lux* genes encode the enzyme luciferase that uses oxygen and stored energy (ATP) to convert luciferin into oxyluciferin. This reaction releases light. *Lux* genes serve as reporter genes, allowing biologists to study gene expression. Expression is indicated by glowing cells.

water. Reporter genes have many valuable roles in research and medicine (Chapter 10). In the next section we explore a range of everyday applications that involve microbes.

5.3   Using Microbes for a Variety of Everyday Applications

Harnessing the great potential of microbes for making foods and for developing and producing new drugs are among the most common everyday applications of microbial biotechnology.

Food Products

Microbes are used to make many foods, including breads, yogurt, cheeses, and sauerkraut as well as beverages such as beer, wine, champagne, and liquors. As a child you probably learned the classic nursery rhyme "Little Miss Muffet." The tale of Miss Muffet, sitting on her tuffet, eating "curds and whey" might seem like an

---

TOOLS OF THE TRADE

The Yeast Two-Hybrid System

The **yeast two-hybrid system** is an innovative technique for studying proteins that interact with each other; it provides a way of understanding protein function. Suppose you identified several proteins that you believed might interact and work together as part of a metabolic pathway for the synthesis of an important hormone in the body. The yeast two-hybrid system is an excellent technique to use to determine if these proteins do, in fact, bind to and interact with each other.

As shown in this figure, the gene for one protein of interest is cloned and expressed as a fusion protein attached to the DNA binding domain (DBD) of another gene. These DNA-binding sites are commonly found on proteins, such as transcription factors that interact with DNA. This protein is often called the "bait" protein because it is used to find other proteins called "prey" that may bind to it. The gene for the second protein of interest is fused to another gene that contains transcriptional activator domain (AD) sequences. AD proteins, also required for the attachment of DNA-binding proteins to DNA, stimulate the binding domain of a bait protein to interact with DNA-binding sites such as a promoter sequence for a reporter gene.

This modified DNA is introduced into yeast cells such as *Saccharomyces cerevisiae*. Neither the DBD fusion protein nor the AD fusion protein alone can stimulate transcription from this reporter gene. In the figure, the *lacZ* gene, which encodes the enzyme β-galactosidase, is a common reporter gene whose activity can easily be measured using simple colorimetric tests. Only a combination of both proteins, a "hybrid" or complex of both proteins, will stimulate expression of the reporter gene. Because scientists can create mutations of both proteins of interest and see how these mutations affect the ability of the two proteins to interact, we can learn a lot about the structure and function of those proteins.

The yeast two-hybrid system is a powerful example of how microbial biotechnology can be used for a research application.

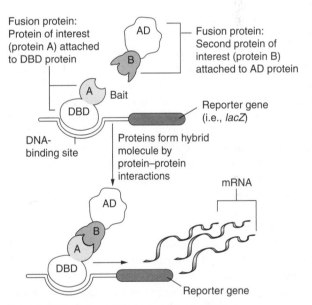

Hybrid protein can now stimulate transcription of reporter gene. Activity of protein from reporter gene mRNA can be measured.

The Yeast Two-Hybrid System for Studying Protein Interactions If a researcher wanted to know if two proteins (A and B) interacted with one another, one protein could be expressed as a fusion protein attached to the DNA binding domain (DBD) of another protein (such as a transcription factor) and the second protein attached to an activator domain (AD) protein. Neither fusion protein alone would be capable of binding to and stimulating transcription (mRNA production) of the reporter gene. But the resulting hybrid protein created by protein–protein interactions of the two fusion proteins would stimulate transcription of the reporter gene, which could easily be measured as an indicator of protein interactions between the A and B proteins.

improbable way to discuss biotechnology, but the treat in Miss Muffet's bowl was the result of biotechnology! Curds and whey are formed from coagulated milk, and milk coagulation is an important step during cheese production. To make cheese, milk from cows, goats, or sheep is treated to help it coagulate (*curd*). The watery liquid that remains after curd forms is called *whey*.

One way to make cheese from coagulated milk is to treat the milk with an acidic solution, but the best-tasting cheeses are typically made using the enzyme **rennin.** In the early days of cheese production, rennin was traditionally obtained by extracting it from the stomachs of calves and other milk-producing species such as goats, sheep, horses, and even zebras and camels. Rennin coagulates milk to produce curd by digesting a family of proteins called *casein,* which is a major component of milk. Digested casein forms an insoluble mixture of proteins that clumps (coagulates) in a process similar to what happens when milk spoils.

In the 1980s, using recombinant DNA techniques, scientists cloned rennin and expressed it in bacterial cells and fungi such as *Aspergillus niger*. Recombinant rennin (now called **chymosin**) from microbes is widely used by cheese makers as an inexpensive substitute for extracting rennin from calves. In 1990, rennin was the first recombinant DNA food ingredient approved by the Food and Drug Administration (FDA). For some types of cheese, certain strains of bacteria called lactic acid bacteria (*Lactococcus lactis, L. acidophilus*) are used for coagulation. These bacteria degrade casein and use an enzyme called *lactase* to break down sugars in the milk that are eventually used by bacteria for **fermentation.**

Fermenting microbes

Fermentation is an important microbial process that produces many food products and beverages including a variety of breads, beers, wines, champagnes, yogurts, and cheeses. One of the earliest applications of microorganisms—the brewing of beer and wine—involves fermentation by yeast (**Figure 5.7a**). To appreciate how making beer, wine, and bread requires microbes, you need to know more about the process of fermentation.

Animal and plant cells and many microbes obtain energy from carbohydrates such as glucose by using electrons from these sugars to create a molecule called **adenosine triphosphate (ATP).** ATP production occurs as a series of reactions. The first major reaction, **glycolysis,** converts glucose into two molecules of *pyruvate*. During this conversion, electrons are transferred from glucose to electron carrier molecules called *NAD+* (nicotinamide adenine dinucleotide), which capture electrons to produce NADH (Figure 5.7b). This molecule transports electrons to subsequent reactions in the process that results in ATP production. For certain

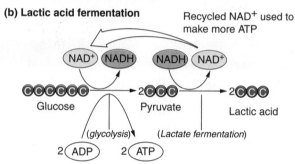
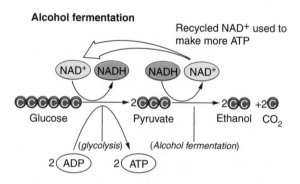

FIGURE 5.7 Fermentation Certain yeasts and bacteria are capable of producing energy from sugars (glucose) through fermentation. (a) Yeast such as *S. cerevisiae* (left), shown here, causes bread dough to rise; other strains of *Saccharomyces* grow on grapevines and are important for making wine (center photo). (b) Anaerobic bacteria that undergo lactic acid fermentation make lactic acid (lactate) as a waste product; alcohol-fermenting bacteria create ethanol and carbon dioxide (CO_2) as waste products.

bacteria and yeasts, oxygen is an important part of these electron transport reactions. Microbes that use oxygen for ATP production are called **aerobes** because they undergo oxygen-dependent (aerobic) metabolism.

Many microbes live in areas where oxygen is rare or absent, such as the intestines of animals, deep water, or soil. Because these organisms must survive without oxygen, they have evolved the ability to derive energy from sugars in the absence of oxygen (anaerobic conditions). This is fermentation, and microbes that use fermentation are called **anaerobes.** Fermentation is similar to glycolysis in that NAD+ is used to capture electrons to make NADH and pyruvate; however, neither the NADH nor the pyruvate has anywhere to go. In aerobic metabolism, oxygen is required to use electrons from NADH and pyruvate to make ATP, but under anaerobic conditions there is little or no oxygen, so the NADH and pyruvate cannot be used to make ATP. All organisms must recycle NADH into NAD⁺. Fermentation enables many anaerobes to do this in the absence of oxygen, and some anaerobes are capable of either fermentation or aerobic respiration depending on the presence or absence of oxygen. In the absence of oxygen, anaerobes have evolved to acquire fermentation reactions as a way to solve the problem of recycling NADH into NAD⁺. Fermenting microbes use pyruvate as an electron acceptor molecule to take electrons from NADH and thus to regenerate NAD⁺ (Figure 5.7b). Two of the most common types of fermentation are **lactic acid fermentation** and **alcohol (ethanol) fermentation.**

In lactic acid fermentation, electrons from NADH are used to convert pyruvate into lactic acid, also called lactate; during alcohol fermentation, electrons from NADH convert pyruvate into ethanol. NAD⁺ is regenerated when electrons are removed from NADH and transferred to pyruvate to make lactate or ethanol in the final step of fermentation. During alcohol fermentation, carbon dioxide gas is also produced as a waste product.

There are many strains of fermenting bacteria and yeast. Other types of fermentation create sauerkraut from cabbage and produce such useful products as acetic acid in vinegar, citric acid in fruit juices, and acetone and methanol, two chemicals often used in laboratories for cleaning glassware. Furthermore, these microbial products have the advantage of being produced more efficiently and cheaply than by other means. Lactic acid fermentation also occurs in human muscle cells during strenuous exercise. The burning you feel in your legs when you run because you are late to class is created by fermentation in muscle cells, creating lactic acid.

So how are fermenting microbes used to make foods and beverages? To make beer and wine, many processes rely on wild strains of yeast that live on grapevines and on such domestic strains of yeast as *Saccharomyces cerevisiae*, which are very good at alcohol fermentation. Large barrels, or fermenters, containing crushed grapes and yeast are mixed together under carefully controlled conditions. Fermenting yeast converts sugars from the grapes into alcohol. Fermentation rates are monitored and carefully controlled by changing both the amount of oxygen in the fermenter and the temperature.

By manipulating fermentation rates, wine makers can control the alcohol content of the brewing wine until the desired alcohol content and flavor are achieved.

Bottles of champagne and other sparkling wines are capped while the yeast is still in the liquid and actively fermenting, thereby trapping carbon dioxide gas in the bottle and releasing it when the bottle is opened, producing the characteristic champagne bottle "pop." Production of some wines relies also on lactic acid bacteria such as *Oenococcus oeni,* which will convert bitter-tasting malic acid that is formed during fermentation into softer-tasting lactic acid and thus give the wine a milder flavor and aroma.

Lactic acid–fermenting bacteria are used to produce cheeses, sour cream, and yogurts; the sharp or sour flavors of these products is primarily due to lactic acid. Yogurt was first created in Bulgaria and has been around for centuries. Yogurt production typically involves a blend of bacteria that often includes strains of anaerobic lactic acid–fermenting microbes such as *Streptococcus thermophilus,* a strain called *Lactobacillus* (*Lactobacillus delbrueckii* and *Lactobacillus bulgaricus*), and another strain called *Lactococcus* (*Lactococcus lactis*). Active cultures of these lactic acid–fermenting microbes are added to mixtures of milk and sugar in a fermenter, which is maintained at carefully controlled temperatures. Microbes in the mixture use sugars to produce lactic acid. Fruit and other flavorings may then be added to the yogurt before it is cooled to refrigeration temperature (4°C–5°C) to prevent changes in its composition. When you enjoy a spoonful of yogurt, you are also eating large numbers of fermenting microbes that are still in the yogurt. Lactic acid and other products of fermentation contribute to the sweet and sour taste of yogurt and assist in the coagulation of the yogurt.

Another lactic acid bacterium, *Lactobacillus sakei,* is found naturally on fresh meat and fish. *L. sakei* serves as a natural biopreservative in meat products such as sausage, where it wards off growth of other undesirable microbes that will spoil the food or cause illness. In addition to lactic acid, this microbe produces molecules called *bacteriocins,* which act as naturally occurring antimicrobial agents to kill other microbes.

Therapeutic Proteins

The development of recombinant DNA technology quickly led to using bacteria to produce such important medical products as therapeutic proteins. Insulin was the first recombinant molecule expressed in bacteria for use in humans. Here we use insulin production as an example of how microbes can be used to make therapeutic proteins.

Producing recombinant insulin in bacteria

Insulin is a hormone produced by cells in the pancreas called *beta cells.* When insulin is secreted into the bloodstream by the pancreas, it plays an essential role in carbohydrate metabolism. One of its primary functions is to stimulate the uptake of glucose into body cells such as muscle cells, where the glucose can be broken down to produce ATP as an energy source. **Type I,** or **insulin-dependent, diabetes mellitus** occurs when beta cells do not produce insulin. The lack of insulin results in an elevated blood glucose concentration, which can cause a number of health problems such as high blood pressure, poor circulation, cataracts, and nerve damage. People with type I diabetes require regular injections of insulin to control their blood sugar levels.

Prior to recombinant DNA technology, insulin used to treat diabetes was purified from the pancreases of pigs and cows before being injected into diabetics. The purification process was cumbersome, expensive, and often produced impure batches of insulin. Also, purified insulin was ineffective in some individuals, and many others developed allergies to insulin from cows. In 1978, insulin was cloned into an expression vector plasmid, expressed in bacterial cells, and isolated by

scientists at Genentech, a biotechnology company near San Francisco, California. In 1982, this recombinant human insulin, called Humulin, was the first biotechnology product to be approved for human applications by the U.S. Food and Drug Administration (FDA).

Bacteria do not normally make insulin, and producing human insulin in recombinant bacteria was a significant advance in biotechnology. It remains an outstanding example of microbial biotechnology in action. Human insulin consists of two polypeptides called the A (21 amino acids) and B (30 amino acids) chains or subunits; they bind to each other by disulfide bonds to create the active hormone (**Figure 5.8**). In the pancreas, beta cells synthesize both insulin chains as one polypeptide that is secreted and then enzymatically cut (cleaved) and folded to join the two subunits. When the human genes for insulin were cloned and expressed in bacteria, genes for each of the subunits were cloned into separate expression vector plasmids containing the *lacZ* gene encoding the enzyme β-galactosidase (β-gal) and then used to transform bacteria (Figure 5.8).

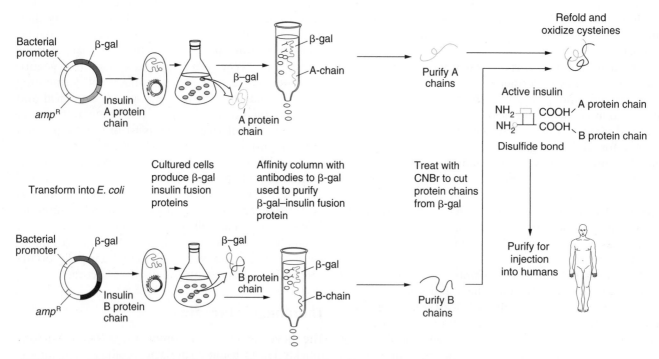

FIGURE 5.8 Using Bacteria for the Production of Human Insulin Insulin was the first protein expressed in recombinant bacteria to be approved for use in humans. Insulin consists of two protein chains (A and B) produced from separate genes. To make recombinant insulin, scientists cloned the insulin genes into plasmids containing the *lacZ* gene, which encodes the enzyme β-galactosidase. Recombinant plasmids were used to transform bacteria, enabling them to produce β-gal–insulin fusion proteins. Affinity chromatography was used to isolate fusion proteins, which were then chemically treated to separate the cloned insulin from β-gal proteins. Purified forms of the A and B protein chains of insulin could then combine to form active insulin, which is given to people with diabetes to control blood sugar levels.

Because the insulin genes are connected to the *lacZ* gene, when bacteria synthesize proteins from these plasmids, they produce a protein that contains β-gal attached to the human insulin protein to create a β-gal–insulin fusion protein. As we saw in Section 5.2, making a fusion protein enables scientists to isolate and purify a protein of interest such as insulin. Bacterial extracts are passed over an affinity column to isolate the β-gal–insulin fusion proteins (Figure 5.8). The fusion protein is chemically treated to cleave off the β-gal, releasing the insulin protein; then, purified A and B chains of insulin are mixed together under conditions that allow the two subunits to bind and form the active hormone. After further purification to conform to FDA guidelines, the recombinant hormone is ready for patient use as an injectable drug.

Shortly after insulin became available, growth hormone—used to treat children who suffer from a form of dwarfism—was cloned in bacteria and became available for human use. A short time later, a wide variety of other medically important proteins that were once difficult to obtain became readily available as a result of recombinant DNA technology and expressing proteins in bacteria. As shown in Table 5.1, many other therapeutic proteins with valuable applications for treating medical illness in humans have been expressed in and isolated from bacteria. A major category of medical products from recombinant bacteria are vaccines. We cover vaccines in Section 5.4.

Using Microbes against Other Microbes

Antibiotics are substances produced by microbes that inhibit the growth of other microbes. Antibiotics are a type of **antimicrobial drug,** a general category defined as any drug (whether produced by microbes or not) that inhibits microorganisms. Penicillin was the first antibiotic to be used widely in humans, and its discovery is an excellent example of how some microbes protect themselves from others by making

TABLE 5.1   THERAPEUTIC PROTEINS FROM RECOMBINANT BACTERIA

Protein	Function	Medical Application(s)
DNase	DNA-digesting enzyme	Treatment of patients with cystic fibrosis.
Erythropoietin	Stimulates production of red blood cells	Used to treat patients with anemia (low number of red blood cells).
Factor VIII	Blood clotting factor	Used to treat certain types of hemophilia (bleeding diseases due to deficiencies in blood clotting factors).
Granulocyte colony-stimulating factor	Stimulates growth of white blood cells	Used to increase production of certain types of white blood cells; stimulate blood cell production following bone marrow transplants.
Growth hormone (human, bovine, porcine)	Hormone stimulates bone and muscle tissue growth	In humans, used to treat individuals with dwarfism. Improves weight gain in pigs and cows; stimulates milk production in cows.
Insulin	Hormone required for glucose uptake by body cells	Used to control blood sugar levels in patients with diabetes.
Interferons and interleukins	Growth factors that stimulate blood cell growth and production	Used to treat blood cell cancers such as leukemia; improve platelet counts; some used to treat different cancers.
Superoxide dismutase	An antioxidant that binds and destroys harmful free radicals	Minimizes tissue damage during and after a heart attack.
Tissue plasminogen activator (tPA)	Dissolves blood clots	Used to treat patients after heart attack and stroke.
Vaccines (e.g., hepatitis B vaccine)	Stimulate the immune system to prevent bacterial and viral infections	Used to immunize humans and animals against a variety of pathogens; also used in some cancer tumor treatments.

TABLE 5.2   COMMON ANTIBIOTICS

Antibiotic	Source Microbe	Common Uses of Antibiotic
Bacitracin	*Bacillus subtilis* (bacterium)	First-aid ointment and skin creams
Erythromycin	*Streptomyces erythraeus* (bacterium)	Broad uses to treat bacterial infections, especially in children
Neomycin	*Streptomyces fradiae* (bacterium)	Skin ointments and other topical creams
Penicillin	*Penicillium notatum* (fungus)	Injected or oral antibiotic used in humans and farm animals (cattle and poultry)
Streptomycin	*Streptomyces griseus* (bacterium)	Oral antibiotic used to treat many bacterial infections in children
Tetracycline	*Streptomyces aureofaciens* (bacterium)	Used to treat infections of the urinary tract in humans; commonly used in animal feed to reduce infections and stimulate weight gain

antimicrobial substances. Alexander Fleming was the microbiologist who, in 1928, discovered that colonies of the mold *Penicillium notatum* inhibited growth of the bacterium *Staphylococcus aureus.* When cultured together on a petri dish, *S. aureus* would not grow in a small zone of agar surrounding mold colonies. A dozen years later, scientists used *P. notatum* to isolate the drug they called penicillin, which was subsequently mass-produced and used to treat bacterial infections in humans.

A majority of antibiotics are isolated from bacteria, and most of these substances inhibit the growth of other bacteria. In the over 60 years since penicillin was discovered, thousands of other antibiotic-producing microbes have been discovered, and hundreds of different antibiotics have been isolated. Table 5.2 shows examples of common antibiotics and their source microbes.

How do antibiotics and other antimicrobial drugs affect bacterial cells? Most of these substances act in a few key ways. Typically they either prevent bacteria from replicating or kill microbes directly, which of course also prevents affected cells from replicating. Antibiotics can damage the cell wall or prevent its synthesis (which is how penicillin acts), block protein synthesis, inhibit DNA replication, or inhibit the synthesis or activity of an important enzyme required for bacterial cell metabolism (**Figure 5.9**).

You sneeze, your body aches, your nose is running, your throat is sore, and you can't sleep, but you still have an exam to take tomorrow afternoon. How will you do it? By going to your doctor and asking for antibiotics of course! But are antibiotics what you really need? Because antibiotics are effective only against

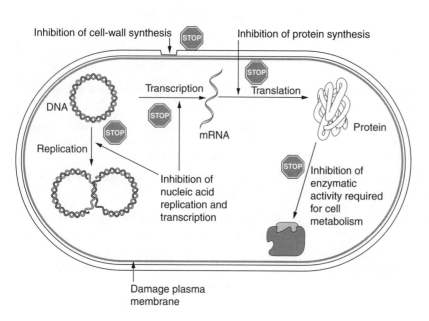

FIGURE 5.9   Antibiotics and Other Antimicrobial Drugs Work against Microbes in a Variety of Ways

bacteria; they do not work against viruses such as those that cause flu. Also, bacterial resistance to antibiotics has become a major problem. Improper use and overuse of antibiotics in humans and farm animals has led to dramatic increases in antibiotic-resistant bacteria, including some strains that do not respond at all to many antibiotics that were effective in the past.

Antibiotic-resistant strains of *S. aureus, Pseudomonas aeruginosa, Streptococcus pneumoniae, M. tuberculosis,* and many other deadly strains of human pathogens have already been detected in hospitals. Because most antibiotics attack a bacterial cell in a limited number of ways (Figure 5.9), resistance to one antibiotic often leads to resistance to many other drugs. Consequently, new antimicrobial drugs that are harmful to bacteria in different ways need to be developed for medical use as well as for treating food animals such as cows, pigs, and chickens. Marine microbiologists are discovering new strains of ocean microbes with novel antibiotics and anticancer compounds. From treetops to polar ice caps, deserts, and the ocean's depths, scientists are bioprospecting many microorganisms as potential sources of new antimicrobial substances.

Another way to create new antimicrobial drugs is to study bacterial pathogens and identify toxins and properties that disease-causing bacteria use to create illness. By understanding the factors involved in causing illness, scientists can develop new strategies to block bacterial replication. For instance, for certain bacteria, their ability to cause disease requires that they attach (adhere) to human tissues. Once attached, bacterial cells can multiply and then produce sufficient toxins to cause illness.

A Florida company called Oragenics received FDA approval to begin clinical trials of a recombinant form of *Streptococcus mutans* to evaluate its safety and efficacy for reducing tooth decay. Unlike naturally occurring strains of *S. mutans* found in the oral cavity, the recombinant strain cannot metabolize sugars to produce lactic acid. Because lactic acid dissolves enamel and dentin in teeth, it leads to cavities and tooth decay. Scientists will attempt to use recombinant *S. mutans* to colonize the oral cavity and replace natural strains of *S. mutans* and then determine if tooth decay is reduced. Recently, scientists at the University of California–Los Angeles (UCLA) created a sugar-free lollipop containing an ingredient in licorice that kills *S. mutans,* and these bacteria-killing lollipops are now available for purchase.

5.4 Vaccines

Antibiotics and vaccines have proven to be very effective for treating infectious disease conditions in humans and animals caused by **pathogens**—disease-causing microorganisms (**Figure 5.10**). However, pathogens with resistance to widely used antibiotics and vaccines have emerged and challenge the effectiveness of vaccines and antibiotics. Infectious diseases caused by microbes affect everyone, and worldwide over 60% of the causes of death among children before age 4 are due to infectious diseases. Without question our ability to prevent, detect, and treat infectious diseases is an important aspect of microbial biotechnology, and vaccines play a key role in this process.

The world's first vaccine was developed in 1796 when Edward Jenner demonstrated that a live cowpox virus could be used to vaccinate humans against smallpox. Smallpox and cowpox are closely related viruses. Smallpox epidemics ravaged areas of Europe, and an estimated 80% or more of Native Americans on the East Coast of the United States died from smallpox infections carried by European settlers in North America. Cowpox produces blisters and lesions on the udders of cows and produces similar skin ulcers in humans. Based on a milkmaid's claim that cowpox infections protected her from smallpox, Jenner prepared his vaccine. He took fluid from cowpox blisters on the milkmaid and used needles containing this fluid to scratch the skin of healthy volunteers. His first "patient" was an 8-year-old boy. A majority of Jenner's volunteers did not develop cowpox or smallpox even when subsequently exposed to persons infected

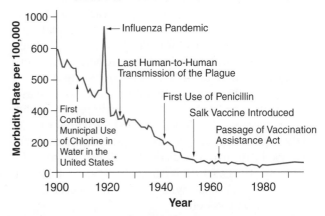

Crude Death Rate for Infectious Diseases

FIGURE 5.10 **The Use of Antibiotics and Vaccines to Combat Infectious Diseases Caused by Microorganisms** Even though the use of antibiotics and vaccines has decreased the incidence of human illness caused by microorganisms in the United States, strains of microbes that show resistance to many popular antibiotics and vaccines are emerging. New antibiotics and vaccines are required to fight these microbes.

*The American Society for Microbiology Report: Congressional Briefing. Infectious Disease Threats, 2001.

with smallpox. Exposure to cowpox fluid had stimulated the immune system of Jenner's volunteers to develop protection against smallpox.

These experiments demonstrated the potential of **vaccination** (named from the Latin word *vacca,* meaning "cow")—using infectious agents to provide immune protection against illness. Although the United States stopped routine vaccinations for smallpox in 1972, by 1980, subsequent widespread applications of the vaccine had eradicated this disease. In the United States many vaccines are routinely given to newborns, children, and adults. Although you may not remember your first vaccination (which usually occurs sometime from 2 to 15 months of age), you were probably vaccinated with the **DPT vaccine,** which provides several years of protection against three bacterial toxins called *d*iphtheria toxin, *p*ertussis toxin, and *t*etanus toxin. Diphtheria can cause breathing problems, paralysis, and heart failure. Pertussis causes whooping cough, which involves episodes of coughing paralysis so severe that it becomes hard for infants to eat, drink or breathe. Tetanus can cause lockjaw, preventing opening of the mouth.

Another childhood vaccine is the **MMR** (*m*easles-*m*umps-*r*ubella) vaccine. As you may know, measles results in a rash, fever, and cough; it can also cause a variety of other complications such as ear infections, seizures, and even death. Mumps causes fever, swollen glands, and headaches, but it can also lead to deafness. Rubella, or German measles, causes a fever, rash, and arthritis.

You were probably also vaccinated with **OPV** (*o*ral *p*olio *v*accine) for the poliovirus, a strain that infects neurons in the spinal cord, causing a devastating paralysis called poliomyelitis (polio). The OPV has dramatically decreased the incidence of polio. It has virtually been eliminated in North America, South America, and most of Europe; however, it still exists in some areas of the world. Polio, once a much more common disease, ravaged millions of children worldwide prior to 1954, when Jonas Salk developed the first vaccine for polio. Salk's original vaccine required injection; Albert Sabin developed the current version, which can be taken by mouth, in 1961. To understand how vaccines work, you must be familiar with the basic aspects of the human immune system.

A Primer on Antibodies

The immune system in humans and other animals is extremely complex. Numerous cells throughout the body work together in intricate ways to recognize foreign materials that have entered our body and mount an attack to neutralize or destroy those materials. Foreign substances that stimulate an immune response are called **antigens.** They may be whole bacteria, fungi, and viruses or individual molecules such as proteins or lipids found on pollen. For instance, people with food allergies have immune responses to proteins, carbohydrates, and lipids in certain foods.

The immune system typically responds to antigens in part by producing antibodies. This response is called **antibody-mediated immunity (Figure 5.11).** When

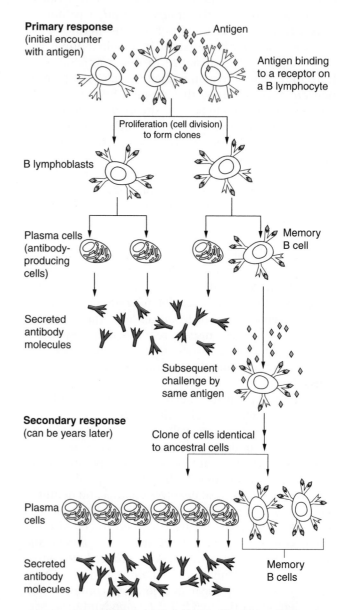

FIGURE 5.11 Antigens Stimulate Antibody Production by the Immune System In response to the initial antigen exposure, which comprise whole cells or individual molecules, B cells divide repeatedly to form many other B cells (clones). B cells differentiate into plasma cells, which produce antibodies specific to the antigen. During this process, memory B cells are formed. If a person is exposed to the same antigen again in the future, even many years later, memory B cells can recognize and produce a stronger and more rapid response to the antigen.

they are exposed to antigens, **B lymphocytes** (simply called **B cells**), which are a type of white blood cell or **leukocyte,** recognize and bind to antigen. **T lymphocytes (T cells)** play essential roles in helping B cells recognize and respond to antigen. After antigen exposure, B cells develop to form **plasma cells,** which produce and secrete antibodies. Most antibodies are released into the bloodstream, but there are also antibodies in saliva, tears, and the fluids lining the digestive system, among others.

Another purpose of antibody production is to provide lasting protection against antigens. During the process of B-cell development, some B cells become "memory" cells, which have the ability to recognize foreign materials years later and in response grow and produce more plasma cells and antibodies, which provide the body with long-term protection against antigens (Figure 5.11).

Antibodies are very specific for the antigen for which they were made, but how do these proteins protect the body against foreign materials? Many antibodies bind to and coat the antigen for which they were made (**Figure 5.12**). After antigens are covered with antibodies, a type of leukocyte called a **macrophage** can often recognize them. Macrophages are cells that are very effective at phagocytosis (which literally means "cell eating," derived from the Greek terms *phago,* "eating," and *cyto,* "cell"). In phagocytosis, macrophages engulf antigen covered with antibody; then, organelles in the macrophage called *lysosomes* unleash digestive enzymes that degrade the antigen (Figure 5.12). When the antigen is a foreign cell such as a bacterium, some antibodies are involved in mechanisms that rupture the cell through a process called *cell lysis.*

We are constantly being exposed to antigens, against which our immune system develops antibodies. But sometimes our natural production of antibodies is not sufficient to protect us from pathogens such as smallpox, viruses that cause hepatitis, and **human immunodeficiency virus (HIV),** the cause of **acquired immunodeficiency syndrome (AIDS).**

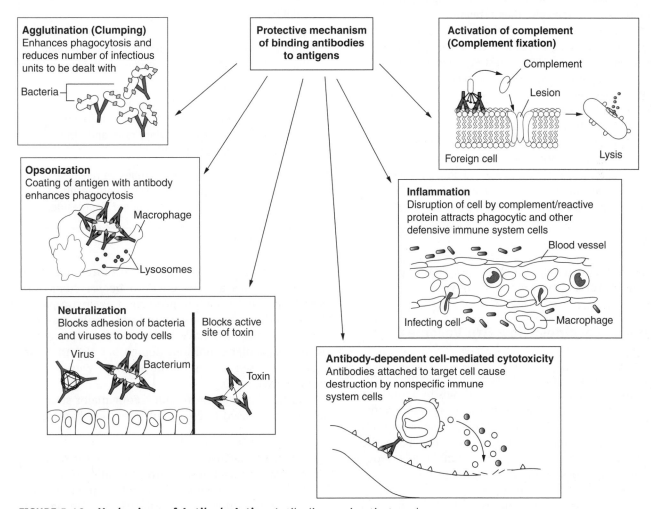

FIGURE 5.12 Mechanisms of Antibody Action Antibodies can inactivate and destroy antigens in a number of ways.

YOU DECIDE

Microbes on the Loose

One of the many controversies surrounding microbial biotechnology is the prospect that recombinant microbes can enter the environment, through accidental introduction or intentional release, as in the ice-minus studies described below. If recombinant microbes are loose in the environment, how can we know what will ultimately happen to these organisms? The first field application of genetically engineered bacteria was developed at the University of California by plant pathologist Steven Lindow and colleagues. They identified a common strain of bacteria called *Pseudomonas syringae*, which makes bacterial proteins that stimulate ice crystal formation. Lindow's group created ice-minus bacteria by removing the ice protein–producing genes from *P. syringae*. They proposed that releasing ice-minus bacteria onto plants would cause the ice-minus bacteria to crowd out normal, ice-forming *P. syringae* and provide frost-sensitive crop plants with protection from the cold, thus extending the growing season and increasing crop yields.

Surrounded by a great deal of controversy, Lindow received approval in 1987 to test *P. syringae* on a crop of potatoes. Around the same time, other scientists received permission to test ice-minus bacteria on strawberries in a small town in California. This was the first time that genetically altered microbes were ever intentionally released into the environment in the United States. In both experiments, a majority of plants were damaged by activists concerned about the release of genetically altered microbes. Ice-minus *P. syringae* have shown some promise for frost protection, but they have not been as effective at crowding out the growth of normal ice-forming *P. syringae* as Lindow and others had hoped.

Because gene transfer between bacteria is a natural process that occurs in the wild, scientists are concerned about horizontal gene transfer, the spread of genes to related microbes. As a result of genetic recombination and the creation of new genes, new strains of microbes with different characteristics based on the genes they inherit may be produced.

- What would happen if recombinant microbes could transfer genetically altered genes into other microorganisms for which they were not originally intended?

- For instance, what would happen if ice-minus bacterial genes were transferred to strains of bacteria that are accustomed to living under cold conditions?

- Once recombinant microbes escape or are released into the environment, we cannot simply call them back into the lab if we do not like what they are doing in the field.

- Can we prevent the escape of genetically altered microbes in field experiments? It is a difficult if not impossible task when wind, rain, and other weather elements are involved.

Should genetically engineered microbes be released even in "controlled" experiments that might result in beneficial applications of biotechnology? You decide.

Biotechnology can help our immune systems by boosting our immunity through the use of vaccines.

Types of Vaccines: How Are Vaccines Made?

Vaccines are parts of a pathogen or whole organisms that can be given to humans or animals by mouth or by injection to stimulate the immune system against infection by those pathogens. When people or animals are vaccinated, their immune systems recognize the vaccine as an antigen and respond by making antibodies and B memory cells. By stimulating the immune system, the vaccine has pressured it into stockpiling antibodies and immune memory cells that can go to work on exposure to the real pathogen in the future,

should such exposure occur. Therefore many but not all vaccines are designed to be *preventative* or *prophylactic* (by providing protection against a pathogen should you be exposed) and not *therapeutic*—that is, a cure once you have become infected or developed a particular condition. Remember also that vaccination is used in pets, farm animals, zoo animals, and even wild animals.

So how are vaccines made? Four major strategies are generally used to create immune responses using vaccines.

1. **Subunit vaccines** are made by injecting portions of viral or bacterial structures, usually proteins or lipids from the microbe, to which the immune system responds. A fairly effective vaccine against hepatitis B virus was one of the first examples

of a subunit vaccine, and vaccines for tetanus, anthrax, and meningococcal disease (which you may have been vaccinated for, because this vaccine is required by many colleges and universities) are also subunit vaccines.

2. **Attenuated vaccines** involve using live bacteria or viruses that have been weakened through aging or by altering their growth conditions to prevent their replication after they are introduced into the recipient. The Sabin vaccine for polio is an attenuated vaccine. So are the MMR, tuberculosis, cholera, and chickenpox (varicella) vaccines as well as many others.

3. **Inactivated (killed) vaccines** are prepared by killing the pathogen and using the dead or inactive microorganism for the vaccine. A mixture of inactivated poliovirus is used in the Salk vaccine against polio. The rabies vaccines administered by injection to dogs, cats, and humans, the DPT vaccine, and the influenza (flu) vaccine, which has become common in recent years, are also examples of inactivated vaccines. Inactivated flu vaccine can also be delivered as a nasal spray.

4. **DNA-based vaccines** have been attempted but so far they have not proven to be widely effective. However, in 2005, the USDA approved the world's first licensed DNA vaccine, a vaccine against West Nile virus (WNV). Developed by Fort Dodge Laboratories of Fort Dodge, Iowa, this vaccine is designed to protect horses from WNV, a mosquito-borne virus. Equine infections of WNV are on the rise, and about a one-third of horses infected with WNV will die or will have to be euthanized. In 2007, the USDA approved the first therapeutic cancer vaccine in the United States for any species, human or animal, for canine melanoma. This vaccine consists of a plasmid containing a gene for a human enzyme (tyrosinase).

Immunity from vaccinations can fade with time, particularly for inactivated vaccines, which often do not produce a strong immune response. As a result, many vaccines require immunization *booster* shots every few years to restimulate the immune system so that it will continue to provide protective levels of antibodies and immune memory cells. For instance, the DPT vaccine is effective for about 10 years, as is the tetanus vaccine, and the flu vaccine that you may have received requires annual injections because new strains of influenza virus are developing each year.

Attenuated and inactivated vaccines were among the first vaccines developed. Some subunit vaccines against bacteria were made prior to recombinant DNA technology by growing bacterial pathogens in liquid culture. Many bacteria release proteins into the surrounding media, and these proteins can be purified and mixed with compounds that will help stimulate an immune response when injected into humans. But as scientists have learned more about the molecular structure of many pathogens, attempts at making recombinant subunit vaccines have become more popular.

For instance, **hepatitis B** is a blood-borne virus transmitted by exposure to body fluids, sexual intercourse, and contaminated blood transfusions. Hepatitis B causes deadly liver diseases. When vaccines for hepatitis B were first prepared, scientists isolated the virus from the blood of infected patients and then used biochemical techniques to purify viral proteins. These proteins were then injected into humans as a vaccine. The hepatitis B vaccine is recommended for international travelers, particularly people visiting Africa and Asia, and health care workers and others who may come in contact with hepatitis-infected persons or their body fluids. Currently a majority of subunit vaccines, including the vaccine for hepatitis B, are made using recombinant DNA approaches in which the vaccine is produced in microbes.

To produce the recombinant subunit vaccine for hepatitis B, scientists cloned genes for proteins on the outer surface of the virus into plasmids. Yeasts transformed with these plasmids are used to express large amounts of viral protein as fusion proteins, which are then purified and used to vaccinate people against hepatitis B infections. This approach is a common strategy for producing subunit vaccines, although sometimes fusion proteins are expressed in bacteria or in cultured mammalian cells. In, 2005, the pharmaceutical company Merck received FDA approval of a recombinant subunit vaccine (Gardasil) against cervical cancer and the first cancer vaccine to be approved by the FDA. Gardasil targets four specific strains of **human papillomavirus (HPV),** which cause about 70% of cervical cancers (HPV strains 16 and 18) and a large percentage of genital warts (caused by HPV strains 6 and 11). Cervical cancer affects 1 in 130 women, nearly half a million women worldwide, and approximately 70% of sexually active women will become infected with HPV during their lifetimes. In the United States, more than 10,000 women each year contract cervical cancer and around 4,000 die of the disease.

Given as a series of three booster shots, Gardasil is designed as a prophylactic vaccine, for use in girls and women aged 9 through 26, which means that it is taken to provide immune protection prior to exposure to HPV. Merck is also seeking FDA approval to expand Gardasil use to women aged 27 to 45 and potentially for use in teen boys. Merck recommends that Gardasil be given to female preteens prior to their becoming sexually active. Several states have pending legislation requiring that schoolchildren be vaccinated with Gardasil.

In Texas, for example, the governor signed an executive order making HPV vaccination compulsory for girls ages 11 to 12. Compulsory vaccination has become a very controversial issue (see problem 8 in the Questions & Activities at the end of this chapter) and Merck has since ceased its efforts to lobby for state laws requiring compulsory vaccination.

Ideally, the immune system can be most effective during the early stages of exposure to an infectious agent, when immune cells can attack the pathogens as soon as they enter the body. Disease-causing viruses use a number of complicated ways to infect cells, replicate, and cause disease. For instance, **human immunodeficiency virus-1 (HIV-1),** the causative agent of AIDS, infects human immune cells by binding to a cell and injecting its RNA genome (**Figure 5.13**). The enzyme reverse transcriptase copies the HIV genome into DNA. HIV and other viruses that transcribe their RNA genomes into DNA are called **retroviruses.** After the viral genome has been copied, it is transcribed to make RNA and translated to produce viral proteins that assemble to create more viral particles that are released from infected cells. We present this brief overview of viral replication because each stage essentially represents a potential target for antiviral drugs, including some vaccines.

Bacterial and Viral Targets for Vaccines

Pathogens are changing all the time, giving rise to both drug- and vaccine-resistant strains and new strains of disease-causing bacteria and viruses. More infectious microbes exist than there are vaccines. As a result, there are many research priorities for improving existing vaccines and producing new ones, and biotechnology companies have in excess of 50 targets for vaccine development. Several of the major vaccine targets include Dengue fever, Hepatitis C and E, sexually transmitted diseases such as herpes, gonorrhea and chlamydial infection, methicillin-resistant *Staphylococcus aureus* (MRSA), pneumococcal disease, rotavirus, and West Nile Virus. In addition, nonpathogenic human diseases such as Alzheimer's disease, multiple sclerosis, allergies, drug addiction, diabetes, and hypertension are targets for *therapeutic* vaccines. Here, we consider a few of the many targets for new vaccines.

The flu is caused by a large number of viruses that belong to the **influenza** family of viruses. Although most people experience flu symptoms that last for a few days and can be readily treated by over-the-counter medications, influenza kills some 500,000 to 1 million people worldwide each year, including about 34,000 deaths in the United States. Because flu viruses mutate so rapidly, no one-size-fits-all vaccine protects against all strains. New flu vaccines are generated each year based on the three main flu virus strains that are expected to be prevalent during the upcoming flu season. Viruses for this vaccine are grown in eggs.

The **World Health Organization (WHO)**—an international group that monitors infectious diseases and epidemics—has established centers in over 80 countries so that it can collect and screen samples of influenza for analysis and then develop vaccine treatment strategies. Infectious disease scientists are considering the development of a "global lab" to monitor strains of influenza viruses around the world, replicate these viruses, and then use recombinant DNA techniques to produce subunit vaccines in response to new viral strains detected. This strategy is a surveillance and rapid-response approach to keeping up with new pathogens and producing vaccines as needed. In the future, it will likely be implemented for many different disease-causing organisms.

Influenza A represents one of the greatest potential threats to human health through a *pandemic,* a global outbreak. Pandemic strains of influenza A have arisen in the past. In 1918, influenza virus killed at least 20 million people. Other pandemics occurred in 1957 and 1968, and epidemiologists predict that a future pandemic could be much worse than previous episodes. Of recent concern was the emerging strain of **avian flu (H5N1),** which received a lot of media attention because of its presence in chickens. Variations in influenza A are due to two glycoproteins called hemagglutinin (HA) and neuraminidase (NA), which project from the surface of the virus. Avian flu is an influenza A subtype called H5N1 because of the variation of these two proteins contained in this strain (abbreviated as H and N when used in strain names). In 2003, strain H5N1 caused a pandemic in chickens in Asia that resulted in the killing of over 200 million birds in an effort to halt the spread. Scientists were concerned that this strain, like other viruses, may mutate and make the jump into humans. Although a few isolated cases of human infection were seen, widespread infection did not occur; however, bird-to-pig transmission of the virus has occurred and mutation of the virus in pigs could produce a strain that would infect humans. Because of the devastation such a virus could cause, the WHO declared development of a vaccine to protect against H5N1 a high priority and vaccines were subsequently produced. In 2009, **swine flu (H1N1)** led to significant public health concerns, but vaccination against this virus has proven effective at controlling its spread in humans.

Tuberculosis (TB), caused by the bacterium *Mycobacterium tuberculosis*, is responsible for between 2 and 3 million deaths each year. Inhaled particles of *M. tuberculosis* can infect lung tissue, creating lumpy lesions called *tubercles*. The spread of TB has been effectively

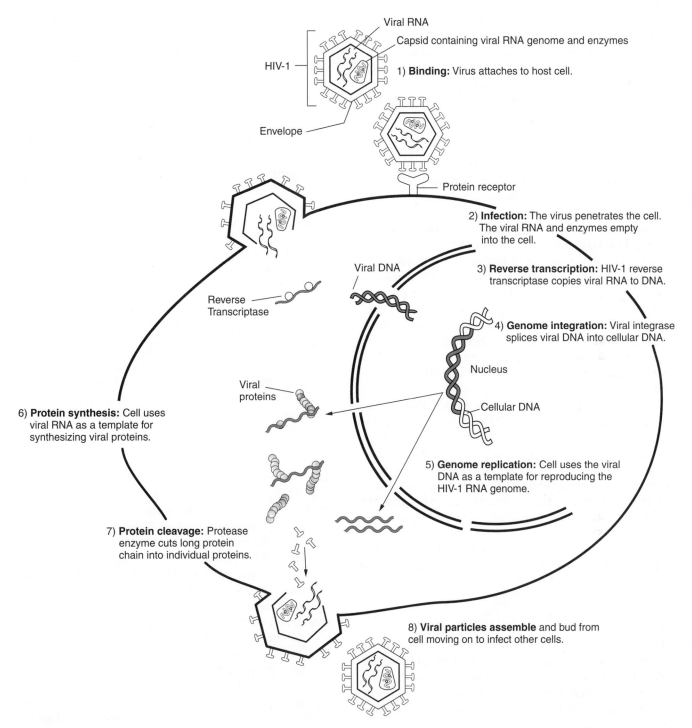

FIGURE 5.13 **HIV-1 Life Cycle** Each stage of HIV-1 replication is a potential target for antiviral drugs and some vaccines.

controlled in many areas of the world by the use of antibiotics and vaccines. But there has been a resurgence of TB because *M. tuberculosis* has proved to be very adept at evolving new strains that are resistant to treatment. Concern over such new strains is so great that in 1993, the WHO declared TB a global health

emergency, and a number of research initiatives were launched. The Bill and Melinda Gates Foundation, together with other organizations, have provided over $30 million for these efforts. The genome for *M. tuberculosis* has been sequenced and, as a result, new proteins have been discovered, leading to the development of new TB vaccines, many of which are currently in clinical trials.

Malaria is caused by the protozoan parasite *Plasmodium falciparum* and transmitted by insects. Worldwide, *Plasmodium* strains are developing resistance to the most commonly used antimalarial drugs. Although these drugs have been effective in parts of the world, the death rate from malaria is still unacceptable. Each year approximately half a billion cases of *Plasmodium* infections occur in children and cause nearly 3 million deaths. Whole-genome microarrays have been made for *Plasmodium* to help scientists identify new gene targets for inhibiting this parasite.

On another front, more than 33 million people are infected by **HIV** worldwide. The need for a vaccine to treat and stop the spread of HIV is critical if we are to curb this devastating disease and eventually eliminate the AIDS epidemic. Several vaccines for the prevention of HIV infections or the treatment of HIV-infected individuals have been tried in humans; most of these vaccines target viral surface proteins, but to date none of these have lived up to their promise. In 2007, a promising vaccine from Merck failed in phase II clinical trials.

One obstacle facing HIV vaccine scientists is the high mutation rate of HIV. Consequently, the multi-subunit vaccines called *cocktails*—those that contain mixtures of many viral proteins, together with antiviral drugs that block viral replication—may be a more effective strategy to combat HIV than using either vaccines or antiviral drugs alone. Similar strategies are being developed for treating other viruses, such as hepatitis B and C.

In the next section, we will consider the many reasons for sequencing microbial genomes and the tools used to carry out this work.

5.5 Microbial Genomes

In 1995, the Institute for Genomic Research, which played a major role in the Human Genome Project, reported the first completed sequence of a microbial genome when they published the sequence for *Haemophilus influenzae*. Since then, over 1,000 microbial genomes have been published, and work is being carried out on the genomes for several hundred other microbes. In 1994, as an extension of the Human Genome Project, the U.S. Department of Energy initiated the **Microbial Genome Program (MGP)**. A goal of the MGP is to sequence the entire genomes of microorganisms that have potential applications in environmental biology, research, industry, and health, such as bacteria that cause tuberculosis, gonorrhea, and cholera, as well as genomes of protozoan

pathogens such as the organism (*Plasmodium*) that causes malaria.

Why Sequence Microbial Genomes?

Streptococcus pneumoniae, the bacterium that causes ear and lung infections, including pneumonia, kills approximately 3 million children worldwide each year. Infections of *S. pneumoniae*, which can also cause bacterial meningitis, have been effectively treated since 1946. But many of these vaccines are ineffective in young children, who are particularly susceptible to infection and serious health consequences. In 2001, the *S. pneumoniae* genome was completely sequenced, and many genes encoding previously undiscovered proteins on the surface of the bacterium were identified. Researchers are optimistic that this new understanding of the *S. pneumoniae* genome will lead to new treatments for pneumonia, including gene therapy approaches to rid children of infections that may persist for years.

This is just one example of the potential power of genomics at work. By sequencing microbial genomes, scientists will be able to identify many secrets of bacteria, from genes involved in bacterial cell metabolism and cell division to genes that cause human and animal illnesses. In addition, researchers will find bacterial genes that may enable scientists to develop new strains of microbes that can be used in bioremediation and to reduce atmospheric carbon dioxide and other greenhouse gases, to find disease-causing organisms in food and water, to detect biological weapons, to synthesize plastics, to make better food products, and to produce genetically altered bacteria as biosensors for detecting harmful substances, among many other examples.

Our ability to sequence microbial genomes is also expected to lead to new and rapid diagnostic methods and ways to treat infectious conditions. For instance, if scientists sequence genes encoding cell-surface proteins that coat a particular bacterial pathogen, they may be able to use these proteins to generate new diagnostic tools, vaccines, and antimicrobial agents.

Bacteria perform a wealth of biochemical activities, which is reflected in their genomes. Of the microbial genomes sequenced to date, approximately 45% of the genes identified produce proteins of unknown function, and approximately 25% of genes discovered produce proteins that are unique to the bacterial genome sequenced. Therefore the potential for identifying new genes and proteins with unique properties that may have important applications in biotechnology is very high.

Selected Genomes Sequenced to Date

Of the millions of different bacteria that have been identified, which ones are of greatest interest to microbial

TABLE 5.3 SELECTED MICROBIAL GENOMES

Bacterium	Human Disease Condition (megabases, mB)	Approximate Genome Size	Approximate Number of Genes
Bacillus anthracis	Anthrax	5.23	5,000
Borrelia burgdorferi	Lyme disease	1.44	853
Chlamydia trachomatis	Eye infections, genitourinary tract infections (e.g., pelvic inflammatory disease)	1.04	896
Escherichia coli O157:H7	Severe food-borne illness (diarrhea)	4.10	5,283
Haemophilus influenzae	Serious infections in children (eye, throat, and ear infections, meningitis)	1.83	1,746
Helicobacter pylori	Stomach (gastric) ulcers	1.66	1,590
Listeria monocytogenes	Listeriosis (serious food-borne illness)	2.94	2,853
Mycobacterium tuberculosis	Tuberculosis	4.41	3,974
Neisseria meningitidis (MC58) infections	Meningitis and blood	2.27	2,158
Pseudomonas aeruginosa	Pneumonia, chronic lung infections	6.30	5,570
Rickettsia prowazekii	Typhus	1.11	834
Rickettsia conorii	Mediterranean spotted fever	1.30	1,374
Streptococcus pneumoniae	Acute (short-term) respiratory infection	2.16	2,236
Yersinia pestis	Plague	4.65	4,012
Vibrio cholerae	Cholera (diarrheal disease)	4.00	3,885

Sources: Sawyer, T. K. (2001). Genes to Drugs. *Biotechniques* 30(1): 164–168. TIGR Microbial Database (www.tigr.org/tdb/mdb/mdbcomplete) and Gold: Genomes OnLine Database (wit.integratedgenomics.com/GOLD).

genome researchers? As shown in Table 5.3, the bacterial genomes that have received the most attention are those from microbes responsible for serious illnesses and diseases in humans. For example, recently the genome for *Pseudomonas aeruginosa* was completed. It is a major human pathogen causing urinary tract infections, a number of skin infections, and persistent lung infections that are a significant cause of death in cystic fibrosis patients. *P. aeruginosa* is a particularly problematic bacterium because it is resistant to many antibiotics and disinfectants commonly used to treat other microbes. Learning more about the genes involved in the metabolism, replication, and the breakdown of compounds (such as antibiotics) in *P. aeruginosa* will be greatly helped by an understanding of its genome.

Another one of the first microbes targeted for genome studies was *Vibrio cholerae,* which is typically found in polluted waters in areas of the world with poor sanitary practices. This bacterium causes the disease **cholera,** which is characterized by severe diarrhea

and vomiting, leading to massive fluid loss, which can cause shock and even death. Strains of antibiotic-resistant *V. cholerae* are causing recurring problems in Asia, India, Latin America, and even areas of the Gulf Coast in the United States. Understanding the genome of *V. cholerae* will help scientists identify toxin genes, genes for antibiotic resistance, and other genes that will augment our current methods for combating this microbe. Genome biologists are also focusing on microorganisms that may be used as biological weapons in a terrorist attack. In Section 5.7, we discuss why and how microbes can be used as bioweapons and what can be learned by studying their genomes.

Scientists have been studying the genetics of lactic acid bacteria for about 35 years helping an effort to understand how these bacteria contribute to the flavor and texture of cheeses, milk, and other products we discussed earlier. Genome projects have been completed for several dozen dairy-related lactic acid bacteria. For example, scientists recently sequenced the genome for

Lactococcus lactis, a strain that is important for making cheese. Such projects have already helped food scientists better utilize different strains to make specific cheeses with enhanced flavor characteristics and to refine culture conditions to maximize the growth abilities of different microbes.

Metagenomic Studies Sequence Genomes from Microbial Communities

Metagenomics involves the sequencing of genomes for entire communities of microbes. Metagenomics projects are sequencing microbial genomes from environmental samples of water, air, and soils as well as from oceans throughout the world, glaciers, mines—virtually every corner of the globe. Estimates also suggest that over 99% of currently known microbial diversity exists in organisms that cannot be cultured. Currently a number of metagenomics projects have been launched around the world, involving international teams of investigators sequencing marine microbes and soil microbes (there is a "terragenome" project under way).

Human genome pioneer J. Craig Venter, who we discussed in Chapter 3, left Celera in 2003 to form the J. Craig Venter Institute (JCVI), and he has played a central role in establishing the field of metagenomics. One of the institute's initiatives is a global expedition to sample marine and terrestrial microorganisms from around the world and to sequence their genomes. On what is called the Sorcerer II Expedition, Venter and his researchers are traveling the globe by yacht on a sailing voyage that has been described as a modern-day version of Charles Darwin's famous treks on the HMS *Beagle.*

A pilot study the institute conducted on the Sargasso Sea off Bermuda yielded around 1,800 new species of microorganisms and over 1.2 million novel DNA sequences. Samples of water from different layers in the water column are passed through high-density filters of various sizes to filter out microbes. DNA is then isolated from the microbes and used for shotgun cloning and then sequenced with computer-automated sequencers that are kept running on board nearly around the clock. This expedition has great potential for identifying new microbes and genes with novel functions, including commercially valuable genes. For example, the Sargasso Sea project identified hundreds of photoreceptor genes. Some microorganisms rely on photoreceptors for capturing light energy to power photosynthesis. Scientists are interested in learning more about photoreceptors to help develop ways in which photosynthesis may be used to produce hydrogen as a fuel source. Medical researchers are also very interested in photoreceptors because, in humans and many other species, photoreceptors in the eye are responsible for vision. The Sorcerer II expedition has

sequenced well over 6 billion base pairs (bp) of DNA from more than 400 uncharacterized microbial species. These sequences contain 7.7 million previously uncharacterized sequences encoding more than 6 million different potential proteins.

The Human Microbiome Project

In 2008 the National Institutes of Health initiated the **Human Microbiome Project,** a $115-million, 5-year metagenomic project to complete the genomes of an estimated 600 to 1,000 microorganisms, bacteria, viruses, and yeasts that live on and inside humans. Microorganisms comprise some 1% to 2% of the human body, outnumbering human cells by about 10 to 1. Many microbes, such as *E. coli* in the digestive tract, have important roles in human health, and of course other microbes make us ill. In addition, 1,200 different bacteriophages (recall from Chapter 3 that phages are viruses that infect bacteria) are found in the gut, and we know virtually nothing about more than half of them.

The Human Microbiome Project has several major goals, including:

- Determining if individuals share a core human microbiome.

- Understanding how we acquire and maintain microbial communities.

- Understanding how changes in the microbiome can be correlated with changes in human health and the conditions (such as stress and diet) that affect the microbiome.

- Developing new methods, including bioinformatics tools, to support analysis of the microbiome.

- Addressing ethical, legal and social implications raised by human microbiome research. Does this sound familiar? Recall that addressing ethical, legal and social issues was a goal of the Human Genome Project (HGP).

It is estimated that the microbiome consists of 100 to 1,000 times the number of human genes. The Human Microbiome Project is still very much in its infancy as a project but it has already sequenced 500 microbial genomes. So far about 3.3 million human gut microbe genes characterized to date appear to be very similar among over 100 individuals. The saliva microbiome is also highly similar from individual to individual, and a Human Oral Microbiome database has been established from these studies; it seeks to develop linkages between the oral microbiome and oral health. More than 700 types of microbes grow in the mouth alone, contributing to oral health issues such as bad breath, plaque formation, and tooth decay. Scientists have characterized the genomes for dandruff-causing bacteria and fungi of the scalp. We have also learned that gut microbes play a role in obesity and that the gut microbial environment changes from childhood to adulthood. Similar projects are under way to study the microbiomes of dogs and other animals. Keep an eye on microbiome and other metagenomics projects, because they will provide very interesting data about the microbes among us.

Viral Genomics

The study of viral genomes is another hot area of research (Table 5.4, on page 149). This is true in part because many deadly viruses mutate quickly in response to vaccine and antiviral treatments. Antiviral drugs are designed to work in several ways. Some antiviral drugs block viruses from binding to the surface of cells and infecting cells; others block viral replication after the virus has infected body cells. Research on viral genomes helps scientists learn how viruses cause disease and leads to the development of new and effective antiviral drugs.

Creating Synthetic Genomes: A Functional Synthetic Genome Is Produced for a Bacterial Strain

In 2010 JCVI scientists published the first report of a functional **synthetic genome.** In this project they designed and had chemically synthesized more than a

thousand 1,080-bp segments covering the entire 1.08-Mb genome of the bacterium *Mycoplasma mycoides* (**Figure 5.14**). To assemble these segments correctly, the segments had 80-bp sequences at each end, which overlapped with their neighbor sequences. These sequences were cloned in *E. coli*. Then, using the yeast *Saccharomyces cerevisiae,* they assembled the sequences into eleven separate 10-kb assemblies, which were eventually combined to completely span the entire *M. mycoides* genome. The assembled genome, called JCVI-syn1.0, was then transplanted into a close relative *M. capricolum* as recipient cells, resulting in a new cell

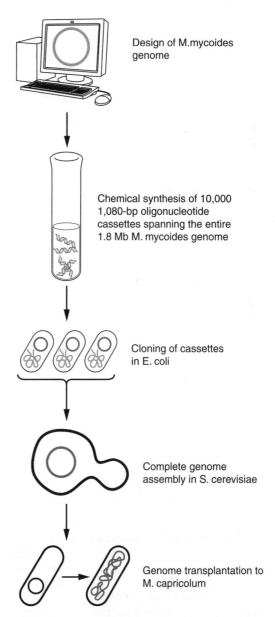

FIGURE 5.14 **Building a synthetic version of the 1.08-Mb *Mycoplasma mycoides* genome JCVI-syn1.0.** Shown here is an overview of the approach used to produce *M. mycoides* JCVI-syn1.0.

Design of M.mycoides genome

Chemical synthesis of 10,000 1,080-bp oligonucleotide cassettes spanning the entire 1.8 Mb M. mycoides genome

Cloning of cassettes in E. coli

Complete genome assembly in S. cerevisiae

Genome transplantation to M. capricolum

YOU DECIDE

Should We Create Life Synthetically?

The JCVI synthetic genome experiments are of limited applicability thus far and will likely be very challenging to apply to other cells types. But if applications of synthetic biology eventually become routine, will we have simplified or demystified life, making it less sacred? Should humans be such "creators" of life? Of course we have been creating life in the lab for a long time; **in vitro** fertilization is one such example. But synthetic biology is a very different approach. Not surprisingly, one fear raised was that bioterrorists could use synthetic genome strategies to recreate deadly bacteria or viruses (read about bioweapons in the next section on bioterrorism) from benign bacteria and viruses. To what extent should we use this technology to remake naturally occurring cells with features we deem better or more desirable? In the future, could synthetic genomes be used to create life from inanimate components? Should this be done if it is technically possible? You decide.

with the JCVI-syn1.0 genotype and the phenotype of a new strain of *M. mycoides*. As shown in **Figure 5.15,** JCVI determined that the recipient cells were taken over to become JCVI-syn.10 *M. mycoides,* in part because they were shown to express the *lacZ* gene, which was incorporated into the synthetic genome. Selection for tetracycline resistance and a determina-

tion that recipient cells also made proteins characteristic of *M. mycoides* and not *M. capricolum* were also used to verify strain conversion.

One particularly impressive accomplishment of these experiments was that the synthetic DNA was "naked" DNA, because it did not contain any proteins from *M. mycoides;* therefore it was capable of transcrib-

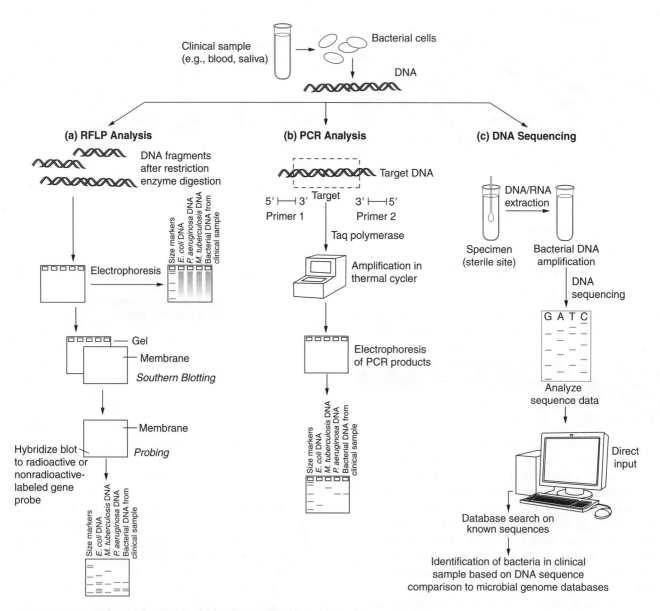

FIGURE 5.15 Using Molecular Techniques to Identify Bacteria Many molecular techniques are available for identifying bacteria. (a) For some pathogens, isolated DNA (which may come from a clinical sample such as blood or saliva) can be digested by restriction enzymes, separated by agarose gel electrophoresis followed by Southern blotting (RFLP analysis). Banding patterns of DNA fragments can be compared to reference strains of known bacteria to allow for a positive identification. In this example, DNA from the clinical sample matches *P. aeruginosa.* (b) PCR has the advantage of being much more sensitive than RFLP analysis; therefore only small clinical samples and small amounts of DNA are required, allowing for early treatment of an infection. (c) DNA-sequencing strategies are also commonly used for microbial identification.

TABLE 5.4   EXAMPLES OF MEDICALLY IMPORTANT VIRAL GENOMES THAT HAVE BEEN SEQUENCED

Virus	Human Disease or Illness	Year Sequenced
Ebola virus	Ebola hemorrhagic fever	1993
Hepatitis A virus	Hepatitis A	1987
Hepatitis B virus	Hepatitis B	1984
Hepatitis C virus	Hepatitis C	1990
Herpes simplex virus, type I	Cold sores	1988
Human immunodeficiency virus (HIV-1)	Acquired immunodeficiency syndrome (AIDS)	1985
Human papillomavirus	Cervical cancer	1985
Human poliovirus	Poliomyelitis	1981
Human rhinovirus	Common cold	1984
Influenza A virus		
• Subtype H5N1 (Avian flu)	Severe flu	2007
• Subtype H5N1 (Swine flu)	Severe flu	2009
Severe acute respiratory coronavirus (SARS-CoV)	Severe acute respiratory syndrome (SARS)	2003
Variola virus	Smallpox	1992

ing all of the appropriate genes and translating all of the protein products necessary for life as *M. mycoides*. This is not a trivial accomplishment. The synthetic genome effectively rebooted the *M. capricolum* recipient cells to change then from one form to another. When this work was announced, J. Craig Venter claimed "This is equivalent to changing a Macintosh computer into a PC by inserting a new piece of PC software."

Venter and others have used recombinant DNA technology to construct synthetic copies of viral genomes. Researchers at Stony Brook University in New York made headlines when they assembled approximately 7,500 bp of synthetically produced DNA sequences to synthesize proteins and lipids; these were assembled into a recreated polio virus: the first synthetically made virus. The genome for the 1918 influenza strain responsible for the pandemic was also assembled in this way.

But Venter's recent work with *M. mycoides* JCVI-syn1.0 9 is hailed as a defining moment in the emerging field of **synthetic biology.** This work did not create life from an inanimate object, since it was based on converting one living strain into another. There are many fundamental questions about synthetic genomes and genome transplantation that need to be answered. But clearly these studies provided key

"proof of concept" that synthetic genomes could be produced, assembled, and successfully transplanted to create a microbial strain encoded by a synthetic genome and bring scientists closer to producing novel synthetic genomes incorporating genes for specific traits of interest.

What are potential applications of synthetic genomes and synthetic biology? JCVI claims that their ultimate goal is to create microorganisms that can be used to synthesize biofuels. Other possibilities exist, such as creating synthetic microbes with genomes engineered for bioremediation, producing alternative fuels, synthesizing new biopharmaceutical products, developing genetically programmed bacteria to help us heal, and making "prosthetic genomes." Work on synthetic genomes and synthetic biology has led to speculation of a future world in which there will be new bacteria and perhaps new animal and plant cells designed and even programmed to be controlled as we want them to be.

5.6   Microbes for Making Biofuels

The United States alone requires about 140 billion gallons of fuel per year to satisfy current needs, and we are producing only about 5 billion gallons per year of ethanol as a **biofuel** produced from grain. The production of biofuels has the potential to provide an alternative energy source and reduce global warming resulting from the burning of fossil fuels. To produce ethanol, alcohol-fermenting microbes convert glucose and other sugars in grain to ethanol, but this process is not cost-effective or efficient. It takes a lot of corn kernels to produce relatively small amounts of ethanol. Although, as you will see, cellulosic biomass such as corn stalks is a readily available and abundant source of sugars for making ethanol, but breaking down glucose from cellulose is no simple matter (Chapter 6). The trick is to break the cellulose down into individual glucose molecules, which can then be used to create ethanol by fermentation processes.

Cellulose in the plant cell wall is somewhat resistant to breakdown naturally in the environment. You also know that you and I cannot digest cellulose; therefore it contributes fiber to out diet. Some techniques use chemical treatments to help loosen the cellulose structure, but these chemicals inhibit many microbes that could be used to make ethanol.

Many companies working on alternative energy sources consider cellulosic sugars for ethanol production as a very sustainable source of biofuels for reducing the world's reliance on fossil fuels. One concept is to create biorefineries in which leftover biomass from conventional crops—such as corn husks and stalks, wood waste such as chips, saw dusts, and yard clippings—would incorporate microbial enzymes to process sugars with great efficiency from biomass into fuels. Researchers are improving strains of bacteria, such as *Zymomonas mobilis*, to help with this process. Key genes encoding enzymes that will convert sugars into ethanol at higher rates than yeasts can do have been engineered into these bacteria, but their effectiveness in scale-up processes for making ethanol is unproven so far.

Recombinant DNA technology is being used to produce *E. coli* with an increased ability to produce ethanol, as well as bacteria with an increased ability to ferment sugars into ethanol by alcohol fermentation. Others have genetically engineered *E. coli* to secrete cellulose-degrading enzymes. In addition, major bioprospecting efforts are ongoing around the world to identify bacteria and algae that produce useful enzymes—enzymes that could help to process biomass into fuel. Later, we will briefly discuss how bioremediation approaches are investigating the use of microbes to degrade components in sediments as a way of generating energy (Chapter 9). Although the future potential of biofuels is unclear, it is expected that research over the next few years will result in significant improvements in biofuel production.

5.7 Microbial Diagnostics

We have repeatedly seen that microorganisms cause a number of diseases in humans, pets, and agriculturally important crops. Scientists can use a variety of molecular techniques to detect and track microbes—an approach called **microbial diagnostics.**

Bacterial Detection Strategies

Before the development of molecular biology techniques, microbiologists relied on biochemical tests and bacteria cultured on different growth media to identify strains of disease-causing bacteria. For example, when doctors take a throat culture, they use a swab of bacteria from your throat to check for the presence of *Streptococcus pyogenes*, a bacterium that causes strep throat. Even though these and other similar techniques still have an important place in microbial diagnosis, techniques in molecular biology allow for the rapid detection of bacteria and viruses with great sensitivity.

Molecular techniques such as restriction fragment length polymorphism (RFLP) analysis, PCR, and DNA sequencing, can be used for bacterial identification (Figure 5.15; see also Chapter 3). If the genome of the pathogen is large and produces too many restriction enzyme fragments, which prevent visualizing individual DNA bands on an agarose gel, DNA may be subjected to Southern blot analysis (Figure 5.15).

Many databases of RFLPs, PCR patterns, and bacterial DNA sequences are available for comparison of clinical samples. For example, if a doctor suspects a bacterial or viral infection, samples including blood, saliva, feces, and cerebrospinal fluid from the patient can be used to isolate bacterial and viral pathogens. DNA from the suspected pathogen is then isolated and subjected to molecular techniques such as PCR (Figure 5.15). PCR is an important tool for diagnostic testing in clinical microbiology laboratories and widely used to diagnose infection caused by microbes such as the hepatitis viruses (A, B, and C), *Chlamydia trachomatis* and *Neisseria gonorrhoeae* (both of which cause sexually transmitted diseases), HIV-1, and many other bacteria and viruses.

Tracking Disease-Causing Microorganisms

Scientists also use molecular biology techniques to track patterns of disease-causing microbes and the illnesses and outbreaks of illnesses they may cause. As you know, microbes play important roles in the production of dairy products such as yogurts and cheeses. But these dairy products are also susceptible to contamination with pathogenic microorganisms. Information about the microbes in milk can be used to determine the quality of milk and milk spoilage. Bacterial contamination of food is a significant problem worldwide. You have probably heard of the bacterium *Salmonella,* which can contaminate meats, poultry, and eggs. *Salmonella* can infect the human intestinal tract causing serious diarrhea and vomiting, symptoms commonly called food poisoning.

After successfully responding to a 1993 outbreak of meat contaminated with *E. coli,* the **Centers for Disease Control and Prevention (CDC)** and the U.S. Department of Agriculture created a network of DNA-detecting laboratories to expand its coverage and boost its response time. This network, called **PulseNet,** enables biologists, using DNA fingerprinting approaches, to rapidly identify microbes involved in a public health condition. Results can be compared with a database to identify outbreaks of microbes in contaminated foods and to decide how to respond so that a minimal number of people are affected.

Approximately 76 million cases of food-borne disease due to microbes occur in the United States each year, causing well over 300,000 hospitalizations and approximately 5,000 deaths. In the United States alone, the *E. coli* strain O157:H7 causes close to 20,000 cases of food poisoning each year. This infectious strain is lethal. PulseNet now monitors *E. coli* O157, *Salmonella, Shigella,* and *Listeria,* and many non-food-borne diseases such as tuberculosis.

Microarrays for Tracking Contagious Diseases

Microarrays have created new approaches for detecting and identifying pathogens and for examining host responses to infectious diseases. For example, microarray pioneer Affymetrics Inc. has developed the SARS-CoV GeneChip, which contains approximately 30,000 probes representing the entire viral genome for the coronavirus that causes **severe acute respiratory syndrome (SARS).** The SARS virus is a highly contagious respiratory virus that has infected approximately 9,000 people and killed nearly 900 since it was first detected in November 2002. Similar chips are being used to detect the flu strains H1N1 and H5N1, which we discussed previously.

Microarray approaches are also being used to study gene expression changes that occur when an organism is infected with a pathogen (**Figure 5.16**), providing a "signature" for infection by a particular organism. With these chips, patterns of genes that are stimulated or inhibited by a pathogen can be analyzed as a hallmark signature unique to that particular pathogen. Notice how the three pathogens used in Figure 5.16 stimulate different sets of genes in mice.

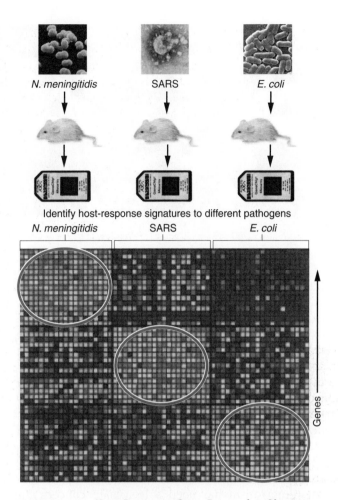

FIGURE 5.16 **Host-Response Gene Expression Signatures for Pathogen Identification** In this animal model, mice were infected with *Neisseria meningitidis* (the bacterium that causes meningitis), SARS, or *E. coli,* and microarray analysis was carried out to examine changes in gene expression following infection by each pathogen. In this example, actively expressed genes are shown as light spots. Notice how each pathogen activates a specific subset of genes (circled), creating a gene expression "signature" that researchers can use to identify the infecting pathogen.

In the next section, we provide a brief glimpse of biological agents that can pose a threat as weapons and discuss how biotechnology can be used to detect and combat bioterrorism.

5.8 Combating Bioterrorism

The tragic events of September 11, 2001, were the most catastrophic attacks of terrorism on American soil. In the weeks that followed these horrific tragedies, America and the world were also served notice of a bioterrorism threat when letters contaminated with dried powder spores of the bacterium *Bacillus anthracis* were

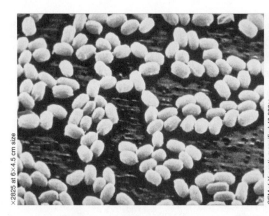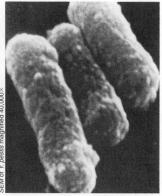

×2825 at 6×4.5 cm size

SEM of *Y. pestis* magnified 40,000×

FIGURE 5.17 Deadly Microbes as Bioweapons Potential bioweapons might include the organisms that cause anthrax (*Bacillus anthracis*, left) and plague (*Yersinia pestis*, right).

mailed to two senators, other legislators, and members of the press. Toxic proteins from *B. anthracis* cause significant damage to cells of the skin, respiratory tract, and gastrointestinal tract, depending on how a person is exposed to the microbe. As a result of this anthrax exposure, 5 people died and another 22 people became ill. These events raised our awareness that bioterrorism activities could cause devastating harm should biological agents be released in large quantities. **Bioterrorism** is broadly defined as the use of biological materials as weapons to harm humans or the animals and plants we depend on for food. Biotechnology by its very definition is designed to improve the quality of life for humans and other organisms. Unfortunately, bioterrorism represents a most extreme abuse of living organisms.

Bioterrorism has been a legitimate concern for centuries. In the fourteenth century, bodies of bubonic plague victims were used to spread the bacterium *Yersinia pestis,* which caused bubonic plague during wars in Russia and other countries. This subsequently played a part in causing the Black Death pandemic that ravaged Europe. Early European settlers of the New World spread measles, smallpox, and influenza to Native Americans. Although the introduction of these diseases may not have been intentional, it led to the deaths of hundreds of thousands of Native Americans because they had virtually no immunity to these pathogens. Between 1990 and 1995, aerosols of toxins from the bacterium *Clostridium botulinum* were released at crowded sites in downtown Tokyo, Japan, although no infections resulted from these attempts. In the last few decades, many other lesser known and, fortunately, unsuccessful incidents have occurred around the world.

Microbes as Bioweapons

New strains of infectious and potentially deadly pathogens are evolving every day all around the world. The threat posed by these disease-causing microorganisms, which could be used as **bioweapons,** may conjure images of science fiction novels; how-

ever, the potential for a bioterrorism attack is real and of significant concern.

Even though thousands of different organisms that infect humans are potential choices as bioweapons, most experts believe that only a dozen or so organisms could feasibly be cultured, refined, and used in bioterrorism (Table 5.5). These agents include bacteria such as *Bacillus anthracis,* the gram-positive bacillus that causes anthrax, and deadly viruses such as smallpox and Ebola (**Figure 5.17**). The possibility that

 **YOU DECIDE**

Should Pathogen Genome Sequences Remain in Public Databases?

Science as a process relies on scientists sharing data and information through presentations at conferences, in publications, and via Web resources such as databases. Now that genomes have been completed for many human pathogens—such as those that cause cholera, anthrax, meningitis, and smallpox—there has been considerable debate about whether genome sequences for potential bioweapon microbes should be publicly available in DNA databases. Media reports on this subject raise concern that terrorists could use such sequence data to develop recombinant proteins for pathogen toxins as a way to produce bioweapons. Others have suggested that terrorists could use genome data to make more effective "superweapon" pathogens resistant to existing vaccines and drugs. However, many genome scientists have spoken in favor of keeping pathogen sequences in public databases, citing the importance of open access to information and claiming that there is so much that we don't understand about the genomes of these pathogens that no human health threat is posed by making their sequences available. What do you think? Should pathogen genome sequences remain in public databases? You decide.

TABLE 5.5   POTENTIAL BIOLOGICAL WEAPONS

Agent	Disease Threat and Common Symptoms
Brucella (bacteria)	Different strains of *Brucella* infect livestock such as cattle and goats. They can cause brucellosis in animals and humans. Prolonged fever and lethargy are common symptoms. The disease can be mild or life-threatening.
Bacillus anthracis (bacterium)	Anthrax. Skin form (cutaneous) produces skin-surface lesions that are generally treatable. Inhalation anthrax initially produces flu-like symptoms leading to pulmonary pneumonia, which is usually fatal.
Clostridium botulinum (bacterium)	Botulism. Caused by ingestion of food contaminated with *C. botulinum* or its toxins. Varying degrees of paralysis of the muscular system created by botulinum toxins are typical. Respiratory paralysis and cardiac arrest often cause death.
Ebola virus or Marburg virus	Both are highly virulent viruses that cause hemorrhagic fever. Symptoms include severe fever, muscle/joint pain, and bleeding disorders.
Francisella tularensis (bacterium)	Tularemia. Lung inflammation can cause respiratory failure, shock, and death.
Influenza viruses (a large, highly contagious group)	Influenza (flu). Severity and outcome depend largely on the strain of the virus.
Rickettsia (several bacteria strains)	Different strains cause diseases such as Rocky Mountain spotted fever and typhus.
Variola virus	Smallpox. Chills, high fever, backache, headache, and skin lesions.
Yersinia pestis (bacterium)	Bubonic plague. High fever, headache, painful swelling of lymph nodes, shock, circulatory collapse, organ failure, and death within days after infection in a majority of cases.

unknown organisms could be used as bioweapons is disquieting because they would probably be very difficult to detect and neutralize. However, a little-known microbe would probably be difficult to deliver as a bioweapon.

As a bioweapon, smallpox, which is a disease caused by the variola virus, is of concern for several reasons. Virtually all humans are susceptible to smallpox infection because widespread vaccination stopped over 20 years ago, when the last case of confirmed smallpox was reported. At that time the WHO declared the disease to be eradicated throughout the world, in large part as a result of vaccines. Recent outbreaks of a monkey smallpox strain, however, may revive smallpox as a concern. Following the anthrax events of 2001, based on concern about the use of smallpox as a bioweapon, the U.S. government started to work with biotechnology companies to mass-produce and stockpile supplies of smallpox vaccines.

Targets of Bioterrorism

Antibioterrorism experts expect that bioterrorists will target cities or events where large numbers of humans gather at the same time. Experts, however, have had a poor track record of predicting where and how such acts might occur. At best, we can speculate that there may be many different potential ways to deliver a bioweapon to injure or kill humans. Bioterrorists are most likely to use a limited number of approaches to achieve quick and effective results. Widespread application of most agents might occur via some type of aerosol release in which small particles of the bioweapon are released into the air and inhaled. The aerosol could be created by grinding the bioweapon into a fine powder and producing a "silent bomb" that would release a cloud of the bioweapon into the air. This aerosol cloud would be gas-like, colorless, odorless, and tasteless. Such a silent attack could go undetected for several days. If exposed to a biological agent, in the days following an attack, a few people might develop early symptoms, which physicians might misdiagnose, or their symptoms might closely resemble common illnesses. Meanwhile, if the biological agent could be spread from human to human, larger numbers of people in other states and even other countries would become infected as infected individuals traveled from place to place and spread their illness. Experts also suspect that biological agents might be delivered by crop-duster planes or disseminated into water supplies.

In addition to the direct threat to humans, experts are concerned with preventing bioweapon attacks that could cause severe damage to crops, food animals, and other food supplies (see Table 5.6). Not only could such an attack affect human health, but this approach could also cripple the agricultural economy of a country if food animals such as cows were infected. If there were general concerns about the safety of beef and milk, many consumers would likely shy away from buying these products for fear of contamination.

Using Biotechnology Against Bioweapons

As was evident during the anthrax incidents of 2001, the United States is generally unprepared for an attack with biological weapons. Numerous agencies including the American Society for Microbiology, the USDA, the CDC, the Department of Health and Human Services, and Congress have worked to develop legislation for minimizing the dangers of bioterrorism and responding to possible strikes. For instance, the U.S. Postal Service has implemented technologies for sanitizing mail by using x-rays or ultraviolet light.

In 2004, the U.S. federal government appropriated approximately $6 billion to be spent over 10 years to combat biological and chemical terrorism through an initiative called Project BioShield. A main goal of BioShield is to develop and purchase substantial quan-

tities of vaccines and drugs to treat or protect Americans from bioweapons. Even with increased budgets, new laws, and international treaties, no measures can guarantee that bioterrorism will never occur or that we would be able to detect and protect people against illness from such an attack if it were to occur. Worldwide efforts to prevent bioterrorism are essential, but how could biotechnology help to detect bioweapons and respond to an attack if it did occur?

Field tests are an essential step in the detection of an attack. Some field tests involve antibody-based tests such as ELISAs to determine whether pathogens, or specific molecules from a pathogen, are present in an air or water sample. Such units were put in place around the Pentagon during the anthrax scare of 2001, the Gulf War, and the wars in Afghanistan and Iraq. These units draw in air and use antibodies to detect pathogens in the air. This technique is flawed because many of these instruments are not very sensitive and cannot detect small quantities of a pathogen. In fact, these sensors have been known to detect harmless microbes that live naturally in the environment. Newer, more sensitive biosensors must be developed. Similar detection tests using PCR have been developed.

Variations of protein-based assays include rapid handheld immunoassays developed by U.S. Navy scientists and protein microarrays used by the military to detect airborne pathogens such as anthrax spores and the smallpox virus (**Figure 5.18**). Designed for use in lab settings or even quick diag-

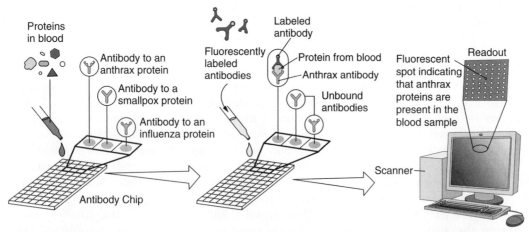

1) Apply blood from a patient to a chip, or array, consisting of antibodies assigned to specific squares on a grid. Each square includes multiple copies of an antibody able to bind to a specific protein from one organism and so represents a distinct disease-causing agent.

2) Apply fluorescently labeled antibodies able to attach to a second site on the proteins recognizable by the antibodies on the chip. If a protein from the blood has bound to the chip, one of these fluorescent antibodies will bind to that protein, enclosing it in an antibody "sandwich."

3) Feed the chip into a scanner to determine which organism is present in the patient's body. In this case, the culprit is shown to be *Bacillus anthracis*, the bacterium that causes anthrax.

FIGURE 5.18 **Protein Microarrays for Detecting Bioweapon Pathogens**

nostic tests in field settings (some field-based micro-arrays use diamond-coated surfaces that make them particularly durable), these arrays detect whole pathogens or pathogen components such as proteins or spores.

Emergency response teams and cleanup crews will also be called to action to evaluate the extent of contamination and determine appropriate cleanup procedures based on the bioweapon released (**Figure 5.19**). Should an attack occur, treatment drugs such as antibiotics will be needed. Countries must build a stockpile of drugs and vaccines that can be widely distributed to large numbers of people if necessary. The basic problem with vaccines is that they must be administered *before* exposure to bioweapons in order to be effective—giving them to infected individuals after an attack is useless. Since 2001, the CDC has increased the U.S. stockpile of smallpox vaccine. Mandatory vaccination of certain military personnel and a voluntary campaign to vaccinate health care workers and those first responders most likely to come in contact with the virus was also instituted. However, even drugs and vaccines may be ineffective if a bioterrorist attack involves organisms that have been engineered against most conventional treatments or if an unknown organism is used as the bioweapon. For instance, you may recall that the anti-

FIGURE 5.19   Battling Bioterrorism Hazardous material workers from the U.S. Coast Guard decontaminate a coworker after working inside the Hart Senate Office Building to clean the building of anthrax spores during the anthrax attacks in 2001. Battling bioterrorism will require the coordinated efforts of many individuals, from scientists, physicians, and politicians to emergency response teams and cleanup personnel.

biotic ciprofloxacin (Cipro) was in high demand during the anthrax threats of 2001.

An unfortunate reality is that somewhere in this world, someone may be working to plan an attack with biological weapons. Will we be prepared?

TABLE 5.6   POTENTIAL BIOLOGICAL PATHOGENS FOR A BIOWEAPONS ATTACK ON FOOD SOURCES

Disease	Target/Vector	Agent
Animal diseases		
Foot-and-mouth disease	Livestock	Foot-and-mouth virus
African swine fever virus	Pigs	African swine fever
Plant diseases		
Stem rust for cereals (fungus)	Oat, barley, wheat	*Puccinia* spp.
Southern corn leaf blight (fungus)	Corn	*Bipolaris maydis*
Rice blast (fungus)	Rice	*Pyricularia grisea*
Potato blight (fungus)	Potato	*Phytophthora infestans*
Citrus canker (bacterium)	Citrus	*Xanthomonas axonopodis* pv. *citri*
Zoonoses		
Brucellosis (bacterium)	Livestock	*Brucella melitensis*
Japanese encephalitis (flavivirus)	Mosquitoes	Japanese encephalitis virus
Cutaneous anthrax (bacterium)	Livestock	*Bacillus anthracis*

Source: Gilmore, R. (2004): U.S. Food Safety Under Siege? *Nature Biotechnology*, 22: 1503–1504.

Microbial Biotechnology Offers Many Career Options

Microbial biotechnology research and development provides opportunities for all education levels. Even though advanced degrees are more common for upper-level positions, people with bachelor's degrees hold a significant number of positions. Owing to the wide variety of products, companies look for people with backgrounds in all areas of biology (molecular biology, protein biochemistry, microbiology, bioengineering, and chemical engineering). These people will design the product and determine how to test it for identification, efficacy, and potency. The biologists and engineers in these groups will develop methods of production on a small scale that can then be used for large-scale production, such as the batch culturing of recombinant microbes in fermenters.

Strict regulation in the biotech industry necessitates that manufacturing be a very precise process. Thus well-trained manufacturing personnel, who can have a variety of educational backgrounds from high school to bachelor's degrees, are needed. Companies typically train personnel for these positions themselves. Quality-control personnel test products and conduct tests on process samples to ensure that the products are produced properly. They collect and test a variety of samples ranging from water and culture media to chemical raw materials and final products. Educational background again can vary from a high school education to a bachelor's degree. Often companies provide training for personnel, but for the more advanced analytical procedures, individuals with an understanding of and experience with bioanalytical methods (e.g., high-performance liquid chromatography [HPLC], gel analysis, blotting techniques, and PCR) will have an advantage.

Quality-assurance personnel are charged with ensuring that products are made according to strict guidelines. They review the manufacturing process record (batch record) before and after manufacturing; they not only must remain current with regulations and be the liaison with the FDA and other regulatory agencies but must also be familiar enough with the production process to serve in this position. A background and understanding of microbiology is very helpful, but an interest in legal matters and regulations and good writing skills are also essential.

Source: Contributed by Daniel B. Rudolph (Process Engineer, Lonza Group LTD, Baltimore, MD).

MAKING A DIFFERENCE

For over 80 years antibiotics isolated from bacteria or antibiotics synthesized on the basis of antimicrobial products of bacteria have had a significant impact on successfully treating infections in humans, pets, livestock, and other animals and limiting the spread of infectious microbes around the world. There is no question that antibiotics as a biotechnology product have made a major difference in improving our quality of life when it comes to combating disease. There are also examples where overuse of antibiotics is causing health concerns in individuals on long-term antibiotic treatments, and such use can also lead to new strains of antibiotic-resistant microbes. Thus bioprospecting continues in our war against infection, as biotechnology companies continue to seek novel antibiotics from microbes.

QUESTIONS & ACTIVITIES

Answers can be found in Appendix 1.

1. Explain how prokaryotic cell structure differs from eukaryotic cell structure by describing at least three structural differences. Provide specific examples of how prokaryotic cells have served important roles in biotechnology.

2. Describe how yeasts differ from bacteria and describe the role(s) of yeast in at least two important biotechnology applications.

3. During what step of the calcium chloride transformation procedure does DNA bind to cells? What step in this procedure causes DNA to enter cells? Describe several advantages of electroporation over the calcium chloride procedure for transformation.

4. Discuss the four major ways in which vaccines can be made to fight a virus or bacterium, and describe how each vaccine works.

5. Why are anaerobic microbes important for making many foods?

6. If you were interested in producing a sweet-tasting wine with low alcohol content, explain how you might attempt to control the oxygen concentration in your bioreactor to achieve the desired results.

7. If you were in charge of protecting citizens from a bioterrorism attack, what biotechnology strategies would you consider for monitoring infectious agents? Do you think that vaccination against a bioweapon such as the *Variola vaccinia* virus, which causes smallpox, should be mandatory for

all citizens even if the vaccine causes life-threatening side effects in some people?

8. How can information gained from studying microbial genomes be used in microbial biotechnology? Provide three examples.

9. Search the FDA website (http://www.fda.gov/) and the CDC website (http://www.cdc.gov/) to determine (1) which influenza strains are approved for vaccine development for the year, (2) which companies have been approved to manufacture the vaccine, and (3) how many doses of the vaccine will be prepared.

10. We have become accustomed to being vaccinated against pathogens such as those that cause measles and polio. However, some states are working on proposals to require the administration of the human papillomavirus vaccine (Gardasil), against the sexually transmitted human papillomavirus, to all teen girls. What do you think about this proposal? Should such vaccination be required? What does this proposal presume about teen sexual activity? This is the most expensive vaccine in history (~$360 for three doses). If such a program becomes mandatory, should it be subsidized by state or federal governments?

11. What is a fusion protein and how can it be used to isolate a recombinant protein of interest?

12. Name three therapeutic recombinant proteins produced in bacteria and explain what they are used for.

13. Most vaccines are designed to be preventative or prophylactic. What does this mean?

14. What is metagenomics? In your answer provide an example of why metagenomics projects can be valuable.

Visit www.pearsonhighered.com/biotechnology

To download learning objectives, chapter summary, "Keeping Current" web links, glossary, flashcards, references, and jpegs of figures from this chapter.

Plant Biotechnology

After completing this chapter, you should be able to:

- Describe the impact of biotechnology on the agricultural industry.

- Discuss the limitations of conventional crossbreeding techniques as a means of developing new plant products.

- Explain why plants are suitable for genetic engineering.

- List and describe several methods used in plant transgenesis, including protoplast fusion, the leaf fragment technique, and gene guns.

- Describe the use of *Agrobacterium* and the Ti plasmid as a gene vector.

- Define antisense technology and give an example of its use in plant biotechnology.

- List some genetically engineered crops.

- Outline the environmental impacts, both pro and con, of biotechnologically enhanced crops.

- Describe the phytochemical opportunities in plant biotechnology.

- Outline several ways in which biotechnology has the potential to reduce hunger and malnutrition around the world.

Bioengineered plants being evaluated for quality control at a plant biotechnology company.

The red juicy tomatoes on sale at the grocery store are a true feat of engineering. Countless generations of selective breeding have transformed a puny acidic berry into the delicious fruit we know today. In the last few decades, conventional hybridization (by cross-pollination) has produced tomatoes that are easy to grow, quick to ripen, and resistant to disease. Pioneering efforts in biotechnological research have created tomatoes that can stay on store shelves longer without losing flavor. The future holds the possibility of an even more amazing transformation for the tomato: it, and other foods, could someday supplement or possibly replace inoculation as a means of vaccination against human disease. For example, researchers have successfully vaccinated volunteer patients against Norwalk virus in clinical trials by having them eat transgenic potatoes that express the vaccine.

In this chapter, we will consider the role of biotechnology in agriculture. First, we'll survey the industry to clarify the motives driving biotech research and development. We will then look more closely at the specific methods used to exchange genes in plants, including how bioengineering can protect crops from disease, reduce the need for pesticides, and improve the nutritional value of foods. Next, we'll examine the future of plant-based biotechnology products, from pharmaceuticals to petroleum alternatives. Finally, we'll consider the environmental and health concerns surrounding plant biotechnology.

FORECASTING THE FUTURE

The transfer of genes to plants has been firmly established as a reliable method to meet the need of future generations for food and energy. Fourteen million farmers in 25 countries currently benefit from genetically engineered (GE) crops, according to the International Service for the Acquisition of Agri-biotech Applications (ISAAA). There are limitations, however, to the entry of new plants to the market, and only a small number of crops have done so. These limitations are due in part to the fact that most of the thousands of existing patents for transgenic plants are held by only three companies: Syngenta, Monsanto, and DuPont. The high royalty fees and restrictions these companies have in place make the costs of developing crops based on these patents prohibitive for most farmers and researchers. This has led several biotech companies to develop their own novel gene transfer methods, and now nonprofit research groups such as the Public Sector Intellectual Property Research for Agriculture (PIPRA) are involved. The Animal and Plant Health Inspection Service (APHIS) of the U.S. Department of Agriculture (USDA) is also assisting. They have launched a pilot program to improve compliance with APHIS field trials and restricted

the movement of regulated organisms by certifying companies through the new Biotechnology Quality Management Service (BQMS). A company that qualifies for BQMS certification can overcome many of the expensive barriers to marketing GE crops by testing the effectiveness of transgenic plants in the field and focusing on products that are safe, affordable, and driven by public need. The future of plant biotechnology is brighter with changes that make producing genetically engineered crops less expensive and improve consumer perceptions of new crops.

6.1 The Future of Agriculture: Plant Transgenics

Over the past 40 years, the world population has nearly doubled while the amount of land available for agriculture has increased by a scant 10%. Yet we still live in a world of comparative abundance. In fact, world food production per person has increased 25% over the past 40 years. How has it been possible to feed so many people with only a marginal increase in available land? Most of that improved productivity has depended on crossbreeding methods developed hundreds of years ago to provide animals and plants with specific traits. Recently, however, the development of new, more productive crops has been accelerated by the direct transfer of genes, as shown in **Figure 6.1**.

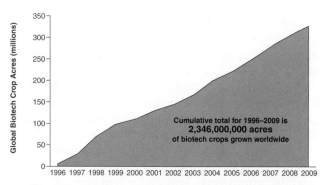

FIGURE 6.1   Total Acres of Biotech Crops Grown in 2009
Between 1996 and 2009, more than 2.3 billion acres of biotech crops were successfully grown as a result of approximately 70 million repeat decisions by farmers to grow these crops. Almost 93% (13 million) of the 14 million farmers growing biotech crops in 2009 were small and resource-poor, living and working in developing countries such as China, India, the Philippines, and South Africa. In terms of total acreage of biotech crops planted, the developed world (with its larger farms) planted 40% of the acreage, while developing countries accounted for 60%. 2009 is the latest year for which figures are available.

Plant transgenesis (the direct transfer of genes to plants) permits innovations that are impossible to achieve with conventional hybridization methods. A few developments that have significant commercial potential are plants that produce their own pesticides, plants that are resistant to herbicides, and even bioproducts like plant vaccines and biofuels. Because producing transgenic plant proteins is relatively easy and the quality of the proteins is reasonably good, the prospects for future research and development in these areas look especially bright. For example, through classic breeding, the average strength of cotton fibers has been steadily increasing by about 1.5% per year. Biotechnology has dramatically accelerated this pace. By inserting a single gene, the strength of one major upland cotton variety was increased by 60%.

The prevalence of transgenic crops in world agriculture is on the rise. In 2005, the billionth acre was planted, and only 2 years later, in 2007, the second billionth was planted. Although the transgenic revolution began in the United States and other developed nations and the United States still dominates the market, other countries have continued to add transgenic crops in recent years. In 2008, some 13.3 million farmers in 25 countries planted transgenic crops—90% of them in developed countries—but just a year later, in 2009, world farmers planted 173 million acres of soybeans (70% of the total planted), 40 million acres of corn (40% of the total planted), 25 million acres of cotton (10% of the total planted), and 15 million acres of canola (5% of the total planted). To these staple crops, herbicide-resistant sugar beets were added in the United States and Canada in 2009 (59% of the beet crop was transgenic in its first year). New biotech crops such as alfalfa, wheat, and potatoes will come to the market within 10 years, as well as high-oleic-acid soybeans and vitamin A–enriched rice. Beyond foods and feed crops, plant-produced vaccines and bioplastics as well as enhanced phytoremediation plants will reach the market in the next 10 years (see Chapter 9).

Although food crops are only one aspect of biotechnology's impact, they have been the focus of considerable controversy worldwide. On the one hand, hunger continues to plague much of the world. This reality is a compelling argument for the rapid development of more productive and nutritious crops. On the other, some sectors are concerned that experimentation could be harmful to the environment and human health.

The debate is far from over. To develop an informed opinion, decision makers must understand the science behind these new products, analyze the products themselves, and be knowledgeable about the regulations that exist to monitor biotechnology

research. In any case, it is unlikely that the revolution in agricultural biotechnology will stop. Protests or not, biotechnology plant products will play a key role in our society. The next section describes the methods used to create new agricultural products.

6.2 Methods Used in Plant Transgenesis

Genetic manipulation of plants is not new. Ever since the birth of agriculture, farmers have selected for plants with desired traits. Since then, technologies have been ever evolving to meet the world's increasing world for agricultural products.

Conventional Selective Breeding and Hybridization

Even though careful crossbreeding has continued to improve plants through the millennia—giving us larger corn cobs, juicier apples, and a host of other modernized crops—the methods of classic plant breeding are slow and uncertain. Creating a plant with desired characteristics requires facilitating a sexual cross between two different plant lines and repeated backcrossing between the hybrid plant's offspring and one of the original parent plants. Isolating a desired trait in this fashion can take years. For instance, Luther Burbank's development of the white blackberry involved 65,000 unsuccessful crosses. In fact, plants from different species generally do not hybridize, so a genetic trait cannot be isolated and refined unless it already exists in a plant strain.

Transferring genes by crossing plants is not the only traditional way to create plants with desirable features. **Polyploids** (multiple chromosome sets greater than normal, usually more than 2N or one set of chromosomes from each parent) plants have been used for many years as a means to increase desirable traits (especially the size) of many crops including watermelons, sweet potatoes, bananas, strawberries, and wheat. The utilization of the drug colchicine (which keeps a cell from dividing after it has double chromosomes) followed by hybridization is a means of introducing commercially important features of related species into potentially new cultivated crops (refer to **Figure 6.2**). This process results in hybrid plants where whole chromosome sets, rather than single genes, from related plants can be transferred. The additional chromosomes often produce plants that produce fruits and vegetable much larger than native wild varieties.

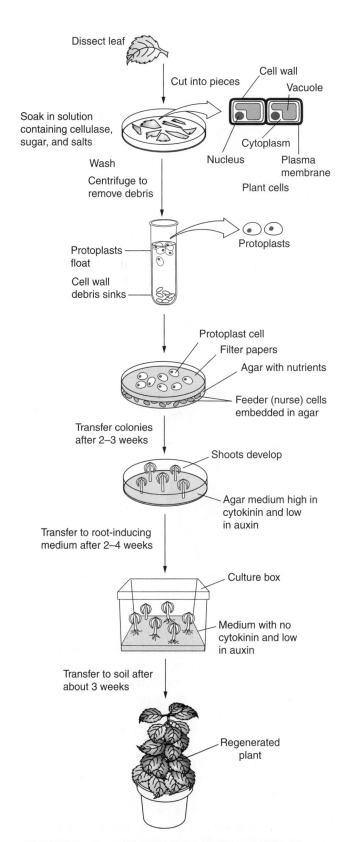

Dissect leaf

Cut into pieces

Cell wall

Vacuole

Soak in solution containing cellulase, sugar, and salts

Cytoplasm

Nucleus

Plasma membrane

Plant cells

Wash

Centrifuge to remove debris

Protoplasts

Protoplasts float

Cell wall debris sinks

Protoplast cell

Filter papers

Agar with nutrients

Feeder (nurse) cells embedded in agar

Transfer colonies after 2–3 weeks

Shoots develop

Agar medium high in cytokinin and low in auxin

Transfer to root-inducing medium after 2–4 weeks

Culture box

Medium with no cytokinin and low in auxin

Transfer to soil after about 3 weeks

Regenerated plant

FIGURE 6.2 Protoplast Fusion and Regeneration of a Hybrid Plant After dissecting a plant leaf, it is possible to create protoplasts by digesting the cell wall with the enzyme cellulase. To create a hybrid plant, protoplasts from different plants are fused by culturing them in a sterile medium that stimulates shoots (cytokinin) and roots (auxin).

Cloning: Growing Plants from Single Cells

Plant cells are different from animal cells in many ways, but one characteristic of plant cells is especially important to biotechnology: many types of plants can regenerate from a single cell. The resulting plant is a genetic replica—or clone—of the parent cell. Animals can be cloned too, of course, but the process is more complicated. (Chapter 7 discusses animal cloning in detail.) This natural ability of plant cells has made them ideal for genetic research. After new genetic material is introduced into a plant cell, the cell rapidly produces a mature plant, and the researcher can see the results of the genetic modification in a relatively short time. Next we consider some of the methods used to insert genetic information into plant cells.

Protoplast fusion

When a plant is injured, a mass of cells called a **callus** may grow over the site of the wound. Callus cells have the capability to redifferentiate into shoots and roots, and a whole flowering plant can be produced at the site of the injury. You may have taken advantage of this capability if you have ever "cloned" a favorite house plant by rooting a cutting.

The natural potential of these cells to be reprogrammed makes them ideal candidates for genetic manipulation. Like any plant cells, however, callus cells are surrounded by a thick wall of cellulose, a barrier that hampers any uptake of new DNA. Fortunately the cell wall can be dissolved with the enzyme cellulase, leaving a denuded cell called a **protoplast.** The protoplast can be fused with another protoplast from a different species, creating a cell that can grow into a hybrid plant. This method, called **protoplast fusion,** as shown in Figure 6.2, has been used to create broccoflower, a fusion of broccoli and cauliflower, as well as other novel plants.

Leaf fragment technique

Genetic transfer occurs naturally in plants in response to some pathogenic organisms. For instance, a wound can be infected by a soil bacterium called *Agrobacterium tumefaciens* (recently reclassified as *Rhizobium radiobacter* by genome analysis, but the name *Agrobactor* is still in common usage; we use it throughout our discussion). This bacterium contains a large circular double-stranded DNA molecule called a *plasmid,* which triggers an uncontrolled growth of cells (tumor) in the plant. For this reason, it is known as a **tumor-inducing (TI) plasmid.** The resulting tumor is known as **crown gall.** If you have

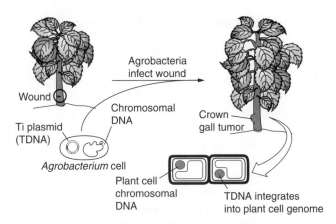

FIGURE 6.3 Process of Crown Gall Formation The Ti plasmid of *Agrobacter* causes "crown galls" in susceptible plants. Through genetic engineering, it is possible to silence the tumor-inducing genes and insert desirable genes into the plasmid, making it a vector for gene transfer.

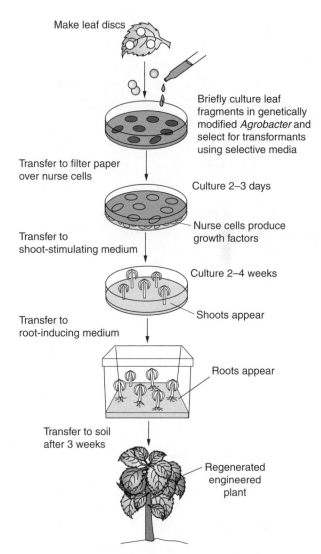

FIGURE 6.4 Transfer of Genetically Modified Ti Plasmid to Susceptible Plants through Tissue Culture

ever seen a swelling on a tree or rose bush, you may have seen *Agrobacter*'s effects (see **Figure 6.3**).

The bacterial plasmid gives biotechnologists an ideal vehicle for transferring DNA. To put that vehicle to use, researchers often employ the **leaf fragment technique.** In this method, small discs are cut from a leaf. When the fragments begin to regenerate, they are cultured briefly in a medium containing genetically modified *Agrobacter,* as shown in **Figure 6.4.** During this exposure, the DNA from the TI plasmid integrates with the DNA of the host cell, and the genetic payload is delivered. The leaf discs are then treated with plant hormones to stimulate shoot and root development before the new plants are planted in soil.

The major limitation to this process is that *Agrobacter* cannot infect **monocotyledonous** plants (plants that grow from a single seed embryo) such as corn and wheat. **Dicotyledonous** plants (plants that grow from two seed halves)—such as tomatoes, potatoes, apples, and soybeans—are all good candidates for the process, however.

Gene guns

Instead of relying on a microbial vehicle, researchers can use a **gene gun** to literally blast tiny metal beads coated with DNA into an embryonic plant cell, as shown in **Figure 6.5**. The process is rather hit or miss—and more than a little messy—but some of the plant cells will adopt the new DNA.

Gene guns are typically used to shoot DNA into the nucleus of a plant cell, but they can also be aimed at the **chloroplast,** the part of the cell that contains chlorophyll. Plants have between 10 and 100 chloroplasts per cell, and each chloroplast contains its own

bundle of DNA. Whether they target the nucleus or the chloroplast, researchers must be able to identify the cells that have incorporated the new DNA. In one common approach, they combine the gene of interest with a gene that makes the cell resistant to certain antibiotics. This gene is called a **marker gene** or reporter gene. After firing the gene gun, the researchers collect the cells and try to grow them in a medium that contains a specific antibiotic. Only the genetically transformed cells will survive. The antibiotic-resistant gene can then be removed before the cells grow into mature plants, if the researcher so desires (for a more detailed explanation of gene markers, see Chapter 5).

Chloroplast engineering

The chloroplast can be an inviting target for bioengineers. Unlike the DNA in a cell's nucleus, the DNA in a

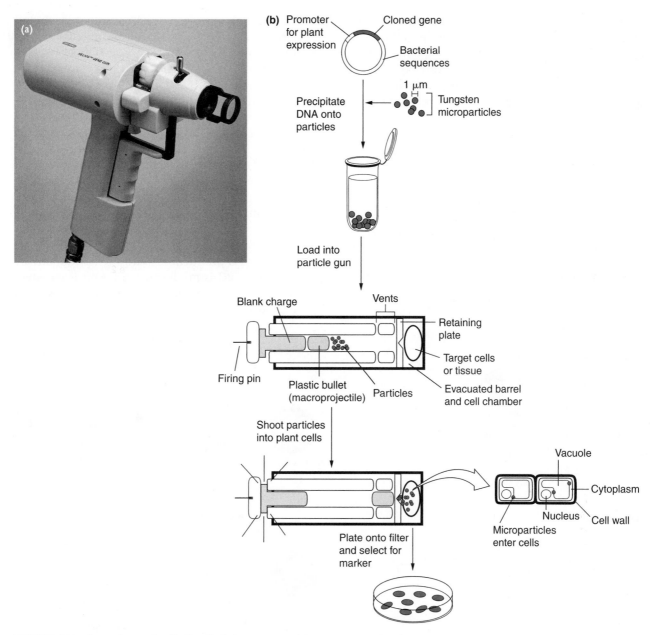

FIGURE 6.5 **Gene Guns** The "bullets" of the gene gun (a) are tungsten beads 1 micrometer in diameter. (b) DNA is coated on the surfaces of the beads and fired from guns with velocities of about 430 meters per second. The targets can include suspensions of embryonic cells, intact leaves, and soft seed kernels.

chloroplast can accept several new genes at once. Also, a high percentage of genes inserted into the chloroplast will remain active when the plant matures. Another advantage is that the DNA in the chloroplast is completely separate from the DNA released in a plant's pollen. When chloroplasts are genetically modified, there is no chance that transformed genes will be carried on the wind to distant crops. This process is shown in **Figure 6.6**, on page 164.

Antisense Technology

Recall the tomato. It is red, juicy, and tasty—and extremely perishable. When picked ripe, most tomatoes will turn to mush within days. But the Flavr Savr tomato, introduced in 1994 after years of experimentation, stayed ripe and fresh for weeks. The Flavr Savr tomato was the first genetically modified food approved by the U.S. Food and Drug Administration

(FDA) and, although it was not an economic success and is no longer available, it is common to find other varieties of genetically modified food, including tomatoes, on the market today. These foods were developed using antisense technology, in which a gene that encodes for a specific trait is removed from plant cells,

used to produce a complementary copy of itself, and transferred back to the original cells using *Agrobacter* as a vector organism.

Ripe tomatoes normally produce the enzyme **polygalacturonase** (or PG), a chemical substance that digests pectin in the wall of the plant. This digestion induces the normal decay that is part of the natural plant cycle. The gene that encodes PG was identified, removed from plant cells, and used to produce a complementary copy of itself. Using *Agrobacter* as a vector, the gene was then transferred into tomato cells. Once in the cell, the gene encoded an mRNA molecule (**antisense molecule)** that united antisense RNA with and inactivated (complementary sequence) the normal mRNA molecule (the sense molecule) for PG production. With the normal mRNA inactivated, no PG is produced, no pectin is digested, and natural "rotting" is slowed. This process is shown in **Figure 6.7**.

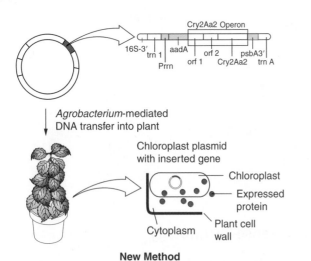

New Method

FIGURE 6.6   Old versus New Method for Cloning Multiple Genes into Plants In the past, when more than one gene was to be expressed, two plants—each with its own inserted gene—had to be produced. (a) Standard crossing by pollen transfer was required to produce the hybrid plant. (b) Now it is possible to insert more than one gene by stacking them in the chloroplast DNA.

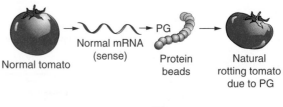

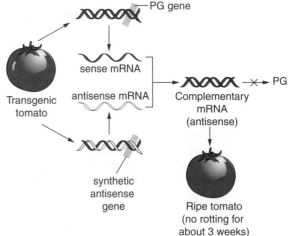

FIGURE 6.7   One of the First Commercial Transgenic Plant Products Tomatoes that rot only slowly are produced by first isolating the gene that encodes polygalacturonase. This gene encodes normal (sense) mRNA, which is translated into PG. After inducing it to produce a cDNA (complementary DNA) counterpart, it is possible to insert the PG into a vector for transfer to tomatoes (see Chapter 3). Transgenic plants with the new insert will produce PG mRNA and antisense mRNA, canceling the production of PG. This slows the rotting process and produces a tomato that will last for about 3 weeks after ripening.

Gene Silencing: An Alternative to Antisense Technology

We can expect to see more gene silencing advancements in the future. Researchers are working on a potato that resists bruising. They have removed a gene responsible for producing an enzyme that promotes color changes in peeled potatoes. It is a subtle improvement, but market analyses have shown that consumers prefer potatoes that are not bruised by handling. Researchers are also working on ways to increase the protein content of plants like potatoes by using genes from chickens. This improvement in the nutritional value of a common food could help many people worldwide to get the protein they need in their diets.

A Purdue University researcher has discovered a fountain of youth for tomatoes that extends shelf life by about a week. He found that by adding a yeast gene, he could increase production of a compound that slows aging and delays microbial decay in tomatoes. The organic compound spermidine is a polyamine found in all living cells. Polyamines such as spermidine enhance the nutritional and processing quality of tomato fruits. Fully ripe tomatoes from those plants, compared with nontransgenic plants, last about 8 days longer before shriveling. Symptoms of decay associated with fungi were delayed by about 3 days. Shelf life is a major problem for any type of produce, especially in countries such as those in Southeast Asia and Africa, which cannot afford controlled-environment storage.

6.3 Practical Applications

Protecting plants from viruses, insects, and weeds while improving their nutrition and properties is the goal of commercial growers, and biotechnology has produced some interesting examples.

Vaccines for Plants

Crops are vulnerable to a wide range of plant viruses. Infections can lead to reduced growth rates, poor crop yields, and low crop quality. Fortunately farmers can protect their crops by stimulating a plant's natural defenses against disease with vaccines. Just like a human vaccine for polio, plant vaccines contain dead or weakened strains of the plant virus that turn on the plant's version of an immune system, making it resistant to the real virus.

Vaccinating an entire field is not easy, but it is no longer necessary. Instead of injecting the vaccine, the vaccine can be encoded in a plant's DNA. For example, researchers have recently inserted a gene from the tobacco mosaic virus (TMV) into tobacco plants. The gene produces a protein found on the surface of the virus and, like a vaccine, turns on the plant's immune system. Tobacco plants with this gene are immune to TMV. The tomato mosaic virus is similar enough to be stopped by this technique. **Figure 6.8** shows this process.

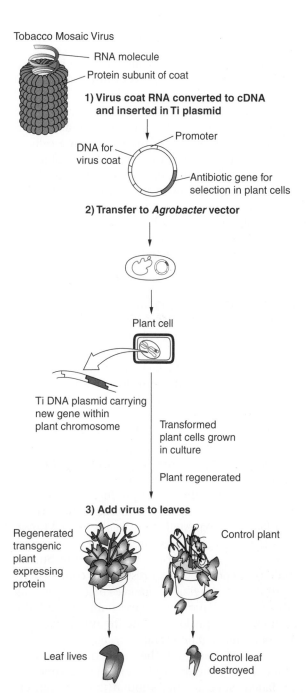

FIGURE 6.8 Plant Vaccines Transgenic plants expressing the TMV coat protein (CP) were produced by *Agrobacterium*-mediated gene transfer. When the generated plants expressing the viral CP were infected with TMV, they exhibited increased resistance to infection. Whereas control plants developed symptoms in 3 to 4 days, transgenic CP-expressing plants resisted infection for 30 days.

Genetic vaccines have already proven themselves in a wide variety of crops. The development of disease-resistant plants has revitalized the once-ravaged papaya industry in Hawaii. In addition, disease and pest-resistant strains of potatoes offer many advantages to both growers and consumers.

Genetic Pesticides

For the last 50 years, many farmers have relied on a natural bacterial pesticide to prevent insect damage to crops. *Bacillus thuringiensis* (**Bt**) produces a crystallized protein that kills harmful insects and their larvae. The crystalline protein (from the *Cry* gene) breaks down the cementing substance that fuses the lining cells of the digestive tract in certain insects. Insects subjected to this protein die in a short time from "autodigestion." The *Cry* gene causing this event is the subject of an expanding market of "insect-resistant" genetically engineered plants. By spreading spores of the bacterium across their fields, farmers can protect their crops without using harmful chemicals.

Now, instead of spreading the bacterium directly across their fields, farmers can grow plants containing Bt genes. Plants that contain the gene for the Bt toxin have a built-in defense. This biotechnologically enhanced pesticide has been successfully introduced into a wide range of plants, including tobacco, tomatoes, corn, and cotton. In fact, most soybean seeds planted today contain the gene for Bt toxin, which effectively kills cotton-infesting insects. Bt with its insecticidal protein is shown in **Figure 6.9a**.

The widespread use of the Bt gene is one of the most remarkable success stories in biotechnology. It is also one of the biggest sources of controversy. Cornell researchers conducted a laboratory experiment in 1999 suggesting that the pollen produced by bioengineered corn could be deadly to monarch butterflies. The results were expected. Researchers had known for years that, in large doses, the toxin naturally produced by *B. thuringiensis* could be harmful to butterflies. Still, the report set off a firestorm of controversy. It was the first tangible evidence that genetically altered food could harm the environment, and the monarch butterfly quickly became the unofficial mascot of opponents of genetic engineering.

When researchers took their experiments out of the laboratory and into the field, many of their concerns quickly faded. Several studies found that few butterflies in the real world would be exposed to enough pollen to cause any harm. In fact, butterflies are unlikely to ingest toxic amounts of pollen even if they feed on milkweed plants less than 1 meter from the typical field of genetically modified corn. Still, scientists speculate that a small percentage of butterflies will

(a)

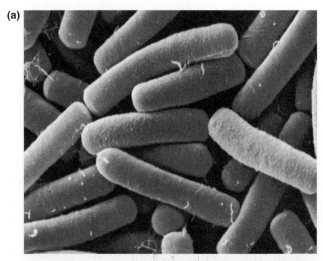

(b)

TEM, × 40,000

FIGURE 6.9 **Bt with Crystal of Insecticidal Protein** Figure (a) shows the crystalline protein as a crystal within the *Bacillus thuringiensis* bacterium. Genetically engineered plants expressing the *Cry* gene are "insect resistant" owing to the production of a small amount of this bacterial protein. Insect larvae (b) that would normally consume plant tissue will die if they consume the crystalline protein (*Cry* gene product).

inevitably get dusted with a lethal amount of pollen. Some of the monarchs that survive the exposure may then be unfit for their long migrations. On the whole, however, concerns that genetically altered corn could devastate monarchs seem to have been disproven: after 2 years of study, the Agricultural Research Service (a division of the USDA) announced in 2002 that Bt toxin posed little risk to monarch butterflies in real-world situations.

Herbicide Resistance

Traditional weed killers have a fundamental flaw: they often kill desirable plants along with weeds. Today, biotechnology allows farmers to use herbicides without threatening their livelihood. Crops can be

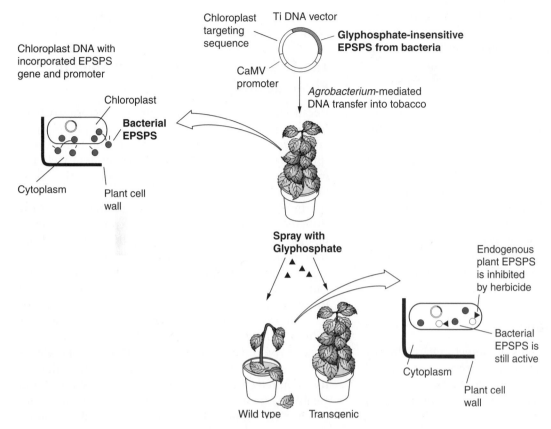

FIGURE 6.10 **Engineering Herbicide-Resistant Plants** Glyphosphate is a compound that blocks a key enzyme in plant protein synthesis. Plants sprayed with this compound (e.g., Roundup) will die because a key enzyme is blocked. Genetically engineered plants that express an alternative gene (*EPSPS*) can bypass this blockage with an enzyme that uses an alternative chemical pathway. Transferring the TDNA vector (from Agrobacter) with the *EPSPS* gene (from bacteria) to plants and activating the promoter (with the help of a chloroplast-targeting sequence) can lead to incorporation of the EPSPS gene in the plants, making them resistant to the glyphosphate. This process has been used in many commercial plant varieties, making them herbicide-resistant and allowing the grower to spray susceptible weeds and resistant plants (and substituting glyphosphate for more harmful herbicides).

genetically engineered to be resistant to common herbicides such as **glyphosphate.** This herbicide works by blocking the enzyme EPSPS, which functions in a biochemical pathway responsible for the synthesis of aromatic amino acids and other compounds vital to plant growth and survival. Through bioengineering, scientists have created transgenic crops that produce an alternative enzyme that is not affected by glyphosphate, meaning that weeds are susceptible while desirable plants are not. This approach has been especially successful in soybeans. Most soybeans grown today contain herbicide-resistant genes. The process is shown in **Figure 6.10**. Glyphosphate is currently the world's most widely used herbicide, effectively controlling of a wide group of unwanted plant species.

Since 1996, the high usage of transgenic glyphosphate-resistant crops in the Americas has led to exclusive use of glyphosphate for weed control over very large areas.

Unfortunately, in regions where transgenic glyphosphate-resistant crops dominate, glyphosphate-resistant populations of damaging weed species have now evolved. A single-site mutation of a proline amino acid at position 101 in *EPSPS* has been implicated in these glyphosate-resistant weeds, which have been found in the United States, Argentina, and Brazil. As more transgenic glyphosphate-resistant crops are planted over the next few years, it is anticipated that other glyphosphate-resistant weed species will evolve. In response to this development,

Monsanto has added a broad-spectrum weed killer to reduce the development of resistant weeds. Future use of glyphosphate-resistant plants will require engineering to combat this event. Other companies are at work on similar products to combat glyphosphate-resistant weeds. Glyphosphate is considered essential for present and future world food production, and any action to secure its sustainability has become a global imperative.

Farmers who plant herbicide-resistant crops are generally able to control weeds with chemicals that are milder and more environmentally friendly than typical herbicides. This development is significant because, before the advent of resistant crops, U.S. cotton farmers spent $300 million per year on harsh chemicals to spray on their fields, exclusive of the large cost in human manual labor necessary to keep cotton plants weed-free, which is no longer necessary.

Enhanced Nutrition

Of all the potential benefits of biotechnology, nothing is more important than the opportunity to save millions of people from the crippling effects of malnutrition. One potential weapon against malnutrition is **Golden Rice**—rice that has been genetically modified to produce large amounts of beta carotene, a provitamin that the body converts to vitamin A. According to recent estimates, 500,000 children in many parts of the world will eventually become blind because of vitamin A deficiency. Currently health workers carry doses of vitamin A from village to village in an effort to prevent blindness. Simply adding this nutrient to the food supply would be much more efficient and, in theory, much more effective.

Biotechnology may not, however, be the magic bullet that ends malnutrition. Although promising, genetically modified foods have their limitations. For instance, the provitamin in Golden Rice must dissolve in fat before it can be used by the body. Children who do not get enough fat in their diets may not be able to reap the full benefits of the enriched rice. Some groups would like to see more conventional breeding techniques used to combat world hunger. For example, although Golden Rice was ready to appear within 2 years of its development, no farmers have, as of 2011, yet planted the rice, largely because of concerns voiced by environmental organizations. These groups endorse programs such as Harvest Plus in place of the introduction of transgenic crops in developing countries. Harvest Plus is a collection of 12 crops that aim to boost levels of vitamin A, iron, and zinc, and it relies on conventional breeding. However, other groups support transgenic crops in these same locations: the Bill and Melinda Gates Foundation, for example, is spending $36 million to support Golden Rice, GM cassava, sorghum, and bananas.

The Future of Plant Biotechnology in Pharmacology

Recall that plants can be ideal protein factories (Chapter 4). A single field of a transgenic crop can produce a large amount of commercially valuable protein. At this time, transgenic corn has the highest protein yield per invested dollar of any bioreactor organism. The possibilities are practically endless.

In the not-so-distant future, farmers will grow human medicine along with their crops. It is already possible to harvest human growth hormone from transgenic tobacco plants. Plants can also manufacture vaccines for humans, as we've seen. Edible vaccines can be produced by introducing a gene for a subunit of the virus or bacterium. The plant expresses this protein subunit, and it is eaten with the plant. When the subunit antigen enters the bloodstream, the immune system produces antibodies against it, providing immunity. The need for inexpensive vaccines that do not require refrigeration was first voiced by the World Health Organization in the early 1990s and has resulted in studies of vaccines in bananas, potatoes, tomatoes, lettuce, rice, wheat, soybeans, and corn. Researchers at Cornell University have recently created tomatoes and bananas that produce a human vaccine against the viral infection hepatitis B. Researchers are actively studying the tomato as another source of pharmaceuticals. Through engineering of the chloroplast (abundant in green tomatoes), scientists hope to create an edible source of vaccines and antibodies, as shown in Table 6.1.

Plants express a wide range of chemical compounds called phytochemicals, and biotechnologists are converting plants into small-scale factories to produce chemicals useful to human health. Biotechnology can alter the production of complex technical therapeutic proteins via plant pathways, with examples including antibodies, blood products, cytokines, growth factors, hormones, and recombinant enzymes. This "molecular farming" will likely bring several products to market in the near future with applications to the treatment of diseases and conditions such as cystic fibrosis, non-Hodgkin's lymphoma, hepatitis, Norwalk virus, rabies, and a range of gastrointestinal illnesses.

Rather than growing human or animal cells on expensive nutrient-rich media, biopharmers insert genes into the cell of plants and the plants do the work of transcribing and folding the protein. Since plants can be grown in larger quantities than cell cultures, they can offer a much greater volume of prod-

TABLE 6.1

Beneficial Crop Product Traits	Crops Available Now	Crops That May Be Available in the Future
Bt crops are protected against insect damage and reduce pesticide use. The plants produce a protein—toxic only to certain insects—found in *Bacillus thuringiensis*.	Corn, cotton, potatoes	Sunflower, soybeans, canola, wheat, tomatoes
Herbicide-tolerant crops allow farmers to apply a specific herbicide to control weeds without harm to the crop. They give farmers greater flexibility in pest management and promote conservation tillage.	Soybeans, cotton, corn, canola, rice	Wheat, sugar beets
Disease-resistant crops are armed against destructive viral plant disease with the plant equivalent of a vaccine.	Sweet potatoes, cassava, rice, corn, squash, papaya	Tomatoes, bananas
High-performance cooking oils maintain texture at high temperatures, reduce the need for processing, and create healthier food products. The oils are either high-oleic or low-linoenic types. In the future, there will also be high-stearate oils.	Sunflower, peanuts, soybeans	
Healthier cooking oils have reduced saturated fat.	Soybeans	
Delayed-ripening fruits and vegetables have superior flavor, color, and texture; are firmer for shipping; and stay fresh longer.	Tomatoes	Raspberries, strawberries, cherries, tomatoes, bananas, pineapples
Increased-solids tomatoes have superior taste and texture for processed tomato pastes and sauces.	Tomatoes	
Nutritionally enhanced foods will offer increased levels of nutrients, vitamins, and other healthful phytochemicals. Benefits range from helping developing nations meet basic dietary requirements and boosting disease-fighting and health-promoting foods.		Protein-enhanced sweet potatoes and rice; high-vitamin-A canola oil; increased antioxidant fruits and vegetables

Source: BIO member survey.

uct than a manufacturing plant would. However, since the first human-like enzyme was first produced in transgenic tobacco plants in 1992 at Virginia Polytechnic Institute, the biopharma industry has had a wave of trials with no approvals. Nevertheless, the first plant-based pharmaceutical products may be on the market before long.

Protalix, a biotech company in Israel, has FDA approval for a drug to treat Gaucher's disease. The disease has no cure, but the product in development breaks down the accumulating fatty substances, and patients would have to take this drug throughout their lives. Since there is no other ready supply, the drug received fast-track approval (see Chapter 12 for a full description of approvals for "orphan" drugs). The drug is manufactured by producing proteins from carrot cell cultures in disposable plastic bioreactors. Another company, Medicago, a U.S. company, is developing flu vaccines in tobacco plants (see **Figure 6.11** on page 170) grown in greenhouses. After the plants are transformed by *Agrobacter*, they produce the necessary protein shell, which is then harvested and made into a vaccine.

The Future of Plant Biotechnology: Fuels

Biofuels (fuels produced from biological products, such as plants) can be produced almost anywhere in the world from homegrown raw materials and may be an important use of plant biotechnology in the future. As the need for alternatives to fossil fuels increases, the U.S. government is looking toward biotechnology

FIGURE 6.11 **Botanical Biopharming** Genetically transformed water plants being transformed to produce pharmaceutical proteins.

line to produce 10 gallons of kernel corn ethanol, which represents a relatively modest net gain. For this reason biotechnology is needed to convert the readily available cellulosic sources into biofuels—perhaps by developing biofuel-producing organisms—and thus making this procedure more economically viable.

Biofuels from Plant Waste

In the future, there may be other opportunities for plants to provide fuels. Specifically, scientists are developing methods to capture energy trapped in plant wastes. The solar energy captured through photosynthesis enables the storage of energy in plant cell wall polymers (cellulose, lignin, and hemicelluloses in straw, husks, hulls, and trees). This energy remains trapped unless plants are burned. Despite the increasing use of starch-based ethanol and biodiesel, fuel produced from plant by-products has been unavailable up to now. If that energy could be released, grasses, wood, and crop residues would offer the possibility of a renewable, geographically distributed source of sugars for conversion to fuel. This process would include collection, destruction of the cell walls (pretreatment), and conversion of the sugars to biofuels.

One such sugar is hemicellulose—a family of polysaccharides composed of five- and six-carbon sugars fibrils. Lignin is the glue that crosslinks these fibrils to provide strength. One challenge in the development of plant wastes as fuel has been finding enzymes that can function in the high acid conditions needed for the pretreatment of hemicellulose in order to break down lignin. Research continues; the refineries developed

to offer solutions. The 2007 Biofuels Initiative increased federal funding 60% over the 2006 budget with the stated intention of replacing 30% of U.S. current fuel with bioethanol by 2030 (see **Figure 6.12**).

Bioethanol refineries have sprung up all over the Midwest (resulting largely from subsidies and incentives) and can be used to convert sugars from any cellulosic source. However it takes 7 gallons of gasoline

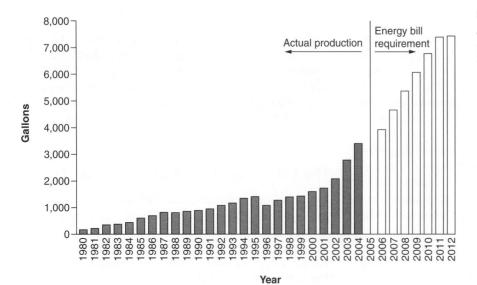

FIGURE 6.12 **U.S. Energy Policy Act of 2005** Calls for 7.5 billion gallons of ethanol and biodiesel to be added to America's fuel supply by 2012.

for bioethanol production from sugars can potentially be converted to convert these plant wastes to fuel.

Biofuels from algae

The most recent group of biofuel producers are looking beyond corn and French-fry grease to microscopic algae for the next alternative to petroleum. Microalgae naturally produce and store lipids (vegetable-like oils). If they can be biochemically altered to become more efficient, biofuels may result. The U.S. Department of Energy (DOE) explored the potential of algae prior to 1996, but when crude oil prices dropped from $50 to $20 per barrel that year, the DOE lost interest. With the passage of the Biofuels Initiative in 2007 and the rising cost of crude oil, interest in algae has returned. The challenge is to produce at least a million gallons of fuel per day. Solazyme, in San Francisco, has developed a closed bioreactor system of proprietary algae that eat anything from waste glycerol to sugar pulp and produce triacylglycerides and methanol. After chemical modification and concentration, the alkanes produced resemble biodiesel. Yields of 50% to 60% of oil per gram of algal cells are considered excellent, and Solazyme's algae are producing 75% oil per gram of dry weight. This company is the first to produce a barrel of microbial fuel oil and is confident that the goal of a million gallons of fuel per day can eventually be achieved.

6.4 Health and Environmental Concerns

Ever since the inception of transgenic plants, people have worried about potentially harmful effects to humans and the environment. In an age when "natural" is often equated with "safe," these decidedly unnatural plants carry an air of danger. Activists have staged protests against companies producing genetically modified plants (GMOs, or genetically modified organisms) (see **Figure 6.13**).

Such fears have the power to shake up an industry. In 2000, potato-processing plants in the Northwest stopped buying genetically modified potatoes. There was never any sign that these potatoes—engineered to be pest-resistant—were inferior or dangerous. They looked and tasted just like non–genetically modified potatoes, and farmers did not need to use gallons of chemicals to get them to grow. They were able to survive aphids and potato bugs but not the tide of public opinion (see **Figure 6.14**).

What are some opponents of plant biotechnology saying, and what are some of the other points of view? What are the pros and cons of GMO crop production?

FIGURE 6.13   A Parade in Boston Protesting Genetically Modified Food at the 2000 BIO Conference

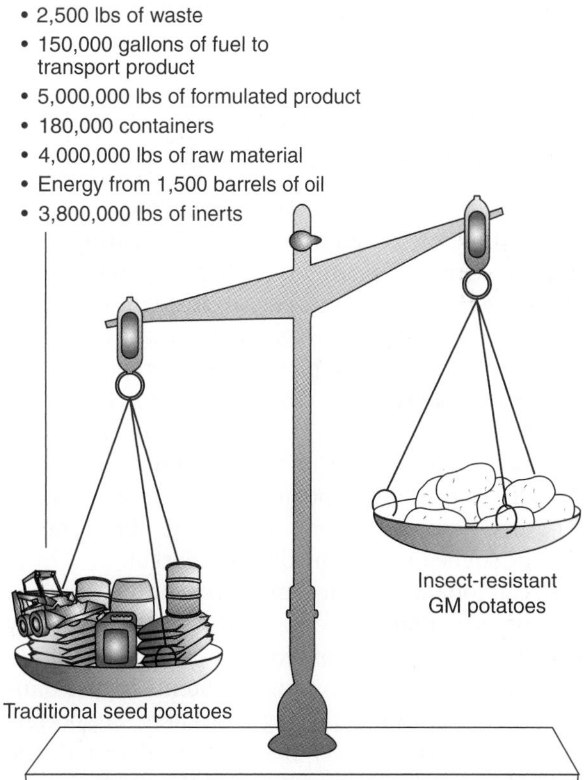

- 2,500 lbs of waste
- 150,000 gallons of fuel to transport product
- 5,000,000 lbs of formulated product
- 180,000 containers
- 4,000,000 lbs of raw material
- Energy from 1,500 barrels of oil
- 3,800,000 lbs of inerts

Insect-resistant GM potatoes

Traditional seed potatoes

Comparative resources from planting to harvest

FIGURE 6.14   Comparison of Resources for Production of GMO and Traditionally Grown Potatoes Considerably more raw materials and energy are expended in the production and application of this insecticide compared with having insecticidal genes in the plant itself.

TOOLS OF THE TRADE

Excision of Reporter Genes

Researchers know that a gene has been transferred to a plant cell because antibiotic-resistant genes (for example, kanamycin antibiotic resistance) are usually used as "reporters" for commercial plant engineering. They allow only those transformed plant cells that can live on the antibiotic medium to be selected. The presence of antibiotic genes (and small amounts of antibiotics) in plants has caused some public concern. However, we know that it is possible to remove specific genes after transformation and selection because four scientists at Rockefeller University have developed the process. It involves the use of a promoter that can be activated to stimulate excision through naturally existing mechanisms in the plant embryo or plant organ tissue *after* the antibiotic selection for the transformed cells has occurred (see Chapter 3 to learn more about the specific scientific tools used in the process described above).

Concerns about Human Health

Every plant contains DNA. Whenever you munch a carrot or a bite into a slice of bread, you're eating more than a few genes. Opponents of genetic engineering have nothing against genes per se. Instead, they fear the effects of *foreign* genes, bits of DNA that would not naturally be found in the plant. A 1996 report in the *New England Journal of Medicine* seemed to confirm at least some of those fears. The study found that soybeans containing a gene from the Brazil nut could trigger an allergic reaction in people who were sensitive to Brazil nuts. Because of this discovery, this type of transgenic soybean never made it to market.

We can look at this incident in two different ways. Opponents say that this case of the soybean clearly demonstrates the pitfalls of biotechnology. They envision many scenarios where novel proteins trigger dangerous reactions in unsuspecting customers. Supporters see it as a success story: the system detected the unusual threat before it ever reached the public.

At this time, most experts agree that genetically modified foods are unlikely to cause widespread allergic reactions. According to a recent report from the American Medical Association, very few proteins have the potential to trigger allergic reactions, and most of them are already well known to scientists. The odds of an unknown allergen "sneaking" into a genetically modified food on the grocery shelf are very small. In fact, biotechnology may someday help prevent allergy-related deaths. Researchers are now working to produce peanuts that lack the proteins that can trigger violent allergic reactions.

Allergies are not the only concern. Some scientists have speculated that the antibiotic-resistant genes used as markers in some transgenic plants could spread to disease-causing bacteria in humans. In theory, these bacteria would then become harder to treat. Fortunately bacteria do not regularly scavenge genes from our food. According to a recent report in the journal *Science,* there is only a "minuscule" chance that an antibiotic-resistant gene could ever pass from a plant to bacterium. Furthermore, many bacteria have already evolved antibiotic-resistant genes.

If you scan the antibiotechnology literature, you'll see many more accusations. Headlines such as "Frankenfoods may cause cancer" are common. To this date, however, science has not supported any of these concerns. The National Academy of Sciences recently reported that the transgenic food crops on the market today are perfectly safe for human consumption.

Concerns about the Environment

Recall from the section on genetic pesticides that recent studies have put to rest fears that bioengineered corn could kill large numbers of monarch butterflies. However, worries about the environment have not disappeared. For one thing, genetic enhancement of crops could lead to new breeds of so-called superweeds. Just as genes for antibiotic resistance could theoretically spread from plants to bacteria, genes for pest or herbicide resistance could potentially spread to weeds. Because many crops—including squash, canola, and sunflowers—are close relatives to weeds, crossbreeding occasionally occurs, allowing the genes from one plant to mix with the genes of another. At this time, however, few experts predict any sort of explosion of genetically enhanced weeds. Further studies are needed to gauge the full extent of this threat and develop ways to minimize the risk.

The potential ecological hazards of biotechnologically enhanced crops must be weighed against the clearly established benefits. First and foremost, biotechnology can dramatically reduce the use of chemical pesticides, as seen in Figure 6.14. One of the key environmental benefits of biotech crops is the reduction in insecticide and herbicide applications to crops. In countries where biotech crops have been planted, pesticide use on four biotech crops—soybeans, corn, cotton, and canola—has fallen by 791 million pounds per year. (8.8%). This has resulted in a 17.2% reduction in the associated environmental impact.

YOU DECIDE

The StarLink Episode

In 2000, traces of genetically modified StarLink corn turned up in taco shells sold in grocery stores. Intended for animal consumption, the corn contained the gene for herbicide resistance, which breaks down in the soil or the stomach of cattle. On the surface, this news was not shocking. Many processed foods on market shelves contain genetically altered corn or soybean products. This particular type of corn, however, had never been approved for human consumption because of lingering concerns over potential allergic reactions. The Environmental Protection Agency (EPA)—the agency that regulates the use of all pesticides—had approved StarLink only for animal feed and industrial use. The discovery of StarLink corn in the food supply triggered a massive recall of potentially "tainted" products. Soon after the recall, Aventis, the company that produced StarLink, struck a deal with the EPA and agreed to stop planting the corn.

Had this "unapproved pesticide" really done any harm? The Centers for Disease Control and Prevention (CDC) took immediate steps to find out. A handful of people had complained of allergic-type reactions after eating the genetically altered corn, and the CDC investigated each case closely. A total of 28 people were found to have had symptoms consistent with an allergic reaction. However, blood tests showed that none of these people were sensitive to the Bt protein.

How should this episode have been handled? Was the public adequately protected from harm? Did the government overreact? What is the best balance between regulations and commercial interests? You decide.

Forests of genetically altered trees could pull billions of tons of carbon from the atmosphere each year and reduce global warming, according to researchers at Lawrence Berkeley National Laboratory and Oak Ridge National Laboratory. They claim that it might be possible to alter trees genetically so that they would send more carbon into their roots, keeping it out of circulation for centuries. These innovations could substantially boost the amount of carbon that vegetation naturally extracts from air. This change would require a modification of the current regulatory climate for producing genetically engineered trees in the United States and require a change in societal perceptions of the issues surrounding the use of genetically altered organisms. The potential exists, but the implementation depends on greater acceptance of genetically modified organisms in our environment.

On the whole, biotechnology does not seem to be taking us to the brink of ecological disaster and may actually offer some solutions to environmental problems. Indeed, the National Academy of Sciences recently reported that biotechnologically enhanced crops pose no greater environmental threat than traditional crops.

Regulations

Biotechnology is not a lawless frontier. As we have already seen, several different agencies regulate the production and marketing of genetically modified foods. The FDA regulates foods on the market, the USDA oversees growing practices, and the EPA controls the use of Bt proteins and other so-called pesticides. The approach of these agencies has changed over the years, especially in the case of the FDA, but they are actively involved in approving plant crops. (Chapter 12 discusses the regulation of plant biotechnology in much greater detail.)

As early as 1992, at the beginning of the biotechnology revolution—the FDA announced that genetically altered food products would be regulated by the same tough standards applied to regular foods—nothing more, nothing less. Even though they were not bound by law, food companies voluntarily consulted with the FDA before marketing any product. In 2001, the agency adopted a stricter, more formal approach. Under these rules, companies must notify the FDA at least 120 days before a genetically altered food reaches the market. The manufacturer must also provide evidence that the new product is no more dangerous than the food it replaces. The determining factor for plant-based foods or products will be the attitude of consumers. The Center for Food Safety and the Grocery Manufacturers Association and Food Products Association have informed the USDA of their "strong opposition to the use of food crops to produce plant manufactured pharmaceuticals in the absence of controls and procedures that assure essentially 100% of the food supply." There are many nonfood plants and contained-growth plant systems that will satisfy this concern, and no manufacturer has violated these rules to date.

Today's Agriculturalist

Agriculturists in the twenty-first century must understand genetic engineering (and the associated regulations), botany, statistics, chemistry, and global imaging systems. In other words, the labor-intensive agriculture industry that we know today will require the tools of science. American farmers and ranchers have already embraced technology, allowing them to become the most efficient producers in the world. Education is the key to the decision making that will ultimately decide the quantity and quality of food and fiber available to the world. A national study sponsored by the USDA and Purdue University projected about 58,000 job openings per year in the field of agricultural science, considerably more than the available number of college graduates in agriculture.

Agricultural biotechnology is one of the newest applications of biotechnology, with industries interested in employees with skills in biotechnology and agriculture. The ability to continue to produce the food needed to feed a human population that is doubling every 45 years from about 2% of the total land surface of the world cannot occur without advancements in technology, and biotechnology has delivered the most change to agriculture in the last decade. Agricultural biotechnology has a great future for those willing to develop the needed skills.

MAKING A DIFFERENCE

The use of petroleum fuels has led to a number of environmental problems, including global warming. The move to commercialize plant- and algae-based biofuels is providing a potential solution to this problem. One company, Eco Solution, is attempting to increase oil production from microalgae through stress. Eco Solution claims that "natural evolution" improves oil production by selecting among genetic variants. Algae normally produce some oil for buoyancy and as a food reserve, and they use CO_2 for photosynthesis. The biomass resulting from harvesting the algae can also be used for methane production. Other companies are utilizing other organisms to produce biofuels, like the Jatropa plant from Central America. SG Biofuels has tested the oil from this plant and found that it burns hotter and produces fewer emissions as a jet fuel than petroleum-based jet fuel. The problems of plant production, fuel extraction, and environmental concerns can be solved by people with the skills and knowledge needed to spur the development of biotechnology.

QUESTIONS & ACTIVITIES

Answers can be found in Appendix 1.

1. How do you detect whether the crops that you eat, say corn or soybeans, have been enhanced through biotechnology?

2. What are the environmental benefits from agricultural biotechnology?

3. What are some of the consumer benefits from agricultural biotechnology?

4. What are the grower benefits from agricultural biotechnology?

5. Are biotechnologically enhanced crops labeled?

6. Will insect-protected crops promote the development of insect resistance?

7. How do you know that there will not be any long-term consequences to humans or animals from the consumption of food or feed that contains products from biotechnologically enhanced plants?

8. Under what circumstances are gene guns primarily used as the means to transfer genes into plants?

9. Golden Rice has been responsible for saving many lives and preventing disease in children born in developing countries. How much has this technology cost these countries?

10. Explain how gene stacks can be used to produce insect-resistant plants through biotechnology.

Visit www.pearsonhighered.com/biotechnology

To download learning objectives, chapter summary, "Keeping Current" web links, glossary, flashcards, and jpegs of figures from this chapter.

Animal Biotechnology

After completing this chapter, you should be able to:

- List some of the medical advances made using animal research models.

- Explain how an animal is chosen as a model for drug studies.

- Describe two alternatives to the use of animal models and discuss their limitations.

- Discuss some of the ethical concerns surrounding animal research.

- Outline the process used to clone an animal.

- Discuss some of the limitations to current animal cloning practices.

- List some biological products that can be produced using transgenic animals as bioreactors.

- Explain how knockout and knock-in animals can be used to provide information about genetic disorders and other diseases.

- Describe the process used to create monoclonal antibodies from hybridomas.

Through biotechnology, pigs may become a source of human organs.

Dolly looked like a sheep and acted like a sheep, but Dolly was not just any sheep. She spent considerable time in front of television cameras because she was a clone: an identical twin animal engineered from a donor cell.

Some people feel uneasy seeing the words *animal* and *engineering* in the same sentence. For them, the words conjure up images of legless cows or featherless chickens. But biotechnologists are not in the business of creating bizarre new animals. Genetically engineered animals are used to develop new medical treatments, improve our food supply, and enhance our understanding of all animals, including humans. Animals can also be used for purposes that may seem inhumane and unethical, but government regulations have largely prevented the widespread introduction of such controversial practices.

Animals offer biotechnological opportunities, but they also present many tough scientific and ethical challenges. Researchers who successfully meet those challenges have the potential to make important contributions to science and society. For example, animal health products generated over $17.4 billion in 2005, a figure that reached $21.7 billion worldwide in 2010. Thus, veterinary health products benefit to both human and nonhuman animals.

This chapter reviews many aspects of animal biotechnology. We begin by looking at the use of animals in research and then visit some of the issues at the cutting edge of biotechnology: cloning, transgenics, and the use of animals as bioreactors.

FORECASTING THE FUTURE

Health care costs are spiraling out of control, with no easy remedy in sight. One way to help reduce costs is to expand the offerings of vaccines to prevent disease. Every dollar invested in vaccination returns between $7 and $20 in averted health care costs. These savings have been made more obvious by the success of human papillomavirus vaccine (Gardisil), against the virus causing cervical cancer, and the H1N1 swine flu vaccine. Although vaccines accounted for only 2% of the world pharmaceutical market in 2009, their growth is expected to outpace the broader drug market in the future. The biotech industry sees the opportunity in vaccine development and is researching the potential of using veterinary products for human health vaccines. A form of total color blindness (congenital achromatopsis) was recently cured in dogs by Andrias Komaromy, using gene therapy, at the Penn School of Veterinary Medicine (University of Pennsylvania). Achromatopsis is a rare inherited disorder that occurs in 1 in every 30,000 to 50,000 people. The development of a vaccine for these individuals could now be possible as a result of the veterinary treatment's success.

There are other veterinary products that have potential for human health care application and health care cost savings.

7.1 Animals in Research

The white mouse is an icon of scientific research; along with the lab coat and microscope, it represents the business of discovery. Although animals have been central to research methods for decades, their use is controversial. Next, we explore the role of animals in research and the resulting issues.

Why Use Animals in Research?

Research using animals has been key to most medical breakthroughs in the past century. Without question, such research has prevented much human suffering. Here are a few examples:

- Without the polio vaccine, which was developed using animals, thousands of children and adults would die or suffer debilitating side effects from the disease each year.
- Without dialysis, which was tested on animals, tens of thousands of patients suffering from end-stage kidney disease would die.
- Without cataract surgery techniques, which were perfected on animals, more than a million people would lose sight in at least one eye this year.

Animal research has also directly benefited animal health:

- Biotechnology has developed 111 USDA-approved veterinary biologics that treat heartworm, arthritis, parasites, allergies, and heart disease.
- Research on animals has also led to vaccines that veterinarians use daily to prevent rabies and feline HIV.

You might wonder why testing on cells in vitro (cells cultured outside the body) would not be sufficient. The answer is that new drugs and medical procedures have effects beyond single cells in tissues and organs and must be tested in animals to determine the impact of the treatment on an entire organism. For example, a drug may be a powerful agent able to kill cancer cells, but it may also be destructive to unrelated organs.

One drug with unexpected side effects is finasteride (Propecia), which is used to encourage hair growth. It carries a clear warning that pregnant women should not handle broken or crushed pills. That warning and a special protective coating on the pills themselves are the direct result of information collected during testing on animal subjects. Those tests revealed serious birth defects (malformations of the reproductive organs) in male offspring born to animal mothers given large oral doses of the drug. Even though merely handling the pills is unlikely to result in similar defects in human infants, the warnings were put in place before the drug was marketed. In this case, experiments using animal subjects helped to prevent unexpected disasters. Thanks to animal testing, pregnant women are never given the drug.

Types of Animals Used in Research

The differences between mice and humans are obvious, but there are many genetic and physiologic similarities between the two. For this reason, research conducted on mice—or other animals, for that matter—can tell us much about ourselves (**Figure 7.1**). The animals used most often in research are pure-bred mice and rats (of which the genetics are known and controlled), but other species are also used. Researchers have developed model animal systems using both vertebrates (zebrafish) and invertebrates (fruit flies) and worms. The tiny zebrafish (*Brachydanio rerio*), an extremely valuable research animal, is a popular and hardy aquarium fish (**Figure 7.2**).

Zebrafish are small (adult length is 3 cm, about the length of a paper clip), making it possible to house large numbers of animals in small spaces. Spawning is virtually continuous by this active warm-water fish, with only about 3 months between generations. The average number of progeny per female per week can exceed 200. Zebrafish eggs complete embryogenesis (early organ development) in about 120 hours; the gut, liver, and kidneys are developed in the first 48 to 72 hours. Because the fish grow so rapidly, scientists can test for

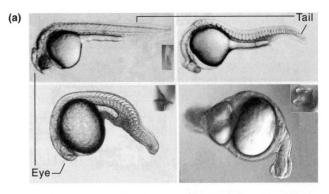

FIGURE 7.2 Zebrafish Are Commonly Used in Animal Research Zebrafish are fast-growing alternatives to other animals for gene studies and toxicity analyses. After transplanting genes into their embryos (a) it becomes possible to study developmental changes (like blood vessel growth) before the fish become adults (b). The transparent embryos, fast growth rate, and ease of study of zebrafish make them common animal models for testing.

toxicity or adverse effects in about 5 days. If a drug is toxic to a zebrafish, it is probably toxic to humans.

Zebrafish are ideal for studies of development and for genetic research. The rapid growth of an easily visible embryo inside the zebrafish's egg, as seen in Figure 7.2a, makes studies efficient. The eggs lend themselves readily to gene transfer, there is no need to place eggs inside a donor mother for gestation, and the embryos are transparent, making it possible to study cell division under the microscope in its first hours.

Rats are superior to mice for many early drug toxicity tests because of their physiology and more human-like responses to drugs. In addition, the relatively large size of rats facilitates surgical and physiologic experimentation. This is particularly important in studies that require dissection or isolation of specific cell types and organ structures, as is the case with pharmacokinetics, which is the study of how drugs interact with tissues and how they are metabolized. This, combined with the fact that most toxicology data have been collected in the rat, has made them the preferred model organism for determining drug metabolism and early toxicology profiles.

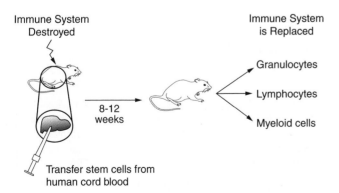

FIGURE 7.1 Mice Tell Us about Ourselves Here, after the immune system of a developing mouse is destroyed, it can be replaced by injecting human stem cells. In this way it is possible to study therapies for humans using mice as models.

There are times when mice, fish, or rats are not the best test animals. For example, the lung and cardiovascular systems of dogs are similar to those of humans, making dogs a better choice for the study of heart disease and lung disorders. HIV/AIDS research is conducted on monkeys and chimpanzees because they are the only known animals that share humans' vulnerability to the virus. Although cats, dogs, and primates like monkeys and chimps are used in specific instances when their particular biology is important to research, they make up less than 1% of the total number of research animals (the majority are mice). In addition, the number of those animals used in experiments has been declining for the past 20 years because of the increased use and availability of alternative methods of testing, which we will discuss shortly.

Regulations in Animal Research

According to regulations enforced by the U.S. Food and Drug Administration (FDA), new drugs, medical procedures, and even cosmetic products must pass safety tests before they can be marketed. Safety testing involves a rigorous scientific methodology known as **phase testing,** as illustrated in Table 7.1.

As part of phase testing, the FDA requires manufacturers to conduct a statistically significant number of trials on cell cultures as well as trials in live animals and human research participants before such products can be made available to the general public.

If initial tests on cell cultures indicate dangerous levels of toxicity, the product is never tested on live animals. The choice of animal to test is usually determined by its genetic similarity to humans or potential to mimic conditions in humans. Two or more species are used in the tests because different effects may be revealed in different animals. Toxicity and unexpected nontarget effects can often be detected by using more than one species with genes homologous to those of humans. The rate of absorption, specific chemical metabolism, and time required to excrete (eliminate from the body) the substance can all be examined in animal models. If significant problems are detected in the animal tests, human research participants are never recruited to participate in trials. Although they are rare, unintended results can occur; participants are informed of this potential as a requirement of informed consent.

Alternatives to the Use of Animals

Whenever possible, researchers use cell cultures and computer-generated models in initial testing. Both of these tools are significantly less expensive than research animals and can save time as well.

TABLE 7.1   FOOD AND DRUG ADMINISTRATION REQUIRED TESTING PHASES FOR DRUG APPROVAL

FDA Phase testing involves the use of animals for pre-clinical testing before allowed in humans. If the new drug candidate has proven to be non-toxic and has benefit, then it can be awarded and Investigational New Drug (IND)status. If it is successful in the three phases of human testing it can receive a New Drug Application (NDA) and likely approval for marketing. The FDA continues evaluating the NDA for another 2.5 years, resulting a total of about 12 years for a successful drug approval.

	Preclinical Testing		Phase I	Phase II	Phase III		FDA	Phase IV
Years	3.5		1	2	3		2.5	12 total
Tested on	Animals in the lab		20–80 healthy volunteers	100–300 patient volunteers	1,000–3,000 patient volunteers			
Purpose	Assess safety and biological activity	File IND at FDA	Determine safety and dosage	Evaluate effectiveness and look for side effects	Verify effectiveness, monitor adverse reactions from long-term use	File NDA at FDA	Review process/approval	Additional testing after approval required by FDA
Success rate	5,000 compounds evaluated		5 enter trials				1 approved	

Source: www.fda.gov/cder/handbook/develop.htm

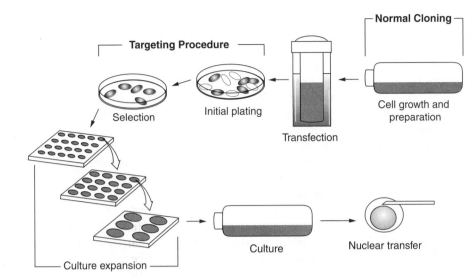

FIGURE 7.3 Selecting and Preparing Tissue for Nuclear Transfer Target cells receive gene insertion by transfection. The small number with the correct insertion are selected and cloned numerous times in culture (in vitro). At each step there is attrition due to the finite life span of somatic (body) cells in culture. Ultimately appropriate somatic cell nuclei can be obtained from the culture for nuclear transfer.

As mentioned earlier, cell culture studies are often used as a preliminary screen to check on the toxicity of substances. In addition, fundamental questions about biology are often best answered by collecting evidence at the cellular level. Clues about which compounds might make the best drugs are often the result of in vitro research (see **Figure 7.3**). But, as already noted, cell cultures cannot provide information about the potential impacts on an entire living organism.

Computers are also used to simulate specific molecular and chemical structures and their interactions. Computer models are becoming increasingly sophisticated, but computers are limited by their programming. They can provide clues, but they cannot provide final answers. Unless we already know the impact on a living system, we cannot create a computer program that can simulate that system. Computers are excellent for processing data and helping us to see the patterns that emerge, but they cannot react unless we have programmed them to do so. However, they can reduce animal usage in research. For example, a computer model to test the efficiency of finasteride would have resulted in a clear message that it was a good drug; the computer model could not be expected to address the problem of unexpected birth defects because programmers would not have known in advance of possible side effects.

Regulation of Animal Research

Animal research is heavily regulated. The federal Animal Welfare Act sets specific standards concerning the housing, feeding, cleanliness, and medical care of research animals. Before a study using animals can begin, researchers are required to prove the need to

employ animals. They also must select the most appropriate species and devise a plan for using as few animals as possible. An oversight committee with the host institution reviews the research while it is in progress to make certain that institutional and federal standards are being upheld. Government agencies regularly monitor the conditions in the laboratories. To receive funding for research from the National Institutes of Health (NIH), the FDA, or the Centers for Disease Control and Prevention, researchers must follow the standards of care set out in *The Guide for the Care and Use of Laboratory Animals,* formulated by the National Academy of Sciences. Given the cost of research, the availability of grant monies from those agencies can be essential; thus most institutions are eager to follow the guidelines. Researchers are tasked with achieving the "Three Rs" of animal research (see Chapter 12 for more on regulations pertaining to animal use in biotechnology):

- *Reduce* the number of higher species (cats, dogs, primates) that are used
- *Replace* animals with alternative models whenever possible
- *Refine* tests and experiments to ensure the most humane conditions possible

Veterinary Medicine: Benefits for Humans and Animals

Not all animal research is conducted in laboratories for the sole benefit of humans. Veterinarians, who devote their lives to the care of animals, also participate in research. Research at veterinary clinics has led to new cancer treatments for pets. Because animal and human cancers are strikingly similar, information gleaned from one species might be used to treat the other. For example, the BRCA1 gene found in 65% of human breast tumors is similar to the BRCA1 gene in dogs. Clinical trials of cancer therapies on pets can pave the way for human trials.

Randomized trials in both dogs and humans have shown that hyperthermia (heating) of a tumor site combined with radiation can be more effective in local tumor control than radiation therapy alone. Researchers at Duke University are using hyperthermia along with gene therapy to destroy solid tumors. They inject directly into the tumors a gene that produces interleukin-12 (IL-12). They then apply heat only to the tumor. The heat turns on the heat-shock control gene within the tumor, promoting the production of IL-12, which helps destroy the cancerous tissue. If successful, this therapy, now being tested on cats, could be phased into human trials faster than it might have been had the early trials been conducted on cultured cells; it might also benefit cats with these tumors.

7.2 Cloning

The production of an animal from the nucleus of another adult animal was a scientific breakthrough in 1997, when Dolly the sheep was produced in this way, but it also created controversy over whether human cloning would ever be feasible. Here, we examine the methods and issues involved in cloning.

Creating Dolly: A Breakthrough in Cloning

Human DNA, like that of all higher animals, contains genes from two parents. The mixture of genes comes down to chance. To prove this, just look at the children in any large family: some may have curly hair and others have straight hair; some may be able to "roll" their tongues and others cannot. One child may have a genetic disorder that the others do not have. This uncertainty is eliminated with cloning. When cloning is used to reproduce an animal, there is only one genetic contributor. If the donor sheep has long soft wool, the lamb clone will too. The offspring have exactly the same genetic programming as the parent.

Embryo twinning (splitting embryos in half) was the first step toward cloning. The first successful experiment produced two perfectly healthy calves that were essentially twins. This procedure is commonly practiced in the cattle industry today, where there is an advantage in producing many cattle with the same strong blood lines. Embryo twinning efforts have also

resulted in the birth of Tetra, a healthy rhesus monkey produced from a split embryo. The embryo-splitting procedure is relatively easy, but it has limited applications. Even though the end result is identical twins, the exact nature of those twins is the result of old-fashioned mixing of the genetic material from two parents. You may end up with animals that have the desired characteristics, but you have to wait until they are grown to find out. This is why Dolly the sheep—who was created from an adult cell, not an embryo—was such a breakthrough. Dolly was the exact duplicate of an adult with known characteristics.

To create a clone from an adult, the DNA from a donor cell must be inserted into an egg. In the first step, cells are collected from the donor animal and placed in a low-nutrient culture solution. The cells are alive but starved, so they stop division and switch off their active genes. The donor cells are then ready to be introduced into a recipient egg, as seen in **Figure 7.4.**

The egg itself is prepared by **enucleation.** A pipette gently suctions out the DNA congregated in the nucleus of the egg. The researcher must then choose a method for delivering the desired genetic payload into the egg. One method is illustrated in Figure 7.4. Another method, known as the Honolulu technique, involves injecting the nucleus of the donor cell directly into the middle of the enucleated egg cell. No matter which technique is used, manipulating the microscopic cells can be extremely difficult.

After the DNA has been delivered to the egg, biology takes over. The new cell responds by behaving as if it were an embryonic cell rather than an adult cell. If all goes well, cell division occurs, just as it would in an ordinary fertilized egg. For the first few days, the embryo grows in an incubator. Then, when the new embryo is ready, it is transferred into a surrogate mother for gestation. The surrogate will give birth to a clone that is genetically identical to the genetic donor. The list of species successfully cloned in this fashion now includes sheep, goats, pigs, cattle, and an endangered cow-like animal called a gaur.

Limits to Cloning

There are limits to the power of cloning technology. First, the donor cell must come from a living organism. Free-ranging dinosaurs à la *Jurassic Park* are likely to remain science fiction. Some people are intrigued by the possibility that cells collected from frozen mammoths discovered in Siberia might be cloned and fostered by a modern-day elephant. Most experts, however, think that the chances of finding a viable mammoth cell, managing to clone it, and succeeding in fostering it in an elephant or other distant relative are laughably remote.

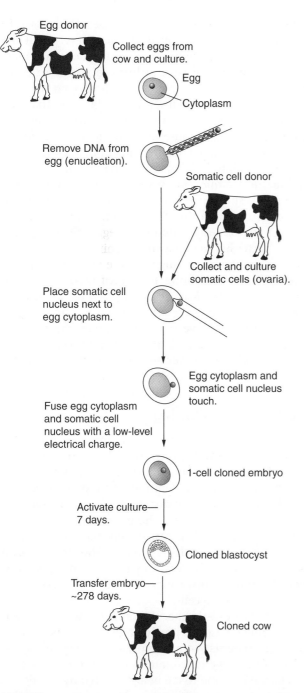

FIGURE 7.4 Nuclear Transfer to Produce a Cloned Animal Cloning usually begins with removal of the egg nucleus from a fertilized egg and replacement with another nucleus (possibly from an adult animal, like a cow). If the process is successful, the desirable characteristics can be obtained with a high degree of certainty. Success is not always easy (especially because, in cattle, it takes 278 days to find out if it worked!).

It is also important to realize that even though clones are exact genetic duplicates, they are not exactly identical to the original in other ways. Animals are shaped by their experiences and environments as well

as their genes in the same way that identical human twins are not exactly the same and have different personalities. For this reason, people who are planing on banking cells in the hope of recreating beloved pets might well be disappointed. For example, a dog's personality is shaped in part by its relationship to its littermates and by the events of its life. Although we may be able to reproduce a genetically identical animal, there is no way to duplicate the early experiences that molded the dog's behaviors. In 2002, the the California company Genetics Savings & Clone (GSC) made headlines when they successfully cloned a cat in an effort to start a viable pet cloning business, despite the uncertainties involved (**Figure 7.5**).

Even racehorses, which are carefully bred to have essential genetic characteristics, are shaped by factors other than genetics. The fun of running and the will to win may not be genetically encoded. For that reason, it is unlikely that there will be a rush to clone a winning horse.

More significantly, the present success rate for cloning is actually quite low, although it is improving. Dolly was the result of 277 attempts. Of these, only 29 implanted embryos lived longer than 6 days, and of

FIGURE 7.5 Cloned Kittens In 2004, at the annual Cat Fanciers' Association/IAMS New York cat show, 4-month-old cloned cats Tabouli and Baba Ganoush and their genetic donor, Tahini, were introduced. The Bengal kittens were cloned by Genetic Savings & Clone of Sausalito, California.

those 29 embryos, only Dolly was born. Cloning failures do not make the evening news, but they are a significant barrier to the feared "animal assembly line" of cloning, which in any case is not the aim of researchers. Scientists are still trying to understand some of the unexpected problems that arise during the cloning process. Many clones are born with defects including enlarged tongues, kidney problems, intestinal blockages, diabetes, and crippling disabilities resulting from shortened tendons. In addition, the mortality rate among the surrogate mothers is much higher than among normal mothers. No clear pattern to these perplexing problems is apparent.

An even more challenging problem is the possibility that clones may become old before their time. This problem came to light when analyses indicated that Dolly's cells seemed to be *older* than she was. This observation bears a little explanation. Every cell in your body contains the chromosomes that determine your genetic makeup. At the end of each chromosome is a string of highly repetitive DNA called a *telomere* (see Chapter 2 for a review). As your cells replicate, the telomere becomes shorter with each division. In this sense, there really is a biological clock inside living cells. A cell's chromosomal age can be read by inspecting the length of its telomeres. As the telomere unravels, signs of aging appear in the organism. Tests showed that Dolly's telomeres were not like those of other lambs born the year she was; instead, they were about the same length as her donor's, or 3 years older than Dolly. Dolly was diagnosed with arthritis, a possible sign of premature aging. She eventually developed a virus-induced lung tumor that led to her being euthanized in 2003 (this particular respiratory virus commonly infects sheep, and scientists at the Roslin Institute found no evidence indicating that the tumor originated because Dolly was a clone). Other clones may be in for a similar fate. Cloned mice, for example, have been found to have much shorter life spans than typical mice. There is some speculation that life span may vary from species to species, but all the evidence seems to indicate that a shorter life span may be part of a clone's destiny.

This cloud over cloning has a potential silver lining. Researchers investigating the problem of shortened telomeres in clones learned more about telomerase, an enzyme that regenerates the telomere. This enzyme could eventually serve as a key to understanding and preventing many age-related diseases as well as possibly providing an answer to the shortened clone (or nonclone) life span.

The Future of Cloning

When Dolly's novel birth was announced in the media, there was a flurry of speculation and concern about the eventual role of cloning, such as the widespread replacement of natural birth processes. This speculation owes a great deal to the fact that cloning is still a young science and therefore subject to much experimentation: some researchers have even managed to make clones from clones. DNA from frozen cells has been used successfully in the cloning process, opening up new possibilities that will depend on cryogenically preserved cells.

Among the early speculation was the notion that clones could be used to provide replacement body parts for their "parents"—the donors of the original cells. Even though a genetic match would probably reduce the chance of clone tissue being rejected by the donor turned transplant recipient, this is not a practical solution to the problem of transplant organ shortages. Aside from the profound ethical questions about such a practice, it would take years for the clone to become mature enough to donate organs. Far more promising biotechnology solutions are on the horizon: xenotransplantation, the use of organs from other species, some of which may have been made more suitable through the use of transgenics, is expected to become a viable source of organs within the next 5 years. (Xenotransplantation is discussed in more detail in Chapter 11.) Other scientists have had early success using cell cultures to grow replacement skin, eyes, and rudimentary kidneys for animals ranging from tadpoles to cattle.

YOU DECIDE

Genetics Savings & Clone

In 2002, the California company Genetics Savings & Clone (GSC) made headlines when they successfully cloned a cat: Cc (Carbon Copy). They then produced two more cloned cats, Baba Ganoush and Tabouli (Figure 7.5).

In 2004, the company again made the news when it revealed that it had sold a cloned kitten to a woman in Texas. What's more, for a mere $50,000, GSC would clone anyone's cat. Within a few years GSC reduced the price to $32,000, and by late 2006 it announced that it would close at the end of the year and give a refund to anyone who had put down a deposit for a cloned cat. After 6 years in business, GSC cloned 5 kittens but sold only 2. Many questioned whether a company could be profitable selling cloned pets, given the costs of cloning. Others questioned the ethics of cloning pets for human enjoyment or to create a genetically identical clone of a deceased pet. What do you think? Should pets be cloned? If a pet in your family died, would you want to buy an identical clone if it was affordable? You decide.

So why are clones necessary? There are many reasons why it is advantageous to have multiple animals with the exact genetic makeup. Clones can be used in medical research, in which their identical genetics makes it easier to sort out the results of treatments without the confounding factor of different genetic predispositions. Clones may also provide a unique window on the cellular and molecular secrets of development, aging, and diseases.

In the case of endangered animals, there is a compelling reason to look to cloning to sustain breeding populations. Steps have already been taken to use cloning to preserve animals. Interspecies embryo transfer has been used successfully to provide surrogate mothers for African wildcats (*Felis silvestria libyca*). Noah, the gaur calf, was a clone born to a domestic cow surrogate. Noah later died of an infection unrelated to the cloning process. Despite that disappointment, efforts are now under way to clone pandas, using more common black bears as hosts.

7.3 Transgenic Animals

We've covered transgenics as it applies to plant biotechnology (Chapter 6). The process of introducing new genetic material to an organism is not limited to plants. It can be used on animals as well. We have seen some of the differences between proteins produced from bacterial (prokaryotic) sources and those coming from other sources (Chapter 4). Because the structural differences among proteins are minimal, cell and animal culture offer good solutions to the need for proper structure. Some of the most exciting developments in biotechnology are the result of transgenic research in animals.

The first experiments in animal transgenics were conducted on the cellular level. A new gene was added to a cell in culture and the effects on that one cell were observed. When cloning technology became available, transgenics took a new course. Instead of manipulating a single cell, genes were added to a large number of cells. These cells were then screened to discover which ones exhibited the desired genes. After these cells were selected, each could be used to grow a complete animal from a single cell, using cloning technology (see Figure 7.5).

Transgenic Techniques

A variety of methods can be used to introduce new genetic material into animals, and new methods continue to be developed and refined. Here is a brief overview of some techniques:

- **Retrovirus-mediated transgenics** is accomplished by infecting mouse embryos with retroviruses before the embryos are implanted. The retrovirus acts as a vector for the new DNA. This method is restricted in its applications because the size of the **transgene** (or *trans*ferred *gene*tic material) is limited and the virus's own genetic sequence can interfere with the process, making transgenics a rather hit-and-miss project (retroviruses are discussed in Chapter 3).

- The **pronuclear microinjection** method introduces the transgene DNA at the earliest possible stage of development of the zygote (fertilized egg). When the sperm and egg cells join, the DNA is injected directly into the nucleus of either the sperm or the egg. Because the new DNA is injected directly, no vector is required; therefore there is no external genetic sequence to muddle the process (see **Figure 7.6**).

- In the **embryonic stem cell method**, ES cells are collected from the inner cell mass of blastocysts. They are then mixed with DNA, which has usually

been created using recombinant DNA methods. Some of the ES cells will absorb the DNA and be transformed by the new genetic material. Those transformed ES cells will then be injected into the inner cell mass of a host blastocyst.

- A similar method, **sperm-mediated transfer,** uses "linker proteins" to attach DNA to sperm cells. When the sperm cell fertilizes an egg, it carries the valuable genes along with it.

- Finally, gene guns can also be used on animal cells (see Chapter 6).

Increasing Agricultural Production with Transgenics

Researchers have used gene transfer to improve the productivity of livestock. By introducing genes that are responsible for faster growth rates or leaner growth patterns, animals can be raised to market more quickly. This process is known as selective improvement. Although selective improvement is not without controversy, it produces an increased food supply at a decreased cost.

Traditional cross-breeding techniques have also been used for years to create chickens that grow from chick to market at an increased rate, which has the potential to raise production levels and decrease cost. Recently, breeders discovered that birds that took less than 42 days to grow from chick to market-ready did not produce eggs. Although eggs are not the goal of those raising chickens for meat, it is impossible to raise chickens without some eggs to hatch. Growing chickens so quickly is not without controversy. However, poverty and food shortages mean that less costly agricultural practices must be explored. Every extra day represents additional costs. Americans consume 10 billion chickens per year, and any reduction in the time required to produce those billions of chickens would embody significant savings for producers, which could then be passed on to consumers. These potential savings are motivating some current research in which genes from fast-growing meat-producing chicken breeds are being swapped into eggs produced by egg-laying breeds. These researchers have met with success on a limited scale. Now the challenge is to develop methods that will work on the scale of industrial farming. In the next few years, researchers hope to devise equipment that will accurately inject the desired traits into 30 or 40 billion chicken embryos a year.

Eggs themselves could even be made healthier for human consumption through transgenics. Eggs are an inexpensive source of high-quality protein, but many people avoid them because they are rich in cholesterol. Tweaking the genes responsible for cholesterol production could result in healthier, lower-cholesterol eggs.

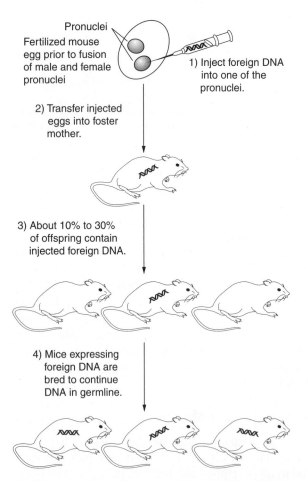

Pronuclei
Fertilized mouse egg prior to fusion of male and female pronuclei

1) Inject foreign DNA into one of the pronuclei.

2) Transfer injected eggs into foster mother.

3) About 10% to 30% of offspring contain injected foreign DNA.

4) Mice expressing foreign DNA are bred to continue DNA in germline.

FIGURE 7.6 Pronuclear Microinjection Cloning The process of injecting DNA into one of the pronuclei (the female pronucleus and male pronucleus combine at fertilization) so that it is incorporated can be used to produce clones that bear the new DNA.

The dairy industry is also a target for genetic improvement. Researchers have used transgenics to increase milk production, make milk richer in proteins and lower the fat content. Herman, a transgenic bull (produced by trangenics and nuclear transfer), carries the human gene for lactoferrin. This gene is responsible for the higher iron content (essential to the healthy development of infants) found in human milk. Herman's offspring also carry the gene for lactoferrin. Consequently, in much the same way that human polyclonal antibodies can be produced in the plasma of cattle (see **Figure 7.7**), Herman's daughters may be the first of many cows able to produce milk that is better suited for consumption by human children.

Another important improvement is the reduction of diseases in animals raised for food. During the epidemic of foot-and-mouth disease in England in 2000, herds of dairy and beef cattle as well as sheep and goats were destroyed. The loss of entire herds was catastrophic to farmers and devastating to that nation's agriculture industry. In the United States, concerns over the spread of foot-and-mouth disease resulted in the confiscation and destruction of sheep that had been imported from England, even though they did not necessarily test positive for the disease. Spurred by the threat of future disease outbreaks, researchers are studying disease-prevention genes and hope to develop animals that are resistant to foot-and-mouth disease. Similar processes may someday help prevent cholera in hogs and Newcastle disease in chickens.

Researchers at the U.S. Department of Agriculture in Maryland turned to genetic engineering to create dairy cows that could resist mastitis, an inflammation of milk glands caused by bacterial infections. Bacteria such as *Staphylococcus aureus* infect mammary glands, resulting in a highly contagious condition that spreads rapidly and results in losses of billions of dollars for the dairy industry. Scientists created transgenic cows containing a gene from *Staphylococcus simulans*, a relative of *S. aureus,* which produces a protein called lysostaphin, which kills *S. aureus*. Preliminary data indicate that these animals do indeed show resistance to mastitis. But researchers must be cautious. They know that these animals will not necessarily be fully protected from infections by *S. aureus* or other microbes and that these cows represent only a first step toward improving the resistance of cows to the bacteria that cause mastitis. Whether the dairy industry will embrace these and other genetically modified (GM) cows remains to be seen.

Researchers at the University of Guelph in Ontario, Canada, recently developed the **EnviroPig,** a transgenic pig expressing the enzyme phytase in its saliva. It is deemed environmentally friendly because it produces substantially less phosphorus in its urine and feces than

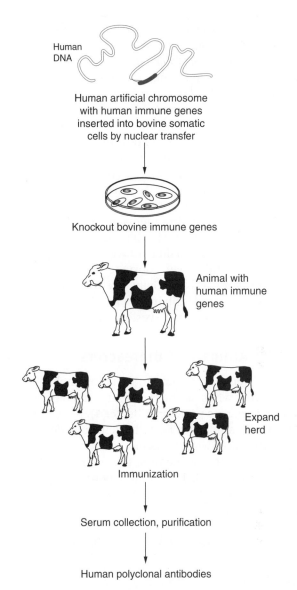

FIGURE 7.7 Human-Like Plasma from Cattle Human plasma for patients is in short supply and could be obtained from cattle cloned to carry human genes for blood factors and antibody genes. Human artificial chromosomes carrying these genes have been successfully transferred by procedures described earlier (Chapter 4).

nontransgenic pigs. Phosphorus is the major pollutant generated on pig farms. Phytase degrades phosphates in the pigs' food, reducing the amounts of phosphorus waste products excreted by about 30%.

The safety of the food that reaches our tables can also be improved by transgenic technology. Human food poisoning may be reduced in the future by transferring antimicrobial genes to farm animals. This application could reduce the thousands of food poisoning deaths in the United States each year and also mean that fewer antibiotics will be used in agriculture. That is good news, because the overuse of antibiotics is leading to the development of resistant strains of *Salmonella, Listeria,* and *Escherichia coli.*

Researchers are also turning to transgenics to create animals that will better serve hungry and poor populations around the world. In many areas, available arable land is limited, and little can be set aside for grazing animals. Researchers are directing their attention to the problem of transforming animals into more efficient grazers, so that more animals can be sustained on the same land. To this end, they are looking at the genetic structure of wild grazers, like African antelopes, that thrive in environments where domestic livestock would fail. It may also be possible to create cattle that are better able to do hard work such as pulling plows even while they consume less food. As researchers work toward improving livestock, they are also concerned with identifying and conserving the genetic heritage of traditional livestock breeds.

Transgenic Animals as Bioreactors

Recall that proteins are an important biotechnological product and that whole animals can serve as "bioreactors" to produce those proteins (Chapter 4). An animal bioreactor works as follows: First, the gene for a desired protein is introduced via transgenics to the target cell. Second, using cloning techniques, the cell is raised to become an adult animal. That adult may produce milk or eggs that are rich in the desired protein, thanks to the introduced gene.

An example of an application of this technology is Biosteel, an extraordinary new product that may soon be used to strengthen bulletproof vests and suture silk in operating rooms. Fundamentally, Biosteel is made from spiderweb protein, which is one of the strongest fibers on earth. In the past, spider silk has never been a viable industrial product because spiders produce too little silk for it to be useful. Thanks to transgenics, spiders are not the only source of spider silk proteins. The gene that governs the production of spider silk has been successfully transferred into goats, and those goats have reproduced, passing the new gene to their offspring. Now there is a whole herd of "silk-milk" goats. There is no need to monitor these bioreactors. These goats, like the ones shown in **Figure 7.8**, eat grass, hay, and grain and naturally produce the milk that now contains the proteins that make up spider silk. The silk protein is then purified from the goats' milk and woven into thread for fabrication into bulletproof vests and other products (see Chapter 4 for a refresher on protein purification methods).

Transgenic animals are also being used to produce pharmaceutical proteins. A human gene called *ATryn,* which is responsible for an anticlotting agent in humans, is needed by patients with a hereditary deficiency to prevent thromboembolism before or after surgery or childbirth. The company that developed the transgenic goat (GTC) linked *ATryn* to mammary-specific promotors so that it would be expressed in milk. The transgenic method is faster, more reliable, and cheaper than other more complex ways of synthesizing pharmaceutical proteins. The product was approved for use by the FDA in 2009, and although the market for patients with hereditary antithrombin deficiency treatable with *ATryn* is not large, it may prove to have

FIGURE 7.8 Goats with Transgenes That Produce Valuable Proteins in Their Milk Transgenic animals (bioreactors)—including goats, cattle, and chickens—offer the advantage of lower material and product costs for transgenic protein production.

other uses in the operating room. This transgenic system is considerably less expensive using bioreactor animals and is under development for other proteins (like monoclonal antibodies) that are needed in large volumes.

You may be wondering why researchers in transgenics so often use goats. Goats reproduce more quickly and are cheaper to raise than cattle. Because they also produce abundant milk, they are an excellent choice for transgenic product development. Many diseases need short-term treatment of larger-quantity recombinant proteins that can be made amply available from transgenic animal sources. Until gene therapy becomes broadly utilized, transgenic-derived therapeutic proteins are the most attractive alternative for replacement therapy in genetic disorders such as hemophilia. Hemophiliacs lack the functional clotting protein antithrombin III. This protein is being produced in transgenic goats and being tested for FDA approval. Using transgenic animal bioreactors for the production of complex therapeutic proteins provides lower production costs, higher production capacities, and safer pathogen-free products.

Milk-producing animals are not the only animal bioreactors, however. Eggs are another efficient method of producing biotech products in animals. Because poultry have an efficient reproductive system, it is easy to generate thousands of eggs from a small flock in a relatively short time. An average hen produces 250 eggs per year. With each egg white containing about 3.5 grams of protein, replacement of even a small fraction with recombinant protein creates a highly productive "hen bioreactor." Some medical products such as lysozyme and attenuated virus vaccines are already derived from eggs.

Knockouts: A Special Case of Transgenics

Adding genes is only one side of the transgenic story; the influence of a gene can also be subtracted. Researchers are working with transgenic mice—so-called knockout mice—to better understand what happens when a gene is eliminated from the picture. **Knockouts** have been genetically engineered so that a specific gene is disrupted. Simply put, an active gene is replaced with DNA that has no functional information. When the gene is knocked out of place by the useless DNA, the trait controlled by the gene is eliminated from the animal.

Knockout mice begin as ES cells with specially modified DNA. The DNA is prepared using recombinant DNA techniques either to clip away a specific portion, leaving a deletion in the genetic code, or to replace one portion of the DNA with a snippet of DNA that has no function. If you think of DNA as an arrangement of letters forming a sentence that provides directions, the knockout process would delete and possibly replace a word. If we begin with "Open the door and place the mail on the table" and knock out a portion of the message, we might end up with "Open the mail on the table." The resulting action would be different from the action that the original instruction was intended to produce. A slightly different knockout and replacement could result in "Open the elephant and place the mail on the table." In this case, nothing would be accomplished because the sequence is nonsensical. Similarly, deleting or replacing a bit of DNA can result in a change in the traits expressed, including termination of the expression of vital proteins.

After the DNA is modified, it is added to ES cells, where it recombines with the existing gene on a chromosome, essentially deleting or replacing part of the genetic instructions in the cell. This process is called **homologous recombination,** as seen in **Figure 7.9**, on the next page. The modified ES cells are then introduced into a normal embryo, and the embryo is implanted in an incubator mother. Interfering RNA is also capable of gene silencing and being utilized as an alternative to homologous recombination in many applications (see Chapter 2 to review RNAi).

A genetically modified organism—generally a mouse pup—is not the end of the story. This offspring of the mouse pup is a **chimera.** Some of its cells are normal and some are knockouts. The activity of the normal cells often conceals the deficit caused by the missing or replaced gene. Two generations of crossbreeding are required to produce animals that are complete knockouts. The reward is worth waiting for in terms of genetic research. When we know what gene has been knocked out and we see the result of that change, we have a clear idea of the function of that gene. Being able to see clearly what happens when a gene is damaged or absent will provide new insights into a host of genetic disorders, including type I diabetes, cystic fibrosis, muscular dystrophy, and Down syndrome. A good example of this is the breast cancer mouse, which was patented in 1988 by Harvard University scientists and has been used extensively to test new breast cancer drugs and therapies (the technology is offered at minimal or no cost to researchers). The ability to test new therapies in these breast cancer mice, rather than human patients, has saved many cancer patients countless hours of unnecessary suffering.

A common approach is now to compare the effect of a potential drug product on knockout animals and **knock-in** animals. Animals can have a human gene inserted to replace their own counterpart by homologous recombination to become knock-ins. In a knockout mouse, the drug would have no or minimal effects because the gene product was not present, but it would

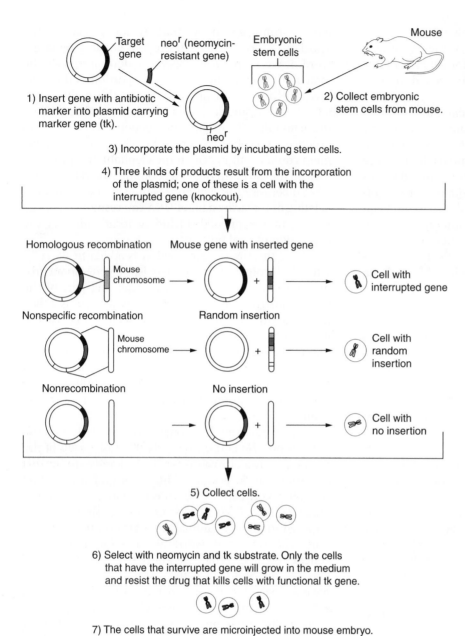

1) Insert gene with antibiotic marker into plasmid carrying marker gene (tk).

2) Collect embryonic stem cells from mouse.

3) Incorporate the plasmid by incubating stem cells.

4) Three kinds of products result from the incorporation of the plasmid; one of these is a cell with the interrupted gene (knockout).

Homologous recombination

Mouse gene with inserted gene

Mouse chromosome

Cell with interrupted gene

Nonspecific recombination

Random insertion

Mouse chromosome

Cell with random insertion

Nonrecombination

No insertion

Cell with no insertion

5) Collect cells.

6) Select with neomycin and tk substrate. Only the cells that have the interrupted gene will grow in the medium and resist the drug that kills cells with functional tk gene.

7) The cells that survive are microinjected into mouse embryo.

FIGURE 7.9 **How to Make a Knockout Mouse** Transferring a marker (neomycin resistance) gene into a mouse target gene by homologous recombination disrupts the gene's function, allowing selection for the disrupted gene. Thymidine kinase (tk) is used to select from antibiotic-resistant transformants because cells with this inserted gene will be killed in the presence of an altered nucleotide during DNA synthesis. After selection, the cells with the disrupted gene can be inserted into mouse embryos to produce mice with this gene "knocked out." By similar methods, it is possible to "knock in" another gene (like a human gene) to make testing of the gene possible in a test animal.

have effects in the presence of the human gene in the knock-in mouse. If effects are seen in the knockout animal, they are off target and can represent potential side effects of the candidate drug (saving human suffering without having to test humans).

Thus far, mice and zebrafish have commonly been used as knockout animals. Research is also being done on knockout primates. The first steps toward developing a potentially invaluable animal model have already resulted in genetically modified monkeys. Researchers at the Oregon Regional Primate Research Center at the Oregon Health Sciences University in Portland have introduced the gene for green fluorescent protein (GFP)

into a rhesus monkey named ANDi (the name stands for "inserted DNA," spelled backward). The GFP gene, which had its origin in a jellyfish, does not cause ANDi to glow in the dark. In fact, he looks exactly like a regular rhesus monkey. It takes a special microscope to see the "glow" in ANDi's cells. Not everyone agrees with the idea of using monkeys for knockout research, even though they have considerable biological similarity to humans (see "You Decide: Monkey Knockouts"). The demonstration that a gene can be inserted into monkeys was a major accomplishment to prove that the method can work in higher animals, even if it never results in a significant medical application.

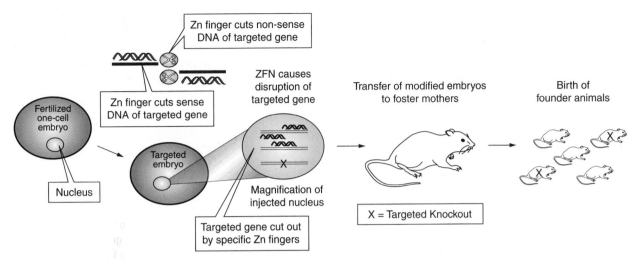

FIGURE 7.10 Zinc Fingers: Knockout Rats for Drug Toxicity Testing The potential to reduce the length and cost of drug toxicity tests and to reduce the number of animals needed to yield reproducible, relevant data has been fulfilled with a new method for creating knockout rats. Site-specific zinc-finger nucleases are used to knock out targeted genes known to be related to drug toxicity. Founder animals are produced from these rat embryos in 6 weeks; they can be used to produce clones in quantity, and the process takes less time than homologous recombination.

Until now, the availability of genetically modified rats as knockouts was limited. The Sigma company reported the first knockout rat in 2009, using the zinc-finger nuclease methodology (see **Figure 7.10**). Knockout rat models offer an alternative that can reduce drug development costs by providing a more human-like model while decreasing time to market. Zinc-finger nuclease-mediated gene targeting in the rat has opened up a whole new world of possibilities for the development of more predictive models of drug metabolism in humans.

7.4 Producing Human Antibodies in Animals

The power of antibodies places them among the most valuable protein products. Here, we look at how antibodies are produced through the use of animals as bioreactors.

Monoclonal Antibodies

Researchers have long dreamed of harnessing the specificity of antibodies for a variety of uses that target particular sites in the body. In the 1980s, the use of an antibody as a targeting device led to the concept of the "magic bullet," a treatment that could effectively seek

YOU DECIDE

Monkey Knockouts

ANDi is the first step on the path to developing a knockout monkey that could be a better animal model for studying diseases and their treatments. Genetically modified primates could potentially be used to study breast cancer, HIV, and any number of other diseases.

Not everyone agrees that creating a knockout monkey ought to be a scientific goal. Some point out that mice, and even zebrafish, share the majority of human genes and have been used effectively in research. Others argue that a knockout monkey could present a threat to public health because a pathogen that could make the leap from the study animals to the human population might develop. They point to the possibility that HIV may be a pathogen that made an unexpected and unassisted leap from wild monkeys to human beings. Others are uncomfortable with the use of primates in any sort of research. They argue that, precisely because these primates are so like us, they may share our self-awareness, which makes such research ethically repugnant.

What do you think? Should we press forward to develop a knockout monkey? Or are primates too closely related to humans to be used in this way? You decide.

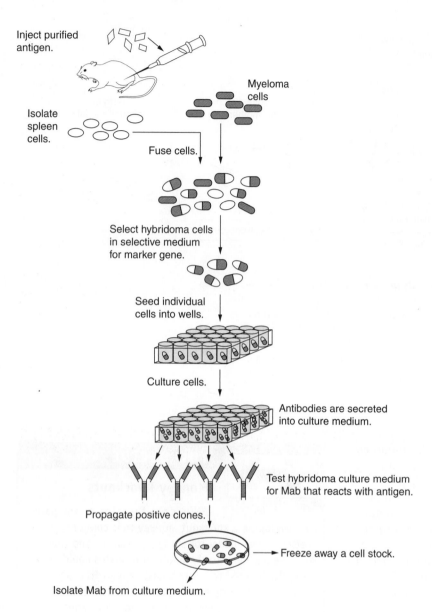

Inject purified antigen.

Isolate spleen cells.

Myeloma cells

Fuse cells.

Select hybridoma cells in selective medium for marker gene.

Seed individual cells into wells.

Culture cells.

Antibodies are secreted into culture medium.

Test hybridoma culture medium for Mab that reacts with antigen.

Propagate positive clones.

Freeze away a cell stock.

Isolate Mab from culture medium.

FIGURE 7.11 Monoclonal Antibody Production An antigen-stimulated mouse will provide spleen cells that can be fused with myeloma (cancer) cells to continuously produce monoclonal antibodies in a culture.

and destroy tumor cells wherever they might reside. One major limitation in the therapeutic use of antibodies is the problem of producing a specific antibody in large quantities. Initially, researchers screened **myelomas,** which are antibody-secreting tumors, for the production of useful antibodies. But they lacked a means to program the myeloma to manufacture an antibody to their specifications. This situation changed dramatically with the development of monoclonal antibody technology. **Figure 7.11** illustrates the procedure for producing monoclonal (made from a single clone of cells) antibodies (Mabs).

First a mouse or rat is inoculated with the antigen (Ag) to which an antibody is desired. After the animal produces an immune response to the antigens, its spleen is harvested. The spleen houses antibody-producing cells, or lymphocytes. These spleen cells are fused en masse to a specialized myeloma cell line that no longer produces an antibody of its own. The resulting fused cells, or hybridomas, retain the properties of both parents. They grow continuously and rapidly in culture like the myeloma (cancerous) cell and produce antibodies specified by the fused lymphocytes from the immunized animal. Hundreds of hybridomas can be produced from a single fusion event. The hybridomas are then systematically screened to identify the clone, which produces large amounts of the desired antibody. After this clone is identified, the antibody will be produced in large quantities.

Monoclonal antibody products are now used to treat cancer, heart disease, and transplant rejection, but the process has not always been smooth. An early limitation to the success of therapeutic antibodies was immunogenicity. The classic method for producing Mabs using mouse hybridoma cells provoked an immune response in human patients. Mouse hybridomas produce enough "mouse" antigens to still evoke an undesirable immune response in humans. Patients quickly eliminated mouse Mabs, so that the therapeutic effects were short-lived. Researchers called this the **human antimouse antibody (HAMA) response.** The solution is to make Mabs more human. Removing mouse antigens or creating chimeric cells is costly and time-consuming, and the resulting Mabs still provoke a HAMA response. Researchers are working on solving this problem and are producing Mabs in different organisms. In 2010, there were 22 therapeutic Mabs that had been approved by the FDA. All but three of these have been engineered to reduce immunogenicity, and most have some human genetic sequence. Transgenic mice producing human-ized Mabs are under study in at least 33 current trials for new drugs.

In 2007, the FDA approved the first monoclonal antibody produced in animals that does not activate the human immune system to rejection. Panitumab, produced from transgenic mice with knock-in human immune genes, was manufactured by Amgen in Thousand Oaks, California. The process involved inactivating the mouse antibody machinery (heavy- and light-chain proteins of antibodies) and introducing the human equivalent of these antibody-producing genes by homologous recombination into the inactivated (deleted) regions. The work was done in mouse ES cells, followed by the introduction of these cells into mouse embryos to produce founder animals. The transgenic mice (xenomice) produce a fully human antibody against epidermal growth factor receptor (as a treatment for people with advanced colorectal cancer), which can be purified and does not produce the HAMA response.

CAREER PROFILE

Animal Technicians and Veterinarians

Animal technicians assist scientists who use laboratory and farm animals in biotechnology research or product testing. They perform medical procedures on animals and care for research animals before and after surgery; they may also perform some surgeries. Technicians not only help to restrain the animal during examinations and inoculations but may also participate in making daily animal observations, breeding animals, and weaning them. Keeping detailed records of animal health and writing standard operating procedures (SOPs) are also important parts of this job.

Animal technicians usually have an associate's degree in veterinary science or a related field. Knowledge of biology and math is also preferred. Many biotechnological positions prefer or require that applicants be certified by the American Association for Laboratory Animal Science (AALAS; for information, visit their website at www.aalas.org).

Salary ranges are skilled entry level, and technicians must be prepared to work nights, weekends, and holidays (because animals need daily care). Most companies offer excellent benefit packages, flexible schedules, and work release for continuing education. Extra hours are commonly available in this position.

Veterinarians play a major role in the health care of all types of pets, livestock, zoo animals, sporting animals, and laboratory animals. Some veterinarians use their skills to protect humans against diseases carried by animals and conduct clinical research on human and animal health problems. Others work in basic research, broadening the scope of fundamental theoretical knowledge, and in applied research, developing new ways to use knowledge. Prospective veterinarians must graduate from a 4-year program at an accredited college of veterinary medicine with a doctorate of veterinary medicine (DVM or VMD) degree and obtain a license to practice.

All of these positions require a significant number of credit hours (ranging from 45 to 90 semester hours) at the undergraduate level. Competition for admission to veterinary school is keen; about 1 in 11 applicants were accepted in 2010. Veterinarians who seek board certification in a specialty must also complete a 2- to 3-year residency program providing intensive training in specialties such as internal medicine, oncology, radiology, surgery, dermatology, anesthesiology, neurology, cardiology, ophthalmology, and exotic small animal medicine. The median annual salary of veterinarians in 2010 was $92,570.

MAKING A DIFFERENCE

The Human Genome Project of 2001 was an international consortium effort. A similar international effort to identify the function of every gene in the mouse has begun. To accomplish this, a known gene is knocked out (of the 20,000 genes making up the mouse genome); then the strain is rigorously compared with a normal mouse strain to determine physical and behavioral differences. Since 95% of mouse genes are similar to those of humans, the database resulting from the project will have relevance to human disease and drug research. The work is expected to take 3 years. If only a handful of useful drugs result (at about $1 billion each in value), the $900 million effort will have paid for itself. Even if the process takes more than the next decade, it will still provide valuable data. The repetitive nature of the study (partially funded by the NIH) of each gene suggests a heavy involvement of biotechnicians and lab analysts, or "full employment" for biotechnologists, although the process may become automated at some point. The opportunity to be part of an international effort with the potential to cure human diseases would be difficult to find in any other field than biotechnology.

QUESTIONS & ACTIVITIES

Answers can be found in Appendix 1.

1. Why are myeloma cells used in hybridomas?

2. Why would a human immune system reject an antibody produced in a mouse?

3. How have some biotechnology companies produced antibodies that do not stimulate a HAMA-like response?

4. Discuss some of the ethical concerns about using animals in research, including regulatory requirements for their use.

5. How do knockout and knock-in animals provide better predictions of how a drug will work in humans?

6. Explain the use of hyperthermia in cancer treatment in animals.

7. Why are zebrafish commonly being used to test new human drugs?

8. Explain how "model animals" are chosen.

9. Explain the concept of "reduce, replace, and refine" in the use of animals in biological research.

10. Explain pronuclear microinjection as a transgenic method.

Visit www.pearsonhighered.com/biotechnology
To download learning objectives, chapter summary, "Keeping Current" web links, glossary, flashcards, and jpegs of figures from this chapter.

DNA Fingerprinting and Forensic Analysis

After completing this chapter, you should be able to:

- Define DNA fingerprinting.
- Outline the process of collecting and preparing a DNA sample to be used as evidence.
- List some factors that can degrade DNA evidence and some of the precautions required to maintain the reliability of DNA evidence.
- Describe the polymerase chain reaction (PCR) method is used to produce a DNA fingerprint.
- Explain how short tandem repeat (STR) profiles are compared in DNA fingerprint analysis.
- Explain how a DNA fingerprint can be used to exclude a suspect.
- Examine two or more DNA fingerprints in the lab to determine if they match.
- Describe the use of DNA fingerprinting techniques in establishing familial relationships.
- List some of the uses of DNA analysis in biological research.
- Discuss some of the ethical issues surrounding DNA fingerprinting.

A biotechnician examines DNA fingerprints looking for similarities and differences between individuals.

orensic science is the intersection of law and science. It can be used to condemn the guilty or exonerate the innocent. Many court cases hinge on scientific evidence provided by forensics. Throughout the years, scientists have developed technologies to uncover facts in criminal investigations, and the law has been quick to embrace these new technologies as they became available.

For example, in the late 1800s, efforts to fight crime were given a boost by a new technology: photography. Thanks to the invention of the camera, it became possible to depict criminals in custody so accurately that images could be used as references. Photographs were certainly an improvement over verbal descriptions and hand-drawn "wanted" posters, but they still had severe limitations. Criminals found many ways to alter their appearance—via haircuts, the addition of facial hair, or use of eyeglasses—that could make identification based on a photograph almost impossible.

A little more than 100 years ago, scientists discovered that the tiny arches and whorls in the skin of human fingertips could be used to establish identity. After a single bloody print on the bottom of a cash box helped solve a murder in England, the process of inking a suspect's fingers and collecting a set of prints became routine. The Federal Bureau of Investigation (FBI), the Central Intelligence Agency (CIA), and other law enforcement agencies amassed huge collections of these prints. At first, clerks were responsible for painstakingly examining the prints visually, looking for matches. With the development of the computer, however, the sorting process became less tedious and more reliable. But fingerprints are not always available: they can be wiped away, and gloves can be worn to avoid leaving them behind.

In 1985, a revolutionary new technology emerged as an important forensics tool. Instead of counting on smudged fingerprints left at the scene of a crime, investigators could look at a new kind of "fingerprint," the unique signature found in each person's genetic makeup. In this chapter, we look at a description of DNA fingerprints and discuss what makes each person's DNA unique. Then we learn the processes used to collect DNA samples and to produce DNA fingerprints for analysis. We compare DNA profiles and learn what it takes for a DNA match to occur. Next we consider examples that illustrate the value—and vulnerabilities—of DNA as evidence. We'll then look at how DNA profiles can be used to establish familial relationships, focusing on the use of mitochondrial DNA to provide proof of kinship. Finally, because DNA profiles are not limited to humans, we look at the uses of DNA analysis to establish the origin of valuable plant crops, help to enforce laws to protect wildlife, and perform biological research.

FORECASTING THE FUTURE

The more that is known about a patient's DNA sequence, the easier it is to make predictions about what types of drugs will be effective in treating that patient's genetically based conditions. Sequencing an individual's DNA to figure out the best course of drug treatment differs from earlier tests, which serve only to diagnose certain genetic diseases (such as prostate cancer, cystic fibrosis, and diabetes). Now genetic sequencing will have to become faster and cheaper in order to develop diagnostic tests that allow for personalized therapies on a large scale. But recent advances are producing second- and third-generation innovations that may reduce the cost of human DNA sequencing by two-thirds in as little as 10 years. Medicare and private insurers have been slow to accept the cost of reimbursing patients for DNA sequencing, but strong demand for personalized medicine in pharmaceutical research could increase acceptance of these new costs. In fact, it is likely that in the near future, the U.S. Food and Drug Administration (FDA) will require companion diagnostics (in the form of DNA sequencing of phase trial participants) for every drug being tested to treat genetic disorders (see Chapter 12). Additionally, in June 2010, the FDA put five genetic test makers on notice that they must get federal approval before marketing their products directly to consumers. The biology of how DNA variations actually cause certain diseases is still being actively researched, but these efforts will undoubtedly lead to the need for more DNA testing on an individual level. Making genetic tests the standard of care has the potential to significantly reduce medical costs and increase safety.

8.1 What Is a DNA Fingerprint?

We know that every human being carries a unique set of genes. The chemical structure of DNA is always the same, but the order of the base pairs in chromosomes differs in individuals. The novel assemblage of the 3 billion nucleotides formed into 23 pairs of chromosomes gives each of us a unique genetic identity.

We also know that every cell contains a copy of the DNA that defines the organism as a whole, even though individual cells have different functions (cardiac muscle cells keep our hearts beating, neurons transmit the signals that are our thoughts, etc.). These two aspects of DNA—its uniform nature in a single individual and the genetic variability between individuals—make DNA fingerprinting possible. Because every cell in a body shares the same DNA, cells collected by swabbing the inside of a person's cheek will be a perfect match for those found in white blood cells, skin cells, or any other tissue.

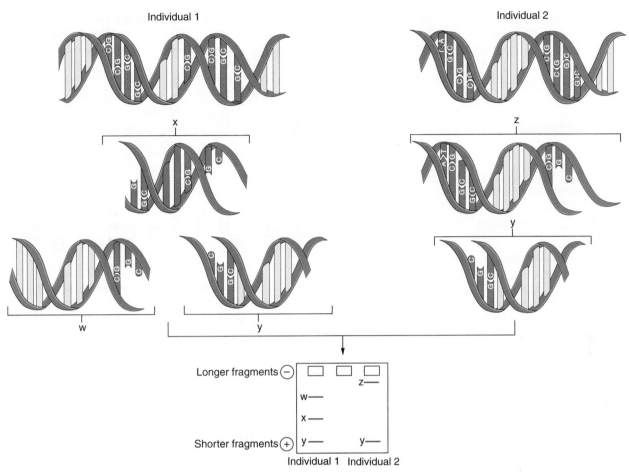

FIGURE 8.1 Restriction Fragment Length Polymorphisms (RFLPs) Can Be Used to Produce DNA Fingerprints Restriction enzymes recognize a short sequence of nucleotides and cut the DNA. Random mutations interfere with enzyme recognition, creating different lengths of DNA (not cut if mutated). The pattern of these cut fragments (when separated on a gel by electrophoresis) can be used to determine similarities and differences. Human DNA fingerprinting has moved to STR analysis largely because of the need for considerably less DNA for the PCR process. Individual 1 in the figure has three cut sites in his DNA, while individual 2 has only 2. The banding patterns resulting from the electrophoretic separation shows the differences.

How Is DNA Typing Performed?

Only a tenth of 1% of DNA (about 3 million bases) differs from person to person. These variable regions can generate a DNA profile of an individual, using samples from blood, bone, hair or other body tissues containing a nucleus with DNA. In criminal cases, this generally involves obtaining samples from crime-scene evidence and a suspect, extracting the DNA, and analyzing it for the presence of a set of specific DNA regions.

There are two main types of forensic DNA testing: one based on restriction fragment length polymorphism (RFLP) and the other on the polymerase chain reaction (PCR) (see **Figure 8.1** for an illustration of an RFLP). Generally, RFLP testing requires larger samples

than does PCR, and the DNA must be intact and undegraded (the FBI accepts RFLPs of four types for case comparison). Crime-scene evidence that is old or present only in small amounts is often unsuitable for RFLP testing. Warm, moist conditions may accelerate DNA degradation, rendering it unsuitable for RFLP testing in a relatively short period of time. Because of this, RFLP is not used as often for DNA testing as it once was.

PCR-based testing requires less DNA than RFLP testing and is still effective if the sample is partially degraded. However, PCR-based tests are also extremely sensitive to contamination by foreign DNA at the crime scene and within the laboratory.

Fortunately it is not necessary to catalog every base pair in an individual's DNA to arrive at a unique

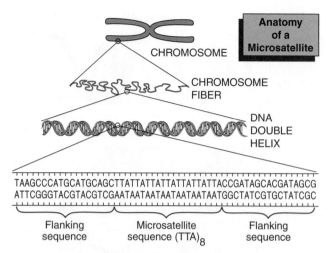

FIGURE 8.2 **Anatomy of a DNA Microsatellite** A microsatellite is a variable number of repeated nucleotides (like TTA) that occur in specific locations in the genome. Using primers for the flanking regions on the ends of the microsatellite allows for amplification using PCR (from both directions). Individual inherit a specific number of these repeats from their parents, but the number of repeats varies for unrelated individuals, forming a distinct pattern or DNA fingerprint.

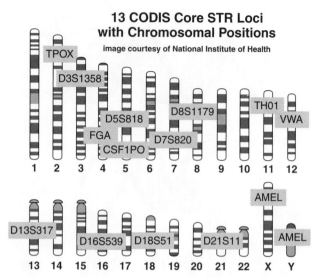

FIGURE 8.3 **Human Short Tandem Repeat Patterns** The FBI has chosen 13 unique STR regions (shown on human chromosomes) for analysis and comparison in its library of DNA fingerprints. DNA primers for the flanking regions of these sites are commonly found in PCR kits that amplify these sites. After DNA sequencing, the results indicate the number of repeats for each homologous chromosome at that site and constitute a DNA fingerprint for that individual.

fingerprint. Instead, DNA profiling depends on only a small portion of the genome. Every strand of DNA is composed of both active genetic information, which codes proteins, and inactive DNA, which does not. Some of these inactive portions contain repeated sequences of between 1 and 100 base pairs (see Chapter 3 for a review). These sequences, called **variable number tandem repeats (VNTRs),** are of particular interest in determining genetic identity. Every person has some VNTRs that were inherited from his or her mother and father. No person has VNTRs that are identical to those of either parent (this could only occur as a result of cloning). Instead, the individual's VNTRs are a combination of repeats of those of the parents' DNA regions in tandem. The uniqueness of an individual's VNTRs provides the scientific marker of identity known as a DNA fingerprint.

DNA fingerprints produced by PCR are usually restricted to detecting the presence of **microsatellites** (**Figure 8.2**), which are one- to six-nucleotide repeats dispersed throughout chromosomes. Because these repeated regions can occur in many locations within the DNA, the probes used to identify them complement the DNA regions that surround the specific microsatellite being analyzed. The small size of these repeated segments has resulted in the term **short tandem repeat**, or **STR**. The FBI has chosen 13 unique STRs to test for DNA profiles contained within their **Combined DNA Index System,** or **CODIS**, as seen in **Figure 8.3**.

8.2 Preparing a DNA Fingerprint

Preparing a DNA fingerprint requires specimen collection, DNA isolation and quantification, and PCR amplification. We discuss these steps in detail on the following pages.

Specimen Collection

Crime scene investigators routinely search for sources of DNA: dirty laundry, a licked envelope, a cigarette butt, or anything else that might be a source for human cells left behind. Tiny blood stains or a trace of saliva is often all it takes to crack a case.

Every living thing has DNA, so every crime scene is full of sources of possible contamination. For this reason, scrupulous attention to detail is required in collecting and preserving evidence. To protect the evidence, workers at a crime scene must take the following precautions (see FBI website for a full list of precautions):

- Wear disposable gloves and change them frequently.

- Use disposable instruments (such as tweezers and swabs).

- Avoid talking, sneezing, and coughing to prevent contamination with microdroplets of saliva.

- Avoid touching anything that might contain DNA (like their own body parts) while handling evidence.
- Air-dry evidence thoroughly before packaging. Microbes can contaminate a sample.
- Store evidence in specifically designed materials: plastic bags can retain damaging moisture.

As this partial list shows, the enemies of evidence are everywhere. Sunlight and high temperatures can degrade DNA. Bacteria, busy doing their natural work as decomposers, can contaminate a sample before or during collection. A small amount of moisture can damage and ruin a sample.

DNA fingerprinting is a comparative process. DNA from the crime scene must be compared with known DNA samples from the suspect. The ideal specimen used to compare evidence is 1 mL or more of fresh whole blood treated with an anticlotting agent called ethylenediaminetetraacetic acid (EDTA). This quantity of whole blood contains an adequate amount of leukocytes with nuclear DNA for testing. Unlike erythrocytes, which have no nuclear DNA, leukocytes contain the normal complement of human chromosomes. If a clean specimen cannot be obtained from the known source, all is not necessarily lost—DNA has been retrieved and successfully analyzed from samples that were a decade old by amplifying the DNA used for comparison by PCR.

Extracting DNA for Analysis

After a sample is collected from the known source, a lab technician is responsible for determining its genetic profile. First, the technician extracts the DNA from the sample. DNA can be purified either chemically (using detergents that wash away the unwanted cellular material) or mechanically (using pressure to force the DNA out of the cell). Once the DNA is extracted, the technician must follow several steps to transform the unique signature of that DNA into visible evidence.

PCR and DNA Amplification

Several thousand cells are required to do an RFLP analysis, more than can often be collected at a crime scene. Imagine a case in which the only evidence is the saliva on an envelope flap or a blood stain invisible to the naked eye. Because of this is, PCR is now normally used to "grow" or amplify the sample into an amount that can be analyzed,

PCR is much like a photocopier for DNA. The first step involves locating the portions of DNA that can be useful for comparison. Primers, or short pieces of DNA, are the tools used to find those portions on each DNA strand in the 5′ to 3′ direction (see arrows in Figure 8.4). The primers behave much like probes and seek out complementary sections of the DNA. Once the complementary segment of DNA is located, copying begins using a device called a *thermal cycle* (see Chapter 3). Figure 8.4 illustrates PCR as used in DNA amplification for fingerprinting. With the aid of the enzyme *Taq* polymerase and DNA nucleotides, cycles of heat and cold cause the DNA to separate and replicate repeatedly. Within hours, the DNA segments can be increased by millions or even billions. That rapid growth presents a possible problem: PCR is very sensitive to contamination, and a small error in field or laboratory procedure can result in the duplication of a useless DNA sample a million times.

STR Analysis

After STRs are amplified by PCR, the alleles are separated and detected using capillary electrophoresis, which separates the lengths of amplified DNA, allowing the number of repeats in each of the two alleles on homologous chromosomes to be determined. An STR contains repeated numbers of a short (typically three- to four-nucleotide) DNA sequence. The number of repeats within an STR is referred to as an allele. For instance, the STR known as D7S820, found on chromosome 7, contains between 5 and 16 repeats of GATA. Therefore, there are 12 different alleles possible for the D7S820 STR sequence. An individual with D7S820 alleles 10 and 15, for example, would have inherited a copy of D7S820 with 10 GATA repeats from one parent and a copy of D7S820 with 15 GATA

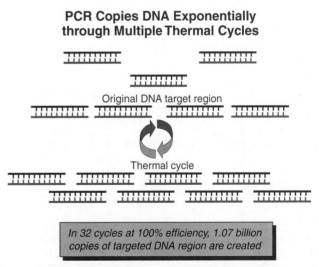

PCR Copies DNA Exponentially through Multiple Thermal Cycles

Original DNA target region

Thermal cycle

In 32 cycles at 100% efficiency, 1.07 billion copies of targeted DNA region are created

FIGURE 8.4 Amplifying DNA at Specific Sites by PCR DNA primers for the flanking regions of CODIS sites results in DNA amplification at specific STR sites. Many PCR kits contain reaction beads with multiple primers, making it possible to amplify DNA for multiple sites simultaneously.

repeats from the other parent. There are 12 different alleles for this STR, creating 78 different possible genotypes, or pairs of alleles, for this one site. The large number of variations for each STR provides the variability that is needed to distinguish different individuals using DNA fingerprints. **Figure 8.5** shows alleles of all 13 STR sites accepted by the FBI for CODIS.

8.3 Putting DNA to Use

DNA fingerprinting is similar to old-fashioned fingerprint analysis in an important way. In both cases, evidence collected from a crime scene is compared with evidence collected from a known source. During evaluation of the samples, analysts are looking for alignment of the bands. If the samples are from the same person (or from identical twins), the bands or dots should line up exactly. All tests are based on exclusion. In other words, testing continues only until a difference is found. If no difference is found after a statistically acceptable amount of testing, the probability of a match is high. See if you agree with the exclusion of one individual as shown in **Figure 8.6**.

As we will see, such clear-cut visual evidence can be invaluable to crime investigators. Next we consider a few examples of DNA profiling in action.

The Narborough Village Murder Case Led to a New Method of DNA Separation

The first reported use of genetic fingerprinting in a criminal case involved the sexual assault and murder of a schoolgirl in the United Kingdom in 1983. Sir Alex Jefferies and his colleagues, Dr. Peter Gill and Dr. Dave Werrett, developed techniques for extracting DNA and preparing profiles using old samples of human tissue. Gill also developed a method for separating sperm from vaginal cells, which allows fingerprints to be run on vaginal cells first and then sperm cells in order to provide comparison samples (**Figure 8.7**). In this method, a detergent breaks open vaginal cells but leaves sperm cells intact. Without these developments, it would be difficult to use DNA evidence to solve rape cases because sperm cells could not be located to compare them against known samples.

In this case, Jefferies and his colleagues compared DNA evidence collected from the murder scene with a semen sample from an earlier similar rape and murder, and the analysis indicated that both crimes were

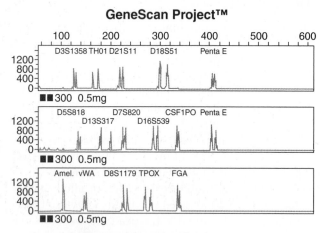

GeneScan Project™

FIGURE 8.5 Gene Scan Display of all 13 STRs accepted by CODIS Utilizing capillary electrophoresis to detect the amplified STR microsatellite DNA sequences, it is possible to determine the number of STRs. By comparing the patterns obtained as evidence, it can be determined if the DNA patterns are exactly the same or different. Note that none of the DNA fingerprints shown are exact matches.

Case Mockup: Results

Sample ID	D3S1358	VWA	FGA	AMEL	D8S1179	D21S11	D18S51	D5S818	D13S317	D7S820
Jane Doe, Reference Sample	15, 16	16, 19	18, 22	X	11, 12	30, 31.2	14	9, 12	9, 12	8, 11
Dorothy Smith, Reference Sample	15, 18	16, 17	20, 21	X	10, 13	31, 31.2	12, 16	10, 12	11, 12	8
Knife Blade Stain A	15, 16	16, 19	18, 22	X	11, 12	30, 31.2	14	9, 12	9, 12	8, 11
Knife Blade Stain B	15, 16	16, 19	18, 22	X	11, 12	30, 31.2	14	9, 12	9, 12	8, 11

Conclusions: Dorothy Smith CAN be excluded from contributing to the DNA found on the knife blade stains A and B. Jane Doe CANNOT be excluded from contributing to the DNA found on the knife blade stains A and B.

FIGURE 8.6 Who Is Excluded by DNA Comparison? The DNA profiles shown below were taken from a crime scene. Knife stains A and B were found on the blade of a knife used in the crime, while stain C was found on the handle of the knife. All three stains were amplified for DNA analysis. Two individuals were tested for a match to the DNA profiles. Read the conclusion from the figure and see if you agree with the analysis.

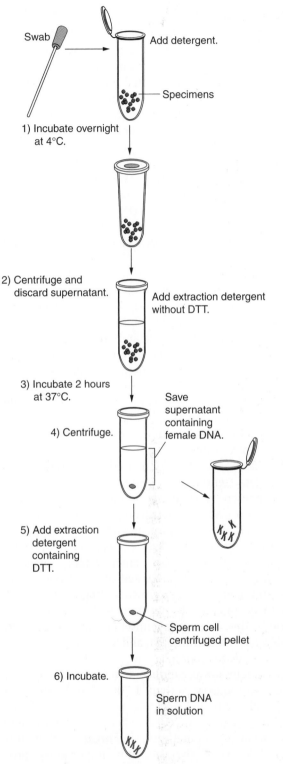

1) Incubate overnight at 4°C.

2) Centrifuge and discard supernatant.

Add extraction detergent without DTT.

3) Incubate 2 hours at 37°C.

4) Centrifuge.

Save supernatant containing female DNA.

5) Add extraction detergent containing DTT.

Sperm cell centrifuged pellet

6) Incubate.

Sperm DNA in solution

Swab — Add detergent. — Specimens

FIGURE 8.7 Isolation of Sperm DNA or Vaginal Cell DNA from Mixed Sources DNA from a sperm cell can be fingerprinted to be compared with the profile gotten from a sperm sample obtained on a vaginal swab. The buffer, containing dithiothreitol (DTT), will break the sperm cell membrane and can be used to detect sperm DNA after vaginal DNA has been analyzed separately in mixed evidence (owing to the ability of the sperm cell membrane to resist disruption until the addition of DTT).

committed by the same man. At this point, the police had a prime suspect. However, when the DNA evidence was compared with a DNA sample from the suspect's blood, the DNA did not match. The prime suspect was therefore excluded from consideration.

The investigation continued. The police conducted the world's first mass screening of DNA, collecting 5,500 samples from the district's male population. Using simple blood-typing tests, all but 10% of the initial group were quickly excluded (using RFLP analysis methods). After many grueling hours of analysis, the investigation had come to a standstill: none of the remaining profiles matched that of the rapist/killer.

Then came the lucky break. A man was overheard saying that he had given a sample in the name of a friend. When the man who had evaded the mass testing was apprehended, his DNA was analyzed and found to be an exact match for the DNA in the semen specimens from the crimes. The suspect confessed and was sentenced to life in prison.

This case highlights one of the important limitations to the use of DNA evidence: unless there is a known sample to be used as a comparison, identity cannot be established. In the example shown in **Figure 8.8**, the blood samples from the victim and the suspect are known. By comparing these two known DNA fingerprints, investigators were able to place the suspect at the scene of the crime based solely on the DNA fingerprints of the blood found on the suspect's clothes.

Is Eyewitness Testimony More Reliable than DNA Testing?

DNA evidence was first used in the United States in 1987 and has since helped to resolve thousands of cases. A particularly important use of DNA evidence is to refute other erroneous evidence that has sometimes led to false convictions. DNA evidence is especially valuable when it is used to expose faulty eyewitness testimony. Eyewitness testimony, which might seem to be the gold standard of evidence, is actually quite fallible.

Sample ID	D5S818
Person A	9, 12
Person B	8 (or 8,8)

FIGURE 8.8 Example of an Exclusion Based on One STR from a Suspect At locus D5S818 (chromosome 5), person A has 9 repeats from his mother and 12 repeats from his father. Person B has 8 repeats from both his parents.

YOU DECIDE

Should DNA Fingerprinting Be Sped Up to Apprehend Repeat Offenders?

Some criminals are arrested, spend less than a day in jail, and then commit crimes while they are out on bail. If police could quickly test suspects' DNA, to see if it matches entries in crime databases, they might be able keep dangerous criminals in jail. Currently, most genetic tests take 24 to 72 hours, and by the time that the results are back, the suspects may have been released. A new test could make it almost as easy as matching fingerprints to compare DNA from people arrested for crimes with other DNA samples from crime scenes stored in forensic databases. With this test, police could check on whether a person's DNA matched that found at past crime scenes while suspects were still being processed. A report on the fast forensic test appears in the August 5, 2010, issue of the journal *Analytical Chemistry*. To increase the speed of forensic DNA testing, the scientists built a chip that can copy and analyze DNA samples taken from a cotton swab. Forensic technicians collect DNA from by swabbing a suspect's mouth, mixing the sample with a few chemicals, and warming it up. The DNA-testing-lab-on-a-chip does the rest. The entire process takes only 4 hours at present.

Is testing a suspect's DNA upon arrest for speedy results a reasonable way to catch repeat offenders, or is it an invasion of privacy? You decide.

For example, in 1988, Victor Lopez, the so-called Forest Hills rapist, was tried for the sexual assault of three women. All three women had described their assailant as a black man when they reported the crime to the police. Because Lopez was not black, concerns were raised that this was a case of mistaken identity. Was Lopez an innocent man falsely accused by the system? The blood of the accused was analyzed; it was then compared with sperm left at the scene and the DNA was a match. Lopez was found guilty of the attacks despite the contradictory eyewitness testimony. Now let's look at how much of what we have learned is admissible as evidence.

World Events Lead to the Development of New Technologies

When major disasters occur and large numbers of people need to be identified, DNA comparisons are not enough. New methods of cataloging evidence have resulted in faster and more efficient comparisons to distinguish individuals.

World Trade Center

Shortly after the twin towers of the World Trade Center in New York City were destroyed by the terror attacks of September 11, 2001, forensic scientists came together and rapidly accelerated efforts to use DNA-based techniques to identify the remains of victims. They were confronted with a tremendous amount of debris at the site, dangerous working conditions, heat and microbial decomposition of remains, and hundreds of thousands of tissue samples from the nearly 3,000 individuals lost. These issues made it evident that new strategies would need to be employed to quickly prepare and organize DNA profiles and compare them with DNA profiles from relatives. How would scientists establish DNA identity for those who perished in over 1.5 million tons of rubble?

State agencies such as the New York Department of Health, and the medical examiner's lab, and several federal agencies including the U.S. Department of Defense, National Institutes of Health, and the U.S. Department of Justice immediately responded to help with this task. Within 24 hours after the disaster, the New York City Police Department had established collection points throughout the city where family members could file missing persons reports and provide cheek cell swabs for DNA isolation. Personal items (combs, toothbrushes) from the missing were also collected for DNA profiling.

Myriad Genetics, Inc., Gene Codes Forensics, DNA-VIEW, Celera Genomics, Bode Technology Group, and Orchid Genescreen were among the companies assisting in this effort. Several other companies were involved in developing new software programs to help match samples submitted by family members to DNA profiles obtained from World Trade Center victims. Available tissue samples primarily consisted of small bone fragments and teeth, which provided fragmented DNA owing to degradation by the intense heat at the site. Because of this, forensic scientists primarily used STR, mitochondrial DNA (mtDNA), and SNP analysis of DNA fragments to develop profiles.

DNA analysis was conducted on over 15,000 tissue samples, although less than 1,700 of the estimated 2,819 people who died at the site were ultimately identified. This tragedy forced the development of new forensic strategies for analyzing and organizing remains recovered from the site and, most importantly, provided closure for a number of families that lost loved ones in the attack.

South Asian Tsunami

The South Asian tsunami of December 2004 was a tragedy that claimed over 225,000 lives and devastated areas of Indonesia, Sri Lanka, and Thailand. The

YOU DECIDE

Crime-Fighting Tool or Invasion of Privacy?

In the Narborough murders, many individuals were tested because they all fit a general description: they were young men who lived in the area of the murders. This strategy led to the conviction of the killer, but mass testing, especially on the basis of a general description, is extremely controversial. Opponents point out that more than 5,000 innocent men were called on to give a sample of their blood during the investigation, and the guilty individual very nearly escaped detection despite these efforts. Such mass testing without due cause is now prohibited both in the United States and the United Kingdom. However, many people still believe that even the routine quest for DNA evidence during a criminal investigation continues to undermine the fundamental right to privacy.

Of particular concern are DNA-profiling databases. Computer-searchable collections of DNA data are now authorized by all 50 states. Many states have also registered the DNA profiles they collect on CODIS, the Combined DNA Index System run by the FBI. In addition to profiles of convicted offenders, CODIS contains a compilation of unidentified DNA profiles taken from crime scenes, and some advocates are calling for adding DNA profiles of all people arrested for any offense. Some states have even proposed that DNA profiles be taken from all individuals at the time of birth to create databases that might be used to identify criminals by matching DNA profiles from their relatives.

In early 2007, the federal government proposed that DNA profiles from more than 400,000 illegal immigrants and detainee captives in the war on terrorism be added to CODIS.

By comparing the databases, law enforcement officers can identify possible suspects when no prior suspect exists. (Visit the CODIS FBI website to see how the information is used in solving crimes.)

Opponents do not argue with the potential usefulness of the databases, but they are concerned that the technology could be abused. They point out that taking a blood sample is a more invasive process than taking a set of fingerprints. In some cases, the courts have agreed and determined that DNA collection is a violation of state and federal laws prohibiting unreasonable search and seizure. There is also the possibility that DNA information could someday be used to discriminate against job seekers or those applying for health insurance. Even though profiles are based on "junk" DNA and provide no information about genetic diseases or other traits, the original sample *does* contain an individual's complete DNA. On this basis, some demand that all samples be destroyed after investigation of a specific case is completed.

Proponents of DNA collection point out that the databases are regulated and secure, and samples are not obviously identified with the name of the source. Only trained professionals collect the blood samples, and the procedure presents no significant health risk to the donor. Is the routine collection of blood and the compilation of DNA databases a reasonable tool in the effort to fight crime or an unwarranted invasion of privacy? Should investigators be allowed to conduct the type of mass screening used in the Narborough case? You decide.

Michigan-based company Gene Codes modified their software system, called **Mass Fatality Identification System (M-FISys),** to help with the Thailand Tsunami Victim Identification effort. Because M-FISys was essentially built in response to the 9/11 tragedy, Gene Codes did not have to write entirely new software and were able to customize M-FISys as necessary. In addition to analyzing mitochondrial DNA (see Section 8.6), M-FISys incorporated male-specific variations in the Y chromosome, called Y-STRs, to aid in the identification of individuals. Within 3 months, approximately 800 victims had been identified.

Gene Codes also established the DNA Shoah Project (*shoah* is the Hebrew name for the Holocaust), an effort to use M-FISys to establish a genetic database of Nazi-era Holocaust survivors with an overall goal to try and reunite an estimated 10,000 postwar orphans around the world.

8.4 DNA and the Rules of Evidence

Before DNA fingerprints could ever be used in a court of law, DNA fingerprinting had to meet the overarching legal standards regarding the admissibility of evidence. Courts use five different standards to determine whether scientific evidence should be allowed in a case. The test used depends on the jurisdiction. When a new technique or method is used to collect, process, or analyze evidence, it must meet one or several of these standards before the evidence is accepted.

- The relevancy test (Federal Rules of Evidence 401, 402, and 403) essentially allows anything that is deemed to be relevant by the courts.
- The *Frye* standard (1923) requires that the underlying theory and techniques used in gathering

evidence have "been sufficiently used and tested within the scientific community and have gained general acceptance." This is known as a general acceptance test.

- The *Coppolino* standard (1968) allows new or controversial science to be used if an adequate foundation can be laid, even if the profession as a whole is not familiar with the new method.

- The *Marx* standard (1975) is basically a common-sense test that requires that the court be able to understand and evaluate the scientific evidence presented. This is sometimes referred to as the no-jargon rule.

- The *Daubert* standard (1993) requires special pretrial hearings for scientific evidence. Under the Daubert standard, any scientific process used to gather or analyze evidence must have been described in a peer-reviewed journal.

The goal is to make certain that the scientific methods and expertise used to provide evidence are trustworthy.

DNA Fingerprinting and the Chain of Evidence

DNA analysis was a relatively new forensic tool when the Los Angeles Police Department used it in the infamous O. J. Simpson trial in 1994. Simpson's ex-wife, Nicole Brown Simpson, and her friend Ronald Goldman were murdered, and Simpson was the primary suspect. Forty-five samples were collected for DNA analysis from the crime scene. During pretrial proceedings, it was announced that the DNA collected at the crime scene matched that of O. J. Simpson.

Defense lawyers immediately attacked the procedures used in collecting, labeling, and testing the evidence. During the trial, the defense showed a videotape of the sample-collection methods and drew on expert testimony to establish doubt about the reliability of the evidence submitted. The defense stressed that contamination could have occurred when a technician touched the ground, when plastic bags were used to store wet swabs, and when sample collection tweezers were rinsed with water between touching samples. While on the stand, one prosecution witness admitted to mislabeling a sample. The possibility that the evidence might be tainted was obvious to both the court and the jury. As a result, the DNA evidence, which had been expected to make the case for the prosecution, was not effective. O. J. Simpson was found not guilty. When the chain of evidence is broken—when the rules of evidence are not followed—DNA samples lose their value in court.

Human Error and Sources of Contamination

One of the greatest threats to DNA evidence is human error. Earlier, we reviewed some of the precautions that crime scene investigators take to preserve and collect DNA evidence. A sneeze, improper storage, failure to label every single sample—these seemingly small mistakes can result in the destruction of evidence. Even if DNA evidence is not degraded by careless handling or bad conditions, it can be disregarded if the "chain of evidence" is suspect, as we saw in the example of the O. J. Simpson trial The chain of evidence requires that the collection of evidence be systematically recorded and access to the evidence must be controlled. Crime scene sample collection presents challenges, but even the collection of samples in a controlled environment, such as a morgue, is also problematic. Studies of morgue tables and instruments have found that the DNA of at least three individuals is often present. That DNA could certainly confuse the results of any analysis undertaken on samples collected in that environment.

DNA evidence is also vulnerable to damage during analysis (for example, DNA from the technician's body or from other sources can be inadvertently added to the sample). Defined standards of laboratory practice and procedure can help guard against these errors, however. When DNA evidence was a new idea, laboratories were not regulated and serious errors were made. Consider the case of *New York v. Castro*, which involved the murder of a woman and her 2-year-old child. The prosecution presented autorads (RFLPs) produced with the Southern blot method of the suspect's DNA and crime scene DNA and claimed that they matched. However, as an expert witness for the defense pointed out, the bands most certainly did not match. In this case, technicians made errors during their analysis and evaluation of the autorads. Fortunately, because the standards for the admissibility of evidence are high and the defense had access to DNA experts as well, these errors did not lead to a fatal miscarriage of justice. The suspect could not be linked to the crime with this evidence.

One step to ensure the reliability of DNA is to make certain that laboratories that process samples adhere to consistently high standards. The American Society of Crime Laboratories Directors (ASCLD), the National Forensic Science Technology Center (NSFTC), and the College of American Pathologists (CAP) all provide accreditation to forensic laboratories. Proficiency testing of technicians has become a basic requirement for employment. These exams include "blind" tests in which the technician is unaware that the sample being processed is actually a test sample submitted by the certifying organization. In addition, the FBI has developed a standard operating procedure for the handling and DNA typing of evidence in criminal cases. With clear

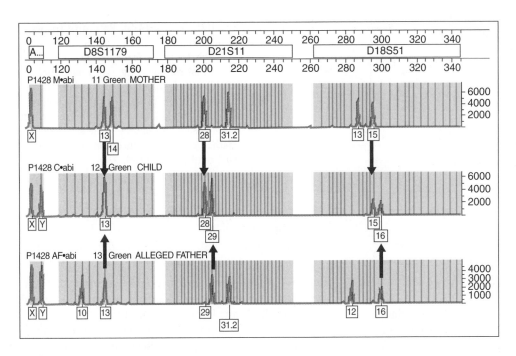

FIGURE 8.9 Paternity Dispute A man has claimed not to be the father of a child, but DNA allelic comparison shows that the child received one allele from each parent. Follow the arrows from the parents to the child to see how each parent contributed one of the two alleles (bands) that show in the DNA fingerprint of the child.

guidelines in place and thorough training of those responsible for collecting and preserving evidence, the reliability of DNA evidence presented in court proceedings should increase.

DNA and Juries

The use of DNA evidence also presents another challenge. To be useful, the science must make sense to the jury evaluating the case. Because DNA evidence is statistical in nature, the results can be confusing to a person without sufficient education, especially when large numbers are involved. When members of a jury panel hear there is "1 chance in 50 billion" that a random match might occur, it is possible they may focus on that one possibility and discount the overwhelming odds against its happening. Oddly enough, the same jurors might accept a conventional fingerprint as more reliable evidence, even though studies have concluded that such evidence is actually *less* statistically valid. Making DNA evidence clear and comprehensible to jurors is not an easy task. And if the DNA evidence is not clearly understood, it may be disregarded.

8.5 Familial Relationships and DNA Profiles

Crime solving is not the only forensic application of DNA fingerprinting. Because family members have a small amount of DNA in common, relationships can be conclusively determined by comparing samples between two individuals. An obvious use of this is in paternity testing (**Figure 8.9**). Every year, roughly 400,000 paternity suits are filed in the United States, Canada, and the United Kingdom. Given samples from the child and adults involved, verifying the child's parentage, and resolving child support or custody disputes is relatively easy. However, reproductive technologies—including in vitro fertilization, artificial insemination, and surrogacy—have introduced a few quirks into custody and paternity. DNA testing has helped untangle cases in the courts that involved sperm mixups in fertility clinics, for example. Thanks to amniocentesis, as shown in **Figure 8.10**, it is possible to verify a child's parentage even before birth.

Mitochondrial DNA Analysis

Mitochondrial DNA analysis can be used to examine DNA from samples that cannot be analyzed by RFLP or STR. Nuclear DNA must be extracted from samples for use in RFLP, PCR, and STR, but mtDNA analysis uses DNA extracted from a cellular organelle called a *mitochondrion*. Older biological samples that lack nucleated cellular material, such as hair, bones, and teeth, cannot be analyzed with STR and RFLP, but they can be analyzed with mtDNA. All daughters have the same mitochondrial DNA as their mothers because the mitochondria of embryos come from the mother's egg cell. The father's sperm contributes only nuclear DNA and no mitochondria. Comparing the mtDNA profile of unidentified remains with the profile of a potential maternal relative can determine whether they share the same mtDNA profile and are related.

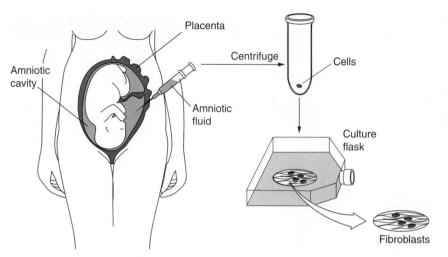

FIGURE 8.10 DNA for Paternity Testing For disputed paternity suits, it is possible to draw a few fetal cells from the fluid that surrounds the fetus (amniotic fluid) without harming the growing fetus. When cultured, these cells can be a source for DNA extraction and fingerprinting. When compared with the DNA of the suspected father, exclusion or inclusion of the individual can be determined.

mtDNA remains virtually the same from generation to generation, changing only about 2% to 4% every million years due to random mutation. Consequently relationships can be traced through the unbroken maternal line, as shown in **Figure 8.11**.

mtDNA evidence was essential in reuniting families torn apart in Argentina during the Falklands war in the early 1980s. During this repressive regime, the military government routinely arrested and questioned anyone suspected of subversive activity. Among those arrested were a number of young pregnant women. Roughly 15,000 of those arrested and questioned simply disappeared. After the collapse of the government in 1983, the mothers and grandmothers of these "disappeared" women became aware that the horrible crime might well be ongoing: babies born to

their daughters while in prison had been taken and were being raised by others—often by the families of those who had murdered their parents.

The mothers sought help from the American Association for the Advancement of Science (AAAS), asking if genetic testing could be used to prove the identity of the stolen children. Through the perseverance of these women and the scientists who volunteered their help, the AAAS was able to prove that at least 51 children had been illegally adopted. DNA evidence was used in court to restore the children to their living relatives.

Y-Chromosome Analysis

The Y chromosome is passed directly from father to son, making the analysis of genetic markers on the Y chro-

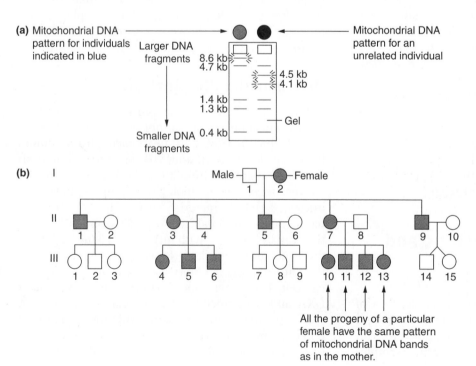

(b)

All the progeny of a particular female have the same pattern of mitochondrial DNA bands as in the mother.

FIGURE 8.11 Maternal DNA Pattern of Inheritance Key genes for mitochondrial function (cell respiration) are located in a small DNA ring in human mitochondria. Because mitochondria are contributed by the egg (only) before fertilization, DNA can be traced through the maternal line with fingerprinting of the mitochondrial DNA. Some genes in mitochondria show variations due to mutations (see top of gel at 8.6, 4.7 vs. 4.5, 4.1 kilobases); others do not.

mosome useful for tracing relationships among males or for analyzing biological evidence involving multiple male contributors. The Y chromosome can be analyzed with nuclear DNA (usually by PCR) methods.

8.6 Nonhuman DNA Analysis

Not every legal case is a matter of human identity. Many legal questions—as well as scientific quandaries—have been answered by examining the DNA fingerprints of plants and animals. Expensive plants, animals, and their biological products can be protected using their DNA profiles.

Identifying Plants through DNA

Ginseng is a valuable herbal product; the market for it is estimated to be $3 million in the United States alone, and the demand is also high in other countries. Currently, two major herbal products are referred to as ginseng. One is native to North America and the other to Asia. Oddly, Asian ginseng is the type most often sold in American health food stores. Most American ginseng—the rarer and more valuable of the two varieties—is exported to Asia. The two types of ginseng look almost identical, but they have very different reputations. Asian ginseng purportedly boosts energy, whereas American ginseng is prized for its supposed ability to calm nerves. Manufacturers are now using DNA sequencing to help make the distinction between the two varieties. Confirming the origin of the product helps ensure quality control and protects the market for American ginseng products.

This same type of analysis is performed for other plants as well. A DNA profile used in quality control of squash seeds is shown in **Figure 8.12**. This quality control process allows the seed supplier to guarantee

TOOLS OF THE TRADE

DNA Profiles Are Advancing Biological Research

In addition to their use in legal proceedings, DNA profiles from animals have been used in direct biological research. In 2002, it was reported that DNA analysis of the remaining bison herds, both public and private, indicated that the majority of bison have some domestic livestock as ancestors. There is no outward sign of the "hybrid" nature of these animals, but their heritage is clearly evident in their DNA. As a result of this research, steps can be taken to preserve the remaining "pure" herds.

Scientists have also found that chickadees, barn swallows, indigo buntings, and other species are not monogamous by examining the DNA of parent birds and their offspring to determine if the babies in the nest actually had different fathers.

In addition, mitochondrial DNA (mtDNA) analysis has proven to be a rich source of new information about evolutionary biology. Mitochondria are estimated to mutate at the rate of 2% to 4% per million years, which allows scientists to trace changes in gene frequency through time. Using frequency changes in mtDNA, scientists are able to differentiate between similar animal species and to examine the role of base substitution during evolution. Only a nanogram of mtDNA is required for direct sequencing identification. In fact, the "Eve hypothesis," which used mtDNA evidence to trace a majority of people now living on earth to a single female ancestor from ancient Africa, is based on mtDNA analysis. The results of that study have now been compared with a similar human origins study following sequence variations in the Y chromosome.

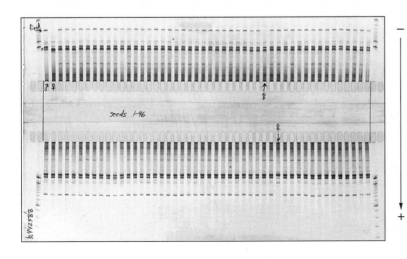

FIGURE 8.12 DNA Profile of Squash Hybrids Used as a Quality Control Analysis The anode is in the center of this gel with cathodes at the top and bottom. When the DNA from 96 seeds is loaded at the center of the gel, it will migrate to the cathodes during electrophoresis. The male parent plant donated the upper DNA band and the female donated the lower of the two bands of the hybrid offspring plant being tested. The hybrid plants all have both bands (one exception gives 98.9% quality assurance of purity in 96 seeds tested). This level of hybridity (quality control) will increase the value of the seed to a grower depending on a high level of hybridity for the growing requirements.

a specific degree of genetic hybridity, thus increasing the seeds' value to growers, who depend on these traits to sell their high-value crops.

Another interesting "paternity" case of the plant world involves the ancestry of cabernet sauvignon grapes, which are prized for wine production. Hybrid grapes are considered inferior by some wine purists and legally excluded from bearing the prestigious distinction *appellation d'origine contrôlée* in France. When scientists examined the DNA of cabernet sauvignon plants, it was evident that two other varieties of grapes, cabernet franc and sauvignon blanc, were the ancestors of cabernet sauvignon. This finding challenges the old notion that varieties derived from crossbreeding are inherently less valuable as wine grapes.

Animal DNA Analysis

DNA evidence has also been used to generate genetic profiles of animals. For example, in Pennsylvania, DNA fingerprinting was used to prove that a hunter had killed a bear illegally. The bear-hunting season in that state is designed to protect pregnant sows. A law making it illegal to kill bears in their dens ensures that a large portion of the breeding female population will be protected from loss during the season. In this case, a witness reported seeing a hunter discharge his rifle into a bear's den. The hunter had registered his kill at the appropriate check station (without disclosing the circumstances under which it was shot), where one of the harvested bear's premolars was removed, as specified by state hunting regulations, to verify the sex and

age of the kill. As part of their investigation of this eyewitness report, authorities collected blood samples from the den and compared the DNA from the den with that from the check station. The tests revealed that the DNA samples were from the same animal, even though the hunter, in an effort to evade prosecution, had claimed to have killed the bear 5 miles from the den. Because of the DNA evidence, he was found guilty.

DNA profiling is regularly conducted by wildlife management authorities in such cases. **Figure 8.13**, for example, shows DNA profiles for four deer that were killed by a hunter with a permit to kill only one deer. Armed with the tools of DNA fingerprinting, wildlife managers have improved their ability to prove that game has been taken beyond the limits set by regulations.

DNA Tagging to Fight Fraud

A highly effective barrier to product counterfeiting and piracy is possible by using DNA as an authentication label hidden in a wide variety of products. Virtually anything of value, from fine art to sports memorabilia, can have DNA incorporated into it; the DNA labels can then be detected using PCR or hybridization techniques to verify authenticity. For example, footballs in the 2011 Super Bowl were encoded with DNA to authenticate them as official products of the National Football League. During the 2010 Sydney Summer Olympics, DNA Technologies of Halifax, Nova Scotia, tagged 34 million labels with

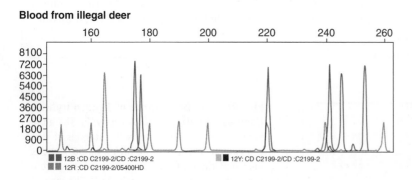

Blood from illegal deer

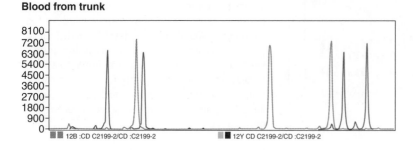

Blood from trunk

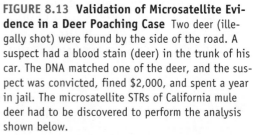

FIGURE 8.13 Validation of Microsatellite Evidence in a Deer Poaching Case Two deer (illegally shot) were found by the side of the road. A suspect had a blood stain (deer) in the trunk of his car. The DNA matched one of the deer, and the suspect was convicted, fined $2,000, and spent a year in jail. The microsatellite STRs of California mule deer had to be discovered to perform the analysis shown below.

specific DNA strands and then the labels were attached to Olympic-licensed merchandise. Random inspections of merchandise vendors in Sydney revealed that nearly 15% of the merchandise sold as officially licensed was in fact counterfeit, contributing to nearly $1 million in lost sales.

Earlier, you learned why DNA is a good forensic marker. These same reasons make DNA a good physical marker that is difficult to counterfeit. Typically, DNA for labels incorporate random single-stranded sequences from 20 to several thousand nucleotides long. These strands are then mixed into ink, cloth, threads, or other materials used in the product to be tagged. To determine whether an item is authentic, the portion thought to contain the DNA tag is removed and analyzed by PCR, or probes are added to the component that will fluoresce if they hybridize with the single-stranded tags in the product. DNA tagging has become a commonly used tool to combat counterfeiting.

CAREER PROFILE

Laboratory Technician

Laboratory technicians are commonly employed in DNA fingerprinting labs to perform tests on evidence collected from crime scenes. These technicians must be able to follow directions with great accuracy. They must take great care to keep the work area clean so as to prevent contamination. In fact, all technicians working in law enforcement labs must pass a quantitative analysis course as a condition of employment. Technicians frequently may be required to work in a special "clean room" environment wearing sanitary gowns. Salaries range from $21,000 to $35,000 for an entry-level worker to $25,000 to $60,000 for more experienced technicians.

Most employers look for individuals with a bachelor's degree in biology, biochemistry, or molecular biology. Some hire applicants with a specialized associate's degree in biotechnology and laboratory experience. Hands-on laboratory experience is a critical skill in this position and may be gained through course work, internships, and on-the-job training. However, technicians also need good math and communication skills. Writing is especially important because lab notebooks are considered legal documents. Strong computer skills are also necessary. Opportunities for advancement to research associate, especially for those with a bachelor's degree, are good.

MAKING A DIFFERENCE

As the cost of sequencing the human genome becomes more affordable, the information gained from comparing multiple human genomes becomes more valuable. The human reference genome, a mosaic of DNA sequences derived from several individuals, was first created in 2001 and has been used since then as the basis for comparison in genomic research. In 2010, researchers at the University of Washington, led by Evan Eichler, Ph.D., have discovered 2,363 DNA sequences that were not charted in the original human reference sequence. Since each segment of the human reference sequence is from a single person, it is possible that they did not possess the newly discovered sequences. This may be not be surprising when you consider that about 80% of the original reference genome came from eight people. The researchers found that some of these newly discovered sequences were common or rare in different populations and that some of them were present in different mammal species (chimpanzee, Bornean orangutan, Rhesus monkey, house mouse, Norway rat, dog and horse). The 1,000 Genomes Project (an international effort to fully sequence the genomes of 1,000 anonymous individuals) and other genome studies are accumulating a considerable amount of data on the human DNA sequence that will be mapped to the original (and incomplete) reference genome. The process of DNA sequencing is largely technical and represents a significant opportunity for those who will be contributing to the "latest reference human genome," which will provide valuable information for DNA technology.

QUESTIONS & ACTIVITIES

Answers can be found in Appendix 1.

1. Explain what is meant by a polymorphic DNA locus.

2. How do polymorphic DNA sequences (e.g., VNTRs) affect someone?

3. How many STR bands found in a child's DNA should occur in the DNA fingerprint of the father? The mother?

4. Why is PCR-based DNA fingerprinting used in forensics more commonly than RFLP?

5. How could contaminants from the technician possibly affect a DNA fingerprint?

6. Why is eyewitness testimony considered less reliable than DNA evidence? (Give an example.)

7. How was it possible to prove that cabernet sauvignon was a cross between cabernet franc and sauvignon blanc?

8. What result would you expect if you performed a quantitative DNA test (slot blot) with twice the amount of DNA? Why would you do it?

9. How many errors can you count when you watch an episode of *Crime Scene Investigation*?

10. Give at least one example of how mtDNA evidence has been used.

Visit www.pearsonhighered.com/biotechnology

To download learning objectives, chapter summary, "Keeping Current" web links, glossary, flashcards, and jpegs of figures from this chapter.

Bioremediation

After completing this chapter, you should be able to:

- Explain what bioremediation is and describe why it is important.

- Describe advantages of bioremediation strategies over other types of cleanup approaches.

- Name common chemical pollutants that need to be cleaned up in different zones of the environment.

- Distinguish between aerobic and anaerobic biodegradation and provide examples of microbes that can contribute to bioremediation.

- Explain why studying genomes of organisms involved in bioremediation is an active area of research.

- Define phytoremediation and understand how it can be used to clean up the environment.

- Discuss how in situ and ex situ approaches can be used to bioremediate soil and groundwater.

- Discuss the roles of bioremediation at a wastewater treatment plant.

- Provide examples of how genetically modified organisms can be used in bioremediation.

Landfills around the world are overflowing with tons of garbage that are generated each day. Bioremediation has great potential for cleaning up chemicals in the environment, reducing waste in landfills, and even producing energy from society's garbage.

Our environment is being threatened with alarming frequency. The air we breathe, the water we drink, and the soil we rely on to grow plants for food are all being contaminated as a direct result of human activities. The average American generates approximately 5 pounds of garbage per day—over 1,700 pounds of trash in a year per person, leading to a total of about 250 million tons of trash every year. Only about one-third (some 83 million tons) of this trash is recycled or composted; the remainder goes into landfills. Yet household wastes are a relatively small part of the problem. Pollution from industrial manufacturing wastes as well as from chemical spills, household products, and pesticides has led to contamination of the environment. An increasing number of toxic chemicals are presenting serious threats to the health of environments throughout the world and to the organisms that live there.

Just as biotechnology is considered to be a key to identifying and solving human health problems, it is also a powerful tool for studying and correcting the poor health of polluted environments. In this chapter we consider how biotechnology can help solve some of our pollution problems and create cleaner environments for humans and wildlife through bioremediation.

FORECASTING THE FUTURE

There are many exciting potential future directions for bioremediation. Among the most likely areas of substantial progress will be global efforts in bioprospecting to find previously undiscovered metabolizing microbes and studying the genomics of these microbes to design pollution-degrading bacteria and plants. For example, in this chapter we discuss how scientists are analyzing oil-degrading microbes discovered in the Gulf of Mexico at the site of the *Deepwater Horizon* oil spill. Creating genetically engineering microbes and plants to degrade particularly toxic and harmful pollutants such as heavy metals and radioactive compounds will continue to be a focus of bioremediation, as will using bioremediation microbes as an energy source. Also, keep an eye out for applications of synthetic biology to artificially create new types of metabolizing microbes with synthetic genomes including genes customized for degrading specific pollutants.

9.1 What Is Bioremediation?

Bioremediation is the use of living organisms such as bacteria, fungi, and plants to break down or degrade chemical compounds in the environment. It takes advantage of natural chemical reactions and processes through which organisms break down compounds to

obtain nutrients and derive energy. Bacteria, for example, metabolize sugars to make adenosine triphosphate (ATP) as an energy source for cells. But in addition to degrading natural compounds to obtain energy, many microbes have developed unique metabolic reactions that can be used to degrade human-made chemicals. Bioremediation cleans up environmental sites contaminated with pollutants by using living organisms to degrade hazardous materials into less toxic substances.

Bioremediation is not a new application. Humans have relied on biological processes to reduce waste materials for thousands of years. In the simplest sense, the outhouse—which relied on natural microbes in soil to degrade human wastes—was an example of bioremediation. Similarly, sewage treatment plants have used microbes to degrade human wastes for decades. But as you will learn in this chapter, modern applications in bioremediation involve a variety of new and innovative strategies to clean up a wide range of toxic chemicals in many different environmental settings.

Taking advantage of what many microbes already do is only one aspect of bioremediation. One of its key purposes is to improve natural mechanisms and increase rates of biodegradation to accelerate cleanup processes. In this chapter, we explore some of the ways in which scientists can stimulate microbes and other organisms to degrade a variety of wastes in many different situations. Another important aspect of bioremediation is the development of new approaches for the biodegradation of waste materials in the environment, which can involve using plants and genetically modified microorganisms.

Why Is Bioremediation Important?

Our quality of life is directly related to the cleanliness and health of the environment. We know that environmental chemicals can influence our genetics and that some chemicals can act as mutagens, leading to human disease conditions. Clearly, there is reason to be concerned about both short- and long-term chemical exposure and the effects of environmental chemicals on humans and other organisms.

According to some estimates, over 200 million tons of hazardous materials are produced in the United States each year. Accidental chemical spills can and do occur, but these events typically are contained and cleaned up rapidly to minimize their impact on the environment. More problematic, however, are illegal dumping practices and sites contaminated through neglect, such as abandoned warehouses, where stored chemicals may leak into the environment. In 1980, the U.S. Congress established the **Superfund Program** as an initiative of the **U.S. Environmental Protection Agency (EPA)** to counteract careless and even negligent practices of chemical dumping and storage as well as to express concern over how these pollutants might affect human health and the environment. The primary purpose of the Superfund Program is to locate and clean up hazardous waste sites in order to protect U.S. citizens from contaminated areas.

One in every five Americans lives within 3 to 4 miles of a polluted site treated by the EPA. In the more than 25 years since Superfund began, the EPA has cleaned up more than 700 sites in the United States. Nevertheless, well over 1,000 sites await cleanup, and Department of Energy estimates suggest as many as 220,000 sites need remediation, with many new sites identified each year. Estimates for the cleanup costs of currently identified polluted areas in the United States are in excess of $1.5 trillion. The extent of contamination of other sites and the number of sites requiring cleanup will likely push that estimate much higher. Clearly environmental pollution in the United States is an important problem that is receiving a lot of attention. In many other countries, environmental pollution is an even greater problem. To learn more about the Superfund Program, visit the Superfund website, which is listed on the Companion Website. Through this site you can also check on contaminated environments close to where you live that are on the Superfund priority list for cleanup.

Through the National Institute of Environmental Health Sciences, a division of the National Institutes of Health, the United States started a program called the **Environmental Genome Project.** A primary purpose of this project is to study and understand the impacts of environmental chemicals on human disease. This includes the study of genes that are sensitive to environmental agents, learning more about detoxification genes, and identifying single-nucleotide polymorphisms that may be indicators of environmental impacts on human health. Ultimately, this project will generate genome data enabling scientists to carry out epidemiological studies that will help them not only to better understand how the environment contributes to disease risk but also how specific diseases are influenced by environmental exposure to chemicals.

We know that pollution is a problem that can affect human health. However, there are many ways to clean up pollutants, so why use bioremediation? We could physically remove contaminated material such as soil or chemically treat polluted areas, but these processes can be very expensive and, in the case of chemical treatments, can create more pollutants that themselves require cleanup. A major advantage of bioremediation is that most such approaches convert harmful pollutants into relatively harmless materials such as carbon dioxide, chloride, water, and simple organic molecules. Because living organisms are used for the cleanup, bioremediation processes are generally cleaner than other types of cleanup strategies.

Another advantage of bioremediation is that many cleanup approaches can be conducted at the site (in situ) of pollution. Because the contaminated materials do not need to be transported to another site, a more complete cleanup is often possible without disturbing the environment. In the next section, we consider some of the basic principles of bioremediation in terms of common chemical pollutants and the environments they pollute. We also discuss some of the microbes and reactions that are important for bioremediation.

9.2 Bioremediation Basics

Naturally occurring marshes and wetlands have excelled at bioremediation for hundreds of years. In these environments, plant life and microbes can absorb and degrade a wide variety of chemicals and convert pollutants into harmless products. Before we discuss how living organisms can degrade pollutants, we will first consider areas of the environment that require cleanup and take a look at some of the common chemicals that pollute the environment.

What Needs to Be Cleaned Up?

Unfortunate as it may be, the answer to this question is that almost everything needs to be cleaned up. Soil, water, and sediment (a combination of soil and decaying plant and animal life located at the bottom of a body of water) are the most common treatment environments that require cleanup by bioremediation, although new bioremediation approaches are being developed to detect and clean up air pollution. Each area presents its own complexities for cleanup because the type of bioremediation approach used typically depends on site conditions. For instance, it should not surprise you that approaches for cleaning up soil can be very different from those used to clean up water. Similarly, surface water often needs to be treated differently than subsurface water (called groundwater).

Pollutants enter the environment in different ways and affect diverse components of the environment. In some cases, pollutants enter the environment through a tanker spill, a truck accident, or a ruptured chemical tank at an industrial site. Of course, depending on the location of the accident, the amount of chemicals released, and the duration of the spill (hours versus weeks or years), different parts or zones of the environment may be

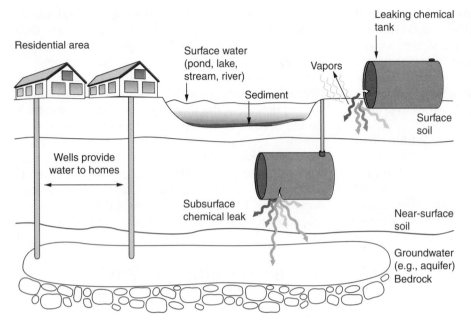

Residential area

Surface water (pond, lake, stream, river)

Sediment

Vapors

Leaking chemical tank

Wells provide water to homes

Subsurface chemical leak

Surface soil

Near-surface soil

Groundwater (e.g., aquifer)

Bedrock

FIGURE 9.1   Treatment Environments and Contamination Zones Chemical spills can create a number of treatment environments and zones of contamination that may be targets for bioremediation. A spill from a leaking chemical tank, which may be located above the ground or below the surface, can release materials that contaminate surface soil, subsurface soil, surface water, and groundwater. In this example, pollution of the aquifer threatens the health of individuals living in houses adjacent to the spill, who rely on the aquifer as a source of drinking water.

affected. **Figure 9.1** provides an example of a leaking chemical tank at an industrial plant. It may initially contaminate only surface and subsurface soils; however, if large amounts of chemicals are released and the leaky tank goes undetected for a long time, chemicals may move deeper into the soil. Following heavy rains, these same chemicals may create runoff that can contaminate adjacent surface water supplies such as ponds, lakes, streams, and rivers. Chemicals may also leak through the ground, creating **leachate**. Leachate can cause contamination of groundwater, including aquifers—deep pockets of underground water that are a common source of drinking water.

Chemicals may also enter the environment through the release of pollutants into the air, which can become trapped in clouds and contaminate surface water and then the groundwater when it rains. This is what causes acid rain, for example. Pollutants from industrial manufacturing, landfills, illegal dumps, pesticides used for agriculture, and mining processes also contribute to environmental pollution. Because the bioremediation approach used to clean up pollution depends on the treatment environment, cleaning up soil is very different from cleaning up water. How bioremediation is used also depends on the types of chemicals that need to be cleaned up.

Chemicals in the Environment

Techniques for using bioremediation to degrade human wastes at a sewage treatment plant are quite different (and somewhat simpler) than those used to degrade the variety of chemicals that exist in the environment. Everyday household materials such as cleaning agents, detergents, perfumes, caffeine, insect repellents, pesticides, fertilizers, perfumes, and medicines appear in our wastewater. Increasingly researchers are also finding that U.S. waterways contain prescription and over-the-counter drugs, including contraceptives, painkillers, antibiotics, cholesterol-lowering drugs, antidepressants, anticonvulsants, and anticancer drugs. Other chemicals that make their way into the environment are the products of industrial manufacturing processes or, as discussed earlier, the result of accidents.

Numerous chemicals from many different sources are common pollutants in the environment. Table 9.1 lists some of the most common categories of chemicals in our environment that require cleanup. Many of these chemicals are known to be potential mutagens and **carcinogens**—compounds that cause cancer. Although we do not discuss the health effects of chemical pollutants in any detail, most of these chemicals are known to cause illnesses ranging from skin rashes to birth defects and different types of cancer, and they can poison animal and plant life. Quite simply, the presence of pollutants in an environment leads to an overall decline of the environment along with the health of the organisms living within it.

In addition to the type of spill and the cleanup environment, the type of chemical pollutant also affects the types of cleanup organisms and approaches that can be used for bioremediation. Throughout this

TABLE 9.1 TWENTY OF THE MOST COMMON CHEMICAL POLLUTANTS IN THE ENVIRONMENT

Chemical Pollutant	Source
Benzene	Petroleum products used to make plastics, nylon, resins, rubber, detergents, and many other materials
Chromium	Electroplating, leather tanning, corrosion protection
Creosote	Wood preservative to prevent rotting
Cyanide	Mining processes and manufacturing of plastics and metals
Dioxin	Pulp and paper bleaching, waste incineration, and chemical manufacturing processes
Methyl t-butyl ether (MTBE)	Fuel additive, automobile exhaust, boat engines, leaking gasoline tanks
Naphthalene	Product of crude oil and petroleum
Nitriles	Rubber compounds, plastics, and oils
Perchloroethylene/ tetrachloroethylene (PCE), trichloroethene (TCE), and trichloroethane (TCA)	Dry cleaning chemicals and degreasing agents TCE is present in some 34% of U.S. water supplies and 60% of Superfund sites
Pesticides (atrazine, carbamates, chlordane, DDT) and herbicides	Chemicals used to kill insects (pesticides) and weeds (herbicides)
Phenol and related compounds (chlorophenols)	Wood preservatives, paints, glues, textiles
Polychlorinated biphenyls (PCBs)	Electrical transistors, cooling and insulating systems
Polycyclic aromatic hydrocarbons (PAHs) and polychlorinated hydrocarbons	Incineration of wastes, automobile exhaust, oil refineries, and leaking oil from cars
Polyvinylchloride	Plastic manufacturing
Radioactive compounds	Research and medical institutions and nuclear power plants
Surfactants (detergents)	Manufacturing of paints, textiles, concrete, paper
Synthetic estrogens (ethinyl estradiol)	Female hormone (estrogen)-related compounds created by a variety of industrial manufacturing processes
Toluene	Petroleum component present in adhesive, inks, paints, cleaners, and glues
Trace metals (arsenic, cadmium, chromium, copper, lead, mercury, silver)	Car batteries and metal manufacturing processes
Trinitrotoluene (TNT)	Explosive used in building and construction industries

chapter we consider strategies for cleaning up many of the pollutants listed in Table 9.1.

Fundamentals of Cleanup Reactions

Microbes can convert many chemicals into harmless compounds either through **aerobic metabolism** (reactions that require oxygen [O_2]) or **anaerobic metabolism** (reactions in which oxygen is not required). Both types of processes involve *oxidation* and *reduction reactions*. You must have a basic familiarity with oxidation and reduction reactions if you are to understand biodegradation.

Oxidation and reduction reactions

Oxidation involves the removal of one or more electrons from an atom or molecule, which can change the chemical structure and properties of a molecule. In the case of a chemical pollutant, oxidation can make the chemical harmless by changing its chemical properties. Oxidation reactions often occur together with **reduction reactions**. During reduction, an atom

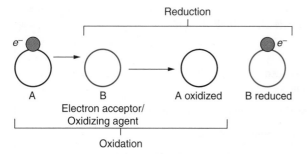

FIGURE 9.2 **Oxidation and Reduction (Redox) Reactions**
Redox reactions are important for the bioremediation of many chemicals. In this oxidation reaction, an electron is transferred from molecule A to molecule B. Molecule B is acting as an electron acceptor or oxidizing agent. In redox reactions, molecule A is oxidized, and molecule B, which gains an electron, is reduced.

or molecule gains one or more electrons. Because oxidation and reduction reactions frequently occur together, these electron transfer reactions are often called **redox reactions** (see **Figure 9.2**).

During redox reactions, molecules called **oxidizing agents**—also known as electron acceptors, because they have a strong attraction for electrons—remove electrons during the transfer process. When oxidizing agents accept electrons, they become reduced. Oxygen (O_2), iron (Fe^{+3}), sulfate (SO_4^{2-}), and nitrate (NO_3^-) are often involved in redox reactions of bioremediation. Redox reactions are important for many cellular functions. For example, human cells and many other cell types use oxidation and reduction reactions to degrade sugars and derive energy.

Aerobic and anaerobic biodegradation

In some environments, such as surface water and soil where oxygen is readily available, aerobic bacteria degrade pollutants by oxidizing chemical compounds. In aerobic biodegradation reactions, O_2 can oxidize a variety of chemicals including **organic molecules** (those that contain carbon atoms), such as petroleum products (**Figure 9.3**). In the process, O_2 is reduced to produce water. Microbes can further degrade the oxidized organic compound to make simpler and relatively harmless molecules such as carbon dioxide (CO_2) and methane gas. Bacteria derive energy from this process, which is used to make more cells; scientists refer to this increase in cell number as an increase in **biomass**. Some aerobes also oxidize **inorganic compounds** (molecules that do not contain carbon), such as metals and ammonia.

In heavily contaminated sites and deep subsurface environments such as aquifers, the concentration of oxygen may be very low. In subsurface soils, oxygen may diffuse poorly into the ground, and any oxygen

that is there may be rapidly consumed by aerobes. Even though it is sometimes possible to inject oxygen into treatment sites to stimulate aerobic biodegradation in low-oxygen environments, biodegradation may take place naturally via anaerobic metabolism. Anaerobic metabolism also requires oxidation and reduction; however, anaerobic bacteria (anaerobes) rely on molecules other than oxygen as electron acceptors (**Figure 9.4**). Iron (Fe^{+3}), sulfate (SO_4^{-2}), and nitrate (NO_3^-) are common electron acceptors for redox reactions in anaerobes (Figures 9.3 and 9.4). In addition, some microbes can carry out both aerobic and anaerobic metabolism. When the amount of oxygen in the environment decreases, they can switch to anaerobic metabolism to continue biodegradation. As you will learn in the next section, aerobes and anaerobes are both important for bioremediation.

The Players: Metabolizing Microbes

Scientists can use microbes—especially bacteria—as tools to clean up the environment. The ability of bacteria to degrade different chemicals effectively depends on many conditions. The type of chemical, temperature, zone of contamination (water versus

Aerobic biodegradation

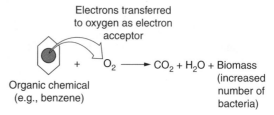

Anaerobic biodegradation

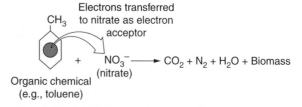

FIGURE 9.3 **Aerobic and Anaerobic Biodegradation**
Aerobic bacteria (aerobes) use oxygen (O_2) as an electron acceptor molecule to oxidize organic chemical pollutants such as benzene. During this process, oxygen is reduced to produce water (H_2O), and carbon dioxide (CO_2) is derived from the oxidation of benzene. Energy from degrading the pollutant is used to stimulate bacterial cell growth (biomass). Similar reactions occur during anaerobic biodegradation, except that anaerobic bacteria (anaerobes) rely on iron (Fe^{+3}), sulfate (SO_4^{2-}), nitrate (NO_3^-), and other molecules as electron acceptors to oxidize pollutants.

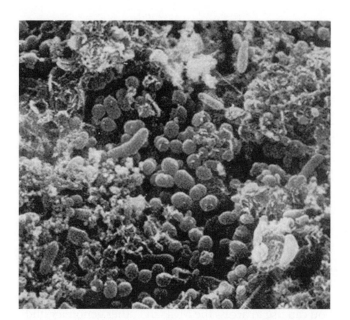

FIGURE 9.4 Anaerobic Bacteria Effectively Degrade Many Pollutants The dry-cleaning agent called perchloroethylene (PCE) is a common contaminant of groundwater; however, anaerobic bacteria can use PCE as food. By growing bacteria on small particles of iron sulfide, which serve as electron acceptors providing the proper chemical environment for anaerobes, bacteria grow rapidly and thrive on PCE.

soil, surface versus groundwater contamination, and so on), nutrients, and many other factors all influence the effectiveness and rates of biodegradation.

At many sites, bioremediation involves the combined actions of both aerobic and anaerobic bacteria to decontaminate the site fully. Anaerobes usually dominate biodegradation reactions that are closest to the source of contamination, where oxygen tends to be very scarce, but sulfates, nitrates, iron, and methane are present for use as electron acceptors by anaerobes. Farther from the source of contamination, where oxygen tends to be more abundant, aerobic bacteria are typically involved in biodegradation (**Figure 9.5**).

The search for useful microorganisms for bioremediation is often best carried out at polluted sites themselves. Organisms living in a polluted site will have developed some resistance to the polluting chemicals and may be useful for bioremediation. **Indigenous microbes**—those found naturally at a polluted site—are often isolated, grown, and studied in a lab and then released back into treatment environments in large numbers. Such microbes are typically the most common and effective "metabolizing" microbes for bioremediation. For instance, strains of bacteria called *Pseudomonas*, which are very abundant in most soils, are known to degrade hundreds of different chemicals. Certain strains of *Escherichia coli* are also fairly effective at degrading many pollutants.

A large number of lesser-known bacteria have been used and are currently being studied for potential roles in bioremediation. For instance, in Section 9.6 we discuss possible applications of *Deinococcus radiodurans*, a microbe that shows an extraordinary ability to tolerate the hazardous effects of radiation. Similarly, researchers at the University of Dublin discovered *Pseudomonas putida*, a strain of bacteria that can convert styrene, a toxic component of many plastics, into a biodegradable plastic.

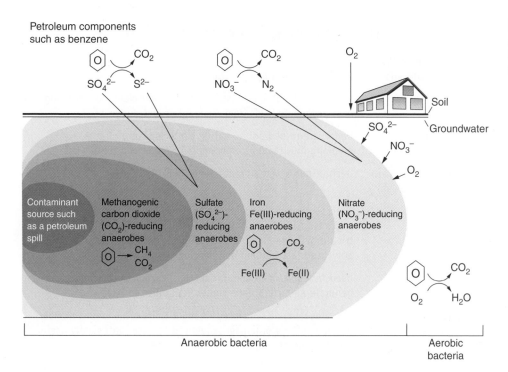

FIGURE 9.5 Anaerobic and Aerobic Bacteria Contribute to Biodegradation of Contaminated Groundwater

Scientists believe that many of the microbes that are most effective at bioremediation have not yet been identified. The quest for new metabolizing microbes is an active area of bioremediation research. In 2010, scientists at the U.S. Geological Survey reported the identification of a bacterium from California's Lake Mono, a lake with high concentrations of arsenic, which appears to metabolize arsenic and even incorporate it into biomolecules such as DNA. Although the accuracy of discovery has been the subject of significant debate in the scientific community.

Scientists are also experimenting with strains of algae and fungi that may be capable of biodegradation. Waste-degrading fungi such as *Phanerochaete chrysosporium* and *Phanerochaete sordida* can degrade toxic chemicals such as creosote, pentachlorophenol, and other pollutants that bacteria degrade poorly or not at all. Asbestos and heavy metal–degrading fungi include *Fusarium oxysporum* and *Mortierella hyaline*. Fungi are also very valuable in composting and degrading sewage and sludge at solid-waste and wastewater treatment plants, polychlorinated biphenyls (PCBs), and other compounds previously thought to be highly resistant to biodegradation.

Bioremediation Genomics Programs

It should not surprise you to learn that many scientists are studying the genomes of organisms that are currently used or may be used in the future for bioremediation. Through genomics, it will be possible to identify novel genes and metabolic pathways that bioremediation organisms use to detoxify chemicals. This will help scientists develop more effective cleanup strategies, including improved strains of bioremediation organisms through genetic engineering (see Section 9.4.). It may also be possible to combine detoxifying genes from different microbes into different recombinant bacteria capable of degrading multiple contaminants at the same time—for example, PCBs and mercury.

There are a variety of ways in which molecular techniques and genomics are being used to analyze genomes of microbes at contaminated sites as a way to identify the genes and proteins involved in bioremediation. The Department of Energy established the Microbial Genome Program (MGP), which has sequenced over 200 microbial genomes, including many microbes that may be useful in bioremediation (Chapter 5). See Table 9.2 for a few examples of recently completed genomics projects involving bioremediation organisms. Be sure to visit the Genomes to Life link on the Companion Website, which you can use to access the MGP and other genomics studies of bioremediation microbes.

Stimulating bioremediation

As discussed previously, bioremediation scientists typically take advantage of indigenous microbes to degrade pollutants. Depending on the pollutant, many indigenous bacteria are very effective at biodegradation. Scientists also use numerous strategies to make microorganisms more effective in degrading contaminants depending on the microorganisms involved, the environmental site being cleaned up, and the quantities and types of chemical pollutants that need to be decontaminated.

TABLE 9.2   EXAMPLES OF BIOREMEDIATION GENOME PROJECTS UNDER WAY OR RECENTLY COMPLETED

Microorganism	Number of Genes (Year Genome Completed)	Bioremediation Applications
Accumulibacter phosphatis	4,790 (2009)	Major microbe used in wastewater treatment plants for removing high phosphate loads from wastewaters and sludge
Alcanivorax borkumensis	2,755 (2006)	Hydrocarbon-degrading marine bacterium very effective at breaking down many components of crude and refined oil
Dehalobacter restrictus	2010 completed (details not fully available yet)	Dechlorinates perchloroethylene (PCE)
Dehalococcoides ethenogenes	1,591 (2005)	Used to degrade hydrogen and chlorine—the only known organism to fully dechlorinate perchloroethylene (PCE) and trichloroethene (TCE); PCE and TCE cleanup in wastewaters; polychlorinated dioxin degradation
Geobacter metallireducens	3,676 (2006)	Used for subsurface metal reduction, carbon cycling, and to generate electricity (see Figure 9.13 on page 222)
Populus trichocarpa (poplar tree)	45,555 (2006)	First tree genome sequenced; thought to have the largest number of genes of any organism sequenced to date; potentially useful for reducing atmospheric carbon dioxide

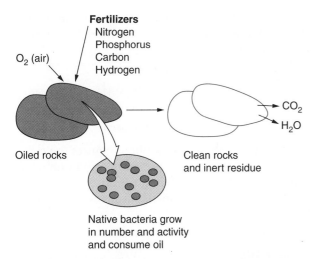

FIGURE 9.6   Fertilizers Can Stimulate Biodegradation by Indigenous Bacteria Bioremediation of chemicals such as those present in oil can be accelerated by adding fertilizers. Fertilizers stimulate the growth and replication of indigenous bacteria, which degrade oil into inert (harmless) compounds such as carbon dioxide (CO_2) and water (H_2O).

Nutrient enrichment, also called **fertilization**, is a bioremediation approach in which fertilizers—similar to the phosphorus and nitrogen that are applied to lawns of grass—are added to a contaminated environment to stimulate the growth of indigenous microorganisms that can degrade pollutants (**Figure 9.6**). Because living organisms need an abundance of key elements such as carbon, hydrogen, nitrogen, oxygen, and phosphorus for building macromolecules, adding

fertilizers provides bioremediation microbes with essential elements to reproduce and thrive. In some instances manure, wood chips, and straw may be added to provide microbes with sources of carbon as a fertilizer. Fertilizers are usually delivered to the contaminated site by pumping them into groundwater or mixing them in the soil. The concept behind fertilization is simple. By adding more nutrients, microorganisms replicate, increase in number (biomass), and grow rapidly, thus increasing the rate of biodegradation.

Bioaugmentation, or **seeding**, is another approach that involves adding bacteria to the contaminated environment to assist indigenous microbes with biodegradative processes. In some cases, seeding may involve applying genetically engineered microorganisms with unique biodegradation properties. Bioaugmentation is not always an effective solution, in part because laboratory strains of microbes rarely grow and biodegrade as well as indigenous bacteria, and scientists must be sure that seeded bacteria will not alter the ecology of the environment if they persist after the contamination is gone.

Phytoremediation

A growing number of approaches are utilizing plants to clean up chemicals in the soil, water, and air in an approach called **phytoremediation.** An estimated 350 species of plants naturally take up toxic materials. Poplar and juniper trees have been successfully used in phytoremediation, as have certain grasses and alfalfa. In phytoremediation, chemical pollutants are taken in through the roots of the plant as they absorb contaminated water from the ground (**Figure 9.7**). As

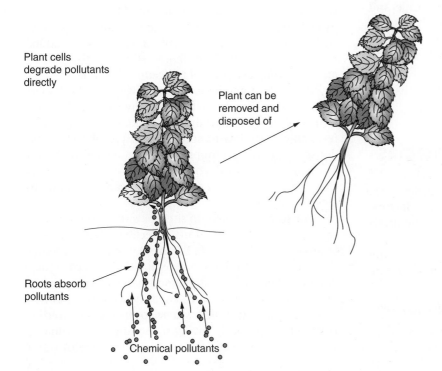

FIGURE 9.7   Phytoremediation Plants can be a valuable addition to many bioremediation strategies. Some plants degrade environmental pollutants directly; others simply absorb pollutants and must later be removed and disposed of.

an example, sunflower plants effectively removed radioactive cesium and strontium from ponds at the Chernobyl nuclear power plant in Ukraine. Water hyacinths have been used to remove arsenic from water supplies in Bangladesh, India. This is a significant technology, considering that arsenic concentrations exceed health standards in 60% of the groundwater in Bangladesh.

After toxic chemicals enter the plant, the plant cells may use enzymes to degrade the chemicals. In other cases, the chemical concentrates in the plant cells so that the entire plant serves as a type of "plant sponge" for mopping up pollutants. The contaminated plants are treated as waste and may be burned or disposed of in other ways. Because high concentrations of pollutants often kill most plants, phytoremediation tends to work best where the amount of contamination is low—in shallow soils or groundwater.

Scientists are also exploring ways in which plants can be used to clean up pollutants in the air, something many plants do naturally: for example, removing excess carbon dioxide (CO_2), the principal greenhouse gas released from burning fossil fuels, which contributes to global warming. Recently the first genome for a tree, the black cottonwood (a type of poplar), was sequenced. Poplars are commonly used for phytoremediation, and genetically engineered poplars have shown promise for capturing high levels of CO_2.

Phytoremediation can be an effective, low-cost, and low-maintenance approach for bioremediation. As an added benefit, phytoremediation can also be a less obvious and more eye-appealing strategy. For instance, planting trees and bushes can visually improve the appearance of a polluted landscape and clean the environment at the same time. Two main drawbacks of phytoremediation are that only surface layers (to around 50 cm deep) can be treated, and cleanup typically takes several years. In the next section, we examine specific cleanup environments and different strategies used for bioremediation.

9.3 Cleanup Sites and Strategies

A variety of bioremediation treatment strategies exist. Which strategy is employed depends on many factors. Of primary consideration are the types of chemicals involved, the treatment environment, and the size of the area to be cleaned up. Consequently, some of the following questions must be considered before starting the cleanup process:

- Do the chemicals pose a fire or explosive hazard?
- Do the chemicals pose a threat to human health including the health of cleanup workers?

- Was the chemical released into the environment through a single incident, or was there long-term leakage from a storage container?
- Where did the contamination occur?
- Is the contaminated area at the surface of the soil? Below the ground? Does it affect water?
- How large is the contaminated area?

Answering these questions often requires the combined talents of molecular biologists, environmental engineers, chemists, and other scientists who work together to develop and implement plans to clean up environmental pollutants.

Soil Cleanup

Treatment strategies for both soil and water usually involve either removing chemical materials from the contaminated site to another location for treatment, an approach known as **ex situ bioremediation**, or cleaning up at the contaminated site without excavation or removal called **in situ** (a Latin term that means "in place") **bioremediation**. In situ bioremediation is often the preferred method of bioremediation, in part because it is usually less expensive than ex situ approaches. Also, because the soil or water does not have to be excavated or pumped out of the site, larger areas of contaminated soil can be treated at one time. In situ approaches rely on stimulating microorganisms in the contaminated soil or water. Those in situ approaches that require aerobic degradation methods often involve **bioventing**, or pumping either air or hydrogen peroxide (H_2O_2) into the contaminated soil. Hydrogen peroxide is frequently used because it is easily degraded into water and oxygen to provide microbes with a source of oxygen. Fertilizers may also be added to the soil through bioventing to stimulate the growth and degrading activities of indigenous bacteria.

In situ bioremediation is not always the best solution. This approach is most effective in sandy soils, which are less compact and allow microorganisms and fertilizing materials to spread rapidly. Solid clay and dense rocky soils are not typically good sites for in situ bioremediation, and contamination with chemicals that persist for long periods of time can take years to clean up in this way.

For some soil cleanup sites, ex situ bioremediation can be faster and more effective than in situ approaches. As shown in **Figure 9.8**, ex situ bioremediation of soil can involve several different techniques, depending on the type and amount of soil to be treated and the chemicals to be cleaned up. One common ex situ technique is called **slurry-phase bioremediation**. This approach involves moving contaminated soil to another site and then mixing the soil with water

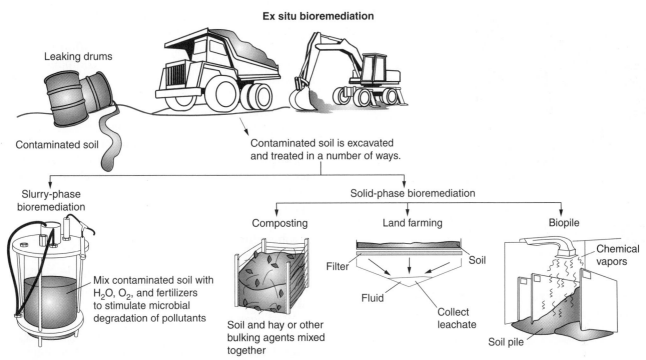

Ex situ bioremediation

Leaking drums

Contaminated soil

Contaminated soil is excavated and treated in a number of ways.

Slurry-phase bioremediation

Mix contaminated soil with H₂O, O₂, and fertilizers to stimulate microbial degradation of pollutants

Solid-phase bioremediation

Composting

Soil and hay or other bulking agents mixed together

Land farming

Filter

Fluid

Soil

Collect leachate

Biopile

Chemical vapors

Soil pile

FIGURE 9.8   Ex Situ Bioremediation Strategies for Soil Cleanup Many soil cleanup approaches involve ex situ bioremediation, in which contaminated soil is removed and then subjected to several different cleanup approaches.

and fertilizers (oxygen is often also added) in large bioreactors, where the conditions of biodegradation by microorganisms in the soil can be carefully monitored and controlled. Slurry-phase bioremediation is a rapid process that works fairly well when small amounts of soil need to be cleaned and the composition of chemical pollutants is well known (Figure 9.8).

For many other soil cleanup strategies, **solid-phase bioremediation** techniques are required. Solid-phase processes are more time-consuming than slurry-phase approaches and typically require large amounts of space; however, they are often the best strategies for degrading certain chemicals. Three solid-phase techniques are widely used: composting, landfarming, and biopiles.

Composting can be used to degrade household wastes such as food scraps and grass clippings; similar approaches are used to degrade chemical pollutants in contaminated soil. In a compost pile, hay, straw, or other materials are added to the soil to provide bacteria with nutrients that help bacteria degrade chemicals. **Landfarming** strategies involve spreading contaminated soils on a pad so that water and leachates can leak out of the soil. A primary goal of this approach is to collect leachate so that polluted water cannot further contaminate the environment. Because the polluted soil is spread out in a thinner layer than it would be if it were below the ground, landfarming also allows chemicals to vaporize from the soil and aerates the soil so that microbes can better degrade pollutants.

Soil **biopiles** are used particularly when the chemicals in the soil are known to evaporate easily and microbes in the soil pile are rapidly degrading the pollutants (**Figure 9.9**). In this approach, contaminated soil is piled up several meters high. Biopiles differ from compost piles in that relatively few bulking

FIGURE 9.9   Soil Biopiles Polluted soil that has been removed from the cleanup site can be stored in piles and bioremediation processes monitored to ensure decontamination of the soil before determining whether the soil can be returned to the environment.

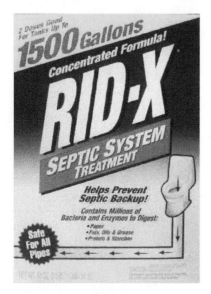

FIGURE 9.10   Septic Tank Additives Stimulate Bioremediation of Household Wastes

agents are added to the soil and fans and piping systems are used to pump air into or over the pile. As chemicals in the pile evaporate, the vacuum airflow pulls the chemical vapors away from the pile and either releases them into the atmosphere or traps them in filters for disposal, depending on the type of chemical. Almost all ex situ strategies for cleaning up soil involve tilling and mixing soil to disperse nutrients, oxygenating the dirt, and increasing interaction of microbes with contaminated materials to increase biodegradation.

Bioremediation of Water

Contaminated water presents a number of challenges. In Section 9.5, we consider how surface water can be treated following large spills such as an oil spill. Wastewater and groundwater can be treated in many different ways, depending on the pollutants that need to be removed.

Wastewater treatment

Probably the best-known application of bioremediation is in the treatment of wastewater to remove human sewage (fecal material and paper wastes), soaps, detergents, and other household chemicals. Both septic systems and municipal wastewater treatment plants rely on bioremediation. In a typical septic system, human sewage and wastewater from a single household move through the plumbing system out to a septic tank buried below ground next to the house. In the tank, solid materials such as feces and paper wastes settle to the bottom to be degraded

by microbes, while liquids flow out of the top of the tank and are dispersed underground across an area of soil and gravel called the septic bed. Within the bed, indigenous microbes degrade waste components in the water.

One commercial application of bioremediation recommended to prevent septic tanks from becoming clogged is to add products such as Rid-X (**Figure 9.10**), which are flushed into the system periodically. These products contain freeze-dried bacteria that are rich in enzymes such as lipases, proteases, amylases, and cellulases, which in turn degrade fats, proteins, sugars, and cellulose in papers and vegetable matter, respectively. Adding microbes this way is an example of the bioaugmentation we discussed in the previous section, and this treatment helps the septic system degrade cooking fats and oils, human wastes, tissue paper products, and other materials that can clog the system.

Wastewater (sewage) treatment plants are fairly complex and well-organized operations (**Figure 9.11**). Water from households that enters sewer lines is pumped into a treatment facility, where feces and paper products are ground and filtered into smaller particles, which settle out into tanks to create a mud-like material called **sludge**. Water flowing out of these tanks is called **effluent**. Effluent is sent to aerating tanks, where aerobic bacteria and other microbes oxidize organic materials in the effluent. In these tanks, water is sprayed over rocks or plastic covered with biofilms of waste-degrading microbes that actively degrade organic materials in the water.

Alternatively, effluent is passed into activated sludge systems—tanks that contain large numbers of waste-degrading microbes grown in carefully controlled environments. Usually these microbes float freely within the water, but in some cases they may be grown on filters through which contaminated water flows. Eventually effluent is disinfected with a chlorine treatment before the water is released back into rivers or oceans.

Sludge is pumped into anaerobic digester tanks in which anaerobic bacteria further degrade it. Methane-producing and carbon dioxide gas–producing bacteria are common in these tanks. Methane gas is often collected and used as fuel to run equipment at the sewage treatment plant. Tiny worms, which are usually present in the sludge, also help break down the sludge into small particles. Sludge is never fully broken down, but once the toxic materials have been removed, it is dried and can be used as landfill or fertilizer.

Scientists have discovered a bacterium called *"Candidatus Brocadia anammoxidans"* that possess a unique ability to degrade ammonium, a major waste

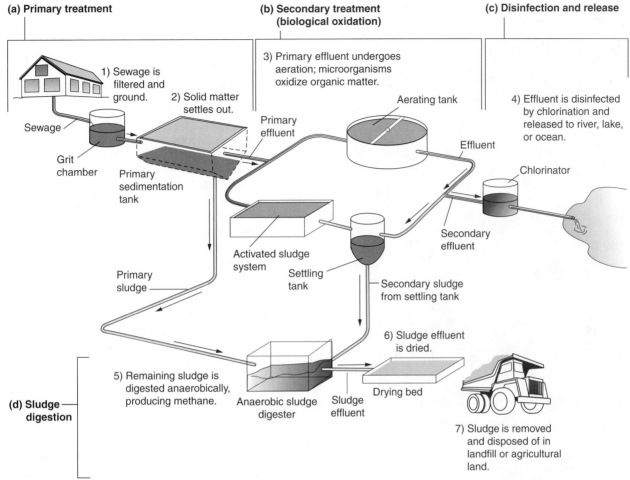

(a) Primary treatment

1) Sewage is filtered and ground.

2) Solid matter settles out.

Sewage

Grit chamber

Primary sedimentation tank

Primary sludge

(b) Secondary treatment (biological oxidation)

3) Primary effluent undergoes aeration; microorganisms oxidize organic matter.

Primary effluent

Aerating tank

Activated sludge system

Settling tank

(c) Disinfection and release

4) Effluent is disinfected by chlorination and released to river, lake, or ocean.

Effluent

Chlorinator

Secondary effluent

Secondary sludge from settling tank

6) Sludge effluent is dried.

Drying bed

(d) Sludge digestion

5) Remaining sludge is digested anaerobically, producing methane.

Anaerobic sludge digester

Sludge effluent

7) Sludge is removed and disposed of in landfill or agricultural land.

FIGURE 9.11 **Wastewater Treatment** Wastewater or sewage treatment facilities are well-planned operations that use aerobic and anaerobic bacteria to degrade organic materials such as human feces and household detergents in both the sludge and water (effluent).

product present in urine (**Figure 9.12** on the next page). Removing ammonium from wastewater before the water is released back into the environment is important because high amounts of ammonium can affect the environment by causing algal blooms and diminishing oxygen concentrations in waterways. Typically, wastewater plants rely on aerobic bacteria such as *Nitrosomonas europaea* to oxidize ammonium in a multistep set of reactions. However, *"Candidatus Brocadia anammoxidans"* is capable of degrading ammonium in a single step under anaerobic conditions, a process called the anammox process. Wastewater treatment plants in the Netherlands are using this strain and treatment plants in other countries may soon be using *"Candidatus Brocadia anammoxidans"* to remove ammonium from wastewater more efficiently.

Groundwater cleanup

With the exception of spills near coastal beaches, most large chemical spills such as oil spills occur in marine environments, often far away from populated areas. However, freshwater pollution typically occurs closer to populated areas and poses a serious threat to human health by contaminating sources of drinking water, either groundwater or surface water such as reservoirs. Groundwater contamination is a common problem in many areas of the United States. Drinking water for approximately 50% of the U.S. population comes from groundwater sources, and, according to some estimates, a large percentage of groundwater supplies in the United States contain pollutants that may have an impact on human health. Polluted groundwater can sometimes be very difficult to clean up because contaminated water gets trapped in soil and rocks and there is no easy way to "wash" aquifers.

Ex situ and in situ approaches are often used in combination. For instance, when groundwater is contaminated by oil or gasoline, these pollutants rise to the surface of the aquifer. Some of this oil or gas can be directly pumped out, but the portion mixed with groundwater must be pumped to the surface and

FIGURE 9.12   Bioreactor Containing *Candidatus Brocadia anammoxidans*, Anaerobic Bacterium That Can Degrade Ammonium Novel metabolic properties enable these anaerobes to degrade ammonium from wastewater in a single step.

passed through a bioreactor (**Figure 9.13**). Inside the bioreactor, bacteria in biofilms growing over a screen or mesh degrade the pollutants. Fertilizers and oxygen are often added to the bioreactor. Clean water from the bioreactor containing fertilizer, bacteria, and oxygen is pumped back into the aquifer for in situ bioremediation (Figure 9.13).

Turning Wastes into Energy

Landfills around the world are stressed to their limits, literally overflowing with trash from homes and businesses. The bulk of our household trash consists of food scraps, boxes, paper waste, cardboard packing containers from food, and similar items. A variety of chemical wastes such as detergents, cleaning fluids, paints, nail polish, and varnishes also make their way into the trash, despite the fact that most states are trying to reduce the amount of chemical waste that can be disposed of as regular garbage.

Scientists are working on strategies to reduce waste, including bioreactors containing anaerobic bacteria that can convert food waste and other trash into soil nutrients and methane gas. Methane gas can be used to produce electricity, and soil nutrients can be sold commercially as fertilizer for use by farms, nurseries, and other agricultural industries. Scientists are also working on seeding strategies that may be used to reduce chemicals in landfills—chemicals which would otherwise seep through the ground and contaminate local ground and surface waters. If successful, these applications of biotechnology may help to reduce the amount of waste and greatly increase the usable space in many landfills.

Bioremediation scientists are also studying polluted sediments in sewage sludge and at the bottoms of oceans and lakes as an untapped source of energy. Sediment in lakes and oceans is rich in organic materials from the breakdown of decaying materials such as

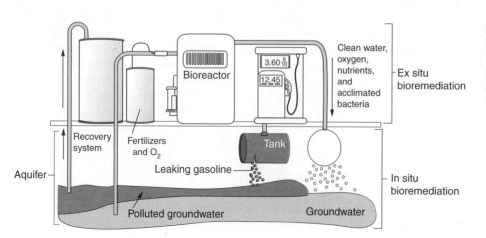

FIGURE 9.13   Ex situ and in situ Bioremediation of Groundwater Gasoline leaking into an underground water supply can be cleaned up using an aboveground (ex situ) system in combination with in situ bioremediation.

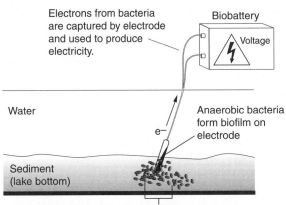

Electrons from bacteria are captured by electrode and used to produce electricity.

Biobattery

Voltage

Water

Anaerobic bacteria form biofilm on electrode

e⁻

Sediment (lake bottom)

Anaerobic bacteria oxidizing organic molecules in sediment transfer electrons to electron acceptor molecules such as iron and sulfur.

FIGURE 9.14   Polluted Sediments May Be an Untapped Source of Energy Scientists have found that anaerobic bacteria in sediment may be a source of energy. Because these bacteria use redox reactions to degrade molecules in sediment, electrons can be captured by electrodes, which can transfer electrons to generators to create electricity.

FIGURE 9.15   Microbe-Powered Fuel Cells Researchers at the University of Massachusetts have demonstrated that *Geobacter* fuel cells can effectively convert sugars into electricity.

leaves and dead organisms. Within this "muck" are anaerobes that use organic molecules in the sediment to generate energy. The term *electrigens* is being used to describe electricity-generating microbes that have the ability to oxidize organic compounds to carbon dioxide and transfer electrons to electrodes. Under certain conditions, electrigens can cluster and interconnect to form nanowires that conduct electrons! Such strains may even be used to make electricity from manure and common household wastes.

Desulfuromonas acetoxidans is an anaerobic marine bacterium that uses iron as an electron acceptor to oxidize organic molecules in sediment. Researchers are exploring ways in which electrons can be harvested from *D. acetoxidans* and other bacteria such as *Geobacter metallireducens* and *Rhodoferax ferrireducens* as a technique for capturing energy in *bacterial biobatteries,* also called *microbial fuel cells,* that can be used to provide a source of electricity (**Figure 9.14**). Researchers at the University of Massachusetts have demonstrated that electrigenic strains of microbes can generate electricity with high efficiency; although preliminary studies suggest that this technique has some promise, more research needs to be done (**Figure 9.15**).

Even though bioremediation strategies have effectively cleaned up many environmental pollutants, bioremediation is not the solution for all polluted sites. For instance, bioremediation is ineffective when the polluted environment contains high concentrations of very toxic substances such as heavy metals,

radioactive compounds, and chlorine-rich organic molecules, because these compounds typically kill microbes. Therefore new strategies will have to be discovered and applied to tackle some of these cleanup challenges. Some of these new bioremediation strategies are likely to involve genetically engineered microorganisms, the topic of the next section.

9.4 Applying Genetically Engineered Strains to Clean Up the Environment

Bioremediation has traditionally relied on stimulating naturally occurring microorganisms; however, many indigenous microbes cannot degrade certain types of chemicals, especially very toxic compounds. For example, some organic chemicals produced during the manufacturing of plastics and resins are resistant to biodegradation and can persist in the environment for several hundred years. Many radioactive compounds also kill microbes, thus preventing biodegradation. To clean up some of these stubborn and particularly toxic pollutants, we may need to use bacteria and plants that have been genetically altered. The development of recombinant DNA technologies has enabled scientists to create genetically modified (GM) organisms with the potential to improve bioremediation processes.

Petroleum-Eating Bacteria

The first effective GM microbes for use in bioremediation were created in the 1970s by Ananda Chakrabarty and his colleagues at General Electric. This work was carried out before DNA cloning and recombinant DNA technologies were widely available. So how did Chakrabarty do this? He isolated strains of *Pseudomonas* from soils contaminated with different types of chemicals, including pesticides and crude oil. He then identified strains that showed the ability to degrade such organic compounds as naphthalene, octane, and xylene. Most of these strains could grow in the presence of these compounds because they contained plasmids that encoded genes for breaking down each component.

Chakrabarty mated these different strains and eventually produced a strain that contained several different plasmids. Together, the combined proteins produced by these plasmids could effectively degrade many of the chemical components of crude oil. For his work, Chakrabarty was awarded the first U.S. patent for a GM living organism. However, this patent decision was very controversial and was held up in the courts for about 10 years. The primary issues being debated were whether life forms could be patented and whether Chakrabarty's recombinant bacterium should be considered a product of nature or an invention. Eventually the U.S. Supreme Court ruled that the development of recombinant *Pseudomonas* was an invention worthy of a patent.

Chakrabarty's approach was not as effective as it might have seemed. Crude oil contains thousands of compounds, and his GM bacteria could degrade only a few of these. The majority of the chemicals in crude oil remain largely unaffected by recombinant organisms. Consequently, developing GM bacteria with different degradative properties is an intense area of research. In the future, a useful approach for cleaning up crude oil may be to release multiple bacterial strains, each with the ability to degrade different compounds in the oil. To date, field applications of GM microbes for bioremediation have been fairly limited, in part because of regulatory obstacles and public concerns over the release of GM bacteria. But GM microorganisms are also often ineffective in the environment because indigenous microbes may outcompete them.

Engineering *E. coli* to Clean Up Heavy Metals

Heavy metals including copper, lead, cadmium, chromium, and mercury can critically harm humans and wildlife. Mercury is an extremely toxic metal that can contaminate the environment. It is used in manufacturing plants, batteries, electrical switches, medical instruments, and many other products. Mercury, and a related compound called methylmercury (MeHg), can accumulate in organisms through a process called **bioaccumulation**. In bioaccumulation, organisms higher up on the food chain contain higher concentrations of chemicals than organisms lower on the food chain. For instance, in a water supply, mercury may be ingested by small fish, which may then be eaten by birds, larger fish, otters, raccoons, and other animals, including humans. Large fish and birds need to eat a lot of small fish; therefore they accumulate more mercury in their systems than small fish and birds that eat less. Similarly, if a person were to eat large fish as a primary source of food, that person would accumulate high amounts of mercury over time. Regular consumption of fish and shellfish contaminated with mercury and methylmercury poses serious health threats to humans, including birth defects and brain damage. For these reasons, in many areas of the United States, health officials suggest that pregnant women and young children eat only small amounts of certain types of fish, such as swordfish and fresh tuna, and restrict these meals to no more than one serving a week.

Because mercury is toxic at very low doses, most current strategies for removing mercury from contaminated water supplies do not remove enough of it to meet acceptable standards. Scientists have developed genetically engineered strains of *E. coli* that may be useful for cleaning up mercury and other heavy metals. They have also identified naturally occurring metal-binding proteins in plants and other organisms. Two of the best-characterized types of proteins—metallothioneins and phytochelatins—have a high capacity for binding to metals. For these proteins to function, however, the metals must enter cells. Scientists have engineered *E. coli* to express transport proteins that allow for the rapid uptake of mercury into the bacterial cell's cytoplasm, where the mercury can bind to metal-binding proteins.

Some of these genetically altered bacteria can absorb mercury directly; others that bind mercury can be grown on *biofilms* to act as sponges for soaking up mercury from a water supply. The biofilms must be changed periodically to remove mercury-containing bacteria. Similarly, genetically engineered single-celled algae containing metallothionein genes and bacteria called cyanobacteria have shown promise for their ability to absorb cadmium, another very toxic heavy metal known to cause many serious health problems in humans.

TOOLS OF THE TRADE

Microcosms Provide Major Benefits

Bioremediation scientists are working on innovative ways to biodegrade different chemical compounds under a myriad of environmental conditions. Industry is continually creating new kinds of chemicals, and many of these will inevitably make their way into the environment. To stay one step ahead of new environmental pollutants, bioremediation researchers must continue to develop new cleanup strategies.

How can scientists study bioremediation of a new chemical that has never made it into the environment? They obviously cannot pollute large areas nor wait for a large-scale disaster like the *Exxon Valdez* spill before testing their theories about how to clean up this new chemical. One of the most practical approaches to learning about new bioremediation strategies is to make a **microcosm,** an artificially constructed test environment designed to mimic real-life environmental circumstances. Some microcosms consist of small bioreactors—about the size of a 5-gallon bucket—containing soil, water, pollutants, and microbes to be tested for their bioremediation abilities. A microcosm may be as small as a few grams of soil in a test tube or vial, but they are more often scaled up to resemble larger environments. For example, large ponds, which may be indoors or outdoors, or soil plots that are prepared to prevent the escape of pollutants outside of the test facility, can be used as microcosms. By carefully designing microcosms,

bioremediation researchers can attempt to simulate, on a small scale, a site that needs to be cleaned up.

When indigenous or genetically engineered organisms with bioremediation potential have been identified, studies in bioreactors or on small isolated areas of land or water can be crucial for determining whether these organisms will effectively clean up pollution in larger settings. By carefully manipulating the environmental conditions in the microcosms, scientists can test the ability of organisms to degrade different pollutants under varying conditions, including different levels of moisture, temperature, nutrients, oxygen, and pH and in different soil types.

Microcosm studies may even involve testing experimental microbes on polluted groundwater or contaminated soil that is placed into the microcosm so that the rates of degradation can be monitored and the cleanup time evaluated. Scientists can also carry out experiments to study the bioremediation of mixtures of pollutants at the same time.

In an attempt to produce the best cleanup results, scientists will analyze the data they have gathered and design new experiments. A cleanup approach that demonstrates success in a microcosm is not guaranteed to succeed in the field. Nevertheless, by testing bioremediation approaches in microcosms, much valuable time and money can be saved before deciding whether a cleanup approach is likely to have any chance of success in cleaning up a polluted environment in the field.

Biosensors

Researchers have developed genetically engineered strains of the bacterium *Pseudomonas fluorescens,* which can effectively degrade complex structures of carbon and hydrogen called polycyclic aromatic hydrocarbons (PAHs) and other toxic chemicals. Using recombinant DNA technology, scientists have been able to splice bacterial genes encoding specific enzymes (which can metabolize these contaminants) to reporter genes such as the *lux* genes from bioluminescent marine bacteria. Recall that *lux* genes are often used as reporter genes because they encode the light-releasing enzyme luciferase (see Chapter 5). As PAHs are degraded, bacteria release light that can be used to monitor biodegradation rates. Similar techniques are being used to develop **biosensors** from recombinant bacteria containing *lux* genes. Such biosensors have proven to be valuable in the assessment of environmental pollutants such as heavy metals. In the future,

GM microbes are expected to play a greater role as biosensors. In the next section, we consider two of the best-studied and most highly publicized examples of bioremediation in action.

Genetically Modified Plants and Phytoremediation

In recent years scientists have been working on genetically modifying plants to improve their phytoremediation capabilities. Currently, using phytoremediation to clean up methylmercury (MeHg) is a very active area of research. Mercury contamination of most fish is a result of MeHg accumulation; throughout the world this has caused many health warnings and restrictions on consuming mercury-contaminated fish. Plants engineered to contain mercury-detoxifying genes from bacteria have shown some potential for accumulating MeHG; eventually they may be used for phytoremediation.

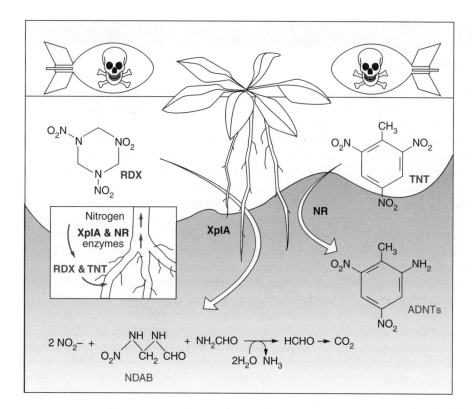

FIGURE 9.16 Phytoremediation of Toxic Explosives Using Transgenic Plants Transgenic plants engineered with the *XplA* or *NR* gene to degrade either RDX or TNT, respectively, absorb the explosive chemicals through their roots and then degrade them into less toxic compounds (aminodinitrotoluenes [ADNTs], or 4-nitro-2,4-diazabutanal [NDAB]), or nontoxic compounds (NH_3, CO_2). Plants engineered with the *XplA* gene use the degradation products from RDX to stimulate their growth.

Another promising area of this research has been the development of plants able to remove chemicals from military explosives and weapons firing ranges that have contaminated soil and groundwater. Hexahydro-1,3,4-trinitro-1,3,5-triazine (commonly called royal demolition explosive, or RDX) and 2,4,6-trinitrotoluene (TNT) are two of the most common chemical contaminants that result from the production, use, and disposal of explosives. These chemicals move readily through soils and contaminate groundwater. Both RDX and TNT are highly toxic to most organisms and pose significant health threats to wildlife and humans.

Notice that TNT was one of the top 20 chemicals listed in Table 9.1 and that the EPA lists both TNT and RDX as priority chemicals to be removed from the environment. These contaminants are a major pollution problem worldwide. Incredibly, there are hundreds of tons of these compounds in sites around the world and tens of thousands of acres deemed unsafe as a result.

A few bacteria and plants have been shown to weakly degrade TNT with low efficiency; degradation of RDX is even less effective because of its chemical structure. In the last few years, however, scientists have used genetic engineering to create transgenic plants that may turn out to be very effective for the phytoremediation of TNT and RDX. A transgenic strain of tobacco containing a nitroreductase gene from

Enterobacter cloacae effectively converts TNT to less toxic chemicals (**Figure 9.16**). Most recently, scientists incorporated a gene called *xplA* from the bacteria *Rhodococcus rhodochrous* into *Arabidopsis thaliana*. The *xplA* gene produces an RDX-degrading enzyme called cytochrome P450, which can degrade RDS once it is absorbed into the plant (see Figure 9.16). It may now be possible to make plants that can degrade both TNT and RDX. Moreover, since the poplar genome has been sequenced, bioremediation scientists working on genetically modified plants are very enthusiastic about the possibility of making transgenic poplars and other fast-growing, deep-rooted trees that can remediate TNT and RDX well below the surface soil. In the next section, we consider several of the best-studied and most highly publicized examples of bioremediation in action.

9.5 Environmental Disasters: Case Studies in Bioremediation

Most bioremediation approaches involve the cleanup of contaminated areas that are relatively small, perhaps several hundred acres in size. Nevertheless, a great deal has been learned about the effectiveness of different bioremediation strategies by studying large-scale environmental disasters that have been treated by bioremediation.

The *Exxon Valdez* Oil Spill

The world relies heavily on crude oil and on petroleum products that can be manufactured from oil. Petroleum products are used not just as gasoline and diesel fuel to power automobiles but also as the basis for hundreds of everyday products including plastics, paints, cosmetics, and detergents. The United States alone uses in excess of 950 billion liters (250 billion gallons) of oil each year, over 350 billion liters of which are imported. When crude oil is transported, some amount of leakage almost always occurs. Tanker accidents spill nearly 400 million liters of crude oil each year, and even larger volumes of oil enter our seas through naturally occurring spills and leaks.

Oil spills have had a tremendous impact on the environment, specifically on large numbers of wildlife (**Figure 9.17a**). Typically large spills do not have a great effect on human life directly. This is because most large spills occur in open oceans or bays far removed from swimming beaches and water supplies. Humans are more affected by small local spills, such as a gas station's leaking underground tank that may threaten local drinking water supplies.

In 1989, the *Exxon Valdez* oil tanker ran aground on a reef in Prince William Sound off the coast of Alaska, releasing approximately 42 million liters (11 million gallons or about 260,000 barrels) of crude oil and contaminating over 1,000 miles of the Alaskan shoreline. Many experimental approaches for bioremediation were implemented to clean up this spill. Prince William Sound became a field laboratory for trying bioremediation cleanup strategies.

As is done at most sizable oil spills, physical cleaning measures were first used to contain and remove large volumes of oil. These measures included the use of containment booms or skimmers—surface nets, fences, or inflatable devices like buoys that float on the surface—which are fixed in place or towed behind a boat to collect and contain oil (Figure 9.17b). Then vacuums were used to pump oil from the surface into disposal tanks. Beaches and rocks were washed with fresh water under high pressure to disperse oil. By diluting and dissolving the oil in the sea, the oil was gradually dispersed. One biotechnology cleanup application used citrus-based products to bind crude oil and allow it to be collected on absorbent pads. But after using all these physical approaches to remove the bulk of the oil, millions of gallons of oil still remained attached to sand, rocks, and gravel both at the surface and below the surface of contaminated shorelines. This is when bioremediation went to work.

(a)

(b)

FIGURE 9.17 Oil Spills Pose Serious Threats to the Environment (a) Oil spills typically have the greatest impact on wildlife. (b) Containment booms are used initially to control the spread of oil and minimize pollution of the surrounding area.

FIGURE 9.18 Applying Fertilizers to Stimulate Oil-Degrading Microbes Cleanup workers spray nitrogen-rich fertilizers onto an oil-soaked beach in Alaska following the *Exxon Valdez* oil spill. The fertilizers greatly accelerate the growth of natural bacteria that will degrade the oil.

As the first step in the bioremediation process, nitrogen and phosphorus fertilizers were applied to the shoreline to stimulate oil-degrading bacteria (mostly strains of *Pseudomonas*) that were living in the sand and rocks (**Figure 9.18**). Indigenous bacteria immediately showed signs that they were degrading the oil. When microbes degrade petroleum products, PAHs are formed and eventually oxidized into carbon chains that can be broken down into carbon dioxide and water. Over time, chemical tests on oil from the shoreline soil showed significant changes in chemical composition, indicating that natural degradation by indigenous bacteria was working. However, oil seeped into sediments and other low-oxygen layers below shoreline rocks, where biodegradation is relatively slow. It may take hundreds of years for the oil spilled by the *Exxon Valdez* to be fully cleared, and some areas of the Alaskan environment may never return to the state in which they were prior to the spill.

Oil Fields of Kuwait

The deserts of Kuwait are literally a living laboratory for studying bioremediation. Ten years after the Gulf War, large areas of Kuwait's desert remain soaked in oil. During the Iraqi occupation of Kuwait from 1990 to 1991, countless oil fields were destroyed and burned, releasing an estimated 950 million liters of oil into the deserts—more than 20 times the amount of oil spilled in the *Exxon Valdez* accident. Kuwaiti scientists studying these spills have found that the spilled oil has severely affected plant and animal life in many contaminated areas. Some plant species have been completely eliminated, and the long-term biological impacts will not be known for many more years.

In contrast to the *Exxon Valdez* spill, bioremediation of desert soils poses a number of different problems. Unlike the spill in Alaska, there are no waves to help disperse and dissolve oil. Dry soil conditions of the desert also tend to harbor fewer strains of oil-metabolizing microbes, and adhesion of oil to sand and rocks slows natural degradation processes. Preliminary studies suggest that novel strains of oil-degrading bacteria are slowly working to break down oil below the surface of the sand.

The Kuwaiti government has developed a $1 billion program to address what may be the world's largest bioremediation project. There are no previously studied sites of this size to use as a model for cleaning up arid desert environments, so bioremediation scientists from around the world are studying Kuwait's deserts in hopes of developing strategies for cleaning up these oil-drenched sands. Information that scientists learn from studying this region will certainly be of value for treating oil spills in other sandy environments.

The *Deepwater Horizon* Oil Spill

On April 20, 2010 British Petroleum's (BP) oil-drilling rig the *Deepwater Horizon* exploded, releasing millions of gallons of crude oil into the Gulf of Mexico about 65 kilometers off the Louisiana coast (**Figure 9.19a**). Although estimates of oil flow have varied widely, more than 600 million liters (4 million barrels) were released into the Gulf of Mexico before the broken well was capped in mid-July, resulting in the largest marine oil spill in history and predictions that it would cause the world's largest environmental disaster. Yet, given the size of the spill, far less oil than many predicted made it to the coast. Oil that was not removed by cleanup crews was dispersed by wave action and through the use of chemical dispersants that broke up the slick. Controlled burns and surface oil evaporation of volatile materials such as PAHs also contributed to oil cleanup, but bioremediation played a big role accounting for the degradation of an estimated 50% of the oil released.

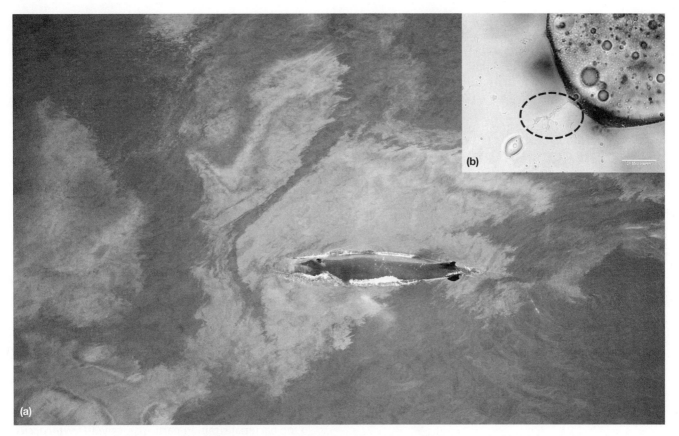

FIGURE 9.19   Microbes Played a Major Role the Bioremediation of Oil from the *Deep-water Horizon* Oil Spill in the Gulf of Mexico   (a) Indigenous microbes that were very abundant in surface and subsurface plumes of oil emanating from the *Deepwater Horizon* spill were essential for bioremediation. Shown here is a whale swimming through a surface plume of oil in the Gulf. (b) An oil droplet magnified 100 times and metabolizing microbes (circled).

Studies by oceanographers from the Woods Hole Oceanographic Institute in Massachusetts reported that during the first 5 days of the spill, microbes had not significantly degraded the oil. Over time, an underwater plume of oil that was at least 22 miles long developed. Within weeks, microbes appeared to be flocking to the plume, replicating rapidly, so that bacteria were twice as dense inside the plume compared to outside the plume (Figure 9.19b). Research on these indigenous microbes in the plume revealed over 1,500 genes encoding proteins designed to degrade hydrocarbons. Clearly hydrocarbon-degrading microbes were highly enriched in the plume, and these microbes were actively breaking down oil. It was also determined that microbes degraded an estimated 200,000 tons of methane that spewed into the Gulf of Mexico during the spill. Ethane and propane were also major hydrocarbons released and these too were degraded by microbes.

The warm waters of the Gulf and components in the plume—such as natural gas, which contains propane and ethane—helped stimulate biodegradation by indigenous microbes. Fertilizers such as iron, nitrogen, and phosphorus were applied at the site to stimulate rates of biodegradation. Gradients in dissolved oxygen levels in the plume also contributed to differences in rates of biodegradation throughout and along the plume. Estimates have suggested that these microbes were reducing oil amounts in the plume by half nearly every 3 days.

Where did they come from? Petroleum-degrading bacteria have existing for eons, thriving on oil that seeps naturally through the seafloor. Each year about 79 million liters of oil leak onto the floor of the Gulf of Mexico through naturally occurring seepage. Within the *Deepwater Horizon* plume, researchers have detected over 900 subfamilies of bacteria, including newly discovered species. It is also likely that chemical dispersants used to break up the stream of oil gushing from the broken well may have created microscopic particles that increased the surface area between oil droplets and water, allowing for greater contact with oil-degrading bacteria. The impact of oil that moved into Louisiana's wetlands and long-term impacts of dispersed oil and chemical dispersants on marine ecosystems in the Gulf region will be closely evaluated in the years to come.

The next section provides a glimpse of how potential applications of bioremediation may be able to solve cleanup problems that have been difficult to treat.

9.6 Challenges for Bioremediation

Biotechnology is a significant tool in our fight to rehabilitate areas of the environment that have been polluted through accidents, industrial manufacturing, and mismanagement of ecosystems. Bioremediation is a rapidly expanding science. Scientists around the world are carrying out research aimed at developing a greater understanding of the microorganisms involved in biodegradative processes, including the identification of novel genes and proteins involved in these breakdown processes. Researchers are studying microbial genetics to create genetically engineered microbes that may be able to degrade new types of chemicals, and they are developing novel biosensors to detect and monitor pollution.

Recovering Valuable Metals

The recovery of valuable metals such as copper, nickel, boron, and gold is another area of bioremediation that has yet to reach its full potential. Through oxidation reactions, many microbes can convert metal products into insoluble substances, called metal oxides or ores, which will accumulate in bacterial cells or attach to the bacterial cell surface. Some marine bacteria that live in deep-sea hydrothermal vents have also shown potential for precipitating precious metals. Using bacteria as a way to recover hazardous metals as part of industrial manufacturing processes is one potential application, but scientists are interested in finding ways in which these bacteria can be used to recover valuable precious metals. For instance, many manufacturing processes use silver and gold plating techniques; these processes produce waste solutions with suspended particles of silver and gold. Microbes may be used to recover some of these metals from the waste solutions. Similarly, microbes may also be used to harvest gold particles from underground water supplies and cave water found in gold mines. Many bacterial strains that live in such environments are actively being studied for these purposes. Plants may also provide a way to harvest metals from the environment. The wild mustard *Thlaspi goesingense*, native to the Austrian Alps, is known to accumulate metals in storage vacuoles.

Bioremediation of Radioactive Wastes

Another area of active research involves developing bioremediation approaches to remove radioactive materials from the environment. Uranium, plutonium, and other radioactive compounds are found in water from mines where naturally occurring uranium is processed. Radioactive wastes from nuclear power plants also present a disposal problem worldwide. The U.S. Department of Energy (DOE) has identified over 100 sites in 30 states that have been contaminated by weapons production and nuclear reactor development. Radioactive waste sites often have a complex mix of radioactive elements such as plutonium, cesium, and uranium along with mixtures of heavy metals and organic pollutants such as toluene.

Although most radioactive materials kill a majority of microbes, some strains of bacteria have demonstrated a potential for degrading radioactive chemicals. For example, researchers at the University of Massachusetts have discovered that some species of *Geobacter* can reduce soluble uranium in groundwater into insoluble uranium, effectively immobilizing the radioactivity. To date, however, no bacterium has been discovered that can completely metabolize radioactive elements into harmless products.

One bacterial strain in particular, called ***Deinococcus radiodurans*** (**Figure 9.20**), is especially fascinating. Its name literally means "strange berry that withstands

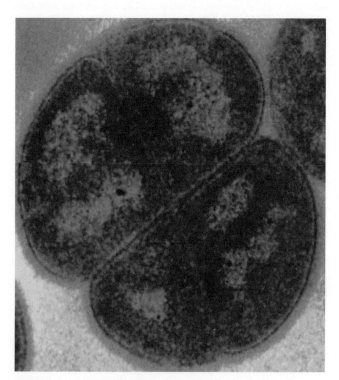

FIGURE 9.20 *Deinococcus radiodurans* These bacteria are highly resistant to damage by radiation.

YOU DECIDE

The PCB Dilemma of the Hudson River

Winding through upstate New York, the Hudson River Valley comprises some of the most beautiful country in the Northeast. But serious problems lurk below the glimmering surfaces of the Hudson's blue water. From 1947 to 1977, the General Electric Company released over 1.2 million pounds of toxic chemicals called polychlorinated biphenyls (PCBs) into the river from facilities in Hudson Falls and Fort Edward, New York. PCBs were commonly used in transformer boxes, capacitors, and cooling and insulating fluids of electrical equipment manufactured before 1977. PCBs are no longer used in manufacturing in the United States, and they have been banned in most of the world.

PCBs are very toxic to humans and wildlife because these fat-soluble chemicals gradually accumulate in fatty tissues. Hudson River fish are contaminated with PCBs at concentrations in excess of safe levels. Consumption of fish from most areas of the Hudson is banned or restricted, but some people ignore these publicized restrictions because the water looks so clear and the fish do not *look* polluted. PCBs present serious health threats to humans, and prolonged exposure to PCBs is known to cause cancer, reproductive problems, and other medical conditions affecting the immune system and thyroid gland. Children are particularly susceptible to the health effects of PCBs.

High concentrations of these chemicals still lurk in the Hudson River and other environments. In the Hudson River, most PCBs are located in the sediment at the bottom of the river. To get rid of these persistent chemicals, environmental groups and the EPA have proposed dredging the 175-mile-long riverbed to remove over 2.5 million cubic yards of sediment, which would remove more than 100,000 pounds of PCBs. Critics of the dredging plan suggest that such operations would only release more PCBs into the water by stirring up the sediment. Instead of dredging, many believe that leaving the sediment in place and letting natural current flows disperse the chemicals, combined with bioremediation through bacterial degradation of PCBs, is the best way to reduce the load of PCBs in the long run.

PCB-degrading anaerobes have been detected in Hudson River sediments. Some anaerobic bacteria are involved in the first step of breaking down PCBs by cleaving off chlorine and hydrogen groups. Aerobic bacteria in the water can further modify PCBs, and others can then convert them into water, carbon dioxide, and chloride. Some predictions suggest that even after dredging, PCB levels in fish will not drop to levels acceptable for human consumption until after 2070. Even if dredging is a good plan, where will the dredged sediments go? How will these chemicals be cleaned up? Some people believe that placing PCB-laden sediments in a sealed landfill, a completely anaerobic environment, will slow down the degradative processes. Should these chemicals be left in the mud to be broken down slowly over time through natural bioremediation, or should humans intervene in an effort to speed up nature's cleanup effort? What are the pros and cons of taking action or doing nothing?

You decide.

radiation." Named the world's toughest bacterium by *The Guinness Book of World Records, D. radiodurans* was first identified and isolated about 50 years ago from a can of ground beef that had spoiled, even though it had been sterilized with radiation. Subsequently, scientists discovered that *D. radiodurans* can endure doses of radiation over 3,000 times higher than other organisms can, including humans. High doses of radiation create double-stranded breaks in DNA structure and cause mutations, yet *D. radiodurans* shows incredible resistance to the effects of radiation. Although its resistance mechanisms are not entirely known, this microbe clearly possesses elaborate systems for folding its genome to minimize damage from radiation, and it also uses novel DNA repair mechanisms to replace damaged copies of its genome. Recent completion of the *D. radiodurans* genome map is expected to provide valuable insight into its unique DNA repair genes.

The DOE is very interested in using *D. radiodurans* and another bacterium, *Desulfovibrio desulfuricans*, for bioremediation of radioactive sites. Recently researchers at the DOE and the University of Minnesota created a recombinant strain of this microbe by joining a *D. radiodurans* gene promoter sequence to the gene encoding the enzyme toluene dioxygenase which is involved in toluene metabolism. This strain demonstrated the ability to degrade toluene in a high-radiation environment. In an effort to immobilize radioactive elements, scientists hope to use this same strategy to equip *D. radiodurans* with genes from other microbes that are known to encode metal-binding proteins.

Finally, it is important that biodegradation scientists continually look ahead so that they can be prepared to predict how new chemicals may affect the environment and thus be able to develop technologies to clean up new pollutants. Bioremediation will not be able to rid all pollutants from the environment, but well-planned bioremediation approaches are an important component of cleanup efforts.

CAREER PROFILE

Validation Technician

Bioremediation is a relatively new discipline of biotechnology, and many smaller companies are involved in varying aspects of the field. Bioremediation is a multidisciplinary industry that requires the collective efforts of many different types of scientists, including microbiologists, chemists, soil scientists, environmental scientists, hydrologists, and engineers. Microbiologists and molecular biologists work hand in hand to carry out research designed to understand the mechanisms that microorganisms use to degrade pollutants. Biologists work together with engineers to design and implement the best treatment facilities for cleaning up the pollutants in question.

One entry-level position in bioremediation is that of the validation technician. Some validation technicians corroborate EPA-filed reports. Knowledge of applicable EPA regulations, the ability to collect pollution data from field sensors (many of which are linked by satellite to monitors), and the know-how to arrange these data in an accessible database are required for this position.

Good computer skills, including familiarity with database programs and satellite geographic information systems, are required. Familiarity with organisms and natural communities is also needed. From entry-level positions, validation technicians with appropriate college coursework (or an applicable degree) can advance to supervisory positions. The genetic engineers who design the bioremediation mechanisms are usually PhD-trained research scientists, but they depend heavily on data collected in field studies by validation technicians.

MAKING A DIFFERENCE

In Hanahan, South Carolina, a suburb of Charleston, a leak of approximately 80,000 gallons of kerosene-based jet fuel occurred in 1975. Despite a number of cleanup measures, the spill could not be kept from soaking through the sandy soil and contaminating the water table. Ten years later, contamination reached a residential area. Excavation of the contaminated soil was impractical because of the large area involved, and removal of the contaminated groundwater would not eliminate the problem. Scientists at the U.S. Geological Survey (USGS) determined that indigenous microbes in the soil were degrading toxic compounds in the jet fuel; they also found that adding fertilizers to these microbes dramatically stimulated rates of biodegradation. This site was one of the first field applications of bioremediation. USGS scientists added nutrients to the groundwater and simultaneously removed some of the contaminated groundwater. By the early 1990s, contamination in the groundwater supply was reduced by 75% and bioremediation had demonstrated its potential worth as an effective cleanup strategy.

QUESTIONS & ACTIVITIES

Answers can be found in Appendix 1.

1. Describe three specific examples of a bioremediation application. Include in your description the organisms likely to be involved in the process, and discuss potential advantages and disadvantages of each application.

2. What is the purpose of adding fertilizers and oxygen to a contaminated site that is undergoing bioremediation?

3. Suppose you learned that a plot of land, which was once used as a chemical dump, had been cleaned up by bioremediation and is now a proposed site for a new residential housing development. Would you want to live in a house that was built on a former bioremediation site? Consider potential problems associated with this scenario.

4. Lead is a very toxic metal that is not easily degraded in the environment. Lead pipes, paint containing lead, car batteries, and even lead fishing sinkers are all potential sources for the release of lead into the environment. Propose a strategy for identifying bacteria that may play a role in degrading lead.

5. Suppose your town was interested in building a chemical waste storage facility in your neighborhood. Consider "lessons learned" from past failures of waste storage, dumping, environmental contamination, and successful applications of bioremediation, and then develop a list of priorities town officials should consider before building the site and questions that should be addressed if the site were to be used.

6. Access some of the websites listed on the Companion Website. Use these sites to search for information about the most prevalent pollutants found in the environment. Make a list of five pollutants and describe where they are found, how they enter the environment, their short- and long-term health effects (in humans or other organisms), their environmental effects, and what types of microbes may degrade these pollutants.

7. Visit the Scorecard website listed on the Companion Website. Scorecard provides pollution data reports for each state in the United States. Enter your zip code and check on the latest reports for the areas closest to where you live.

8. Visit the EPA Superfund site and search for environmental cleanup violations reported in your state. Who was responsible for these violations? What fines (if any) were levied? What efforts are planned to remediate the site or sites?

9. In November 2002, the oil tanker *Prestige* split in half and sank off the northwestern coast of Spain. At least 30,000 tons of fuel oil were spilled. Conduct a Web search to see if you can find any information about bioremediation at this site.

10. Scientists monitoring bioremediation at the *Deepwater Horizon* oil spill in the Gulf of Mexico were concerned that microbial degradation of oil could lead to zones of hypoxia, which might result in massive fish kills. Why might this be a concern? As it turns out, this did not appear to be major problem. Propose reasons why these predictions may have been incorrect.

11. Bioremediation is yet another rapidly changing area of biotechnology. Visit Google Scholar (http://scholar.google.com/) and do a search using the word *bioremediation* (or search for a specific topic in bioremediation such as *phytoremediation*) to find recent publications in bioremediation.

12. In late 2010, a paper was published online in *Science* describing a potentially unprecedented microbe that was reported to metabolize arsenic and incorporate arsenic into its DNA. Many scientists have been skeptical and highly critical of this work. Do a web search for articles on the latest viewpoints of this work.

Visit www.pearsonhighered.com/biotechnology

To download learning objectives, chapter summary, "Keeping Current" web links, glossary, flashcards, references, and jpegs of figures from this chapter.

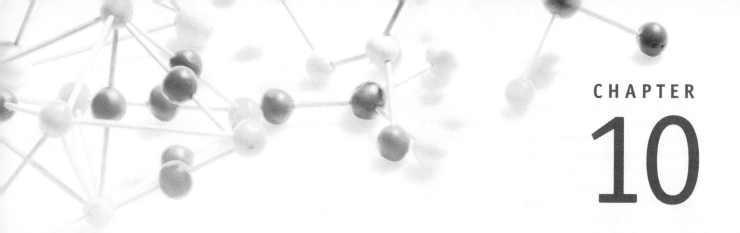

CHAPTER

10

Aquatic Biotechnology

After completing this chapter, you should be able to:

- Discuss important goals and benefits of aquaculture as well as commonly used fish-farming practices and recognize and appreciate the worldwide impact of fish farming.

- Discuss controversies surrounding aquaculture and describe its limitations.

- Understand how novel genes from aquatic species may be beneficial to the biotechnology industry.

- Provide examples of transgenic finfish and their uses.

- Discuss how triploid species are created.

- Explain why scientists are actively bioprospecting aquatic organisms around the world.

- Provide examples of medical and nonmedical products and applications of aquatic biotechnology.

- Define biofilming and explain how scientists are looking to marine organisms as a natural way to minimize biofilming.

- Describe how marine organisms may be used for biodetection and the remediation of environmental pollutants.

Horseshoe crabs (*Limulus polyphemus*) are a source of blood cells for an important test used to ensure that foods and medical products are free of harmful contaminants.

Given that water covers nearly 75% of the earth's surface, it should not surprise you that aquatic environments are a rich source of biotechnology applications and potential solutions to a range of problems. Aquatic organisms exist in a range of extreme conditions, such as frigid polar seas, extraordinarily high pressures at great depths, high salinity, exceedingly high temperatures, and low light conditions. As a result, aquatic organisms have evolved a fascinating number of metabolic pathways, reproductive mechanisms, and sensory adaptations. They harbor a wealth of unique genetic information and potential applications. In this chapter, we consider many fascinating aspects of **aquatic biotechnology** by exploring how both marine and freshwater organisms can be used for biotechnology applications.

FORECASTING THE FUTURE

There are many exciting potential future directions for aquatic biotechnology. Among the most likely areas of substantial progress will be global efforts in bioprospecting to find previously undiscovered aquatic species that may have commercial value. Aquatic environments such as the world's oceans are incredibly rich sources of biological diversity; to survive many marine organisms have to cope with extremes in pressure, temperature, salinity, and other environmental conditions. Hence the potential for finding novel proteins and other bioactive compounds of commercial value is very high. For example, over 40 companies are bioprospecting in the Arctic alone, hoping to exploit novel properties of Arctic organisms in the ice and the waters and soils below the ice. Aquaculture, particularly practices involving genetically modified species, will continue to be a rapidly developing area of aquatic biotech as countries strive to provide protein-rich food sources for their populations, including areas of the world where finfish and shellfish species are nonexistent or overharvested.

10.1 Introduction to Aquatic Biotechnology

Oceans have been a source of food and natural resources for millennia. However, human population growth, overharvesting of fish and other marine species, and declining environmental conditions have caused the collapse of some fisheries, which puts pressure on many species and strains ocean resources. Although scientists have learned a great deal about ocean biology, the vast majority of marine organisms—particularly microorganisms—have yet to be identified.

It has been estimated that greater than 80% of the earth's organisms live in aquatic ecosystems.

The obvious need to utilize the potential wealth of the majority of the earth's surface combined with increasing populations, human medical needs, and environmental concerns about our planet makes aquatic science an emerging frontier for biotechnology research. Aquatic ecosystems may contain the answers to a variety of global problems. In the United States, less than $50 million is spent annually for research and development in aquatic biotechnology. In contrast, Japan spends between $900 million and $1 billion annually. The success of Asian countries that have invested in basic science research on aquatic biotechnology and the financial value of the resulting products have encouraged other countries to invest significant amounts of time and resources in aquaculture.

In the United States, several research priorities have been identified to explore the seemingly endless possibilities of utilizing aquatic organisms. These include the following:

- Increasing the world's food supply
- Restoring and protecting marine ecosystems
- Identifying novel compounds for the benefit of human health and medical treatments
- Improving the safety and quality of seafood
- Discovering and developing new products with applications in the chemical industry
- Seeking new approaches to monitor and treat disease
- Increasing knowledge of biological and geochemical processes in the world's oceans

From fish farming to isolating new medical products from marine organisms, these aspects of aquatic biotechnology are among the range of issues we consider in this chapter. The next section discusses aquaculture, a large and rapidly expanding application of aquatic biotechnology.

10.2 Aquaculture: Increasing the World's Food Supply through Biotechnology

Two fishermen strain to hoist a net overflowing with catfish that are the ideal size and health for human consumption. Every fish in the net is a keeper. When prepared for market, these catfish will possess a highly desired consistency, smell, color, and taste. This scenario does not take place on a fishing dock or a com-

FIGURE 10.1 A Net Full of Farm-Raised Catfish Ready for Market Aquaculture is an effective way of raising large quantities of fish or shellfish that are ready for market consumption in a relatively short period of time. The catfish shown here are market-size catfish, called 103 by the U.S. Department of Agriculture (USDA), which grow significantly faster than other catfish.

mercial fishing boat but instead describes fish farmers clearing nets from a fish-rearing tank (**Figure 10.1**). The cultivation of aquatic animals, such as finfish and shellfish, and aquatic plants for recreational or commercial purposes is known as **aquaculture**. Specifically, marine aquaculture is called **mariculture**. Although aquaculture can be considered a form of agricultural biotechnology, it is typically considered a form of aquatic biotechnology. In this section, we primarily discuss farming of both marine and freshwater species of finfish and shellfish.

The Economics of Aquaculture

The global population of 6.9 billion people is projected to increase to 9.3 billion by 2050. And people with higher standards of living have a tendency to eat more meat and seafood. Worldwide demand for aquaculture products is expected to grow by 70% during the next 30 years. In fact, it has been estimated that, in the near future, worldwide demand for seafood of all kinds will exceed what wild stocks can provide by approximately 50 to 80 million tons. Overharvesting, loss of habitat, and depressed commercial fishing industries are all contributing to the decline in production of wild seafood stocks. According to the Food and Agriculture Organization (FAO) of the United Nations, approximately half of all fish stocks have been deemed fully exploited and another 30% or so deemed overexploited, depleted, or recovering. If demand continues to rise and wild catches continue to decline, we will see a deficit of consumable fish and shellfish. Aquacul-

ture together with better resource-management practices will, in part, overcome this problem.

Worldwide aquaculture production in the year 2000 was approximately 33 million metric tons, having more than doubled since 1984. Recent estimates suggest that close to 50% of all fish that humans consume worldwide are produced by aquaculture and farmed fish are expected to eventually overtake beef products as food. The market for fish captured by conventional commercial fishing methods is approximately 92 million metric tons per year. Total world fisheries, aquaculture plus capture fisheries, now produce more than 125 million metric tons annually. Some sources estimate that aquaculture has already surpassed cattle ranching in terms of the volume of food produced. China, where aquaculture has been practiced for thousands of years, is the single largest producer, accounting for approximately two-thirds of the world's seafood supply, much of it through aquaculture.

Aquaculture in the United States is an industry grossing more than $1 billion a year, which provides only about 5% of the world's seafood supply. Aquaculture production in the United States has nearly doubled over the last 10 years and produces more than 100 different species of aquatic plants and animals. This increase is expected to continue, and similar increases in aquaculture are occurring globally. For example, over half of the salmon sold worldwide is produced by aquaculture, with a nearly 50-fold increase in production of fish-farmed salmon over the past two decades. Aquaculture in the United States became a major industry in the 1950s, when catfish farming was established in the Southeast; today, there are aquaculture facilities in every state. Some of the most successful examples of the business potential of U.S. aquaculture include the Alabama and Mississippi Delta catfish industry, salmon farming in Maine and Washington, trout farming in Idaho and West Virginia, and crawfish farming in Louisiana. Similarly, Florida, Massachusetts, and other states have established successful shellfish farms that have brought benefits to struggling commercial fishers. For instance, Cape Cod, which houses approximately 75% of the Massachusetts aquaculture industry, has successfully raised a number of different shellfish species, including quahogs, soft-shelled clams, blue mussels, and oysters.

Many other countries are actively engaged in aquaculture practices. The United States ranks 13th in total aquaculture production. China is the world's leading producer followed by India, Vietnam, Indonesia, Thailand, Bangladesh, Norway, Chile, the Philippines, Japan, Egypt, and Myanmar, Greek farms are the leading producers of farmed sea bass in the world, producing 100,000 tons each year, which account for over 90% of aquaculture production in Europe. Norway is a leading producer of salmon. Canada produces more than 70,000 tons of Atlantic and Pacific salmon with a yearly crop value of approximately $450 million. The rapid growth and success of the Chilean aquaculture industry has many countries asking, "If Chile can do it, why can't we?" As a result, expanding markets are under way in Algeria, Argentina, Scotland, Iceland, the Faroe Islands, Ireland, Russia, Indonesia, New Zealand, Ecuador, Colombia, Peru and many other nations.

Many of the countries most actively engaged in developing aquaculture industries are doing so because local waters have been overfished to the point where natural stocks of finfish and shellfish have been severely depleted. Through aquaculture, markets can be created in areas where the natural resources have been lost. Aquaculture also provides the benefit of creating seafood industries in areas of the world that, because of geographic location, are not normally known for their fisheries. For instance, a successful fish-farming industry has been developed in the deserts of Arizona, where recycled water from aquaculture is also being used to irrigate crop fields. Fish are even being raised in irrigation ditches. Similar efforts are under way in the arid areas of Australia.

In theory, increased productivity in the raising of fish should lead to decreased retail prices for consumers. In some ways fish farming is more economical than animal farming or commercial fishing. For example, it takes approximately 7 pounds of grain to raise 1 pound of beef, but less than 2 pounds of fishmeal are needed to raise approximately 1 pound of most fish.

TOOLS OF THE TRADE

Fishing for Fish Genes

Studies are under way to learn about the genomes of many different species of commercially valuable fish that are desirable food sources. From tuna to salmon and a variety of shellfish, scientists are attempting to identify genes that contribute to properties such as growth rates, taste, color, and disease resistance. Selective breeding approaches have been popular for producing fish with the desired characteristics. But molecular biology techniques are being used as "tools" to identify specific genes that scientists can use to produce transgenic finfish and shellfish with enhanced properties that will make certain species most attractive to seafood lovers. So how do scientists "fish" for these genes? In some cases, whole genome sequencing studies are under way, but one common approach to this question involves a technique called **differential display PCR.**

This technique is based on comparing DNA or RNA from two different tissue samples or organisms using primers that bind randomly to sequences within a genome. It allows scientists to identify categories of genes that are expressed "differentially"—that is, genes produced in one tissue (or organism) and not another. For example, suppose scientists were trying to identify genes that might contribute to rapid growth rates in shrimp. You might hypothesize that larger shrimp have different genes or greater levels of gene expression for one or more genes involved in growth rate. To discover these genes, you would isolate mRNA from small shrimp and large shrimp, reverse transcribe the RNA to make complementary DNA, and then amplify these samples by differential display PCR. The PCR products are separated by gel electrophoresis. If large shrimp expressed mRNA for unique genes involved in rapid growth, then PCR products for these genes would be detected in samples from large shrimp but not small shrimp. Scientists can then determine the DNA sequence of these PCR products to figure out which genes they have identified. Some of these genes may then be used to create transgenic animals with enhanced growth capabilities.

This method has been used by molecular biologists in many areas of research. For instance, cancer researchers use this approach to identify genes expressed in cancer cells that are not expressed in normal cells.

As another example, farm-raised catfish grow nearly 20% faster in fish farms than catfish in the wild, and they are ready for market sale in approximately 2 years. For fish species that are fed genetically engineered feed at around 10 cents a pound, the return is often 70 to 80 cents a pound on the raised fish—a good return on investment. But as you will learn in this chapter, the economics of aquaculture are not always so favorable.

Although worldwide increases in aquaculture are likely to continue, it is unlikely that fish farming will fully replace wild catches in the near future. One reason is that many species are not amenable to fish farming. Because open-water species such as tuna and swordfish roam over large areas of water and require lots of food, they cannot be raised in fish farms. We discuss other barriers to aquaculture later in this chapter. The purpose of presenting the statistics in this section is not to provide you with a remedy for insomnia but rather to help you develop a perspective on the current value and importance of aquaculture and its future potential.

Fish-Farming Practices

In many ways, aquaculture is an extension of the conventional land-based agricultural techniques that have been practiced for decades. But culturing organisms for human consumption is only one purpose of aquaculture. Organisms are raised for many other reasons, such as providing bait fish for commercial and recreational fishing. Small fish such as anchovies, herring, and sardines are harvested to make fishmeal and oils used in animal feed for poultry, cattle, swine, and other fish. Growing pearls, culturing species to isolate pharmaceutical agents, breeding ornamental fish including goldfish and rare tropical fish, and propagating fish to stock recreational areas are among the many applications of aquaculture.

For example, in New Jersey, fisheries biologists breed hatchery-raised trout to create a "put-and-take" fishery for trout. Each spring the state stocks over 700,000 rainbow trout, brown trout, and brook trout in streams, rivers, ponds, and lakes throughout the state. Many streams and rivers will not sustain trout throughout the year because they get too warm in the summer or because they are of poor water quality for trout. But through stocking ("put"), anglers are encouraged to keep what they catch ("take") for table fare. Many other states have similar stocking programs for a number of warm- and cold-water species, including panfish, bass, catfish, muskellunge, and hybrid striped bass, the last of which were created by breeding freshwater white bass and saltwater striped bass. Such programs also provide angling opportuni-

ties for people who live in urban areas with relatively little access to rural fisheries.

As shown in Table 10.1, a variety of fish and marine organisms have been cultured successfully. In many countries, this list will soon be expanded. For example, marine species such as lobsters and crabs have been farmed on a limited basis but are not yet widely available commercially.

Aquaculture practices vary widely depending on factors such as the species being farmed, life cycles of the species, environmental requirements for growth, and the length of time needed to achieve maturity for market sale. For fish such as salmon and trout, eggs (roe) and milt (sperm) are manually harvested from breeder stocks of adult fish. In an attempt to control the quality and health of the fish produced, parents used for breeder stocks are often fish that display the desired growth characteristics and physical appearance (**Figure 10.2**). Fish growth rate, health, quality, and the color of the meat are among the many features considered in the selection of breeder fish.

Eggs are fertilized in small tanks or containers and hatched embryos, called *fry*, are eventually transferred to larger aquarium tanks with flowing water. Most of

TABLE 10.1   IMPORTANT AQUACULTURE ORGANISMS IN THE UNITED STATES

Freshwater Organisms*	Marine Organisms*
Catfish	Oysters
Crawfish	Clams (hard- and soft-shell)
Trout	Marine shrimp
Atlantic salmon	Flatfishes (turbot, flounder)
Pacific salmon	Sea bass
Bait fish (fathead minnows, golden shiners, mullet, sardines)	Mussels
	Abalone
Tilapia	Quahogs
Hybrid striped bass	
Sturgeon	
Carp (silver, grass)	
Yellowtail	
Bream	
Goldfish	

*Species are ranked approximately according to total production in metric tons from highest to lowest quantity. Production of freshwater organisms exceeds that of saltwater organisms; therefore, the listing of species on the same row does not indicate equal production.

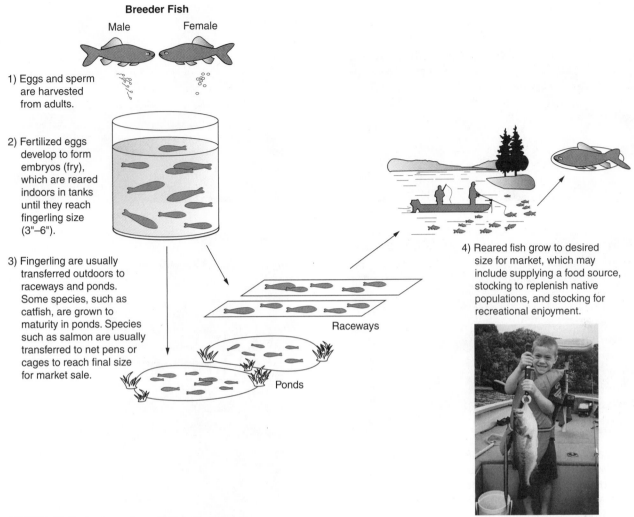

1) Eggs and sperm are harvested from adults.

2) Fertilized eggs develop to form embryos (fry), which are reared indoors in tanks until they reach fingerling size (3"–6").

3) Fingerling are usually transferred outdoors to raceways and ponds. Some species, such as catfish, are grown to maturity in ponds. Species such as salmon are usually transferred to net pens or cages to reach final size for market sale.

4) Reared fish grow to desired size for market, which may include supplying a food source, stocking to replenish native populations, and stocking for recreational enjoyment.

Breeder Fish

Male   Female

Raceways

Ponds

FIGURE 10.2 An Overview of Fish Farming

this culturing initially occurs indoors in a "nursery." Fish typically leave the nursery when they reach *fingerling* size (several inches in length, about the length of a human finger). Fingerlings, sometimes called *smolts*, are usually moved into concrete tanks called *raceways,* which may be indoors or outdoors (**Figure 10.3**). Raceways are often designed as a series of interconnected tanks that allow for the continuous flow of water through the tanks. This is particularly important for species like salmon and trout, which are accustomed to battling the current to survive in streams and rivers. By mimicking environmental conditions, aquaculturists attempt to allow these species to develop physical characteristics similar to those of wild fish, such as the development of musculature and survival instincts.

Some fish species are raised in raceways until they reach market size, or they may be transferred to small shallow ponds for further growth before being harvested. For certain species including salmon, after they reach a modest size, they can be raised to maturity in

net pens, cages, and other enclosures that are placed in lakes, ponds, or estuaries. Shellfish farmers often employ rack or cage systems that are "seeded" with large numbers of tiny nursery-raised immature shellfish such as oysters or clams that adhere to the racks. Racks are then placed in estuarine environments to allow the shellfish to grow to maturity in a natural marine environment. Typically such racks allow for much greater production of shellfish than natural habitat. Most aquaculturists are constantly modifying culturing techniques in an effort to increase the yield of a species in a given area and to minimize the expense-to-profit ratio.

Salmon farming is an excellent example of the fish-rearing process. Eggs and sperm milked from adults bred to grow fast are used for fertilization, and the resulting embryos are raised for 12 to 18 months in tanks until they reach the fingerling stage. Fingerlings are usually transferred to mesh-enclosed pens placed in oceans or bays close to the shoreline. Throughout this process, the salmon diet often consists

(a)

(b)

(c)

(d)

FIGURE 10.3 Raceways, Ponds, Net Pens, and Oyster Racks A wide variety of approaches are used for farming fish and shellfish, including (a) raceways, (b) ponds, (c) net pens for raising salmon, and (d) rack systems for oysters and other shellfish. Many of these approaches are often used in sequence as fish grow from small fingerlings to market size.

of pellets made from ground-up herring, anchovies, fish oil, and animal and poultry by-products (bone, excess meat, and the like). The food may also contain antibiotics, as well as additive dyes to make the salmon flesh pink, and fish may be placed in pesticide baths to remove sea lice and other parasites. Throughout the culturing process, aquaculturists can change the taste of farm-raised species by experimenting with food sources such as vegetable-based foods versus foods derived from fish products. Ultimately, unblemished fish of the appropriate size are packaged for market sale. Figure 10.2 provides an overview of fish farming.

Innovations in fish farming

Many innovative approaches for raising fish are under development. For example, in West Virginia, fisheries biologists are taking advantage of an abundance of abandoned coal mines in the mountain state. The cool mountain groundwater and water from freshwater springs seeps into and fills many mines, providing high-quality environments for rearing cold-water species such as rainbow trout and arctic char. This approach also provides an important use of abandoned mines that would typically serve no purpose otherwise.

Some companies have experimented with **polyculture,** also called *integrated aquaculture;* that is, raising more than one species in the same controlled environment. Raising species that have different nutrient requirements and feeding habits is one way to optimize water resources. Polyculture can involve different fish species cocultured, that is, fish and shellfish raised together, as well as animal and plant polyculture. For example, raising carp in the presence of plants such as lettuce is an effective approach. Roots from the lettuce absorb nutrients and fish waste products (such as nitrogen) from the water and use it as fertilizer to support lettuce growth. By minimizing the cost of artificial filtration and removal of wastes, polyculture limits the amount of wastewater effluent for discharge into the surrounding environment. Another aquaculture approach involves the use of **hydroponic systems**. These systems are small-volume water-flowing systems in which vegetables (like tomatoes and broccoli) or herbs (like basil and chives) are cultured in racks through which wastewater from fish tanks can flow. Polyculture and hydroponic systems are frequently used together when fish such as catfish, carp, and tilapia are being raised (**Figure 10.4**).

Improving Strains for Aquaculture

Many fish-farming efforts involve activities designed to improve certain qualities, such as growth rate, fat content, taste, texture, and color of the finfish or shellfish being raised. As we discuss in Section 10.3, some of the strategies used to improve strains for aquaculture involve molecular alterations in fish genetics. Scientists are studying the mechanisms of gene expression necessary for fish reproduction, growth, and development; techniques for the **cryopreservation** and storage of tissue samples and cells at ultralow temperature (usually between –20°C and –50°C); and delivery systems for hormones designed to induce spawning and improve the growth of fish.

Here we consider a few examples of how strains of finfish or shellfish can be improved for aquaculture by discussing how selective breeding may be used to raise species with desirable characteristics. Fish farmers and scientists work together to identify individuals of a given species with desirable characteristics. For instance, in a population of catfish, not all individuals grow at the same rate or have the same body characteristics, such as muscle mass compared with the percentage of body fat. Scientists can use a number of techniques to identify fish that show fast growth rates and heavy muscle mass. They can use ultrasound machines to measure muscle mass as a means to estimate fillet yield. Fish that show the highest yield are then bred to produce new generations with increased muscle mass.

Scientists at the U.S. Department of Agriculture Catfish Genetics Research Unit in Mississippi have selectively bred a variety of catfish called USDA 103 (a name worthy of a government-issue catfish!) that grows 10% to 20% faster in ponds, trimming the time from birth to market to between 18 and 24 months (refer to Figure 10.1). It has been reported that some USDA 103 catfish can reach sexual maturity a full year faster—by 2 years of age—than other species. On the negative side, USDA 103 catfish consume more feed than other varieties of catfish.

In Arizona, researchers have used selective breeding to identify shrimp that are acclimated to growing

FIGURE 10.4   Polyculture of Lettuce and Tilapia in a Hydroponic System The hydroponic system of lettuce growing in racks above a tank containing tilapia (inset). Water from the tilapia tank, which contains nitrogenous wastes from the tilapia, is cycled through the racks of lettuce. Lettuce will use wastes in the water as fertilizer, thus making efficient use of the water resource.

in low-salinity water. Starting with larval shrimp grown in fresh water with added salt, scientists culture these shrimp through tanks of water with progressively lower salt concentrations. This allows for the farming of high-quality shrimp without the need for saltwater, and low-salt water can then be recycled to irrigate fields of fruits and vegetables. Thus shrimp can be raised in the desert of Arizona hundreds of miles from natural sources of saltwater.

Enhancing the Quality and Safety of Seafood

Aquaculture can be combined with cutting-edge techniques in molecular biology to create finfish and shellfish species that have the color, taste, and texture consumers want. In addition, scientists are working to make seafood safe, so that it is free of pathogens and chemical contaminants.

Igene Biotech of Columbia, Maryland, has used gene-cloning techniques to mass produce **astaxanthin**, the pigment that gives shrimp their pink color. By including recombinant astaxanthin in fish feed, scientists can create salmon and trout with a rainbow of hues in flesh color. Astaxanthin is also thought to have potential value as an antioxidant to be used in nutritional supplements for humans. Most people like salmon with a pink color. In fact, consumer polls suggest that many people believe the redder the salmon, the higher its quality. Dark red–colored salmon is highly prized in the best sushi bars in Japan.

Astaxanthin has no effect on the taste of salmon, but farmers want to grow what consumers prefer. To help fish farmers produce the fish their consumers demand, the Swiss drug company Roche Holding AG, a leading producer of astaxanthin, distributes a "salmofan," which resembles a paint-color chart. Aquaculturists can pick shades ranging from light pink to dark crimson and then purchase food with the concentration of astaxanthin that will produce fish with the flesh color they desire.

A variety of approaches to enhance seafood quality and safety are currently being used, and many more are in development. From detecting contaminated seafood to enhancing the taste and shelf life of seafood, aquatic biotechnologists are working to apply innovative technologies. Agilent Technologies Inc. produces a restriction fragment length polymorphism (RFLP)-based DNA chip that can be used to identify fish species in food products. Port inspectors, government agencies, seafood processors, distributors, and others can use this technology to determine the fish species present in frozen and fresh samples of fish. Similar techniques are used to fight illegal fish-catching poachers by identifying species that were illegally harvested or intentionally mislabeled.

Marine scientists are using molecular biology techniques to identify and learn more about genes encoding toxins produced by marine organisms; this can help them to understand how these toxins cause disease and to minimize the negative health impacts of toxin exposure. Many molecular probes and PCR-based assays are being developed for the detection of bacteria, viruses, and a host of parasites that infect finfish and shellfish. Gene probes have been developed for detecting viral diseases in shrimp and companies are developing gene probes for detecting and assessing the effects of environmental stresses on fish and shellfish. A handheld antibody test kit has been developed to detect *Vibrio cholerae* in oysters. It uses antibodies to detect proteins from *V. cholerae*, the bacterium that causes **cholera**—a very serious illness characterized by severe diarrhea; in severe cases, this can lead to dehydration and death. Cholera is particularly a problem in developing nations with water supplies that are contaminated because of inadequate sewage treatment facilities. This kit has been widely and successfully used in South America to detect contaminated seafood and minimize cholera infections. Similar approaches have been used in Hawaii to develop a dipstick monoclonal antibody test for ciguatoxin, which affects finfish from tropical areas that are sold in the aquarium industry.

Aquatic biotechnologists are interested in developing detection kits or vaccines for a number of pathogens that pose serious threats to fish raised by aquaculture. Several aquaculture companies are working to develop a vaccine for infectious salmon anemia, a deadly condition that causes internal bleeding and destroys internal organs in salmon and leads to the death of hundreds of thousands of salmon worldwide each year. Similar work is being carried out to fight against sea lice, such as *Caligus elongatus*. This tiny parasite attaches to salmon, feeding on salmon blood and creating exposed lesions that render salmon susceptible to deadly infections. Scientists in California, Idaho, Oregon, and Washington have developed a subunit vaccine against the virus causing infectious hematopoietic necrosis (IHN), which is responsible for the loss of large numbers of commercial trout and salmon each year. The vaccine was developed from an IHN protein expressed in bacteria. This vaccine, which protects fish against infection by IHN, is injected into these animals to stimulate the production of antibodies against the IHN virus.

Barriers and Limitations to Aquaculture

Although aquaculture is firmly established as a worldwide application of aquatic biotechnology with enormous potential to provide food in unique ways, there are concerns and obstacles. Not all species are ideally suited for aquaculture. This is particularly true of many highly prized marine species. Water quality issues can be a problem. For some species, it is difficult to maintain water with the proper flow rate to deliver adequate concentrations of nutrients and remove rapidly accumulating waste products properly. This is especially true for marine species that require large areas of ocean for roaming and do not survive when confined to small living spaces. In addition, marine organisms often have long, complicated life cycles involving a series of larval stages, each of which may have different food requirements, before marketable size can be achieved. To raise adults of these species, the loss of young fingerlings can be very high and thus prohibitively costly. Aquaculture is easiest with species that have few larval stages.

Aquaculturists are constantly faced with the problem of minimizing the effects of disease on their fish populations. Given the dense, crowded conditions in which many fish are raised, farmed fish are often more susceptible than native stocks to stress and disease caused by bacterial and viral pathogens as well as lice. Because there is less genetic diversity and disease resistance in farmed fish, infections can spread rapidly. Most fish in a farmed population will have the same resistance and susceptibility to disease; therefore a large supply of fish can be wiped out quickly if disease is not controlled.

Some are concerned that the aquaculture industry may be consuming excessive amounts of wild bait fish, such as anchovies and herring. Although some bait fish are raised by aquaculture, in many areas of the world the bait fish used to make fishmeal are still netted from wild stocks. For every pound of salmon, several pounds of other typically wild-caught fish—such as anchovies, herring, and mackerel—are used to raise the salmon. Growing 1 pound of salmon can require 3 to 5 pounds of wild-caught fish. Scientists are looking for ways to change the carnivorous feeding habits of some farmed fish by switching them to vegetable-based diets, but many aquaculturists are concerned about the quality of these fish. Supporters of vegetable-based diets claim that some people may like salmon that tastes less fishy.

Runoff of animal feces and wastes from traditional land farming is a problem that has significant impacts on the environment, long of concern to environmentalists. Similarly, they are concerned about pollution from fish-farming industries. The waste-laden effluent water from aquaculture operations—which contains untreated feces, urine, and uneaten food—is typically discharged into natural waterways. These wastes are rich in nitrogen and phosphorus and can lead to algal blooms and other problems. Dying algae rob water of oxygen, which can lead to fish kills. Fish wastes also harbor strains of bacterial pathogens such as *Salmonella* and *Pseudomonas*, which can affect human health; however, the likelihood of disease transmission from fish wastes to humans has not been well established. At present, it is generally thought that waste production from aquaculture has a small overall effect on water quality, but this may change as aquaculture efforts increase worldwide.

The extermination of "pest" species at aquaculture facilities is another concern. A number of fish-eating birds (such as cormorants, pelicans, herons, egrets, and gulls) as well as mammals (such as seals and sea lions) can be subjected to authorized and unauthorized capture and extermination. However, many facilities employ nonlethal methods (including visual harassment with the use of lights, reflectors, scarecrows, human presence, and auditory devices such as predator distress calls, sirens, and electronic noisemakers) to deal with these "pest" species. Underwater acoustic and explosive devices may also be used, along with perimeter fencing and protective netting to discourage predators from feeding at fish farms.

Concern has also been raised about the discharge of chemicals commonly used in many aquaculture operations. These include antibiotics, pesticides, herbicides (plant killers), algicides (algae killers), and chemicals used to control parasites. In addition, residual chemicals from other antifouling compounds (such as those used to reduce the growth of barnacles, algae, and other organisms) and anticorrosants may be absorbed by fish and shellfish and passed to other marine organisms and humans. A comprehensive survey of farmed salmon has found that they contain significantly higher levels of polychlorinated biphenyls (PCBs) and other, similar compounds correlated with increased risk of cancer than their wild relatives. This is a major concern for the industry, and fish farmers are working on ways to reduce pollutants in farmed salmon.

A number of federal laws regulate aquaculture and its potential effect on the environment, including the Clean Water Act, which regulates discharges including aquaculture effluent; the Migratory Bird Treaty Act, which protects birds that may pose a threat to aquaculture; the Marine Mammal Protection Act, which prohibits the killing of marine mammals that may be predators at aquaculture facilities; the Federal Insecticide, Fungicide, and Rodenticide Act, which governs use of these substances on crops, including fish; and the Food, Drug, and Cosmetic Act, which oversees approvals for drug use in fish farming and governs seafood safety.

The visual effects of aquaculture on the landscape are also problematic. For instance, in some areas of Scotland, there are concerns that the abundance of shellfish cages along the Scottish coastline is damaging the natural aesthetic appeal of coastal landscapes. A very serious issue raised by both aquaculturists and naturalists is the potential threat to native species by farm-raised fish that accidentally escape into the wild (refer to the "You Decide" box, on the next page).

The Future of Aquaculture

What lies ahead for the aquaculture industry? Much of its future direction will involve overcoming some of the barriers and limitations described previously. To reduce concerns about overfishing baitfish populations for use in fishmeal, many aquaculturists are exploring ways to grow species on different food sources that require little or no wild fishmeal and oils. Advances in fish farming will likely minimize effluent discharges from aquaculture facilities and reduce the use of antibiotics, pesticides, and other chemicals used to treat farmed species. It is also likely that polyculture approaches will become more popular as aquaculturists attempt to maximize water resources.

As we discuss in the next section, molecular genetic approaches will have a powerful impact on aquaculture in the future. Using many of the strategies presented in this chapter, along with recombinant

YOU DECIDE

The Risks of Fish Pollution and Genetic Pollution: Controversies of Aquaculture and Genetically Engineered Species

The term *fish pollution* describes the escape of aquaculture species that can harm natural ecosystems by altering or reducing biodiversity; introducing new parasites; competing with native species for food, habitat, and spawning grounds; interbreeding with native stocks; and destroying habitats. Farm-raised fish do escape. Many examples of fish pollution and its effects have been documented. For example, in some waterways in Florida, tilapia—a rapidly growing aquaculture panfish with a voracious appetite—have outcompeted native species such as bream for spawning areas and food. As a result, some native fish in tilapia-polluted areas have virtually disappeared. Similarly, nonnative species of Pacific white shrimp farmed along the Gulf Coast of Texas and the Atlantic Coast of South Carolina have escaped and are free swimming in these areas.

Evidence suggests that farm-raised Atlantic salmon are breeding in waters of the Pacific Northwest. Twenty-six stocks of Pacific salmon and steelhead are currently listed as endangered or threatened. Many scientists believe that the loss of these native species is due to the spread of lice and other parasites from farmed fish. In Norway, escaped farmed salmon comprise approximately 30% of the salmon in local rivers, and they are thought to outnumber resident salmon in local streams. A December 2000 northeaster in Maine—where salmon farming is the second-largest fishery behind lobstering—caused the uprooting of steel cages containing farm-raised salmon and released over 100,000 fish into Machias Bay. This accident is thought to have been the largest known escape of aquaculture fish in the eastern United States. Many are concerned that this spill will weaken the genetic potential of future generations of wild salmon in Maine rivers. The U.S. government has already listed wild salmon in eight Maine rivers as endangered as a result of concerns about the impact of aquaculture in the area. In some spawning rivers in New Brunswick and Maine, farmed escapee salmon greatly outnumber wild salmon.

The escape of farmed fish has caused environmentalists and other groups to rally for a moratorium on new fish farms in many areas of the world and for regulation of the aquaculture industry. They see fish pollution as a problem because farmed fish can threaten native fish species and affect natural aquatic ecosystems. Many aquaculture species are similar to domestic farm animals in that they have become reliant on humans and are poorly adapted to life in the wild. Escaped farmed fish can affect the gene pool of wild stocks by competing with native stocks during breeding and interbreeding with native species. Evidence also indicates that farmed salmon produce smaller and perhaps less fertile eggs than wild stocks.

Even stronger concerns have been raised about the escape of genetically modified species such as transgenics and triploids. Although most of these species cannot reproduce, no technique ensures that all modified fish will be sterile. Once in the wild, oceans cannot be drained to retrieve them. Transgenic stocks that escape and grow in areas outside of their intended growing range often do not grow as well as they do in their intended environment. As a result, if they breed with native species, mixed stocks often lose biological fitness (the ability to reproduce) and grow more slowly. By introducing transgenic species, it is possible to erode adaptations that have occurred over thousands of years and reduce the fitness of native stocks. Fish with enhanced growth capabilities may be able to outcompete native fish for food and spawning sites. In 2001, Maryland became the first state to ban the raising of genetically modified fish in ponds and lakes connected to other waterways.

Critics also cite historic examples of the spread of nonnative species as reason for concern. For example, European zebra mussels—thought to have entered the United States in ballast water of oceangoing ships entering the Great Lakes during the 1980s—have created significant problems. Colonies of these prolific shellfish have smothered habitat for other species; through filter feeding, they have caused the decline of native plankton in the Great Lakes, blocking pipes and growing on the hulls of ships. Costs associated with controlling zebra mussels in the Great Lakes region alone are estimated to exceed $400 million annually.

Even though the U.S. Food and Drug Administration (FDA) is involved in regulating the safety of transgenic fish as a food source, no clear policies exist for regulating the release of aquaculture species (including transgenics) in the United States. This must change if some of the concerns just described are to be addressed. The USDA is looking closely at ways to assess the risk of bioengineered species and the safety of introducing genetically modified organisms into the environment in both marine and nonmarine environments. Most scientists believe that aquaculture and genetic engineering are necessary to produce enough food for an increasing human population, but are the risks of these technologies greater than the risks of doing nothing? You decide.

DNA technology, aquatic biotechnologists will be working to increase growth and productivity of farmed species while improving disease resistance and the genetic composition of important food species. Molecular biology will have a strong impact on aquaculture, from identifying genes that control reproduction and the spawning of fish and shellfish to the creation of genetically modified species with desired characteristics.

Work on understanding conditions that affect the mortality of marine organisms during critical times in early development will lead to new rearing practices that can raise the productivity of farmed species. Many countries are actively involved in research related to the farming of other species that are difficult to raise by aquaculture. For instance, the South Carolina Department of Natural Resources is experimenting with aquaculture techniques for raising cobia, a very popular game fish of high commercial value. Cobia roam deep open-water areas of the ocean, making them difficult to raise under confined conditions.

So far we have taken a comprehensive look at the aquaculture industry—one of the oldest applications of aquatic biotechnology. In this section, we have seen that aquaculture is not unlike many industries: it benefits society but also poses some problems. In the next section, we explore an area that is in its infancy: the molecular genetics of aquatic organisms.

10.3 Genetic Technologies and Aquatic Organisms

Basic knowledge of gene expression and regulation in aquatic organisms and an understanding of genes involved in processes such as reproduction, growth, development, and survival in extreme environmental conditions will be critical for applications such as maintaining populations of endangered marine species, limiting populations of harmful species, and advancing the genetic manipulation of aquatic organisms.

In addition, pathogens that affect finfish and shellfish result in large financial losses each year. Scientists are working to improve the survival and growth of aquaculture species, and molecular biology is being used to learn more about the immune systems of aquatic species and their susceptibility and resistance to disease-causing organisms, including disease transmission and the life cycles of the pathogens themselves.

As you will soon learn, one of the ultimate applications of improving our understanding of the genomes of aquatic organisms involves manipulating the genetics of marine species.

The Discovery and Cloning of Novel Genes

The gene discovery process in aquatic organisms covers many interesting areas. For example, a great deal of research is dedicated to identifying new genes; learning about the environmental factors that control gene expression, such as the effects of extreme temperature and deep ocean pressures; identifying the molecular basis of unique adaptations; and studying genetic and molecular factors such as hormones that control the reproduction, growth, and development of aquatic organisms.

In addition to identifying novel genes, many research groups are involved in identifying mutations associated with diseases in fish. Eventually, such information will be used in the development of disease-free breeder stocks of fish with selected characteristics for U.S. hatcheries. By identifying deleterious genes that may have negative influences on fish growth, health, and longevity, scientists anticipate that many of these genes can be modified or removed to produce improved species for aquaculture. Discovering genes that can be used as probes for the identification of microscopic marine organisms such as phytoplankton and zooplankton—important food sources for many marine species—will help scientists look at environmental effects on the genetics of these microorganisms. This is a critically important issue for understanding how these microscopic organisms affect organisms higher up in the food chain.

Genes are being identified as DNA markers that can be used to distinguish wild stocks of fish from hatchery-reared stocks. Biologists are interested in using these markers to identify strains that are more resistant to disease and receptive to fish farming. Such markers will also play important roles as scientists attempt to determine the effects of farm-raised escapees on native stocks. Many of these marker genes are species-specific, and they have enabled fish and wildlife officers to apply PCR techniques in investigating cases of illegal harvest and retail of fish that are subject to quotas. For example, in the northeastern United States, blackfish are prized table fare. But the overharvesting of blackfish has resulted in strict size limits and restricted seasonal dates for legally harvesting blackfish. If wildlife officials encounter someone they suspect to be a poacher harvesting blackfish during the closed season but the poacher has only fillets in the boat and not intact carcasses, how can officials determine what species the fillets came from? Using blackfish-specific primers, PCR assays can be run on DNA isolated from fillets to determine if the fillets are from blackfish or not.

Cloning the gene for **growth hormone (GH)** is an excellent example of how the gene discovery process can lead to advances in marine biotechnology. We have discussed how the cloning of human GH led to treatments for dwarfism (Chapter 3). Recall that GH is a hormone produced by the pituitary gland that stimulates the growth of bone and muscle cells during adolescence. Cloning of the salmon GH gene has led to the development of **transgenic** species of salmon that demonstrate greatly accelerated growth rates compared with native strains. We consider the GH example in more detail in the next section. As another example, University of Alabama-Birmingham researchers have cloned the gene for molt-inhibiting hormone (MIH) from blue crabs. Molting (shedding) is triggered by a decrease in MIH, leading to the production of soft-shelled crabs that can be eaten whole. Current research is under way to block the release of MIH as a way to produce soft crabs on demand for the seafood industry.

Antifreeze proteins

One of the most successful examples of the identification of novel genes with promising applications involves genes for **antifreeze proteins (AFPs)**. A/F Protein, Inc., of Massachusetts, is a leader in the production of antifreeze proteins. Many of the first AFPs were isolated from bottom-dwelling fish species such as northern cod, which live off the coast of eastern Canada, and Antarctic fish called teleosts, which live in extremely cold ice-laden waters—some of the most severe environments on earth. Subsequently, AFPs have been isolated from a number of other cold-water species, including winter flounder (*Pleuronectes americanus*), sculpin (*Myoxocephalus scorpius*), ocean pout (*Marcrozoarces americanus*), smelt (*Osmerus mordax*), and herring (*Clupea harengus*). Interestingly, similar proteins have been discovered in insects, bacteria, and other organisms that can survive cold temperatures.

Several types of AFPs have been discovered. Structurally, most AFPs have extensive alpha helices and are held together by large numbers of disulfide bridges. AFPs function to lower the freezing temperature of fish blood and extracellular fluids, thus protecting fish from freezing in frigid marine waters. Seawater freezes at approximately −1.8°C. AFPs typically lower the freezing point of fish body fluids by approximately 2°C to 3°C. Currently, the majority of AFPs are isolated from fish blood. To meet the heavy demands that are expected for applications of AFPs, scientists are working on recombinant AFP production in bacterial and mammalian cells to accommodate anticipated future needs for AFPs worldwide.

AFPs protect living organisms from freezing in a variety of ways. They can bind to the surfaces of ice crystals to modify or block ice crystal formation, lower the freezing temperature of biological fluids, and protect cell membranes from cold damage. Because of these unique abilities, a number of innovative applications for these cryoprotective proteins are being developed. AFPs

are being used to create transgenic fish and plants with enhanced resistance to cold temperatures and freezing. For instance, salmon cannot produce antifreeze molecules; thus they die when exposed to near-freezing water. Waters off the coast of eastern Canada, too cold for wild species of salmon, are being considered as potential aquaculture habitat for freeze-resistant species of transgenic salmon containing AFP genes.

AFP gene promoter sequences are also being used in recombinant DNA experiments to stimulate the expression of transgenes, including GH in salmon. Transcription from AFP promoter sequences is stimulated by cold temperatures. By ligating genes of interest to the 3′ end of AFP promoter sequences, AFP promoters can be used to stimulate the transcription of these "downstream" genes under cold-water conditions (**Figure 10.5**). Thus AFP promoter gene constructs can serve as very effective transcription vectors to transcribe foreign genes, leading to the increased production of protein. Such constructs may be very effective for the genetic engineering of fish, as we discuss in the next section.

AFPs have been introduced into plants such as tomatoes to produce cold-hardy transgenic strains, but these are not widely available yet. Few commercial crop plants produce cryoprotective proteins as effective as AFPs from fish. For many popular crops such as wheat, coffee, fruit (e.g., citrus, apples, pears, cherries, and peaches), soybean, corn, and potatoes, crop damage as a result of cold temperatures is a problem worldwide.

Cryoprotection of human cells, tissues, and organs is a promising medical application of AFPs. As shown in **Figure 10.6**, cold storage of oocytes used for in vitro fertilization is one potential application. AFPs may prove useful for the storage of a number of human tis-

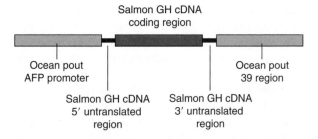

FIGURE 10.5 AFP Promoters Stimulate Gene Expression in Transgenic Fish A recombinant DNA plasmid construct containing an antifreeze protein gene promoter from ocean pout attached to the cDNA for salmon growth hormone can be used to produce rapidly growing fish for aquaculture. Transcription from the AFP promoter is stimulated by cold conditions; therefore transgenic fish containing this construct synthesize large amounts of growth hormone when they are raised in cool water. Increased production of growth hormones causes transgenic fish to grow faster than native nontransgenic strains.

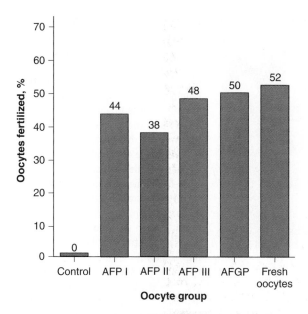

FIGURE 10.6   AFPs as Cryoprotectants of Human Tissues Bovine oocytes were incubated for 24 hours at 4°C in the absence (control) or presence of different types of AFPs or antifreeze glycoprotein (AFGP), a carbohydrate-rich type of AFP, and then fertilized with bovine sperm. Notice how the AFP- and AFGP-treated oocytes show fertilization rates similar to those of fresh oocytes, thus indicating that AFPs can provide cold protection of oocytes that will maintain their ability to be fertilized.

sues, including blood, and for the development of new protocols for the cryogenic storage of human organs such as the heart and liver prior to their use in transplantation surgery.

Finally, scientists are investigating ways that AFPs may be used to improve the shelf life and quality of frozen foods, including ice cream. Changes in the quality of frozen foods often occur during thawing and refreezing, as when you bring home food from the store and put it in your home freezer. It is possible that AFPs can be used to alter the ice crystallization properties of frozen foods to improve their overall quality. Scientists have even proposed using AFPs to control ice formation on aircraft and roadways.

As you can see, AFPs may be useful for a number of interesting applications. Undoubtedly the ocean harbors many species containing many other novel genes that may be exploited to benefit biotechnology in the future.

"Green genes"

An outstanding example of a research application involving a novel gene from an aquatic organism involves the bioluminescent jellyfish *Aequorea victoria*. *A. victoria* can fluoresce and glow in the dark because of a gene that codes for a protein called **green fluorescent protein (GFP)**. GFP produces a bright green glow when exposed to ultraviolet light. It has been estimated

that nearly three-fourths of all marine organisms have bioluminescent abilities. Bioluminescence is often used as a way for fish and other organisms to find each other in the dark environments of the ocean during mating activities. Genes for red, orange, and yellow fluorescent proteins have been cloned from sea anemones, expanding the color palette of proteins available for researchers. Previously, we mentioned that the fluorescence of some marine species is created by bioluminescent bacteria such as *Vibrio harveyi* and *Vibrio fischeri* (Chapter 5). Osama Shimomura, Martin Chalfie, and Roger Tsien shared the 2008 Nobel Prize in Chemistry for the discovery and development of GFP.

Scientists have taken advantage of the fluorescent properties of GFP and other similar proteins to create unique **reporter gene** constructs. Reporter genes allow researchers to detect the expression of genes of interest in a test tube, cell, or even a whole organ. As shown in **Figure 10.7** on page 248, reporter gene plasmids are created by ligating the GFP gene to a gene of interest and then introducing the reporter plasmid into a cell type of choice. Once inside cells, these plasmids are transcribed and translated to produce a fusion protein that fluoresces when exposed to ultraviolet light. Only cells producing the GFP fusion protein fluoresce. In this manner, these plasmids serve to detect or "report" where the gene of interest is being expressed.

This approach has been widely used to study basic processes of gene expression and regulation. Developmental biologists have used GFP-expressing embryonic stem cells to track their differentiation into different cell types during the development of embryos. Scientists have also used GFP reporter gene constructs in many innovative ways that promise to advance medical diagnostics and disease treatment (Figure 10.7b). For example, GFP genes have been used to pinpoint tumor formation in mice, to follow the progress of bacterial infections in the intestines, to monitor the death of bacteria following antibiotic treatment, and to study the presence of food-contaminating microorganisms in the human digestive tract (Figure 10.7c).

Cloning the genomes of marine pathogens

Marine biologists are interested in cloning the genomes of a variety of marine pathogens that affect wild and farmed species as a way to learn about genes these organisms use to reproduce and cause disease. In 2001, Chilean scientists deciphered the genome of the bacterium *Piscirikettsia salmonis*. *P. salmonis* infects salmon and causes a disease called rickettsial syndrome, which affects the livers of infected salmon, leading to their death. Combating this syndrome is a worldwide problem for salmon farmers. Currently, no effective treatment exists. In Chile's salmon industry alone, this

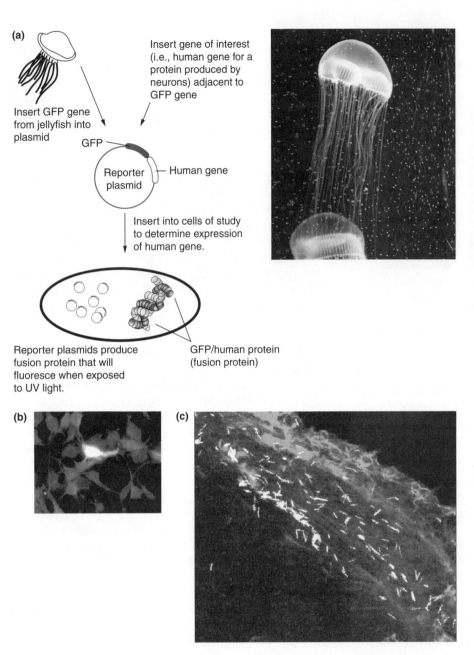

(a)

Insert GFP gene from jellyfish into plasmid

Insert gene of interest (i.e., human gene for a protein produced by neurons) adjacent to GFP gene

GFP

Reporter plasmid — Human gene

Insert into cells of study to determine expression of human gene.

Reporter plasmids produce fusion protein that will fluoresce when exposed to UV light.

GFP/human protein (fusion protein)

(b) **(c)**

FIGURE 10.7 The GFP Gene Is a Valuable Reporter Gene (a) GFP reporter gene constructs can be created in plasmids by ligating the GFP gene to a gene of interest. In this example, a human gene encoding a protein produced by neurons is attached to the GFP gene and the plasmid is introduced into cells in culture. These cells will then express mRNA molecules, which will be translated to produce a fusion protein in which GFP is attached to the cloned human protein. Cells producing this fusion protein will fluoresce when exposed to ultraviolet light, "reporting" the location of the expressed protein. (b) A GFP reporter gene is being used to detect a protein in human neurons that may be responsible for the neurodegenerative condition called Huntington's disease. (c) GFP reporter plasmids have also been used to detect disease-causing bacteria in the human digestive tract, such as the *Campylobacter jejuni* species shown here infecting human intestinal cells. *C. jejuni* is a common contaminant of chicken, causing approximately 2.4 million cases of food poisoning in the United States each year.

syndrome causes financial losses of an estimated $100 million per year. Armed with genome information about pathogens, scientists are trying to bolster the immune systems of farmed species to improve their resistance to pathogens. For example, it may be possible to inject genes or proteins from *P. salmonis* into salmon as vaccines to stimulate their immune systems to defend against infection by these bacteria.

Similar genome studies are under way to clarify the genetics of *Pfiesteria piscicida*, a toxic dinoflagellate that some scientists believe is responsible for major fish kills and shellfish disease in North Carolina estuaries. *P. piscicida* has also caused fish kills and disease in aquaculture facilities from the mid-Atlantic to the

Gulf Coast. In fish, *P. piscicida* causes lesions, hemorrhaging, and other symptoms that can lead to the death of infected fish. One reason this pathogen has attracted so much attention, bordering on hysteria in some coastal communities affected by large fish kills, is because evidence indicates that *P. piscicida* toxins can also have serious health effects on humans.

Scientists are working on ways to use molecular approaches in the battle against diseases and parasites that have essentially eliminated commercial fishing for oysters in several areas of the United States. Across the country, oysters have been under siege. As a result of overharvesting, pollution, habitat destruction, and parasitic and viral diseases, oyster stocks are nonexistent in many areas formerly known as rich sources of oysters for human consumption. Protozoans have caused substantial damage to oyster populations in the eastern United States. The parasites Dermo (*Perkinsus marinus*), SSO (*Haplosporidium costale*), and MSX (*Haplosporidium nelsoni*) have devastated eastern oyster (*Crassostrea virginica*) populations in many areas of the country, such as the Chesapeake Bay in Maryland and the Delaware Bay in New Jersey. Similar problems have occurred along the Gulf Coast and California.

Until recently, part of the difficulty in combating these diseases was a lack of sufficient numbers of parasites to study. Cell culture techniques have been used to overcome this problem, and researchers are now learning a great deal about the molecular biology of Dermo, SSO, and MSX. Molecular techniques have been used to learn about the life cycle of these parasites, and molecular probes are now available for their detection in the field and in aquaculture facilities. For instance, PCR-based approaches have proven very effective for the rapid and sensitive detection of Dermo, SSO, and MSX, allowing aquaculturists to screen and identify diseased oysters before widespread infections occur (**Figure 10.8**).

A greater understanding of genes involved in the oyster's immune system has also led to new strategies for transferring genes for disease resistance from species such as the Pacific oyster (*Crassostrea gigas*) into the more susceptible species of eastern oysters, such as *C. virginica*. In the future it may be possible to create transgenic species of oysters with genes that will provide resistance to damage by these and other parasites.

Genomics and aquatic organisms

Not surprisingly, genomics approaches are being used to analyze the genomes of a number of aquatic species. An international team of scientists has completed the sequencing of the sea urchin's genome. Sea urchins are echinoderms, a group of marine animals that include starfish and sea cucumbers. Urchins are believed to have originated over 540 million years ago. Their genome has been of great interest to molecular biologists because urchins share a common ancestor with humans. From this ancestor a superphylum of animals called the Deuterostomes arose; included in this phylum are echinoderms, humans, and other vertebrates. Because Deuterostomes are more closely related to each other than to animals not in this superphylum, genome scientists expected that sea urchins would be genetically similar to humans. Comparative genomic studies have shown that we share approximately 7,000 genes with urchins, including genes involved in hearing, balance, immunity, and development. We therefore expect that

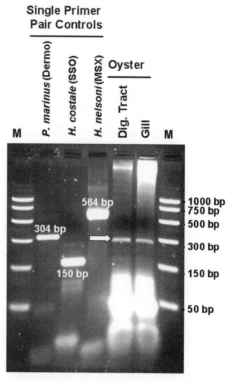

FIGURE 10.8   PCR Can Be Used to Detect DNA from Oyster Pathogens Dermo, SSO, and MSX are significant causes of mortality in oysters. PCR can be used to detect DNA from these pathogens in oyster tissues, helping scientists determine when these parasites are present in an effort to control oyster infection that can decimate large populations of these shellfish. In this example, notice that DNA samples from the digestive tract and gill of an oyster show the presence of DNA for Dermo (indicated by the arrow). M, DNA size markers.

urchins will continue to be very valuable experimental model organisms for molecular biologists studying both aquatic genomes and the human genome.

Genetic Manipulations of Finfish and Shellfish

Biologists throughout the world are using genetic techniques to create breeds of finfish and shellfish with such desired characteristics as improved growth rates and disease resistance. Table 10.2 shows examples of different species that have been genetically engineered for a variety of purposes.

Genetically engineered species: transgenics and triploids

We have discussed how *transgenic* organisms (transgenics) can be created. Transgenic fish, like other transgenics, contain DNA from other species, and interest

TABLE 10.2	GENETICALLY ENGINEERED FISH AND SHELLFISH

Fish Species	Shellfish
Atlantic salmon	Abalone
Bluntnose bream	Clams
Channel catfish	Oysters
Chinook salmon	
Coho salmon	
Common carp	
Gilthead bream	
Goldfish	
Killifish	
Largemouth bass	
Loach	
Medaka	
Mud carp	
Mummichog	
Northern pike	
Penaeid shrimp	
Rainbow trout	
Sea bream	
Striped bass	
Tilapia	
Walleye	
Zebrafish	

Source: Adapted from Goldberg, R., and Triplett, T. (2000): *Something Fishy*. Environmental Defense, www.environmentaldefense.org.

in transgenic fish has increased along with the aquaculture industry. Aquaculturists are interested in using recombinant DNA techniques to genetically engineer fish designed to grow faster and healthier.

As shown in Table 10.3, many species have been genetically modified for potential applications in the aquaculture industry. Foreign genes have been inserted into finfish and shellfish to accelerate growth, increase cold tolerance and disease resistance, and alter flesh qualities to improve table quality. Although transgenic strains of corn, soybeans, and tomatoes have been in use in the United States for years, no transgenic fish have been approved by the FDA for human consumption. A/F Protein, Inc., has requested

approval from the agency to market transgenic fish containing AFP genes. Many other companies are also seeking FDA approval for the sale of transgenic fish. Cuba is the most progressive country in the world when it comes to the use of genetically modified foods, particularly seafood. Genetically modified tilapia are already sold for human consumption in Cuba. Similarly, Norwegian companies have developed genetically modified farm-raised tilapia that grow nearly twice as fast as wild tilapia.

Aqua Bounty Technologies in Massachusetts has produced transgenic Atlantic salmon containing a GH gene and a gene regulatory sequence from Chinook salmon (*Oncorhynchus tshawytscha*)—a species that demonstrates more rapid growth and larger adult-size capabilities than most Atlantic salmon, as well as a regulatory sequence from an eel-like organism called ocean pout. Atlantic salmon normally stop growing in the winter; not so with these transgenics which grow nearly 400% to 600% faster than nontransgenic salmon (**Figure 10.9**). Such fish reach market size in 18 months instead of the traditional 30 months. Aqua Bounty has been trying for more than 10 years to get FDA approval for these salmon despite data indicating no difference in nutritional value, appearance, or taste of these fish compared with wild salmon. In 2009, the FDA classified genetically modified changes in animals as veterinary drugs. As of early 2011, the FDA was still considering approval of these transgenic salmon. A main risk assessment concern continues to be the ecological impacts of these salmon. Aqua Bounty says that all of these transgenic salmon are female and approximately 99.8% of them are sterile *triploids* (see the discussion of triploids beginning on the next page).

FIGURE 10.9 Transgenic Salmon Transgenic salmon, which overexpress growth hormone genes, show greatly accelerated rates of growth compared with wild strains and nontransgenic domestic strains of salmon. On average, these transgenic strains of salmon weigh nearly 10 times more than nontransgenic salmon.

TABLE 10.3 **EXAMPLES OF GENETICALLY MODIFIED SPECIES BEING TESTED FOR USE IN AQUACULTURE**

Species	Foreign Gene	Desired Effect and Comments	Country
Atlantic salmon	AFP	Cold tolerance	United States, Canada
	AFP salmon GH	Increased growth and feed efficiency	United States, Canada
Coho salmon	Chinook salmon GH + AFP	After 1 year, 10- to 30-fold growth increase	Canada
Chinook salmon	AFP salmon GH	Increased growth and feed efficiency	New Zealand
Rainbow trout	AFP salmon GH	Increased growth and feed efficiency	United States, Canada
Cutthroat trout	Chinook salmon GH + AFP	Increased growth	Canada
Tilapia	AFP salmon GH	Increased growth and feed efficiency; stable inheritance	Canada, United Kingdom
	Tilapia GH	Increased growth and stable inheritance	Cuba
	Modified tilapia insulin-producing gene	Production of human insulin for diabetics	Canada
Salmon	Rainbow trout lysosome gene and flounder pleurocidin gene	Disease resistance; still in development	United States, Canada
Striped bass	Insect genes	Disease resistance; still in early stages of research	United States
Channel catfish	GH	33% growth improvement in culture conditions	United States
Common carp	Salmon and human GH	150% growth improvement in culture conditions; improved disease resistance; tolerance of low oxygen levels	China, United States
Goldfish	GH AFP	Increased growth	China
Abalone	Coho salmon GH + various promoters	Increased growth	United States
Oysters	Coho salmon GH + various promoters	Increased growth	United States
Rabbit	Salmon calcitonin-producing gene	Calcitonin production to control calcium loss from bones	United Kingdom
Strawberries and potatoes	AFP	Increased cold tolerance	United Kingdom, Canada

Note: The development of transgenic organisms requires the insertion of the gene of interest and a promoter, which is the switch that controls expression of the gene. AFP, antifreeze protein gene (Arctic flatfish); GH, growth hormone gene.

In 2004, Yorktown Industries of Austin, Texas, made popular news headlines when they announced they had created the GloFish, a transgenic strain of zebrafish containing the red fluorescent protein gene from sea anemones. When ultraviolet light illuminates GloFish, they fluoresce bright pink; they were advertised as the first genetically modified pet sold in the United States. A number of antibiotechnology groups voiced their protests about this novelty use of genetic engineering.

Although transgenic species are the most common type of genetically modified marine species, a number of **polyploid** species have been created. Polyploids are organisms with an increased number of *complete sets* of chromosomes. As we have already discussed, most animal and plant species are **diploid** (abbreviated 2*n*, where *n* = number of chromosomes), meaning that they have two sets of chromosomes in their somatic cells and a single, or **haploid** number (*n*), of chromosomes in their gametes. In humans, the haploid number of chromosomes is 23; therefore the diploid number of chromosomes in human somatic cells is 46 (2*n* = 46). **Figure 10.10** is a simple representation of the differences among haploid, diploid, and polyploid species.

The majority of polyploid marine species created to date are **triploid** species. Triploid organisms contain three sets of chromosomes (3*n*). A number of different techniques can be used to create triploids. Triploids are usually derived by subjecting fish eggs to a temperature change or chemical treatment to interfere with egg cell division. Eggs treated in this way mature with an extra set of chromosomes.

For instance, treating eggs with **colchicine**, a chemical derived from the crocus flower (*Colchicum autumale*), is one common approach to creating polyploids (**Figure 10.11a**). Colchicine blocks cell division by interfering with the formation of microtubules necessary for cell division. As a result, the chromosomes replicate in treated egg cells, but these cells are incapable of dividing. Therefore these eggs have a diploid number of chromosomes, twice as many chromosomes as normal. Fertilization of such an egg cell by a normal haploid sperm cell results in a triploid organism. Another approach to producing triploids involves fertilizing a normal haploid egg with two spermatozoa (Figure 10.11b). The resulting embryo is a triploid containing one set of chromosomes from the egg and one set from each of two different sperm.

Polyploidy usually influences the growth traits of an organism. For instance, triploids typically grow more rapidly, in most species 30% to 50% faster, and larger than their normal diploid cousins. But most triploids are sterile because they produce gametes with an abnormal number of chromosomes; in some cases triploids may not produce any gametes. One of the first widely used triploid strains of fish was the triploid grass carp (*Ctenopharyngodon idella*). Grass carp have a voracious appetite for many types of aquatic vegetation. In the early 1960s, the U.S. Fish and Wildlife Service imported diploid grass carp, which are native to Malaysia. In the early 1980s, triploid grass carp were developed by temperature-shocking eggs to create diploid eggs and fertilizing these eggs with normal haploid sperm. Triploids grow rapidly and can exceed 25 pounds. Because of their ability to consume large amounts of vegetation, grass carp became very popular for controlling the growth of aquatic weeds in freshwater ponds and lakes throughout the United States. Triploid grass carp became an instant favorite among pond and lake managers, who could stock these fish in lakes as a "natural" way to control weed growth, minimizing the need to use large doses of chemical herbicides.

Not everyone has been thrilled with triploid grass carp. In some waterways, because of grass carp over-

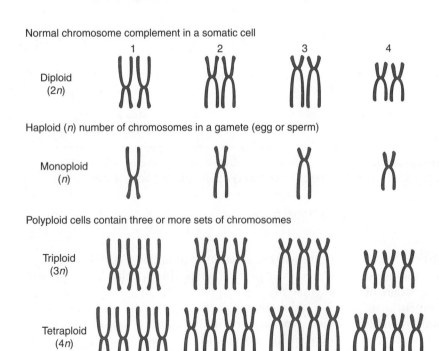

Normal chromosome complement in a somatic cell

Haploid (*n*) number of chromosomes in a gamete (egg or sperm)

Polyploid cells contain three or more sets of chromosomes

FIGURE 10.10 Polyploid Species Have Variations in Complete Sets of Chromosomes Somatic cells from many animal and plant cells have a diploid number of chromosomes, whereas gametes have a single set, or haploid number, of chromosomes. Polyploids contain three or more sets of chromosomes. Chromosomes from an organism with a diploid number (2*n*) of eight chromosomes are shown. Although tetraploid (4*n*) and pentaploid (5*n*) organisms can be created, the vast majority of marine polyploid species are triploids (3*n*).

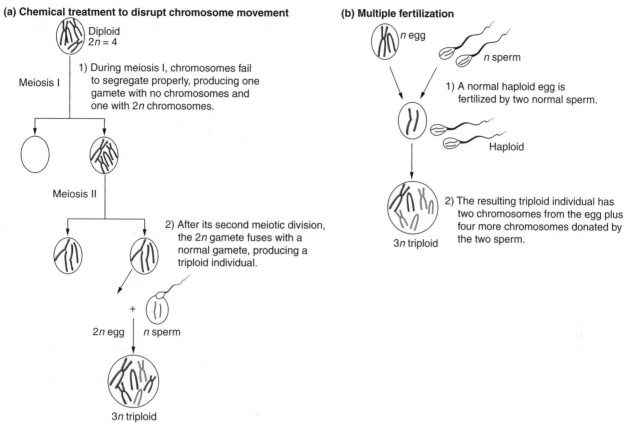

(a) Chemical treatment to disrupt chromosome movement

Diploid
$2n = 4$

Meiosis I

1) During meiosis I, chromosomes fail to segregate properly, producing one gamete with no chromosomes and one with $2n$ chromosomes.

Meiosis II

2) After its second meiotic division, the $2n$ gamete fuses with a normal gamete, producing a triploid individual.

$2n$ egg n sperm

$3n$ triploid

(b) Multiple fertilization

n egg

n sperm

1) A normal haploid egg is fertilized by two normal sperm.

Haploid

$3n$ triploid

2) The resulting triploid individual has two chromosomes from the egg plus four more chromosomes donated by the two sperm.

FIGURE 10.11   Producing a Triploid (a) Egg cells can be chemically treated to block chromosome movement during cell division to produce diploid eggs. (b) When these eggs are fertilized by a normal haploid sperm cell, triploid offspring are produced. A triploid can also be created by the fertilization of a normal haploid egg by two normal haploid sperm.

population (grass carp have few natural predators), they have been so effective at eating their way through aquatic vegetation that water quality has changed substantially. Dramatic decreases in weeds used as habitat by many sport fish, such as trout and bass, have led to a decline in fishing in previously productive fishing waters. Also, many triploids have escaped into waters adjacent to their original stocking, leading to the loss of vegetation in areas such as marshes and natural lakes not originally targeted for weed control. Finally, reproduction from these presumably sterile fish has also been documented.

Another polyploidy success story involves the triploid oyster, credited with reviving a diminishing oyster industry on the West Coast. Oyster production is seasonal because it is strongly influenced by weather, breeding patterns, and habitat. Not only has the development of a triploid strain of the Pacific oyster provided for year-round harvesting, but the triploid oysters grow substantially larger than their diploid cousins. Diploids normally store sugars in their tissue and then become lean in the summer while expending energy

to spawn. Lean oysters are undesirable for market sale. Because triploids do not spawn, they grow fat all summer, making them market-ready.

In the next section we introduce you to some of the valuable ways in which aquatic organisms are being used for important medical applications.

10.4 Medical Applications of Aquatic Biotechnology

Relatively few medical products have been derived from aquatic organisms, but this is rapidly changing. Aquatic scientists believe that oceans and freshwater habitats possess near limitless opportunities for the identification of medical products. In the future, it is anticipated that new and important classes of drugs will be derived from aquatic organisms and used for human benefit, and marine organisms may be used as biomedical models to understand, diagnose, and treat human diseases. Bioprospecting research on marine species is ongoing around the world, and we have many reasons to be optimistic that the waters of the world will yield novel medical treatments.

Bioprospecting to Isolate Medicines from the Sea

A wide number of marine species contain or are suspected to contain compounds of biomedical interest, including antibiotics, antiviral molecules, anticancer compounds, and insecticides. These species include sea sponges, members of the phylum Cnidaria (hydras, jellyfish, sea anemones, and a variety of corals), members of the phylum Mollusca (snails, oysters, clams, octopuses, and squid), and sharks, among many others. The wealth of marine organisms under study is broad and impressive. Let's consider a few potential medical applications of aquatic products.

Osteoporosis, a condition characterized by a progressive loss of bone mass, creates porous, brittle bones that can lead to fractures of the hip, legs, and joints, thus severely hindering an individual's lifestyle. Over 90% of the roughly 25 million Americans affected by osteoporosis are women. A common treatment for osteoporosis is estrogen therapy. This medication is ineffective for many women, and the long-term health effects of estrogen are a concern. Other individuals are treated with human recombinant **calcitonin**, a thyroid hormone that stimulates calcium uptake and bone calcification and inhibits bone-digesting cells called osteoclasts. Researchers have discovered that some species of salmon produce a form of calcitonin with a bioactivity that is 20 times higher than that of human calcitonin. Cloned forms of salmon calcitonin are now available for delivery as an injection form and a nasal spray.

The exquisite patterns of coral reefs around the world are created by the skeletons of corals. These structures partially consist of **hydroxyapatite (HA)**, an important component of the matrix that constitutes bone and cartilage in animals, including humans. The biotechnology company Interpore International has developed technology that allows HA implants to be cut into small boxes and used to fill gaps in fractured bones. These boxes are ultimately invaded by local connective tissue cells that speed repair. As a result, patients avoid the necessity of bone grafts derived from other parts of their bodies. These implants may also serve to fill bone material lost around the root of a tooth.

Similarly, a number of adhesives have been identified in glue-like resins produced by mussels and other shellfish. A favorite dish in many seafood restaurants consists of mussels (*Mytilus edulis*); these are hinged-shelled mollusks that live in harsh, physically demanding environments. They typically adhere to rocks or

pilings at the edges of oceans and bays. Day after day, these creatures are pounded by waves. They dry out during low tides, then as the tide rises, are submerged again and pounded by waves. How do they maintain their contacts to rocks and other structures without being crushed or pulled off the rocks? The answer lies in a unique form of protein-rich superadhesive called **byssal fiber** (**Figure 10.12**).

Byssal fibers are several times tougher and more extensible than human tendons, which themselves are tougher than steel. The adhesive and shock-absorbing elastic properties of byssal fibers absorb energy and stretch as waves tug away at the mussels. Although it would be cost-prohibitive to isolate byssal fibers from mussels directly (nearly 10,000 mussels would be required for 1 gram of adhesive), scientists are using recombinant DNA techniques to express the byssal fiber genes in bacteria and yeast to produce these adhesive proteins on a large scale.

Although still several years from development, byssal fiber proteins are being considered for a wide variety of diverse applications from automobile tires to shoes and from bone- and tooth-repair strategies to soft body armor for soldiers. Other potential uses include surgical sutures and artificial tendons and ligaments to be used as grafts.

Byssal fibers

FIGURE 10.12   Mussels Produce Unique Adhesives Mussels and other shellfish cling tightly to rocks and other surfaces by producing a unique type of adhesive called byssal fiber. Byssal fibers are remarkably strong. They can withstand tremendous physical forces such as those created by waves in the water where mussels live.

One category of drugs from the sea that appears to provide promising medical applications of aquatic biotechnology is the identification of anti-inflammatory, analgesic (pain-killing), and anticancer compounds that may be used to treat humans. Over a dozen different anticancer compounds have been isolated from marine invertebrates, particularly sea sponges, tunicates, and mollusks. Many of these compounds are in various stages of clinical trials that will ideally lead to new and effective drugs on the market. Several groups of researchers are studying venomous marine creatures in the hope of identifying substances that may be used to treat nervous system disorders.

Marine cone snails, a potentially lethal species, produce conotoxins, molecules that can target specific neurotransmitter receptors in the nervous system. In 2004, the FDA app roved the drug **Prialt** (ziconotide), a peptide conotoxin purified from the marine cone snail *Conus magus* by the Elan Corporation of Ireland. Conotoxins such as Prialt represent a promising new source of neurotoxins with the ability to act as strong painkillers by blocking neural pathways that relay pain messages to the brain. Prialt has been successfully used to treat chronic, severe forms of pain, such as back pain.

Yondelis (trabectedin or ecteinascidin), an antitumor drug isolated from the sea squirt *Ecteinascidia turbinate*, has been granted orphan drug status by the European Commission and the FDA for the treatment of soft tissue sarcomas and ovarian cancer. Yondelis is also in phase II trials for the treatment of prostate and breast cancer. Yondelis binds in the minor groove of DNA and inhibits cell division by blocking transcription and DNA repair proteins.

Researchers are also examining anti-inflammatory compounds found in coral extracts. Such compounds may lead to new treatment strategies for skin irritations and inflammatory diseases such as asthma and arthritis. Table 10.4 lists examples of medical compounds isolated from aquatic organisms.

Other researchers are developing culturing systems to provide adequate supplies of marine organisms such as single-celled plankton called dinoflagellates, which contain components useful in the treatment of tumors and cancers. Recently a marine invertebrate called *Bagula neritina* was shown to contain minute amounts of a compound that is active against certain types of leukemia. Because large amounts of this compound will be needed for human studies, molecular cloning techniques will be important if this compound is to be produced in mass quantities.

Similarly, the Japanese pufferfish, or blowfish (*Fugu rubripes*), has been getting a lot of attention lately (**Figure 10.13**). *Fugu* is famous for its ability to swallow water and "puff up" when threatened and to produce a potent nerve cell toxin called tetrodotoxin (TTX). TTX is one of the most toxic poisons ever discovered (nearly 10,000 times more lethal than cyanide). Hollywood has often depicted *Fugu* in movies, showcasing its ability to produce its deadly toxin. In Japan, *Fugu* is a prized and very expensive delicacy for many sushi lovers who enjoy the food quality of this tasty fish despite the risk (eating *Fugu* kills nearly 100 people, mostly in Japan, each year).

Scientists have used TTX to develop a greater understanding of how proteins called sodium channels help neurons to produce electrical impulses. TTX is a deadly poison because it blocks sodium channels and prevents nerve impulse transmission.

FIGURE 10.13 Marine Organisms Such as Pufferfish are Helping Scientists Discover New Ways to Treat Cancer and Chronic Pain in Humans

TABLE 10.4   EXAMPLES OF MEDICAL COMPOUNDS FROM AQUATIC ORGANISMS

Organism	Medical Product	Application
African clawed frog (*Xenopus laevis*)	Magainins	Antimicrobial peptides first discovered in frog skin. Used to treat bacterial infections.
Coho salmon (*Oncorhynchus kisutch*)	Calcitonin	Hormone that stimulates calcium uptake by bone cells. Used to treat osteoporosis.
Hammerhead shark (*Sphyrna lewini*)	Neovastat	Antiangiogenic compound (blocks blood vessel formation). Used to treat cancer.
Leech (*Hirudo medicinalis*)	Hirudin	Saliva peptide from leeches used as an anticoagulant to thin blood.
Marine cone snail (*Conus magus*)	Prialt (ziconotide)	Synthetic peptide toxin used as a pain reliever to treat chronic severe pain and arthritis.
Sea sponge (*Halichondria okadai*)	Eribulin	Chemotherapy compound for breast cancer treatment.
Sea squirt (*Ecteinascidia turbinata*)	Yondelis (trabectedin)	Antitumor agent developed for the treatment of advanced soft tissue sarcoma.

An understanding of how TTX affects sodium channels has led to the development of new drugs that are being tested not only as anesthetics to treat patients with different types of chronic pain but also as anticancer agents in humans. Researchers are also working on sequencing the pufferfish genome, which contains nearly the same number of genes as that of humans but is much smaller. *Fugu* also contains far less noncoding DNA (introns) than the human genome, so it is also considered an ideal model organism for studying the importance of introns.

A steroid called squalamine, first identified in dogfish sharks (*Squalus acanthias*), appears to be a potent antifungal agent that may be used to treat life-threatening fungal infections that can kill patients with conditions such as AIDS and cancer. Sharks rarely develop cancer (although it is a myth that sharks do not get cancer), and shark cartilage has been proposed to be a rich source of anticancer agents. Although no compounds from shark cartilage have demonstrated effectiveness in controlled clinical trials, shark cartilage extracts possess *antiangiogenic* compounds, and cartilage extracts are in various stages of clinical trials. **Angiogenesis,** or the formation of blood vessels, is a process that is often required for growth and development of many types of tumors. By blocking blood vessel formation, antiangiogenic compounds derived from marine species show promise for inhibiting the growth of certain tumors.

In addition, because many aquatic organisms live in harsh environments, scientists are optimistic that we can learn from the adaptations these organisms

have developed. For example, researchers are currently studying marine organisms that show tolerance to ultraviolet light; these may prove to be a source of natural sunscreens. As another example, alligator wounds are remarkably resistant to infection, even though alligators swim in microbe-infested waters. Scientists have found that alligators produce antimicrobial peptides capable of killing such microbes as MRSA (methicillin-resistant *Staphylococcus aureus,* which is responsible for about 70% of lethal infections in hospitals), which have become immune to modern antibiotics.

Even discarded crab shells from the commercial crabbing industry play a role in medical applications of aquatic biotechnology. The outer skeleton, or exoskeleton, of members of the phylum Arthropoda—which includes crabs, lobsters, shrimp, insects, and spiders—is a rich source of **chitin** and **chitosan**. These complex carbohydrates are structurally similar to cellulose, which forms the tough outer layer of the cell wall in plants. Cellulose is widely known as a source of dietary fiber. Similarly, chitin and chitosan are also sources of fiber. Eating vegetables and fruits to get fiber is much easier on your digestive tract than eating crab shells. Nonetheless, ground-up extracts of crab shell can be purchased as a powder in many nutrition stores. Many skin creams and contact lenses also contain chitin, and chitin has been used to create nonallergenic dissolvable stitches that appear to stimulate healing when surgeons place them in humans.

To date, few drugs from the sea have widespread use in the medical market; however, in the future, recombinant DNA technologies will lead to enhanced abilities to produce bulk quantities of the bioactive compounds typically found in very low concentrations in aquatic organisms.

10.5 Nonmedical Products

To further appreciate the potential of our world's waters, we now take a look at a number of aquatic products, from research reagents to food supplements, which have had an impact on the biotechnology industry.

A Potpourri of Products

We have discussed the importance of *Taq* polymerase, isolated from the archaebacteria *Thermus aquaticus*, found in hot springs, which allowed for the development of the PCR as a powerful tool in molecular biology (Chapter 3). The ocean also has proved to be an excellent source of enzymes and other products that have played an important role in basic and applied research. For example, bacteria living near hydrothermal vents (hot-water geysers on the ocean floor) have yielded a second generation of heat-stable enzymes for use in PCR and DNA-modifying enzymes, including ligases and restriction enzymes.

Other enzymes produced by marine bacteria possess a variety of interesting properties that may result in important applications in the future. Some enzymes are salt-resistant, which renders them ideal for industrial scale-up procedures involving high-salt solutions. We have discussed the role of the bioluminescent bacterium *Vibrio harveyi* in detecting environmental pollution (Chapter 5). Marine species of *Vibrio* produce a number of proteases, including several unique proteases that are resistant to the detergents used in many manufacturing processes. As a result, these detergent-resistant proteases may have potential applications for degrading proteins in cleaning processes, including their use in laundry detergent for removing protein stains in clothes. *Vibrio* is also a good source of **collagenase**, a protease used in tissue culturing. When scientists are looking to grow cells, such as liver cells in culture, they can use collagenase to digest the connective tissues holding cells together so the individual cells can be dispersed into cell culture dishes.

Another product of the sea is **carrageenan**, an ingredient in many preserved foods, toothpaste, and cosmetics. This sulfate-rich polysaccharide, extracted from red seaweeds, has been used in many products for over 50 years. There is a large family of carrageenan polysaccharides. They have the ability to form gels of varying densities at different temperatures. As a result, carrageenans have been used as thickening agents and for improving the way foods "feel" in our mouths when we eat them. Some of the most common applications of carrageenans include their use as stabilizing and bulking agents in chewing gum, chocolate milk, beers and wine, salad dressings, syrups, sauces, processed lunch meats, adhesives, textiles, polishes, and hundreds of other products. The Philippines is the world's largest producers of red seaweeds (Rhodophyta), from which many carrageenans are derived. Red algae are also prized for their use in nori, a paper-thin seaweed product used to wrap sushi. Improvements in farming seaweeds have allowed for increased production of other algal polymers, including agar and alginic acids, which are important research materials used, respectively, to make agar gels for plating bacteria and to create agarose gels for DNA electrophoresis (Chapter 3).

Biomass and Bioprocessing

One newly emerging area of marine biotechnology involves the exploration of marine biomass. A mat of aquatic weeds of algae represents biomass, as does a school of fish. As you know, plants are responsible for the production of oxygen through the process of photosynthesis, in which carbon dioxide and water are converted into carbohydrates and oxygen. Marine plants (including seaweeds, grasses, and planktons) use photosynthesis to capture and convert a tremendous amount of energy (nearly 30% of all energy produced worldwide) from the sun into chemical energy. Can chemical energy from such biomass be harvested?

Scientists are examining ways in which algae and plants may be used to produce alternative fuels. For example, it may be possible to take advantage of the rapid biosynthetic capabilities of marine algae—with their ability to mass-produce hydrocarbons and lipids in extraordinary quantities—to provide alternative sources of materials that are normally cost-prohibitive to produce or to isolate from natural materials. Similarly, it may be possible to convert marine biomass into fuels such as ethanol.

The U.S. Naval Research Lab has investigated potential ways to use plankton as underwater "fuel cells." Plankton at the water's surface release energy as they undergo photosynthesis, whereas plankton closer to the sediment at the bottom of the ocean (where there is less oxygen) use other reactions to generate energy. As a result, scientists have found that these plankton create a natural voltage gradient from the surface to the ocean floor that can be harvested to produce an indefinite source of electricity. In the future, the ocean may turn out to be a valuable resource for providing energy. Last, scientists are exploring ways in which biomass of marine algae may be used to increase the absorption of carbon dioxide and decrease greenhouse effects on the earth.

Marine scientists are exploring ways in which **bioprocessing** may involve marine products to produce a biological product such as a recombinant protein. Algae may potentially b e very valuable for expressing recombinant proteins. Researchers have found that they can make an abundance of proteins, such as antibodies, in marine algae because they can be grown on a very large scale. Marine biologists are exploring how marine organisms may be used to synthesize a variety of polymers and other biomaterials, which may be used for industrial manufacturing processes. For instance, oyster shell proteins are being considered as additives in detergents and other solvents as nontoxic, biodegradable alternatives to currently used materials.

We conclude this chapter by discussing how aquatic organisms are being considered for use in applications to clean up the environment.

10.6 Environmental Applications of Aquatic Biotechnology

Unfortunately the world's oceans have long served as dumping grounds for the wastes of humanity and industrialization, with too little thought given to the effect of pollution on fish stocks, marine organisms, and the environment. Clearly, oceans do not have an infinite ability to accept waste products without consequences, and we are seeing the results of this as wetlands and other estuarine environments—which are critical habitats for the spawning of many marine species and the growth of young marine organisms—are showing signs of severe decline due to pollution and human impact.

Bacteria can be used to detect and degrade environmental pollutants (discussed in Chapters 5 and 9). In this section, we consider how aquatic organisms can be used to help clean up and control pollution in the environments in which they live.

Antifouling Agents

Biofilming, also called biofouling, refers to the attachment of organisms to surfaces. These surfaces include the hulls of ships, inner linings of pipes, cement walls, and pilings used around piers, bridges, and buildings. Biofilming also occurs on the surfaces of marine organisms, especially shellfish. If you have spent any time around boat docks and piers located in marine environments, you have undoubtedly seen the effects of biofilming on structures that are in the water.

In marine environments, barnacles, algae, mussels, clams, and bacteria are among the most common organisms responsible for biofilming (**Figure 10.14**). As the term *biofilming* suggests, these organisms literally grow to create a layer, or usually multiple layers, that create a "film" over the surfaces to which they adhere. However, biofilming is not exclusively a marine problem. For example, similar biofilming occurs in the plumbing of your home, on contact lenses, and in your mouth. Bacteria that coat your teeth and bacteria that adhere to implanted surgical

(a)

(b)

FIGURE 10.14 Unwanted Invaders Create Biofilms That Have Serious Economic Impacts (a) Uncontrolled growth of unwanted species, such as that of zebra mussels, presents many problems. (b) Mussels, barnacles, and other organisms can create hard biofilms shown here covering a pipe.

devices and prostheses are examples of biofilming. Although brushing your teeth regularly minimizes biofilming in your mouth, such simple remedies are not available for deterring biofilming in marine environments.

Biofilming can create significant problems. The attachment of marine organisms to the hull of a ship can greatly increase the resistance of the ship as it moves through the water, slowing its travel time and reducing its fuel efficiency. A progressive accumulation of biofilms can clog pipes, block water intake and filtration systems for ships, and corrode metal surfaces. Furthermore, the colonization of surfaces with invertebrates and mollusks creates "hard" biofouling that is difficult and costly to remove (Figure 10.14).

Traditionally, most antifouling agents have employed toxic chemicals such as copper- and mercury-rich paints to minimize the growth of fouling organisms. However, the metals that leach from these paints can contribute to environmental pollution. As a result, researchers are investigating the natural mechanisms that many organisms use to prevent biofouling on their own surfaces. If biofilming is a problem for both manufactured surfaces and the surfaces of marine organisms, how do clams, mussels, and even turtles minimize biofilming and thus prevent their shells from being completely closed by biofilming organisms? Scientists are using molecular techniques to find the answer. Some organisms are thought to produce repelling substances; others appear to produce molecules that block the adhesion of biofilming organisms (**Figure 10.15**). Many marine sea grasses—such as the eelgrass *Zostra marina*—and algae produce compounds that prevent bacteria, fungi, and algae from adhering or that neutralize them. In the near future, these compounds may be used to produce protective coatings for covering ship hulls, aquaculture equipment, and other surfaces susceptible to biofilming.

Biosensors

Scientists are working to explore the use of aquatic organisms as biosensors to detect low concentrations of pollutants and toxins in waterways. Bioluminescent strains of bacteria can be used as biosensors (Chapter 5). Some species use bioluminescence to illuminate their environment; others use it to find mates in the dark depths of the ocean. Most bioluminescent deep-sea fish are involved in symbiotic relationships with bioluminescent bacteria such as *Vibrio fischeri*.

V. fischeri and other bioluminescent strains (such as *V. harveyi*) express *Lux* genes, which code for the light-emitting enzyme luciferase. In response to changing environmental conditions, the intensity of light emitted by *Vibrio* organisms can change. Because

of this ability, *Vibrio* bacteria have been used as biosensors to detect pollutants such as organic chemicals and nitrogen-containing compounds in marine environments.

Not only are some marine organisms useful for detecting environmental pollutions, but many marine species are also believed to possess metabolic pathways for degrading a range of substances both natural and manufactured. Much research is dedicated to characterizing the biochemical pathways involved in degradative processes to determine how marine organisms may be used to remediate, or clean up, the environment of a variety of hazardous substances that enter marine environments.

Environmental Remediation

Recall how native microorganisms or genetically engineered strains have been used to degrade chemicals (Chapters 5 and 9). In much the same way, marine organisms possess unique mechanisms for breaking down substances including toxic organic chemicals such as phenols and toluene, oil products found in harbors and adjacent to oil rigs, and toxic metals.

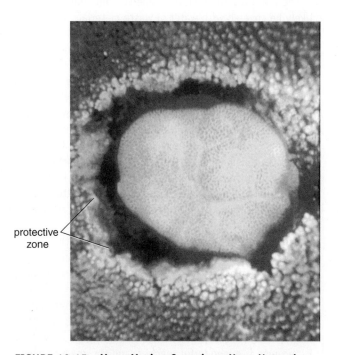

protective zone

FIGURE 10.15 **Many Marine Organisms Have Natural Antifouling Mechanisms** Many stationary marine organisms release defensive chemicals to create a protective zone (indicated by the dark-colored band around this marine coral) around them. These chemicals may deter fouling microorganisms and protect the organism from predators.

One of the earliest techniques used in marine remediation involved increasing the quantity of shellfish in polluted areas. Scientists introduced racks of shellfish into polluted waters to take advantage of the natural feeding method of bivalves such as clams, oysters, and mussels. Because these organisms strain the water, they act as a form of estuarine filters to remove wastes such as nitrogen compounds and organic chemicals. These chemicals, in turn, are absorbed in the tissues of the shellfish. After periods of time, these shellfish can be harvested, thus removing wastes from the water.

Heavy metal contamination of marine waters is the result of many industrial manufacturing processes.

CAREER PROFILE

Careers in Aquaculture

Aquaculture offers a varied range of exciting career opportunities. Approximately 36 million people worldwide are employed in aquaculture and related areas. Those who hold entry-level or technician positions maintain appropriate environmental conditions for the species that are being cultured, feed or supply nutrients in some form, and dispose of the fecal material or uneaten food. Although computer monitoring of many systems is gaining acceptance, visual inspection of farmed organisms (the "crop") cannot be underestimated. For example, changes in growth patterns, feeding, or behavior can foreshadow problems with the crop that must be addressed. Nevertheless, the performance of manual processes is usually an important part of the job. Even the facility foreperson and manager will likely be covered with water or waste materials from time to time—this is part of the business.

There are many potential career paths in aquaculture. Engineers design aquaculture production systems. Biologists and food scientists develop feeds for optimal growth. Plant and animal geneticists breed new hybrids of species presently under culture that may grow faster or have characteristics that are more suitable for sales or hardiness. Veterinarians must attend to sick fish and nurse them back to health. There are also opportunities for sales positions in aquaculture because eventually someone has to sell the end product. If the cultured product is not sold at a profit, considerable effort will have been wasted.

Those entering into the aquaculture field at the technician level should have a background in the sciences or animal husbandry. A B.S. degree in biology or animal science and experience in an aquaculture facility is preferred, and those going into research will need at least a master's degree if not a Ph.D. Most employers look for a motivated person with at least an associate's degree. Experience is a wonderful teacher, and those who have an interest in fish or plant husbandry should make every attempt to garner experience wherever possible. In addition, if there is one thing that every aquaculturist becomes familiar with, it is plumbing! If a plumbing course ever presents itself, take it. Knowledge of both air and water pumps and basic hydrodynamics is critical because it is integral to aquaculture.

Fish and plants are living organisms. If they are ignored, they will die. Therefore the water crop must be monitored every day, several times a day. Fish must be fed, waste material must be removed, oxygen and nitrogen levels must be monitored and adjusted, the problem of predators must be addressed, and the overall condition of the system must be evaluated. Without a person to monitor and maintain a crop, the process will fail—very quickly in most cases.

Working conditions vary because aquaculture does not take place in one specific type of place. Fish and shellfish can be grown in cold or very warm climates. Species can be grown outside or inside, and plants are usually grown in greenhouses. The working conditions may be wet, but they are usually fairly comfortable except for net pen cultures of salmon, which might place a person on a floating net pen in cold, inclement weather, chipping ice off the structure.

Salaries in aquaculture, as in most agricultural sciences, lag behind those in other areas of biotechnology. Those who go into the aquaculture business for themselves will face the same wage challenges as farmers who must wait 2 to 3 years before they can sell their crops. However, the interaction with the crop and freedom to work in often beautiful outdoor situations are appealing. People take more than a passing interest in the fish or shellfish that are being grown, and a sense of ownership and pride is attached. The same is true of the ornamental plants that are grown.

The job market outlook in aquaculture is strong because people worldwide are eating more fish. If the world's oceans can produce only a finite amount of fish and seafood, then aquaculture will have to step up and meet this demand. This area of biotechnology will definitely expand in the future, as will all the jobs associated with it.

Contributed by Gef Flimlin, Rutgers Cooperative Extension, Toms River, New Jersey.

As a solution to this problem, scientists have isolated marine bacteria that oxidize metals such as iron, manganese, nickel, and cobalt. Some of these bacteria can also be used to extract important metals from low-grade ores. Additionally, some marine bacteria and single-celled algae express metallothioneins, a family of metal-binding proteins. These species thrive in water contaminated with cadmium and other heavy metals, where they literally mop up cadmium from the surrounding environment and then degrade toxic metals into harmless by-products. Scientists are looking at ways to use these organisms to extract, recover, and recycle important and expensive metals such as gold and silver from manufacturing processes.

Many marine organisms produce substances that are valuable for degrading and processing a variety of waste materials. By using aquatic organisms or their products, it is possible to stimulate waste degradation in natural environments or in bioreactors seeded with these organisms in much the same way that sewage treatment plants rely on bacteria to degrade fecal wastes. For example, microbiologists at the USDA have experimented with growing nitrogen-metabolizing algae on large mats, called scrubbers, so that they can be used as natural filters. These scrubbers work like charcoal filters in an aquarium in that they bind nitrogenous wastes. Water contaminated with farm animal wastes is passed over the scrubbers, and the algae absorb and metabolize the wastes. These wastes provide nutrients for the algae, which grow into thick mats. The algae are periodically harvested by cutting them back, but they are allowed to regrow like grass. Water cleaned in this way has been used for irrigation, and some of the harvested algae have even been used as livestock food. Similar experiments have been done in Florida using water hyacinths to clean water rich in phosphorus, nitrogen, and other nutrients.

MAKING A DIFFERENCE

During early spring and summer along many beaches on the East Coast of the United States, a common scene has repeated itself for decades. Large numbers of horseshoe crabs (*Limulus polyphemus*) invade shallow bays to mate (see chapter-opening photo). Horseshoe crabs preceded dinosaurs on the earth. In some ways, these "living fossils" have changed very little from their initial appearance over 300 million years ago. *L. polyphemus* was one of the first marine organisms to be used successfully for medical applications. In the early 1950s, it was discovered that *Limulus* blood contains cells that kill bacteria. From this simple observation, a very powerful medical application of marine biotechnology was developed. The **limulus amoebocyte lysate (LAL) test** is an extract of blood cells (amoebocytes) from the crab that is used to detect bacterial **endotoxins**. Endotoxins, also called lipopolysaccharides, are part of the outer cell wall of many bacteria such as *Escherichia coli* and *Salmonella*. Endotoxins can cause instant death to many cells. In humans, exposure to endotoxins from certain bacteria can result in mild symptoms ranging from joint pain, inflammation, and fever to more severe conditions such as a stroke. Certain endotoxins can be lethal.

Researchers discovered that horseshoe crab blood would clot when exposed to whole *E. coli* or purified endotoxins. They later determined that amoebocytes—which are similar to human white blood cells—in horseshoe crab blood could be lysed, centrifuged, and freeze-dried to create a lysate that can be used in an LAL test. The LAL test, a rapid and very effective assay for endotoxins in human blood and fluid samples, is also used to ensure that endotoxins are not present in biotechnology drugs such as recombinant therapeutic proteins. It is also used to detect bacteria in raw milk and beef. In addition, many medical companies and hospitals use the LAL test to make sure that surgical instruments, needles used for drawing blood and cerebrospinal fluid, and implanted devices such as pacemakers are endotoxin-free.

Amoebocytes are collected from crabs, which are then released. The destruction of the crabs' nesting habitat, predation, and harvesting are reducing the *Limulus* population, causing concern about the long-term sustainability of this species. But without question the LAL test is an outstanding example of the power of marine biotechnology to benefit humanity, and it has "made a difference" by ensuring the safety of medical products given to millions of humans, pets, farm animals, and others.

QUESTIONS & ACTIVITIES

Answers can be found in Appendix 1.

1. In 2011, the U.S. National Oceanic and Atmospheric Administration (NOAA) announced a new national initiative to help restore shellfish habitat and increase production of clams and oysters through aquaculture. Propose potential policy initiatives that would support this effort.

2. Describe the differences between transgenic fish and triploids, and discuss how each type of genetically engineered fish can be created.

3. Suggest ways to assess the risk of genetically engineered marine species. Consider this from the standpoint of risks associated with the consumption of genetically engineered species as well as risks associated with introducing these species into natural environments.

4. Create a list of benefits and problems associated with aquaculture.

5. Describe the most interesting aspect of aquatic biotechnology you learned about in this chapter. What topic did you choose? Why did you find it interesting?

6. Name some properties of an aquatic organism that might be attractive to aquatic biotechnologists interested in identifying novel compounds that might be valuable for a biotechnology application.

7. Provide examples of finfish and shellfish species that are important for aquaculture.

8. In this chapter, we have primarily focused on applications of aquatic biotechnology that benefit humans. Many aquatic biotechnologists are interested in using biotechnology to improve native populations of aquatic organisms (finfish or shellfish) around the world. Suggest at least three ways in which biotechnology may be used to restore fish stocks.

9. The GloFish has been marketed as the world's first genetically modified pet. Even though the GloFish was not designed for human consumption, many different groups (including scientists) have voiced strong concerns against this use of transgenic animals. According to Yorktown Technologies, the company that produced the GloFish, the Environmental Protection Agency (EPA), USDA, U.S. Fish and Wildlife Service, and FDA all claimed that there was no need for any of these agencies to regulate or approve sale of the GloFish. Why do you think federal agencies decided that it was not necessary to regulate the GloFish? Describe possible reasons why the GloFish is so controversial. What might some of the primary concerns be about this fish?

10. What are biofilms? Give examples of biofilms in aquatic environments as well as in humans. How may aquatic organisms be used to help scientists combat biofilming?

11. What is the limulus amoebocyte lysate (LAL) test? Briefly describe how it works and its uses.

12. Access the Marinebiotech.org website from the Companion Website. Use the "resources" tab to learn more about dozens of compounds derived from marine organisms. This site is also a good resource for many topics in marine biotechnology.

Visit www.pearsonhighered.com/biotechnology
to download learning objectives, chapter summary, "Keeping Current" web links, glossary, flashcards, and jpegs of figures from this chapter.

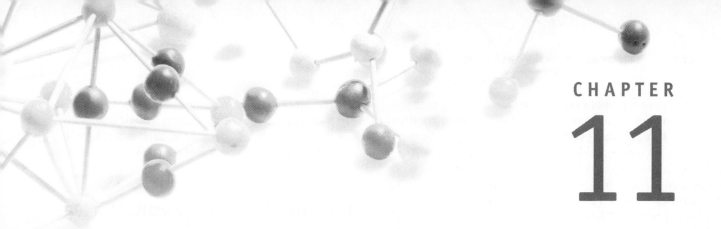

Medical Biotechnology

After completing this chapter, you should be able to:

- Explain why model organisms are important for medical biotechnology.

- Describe different molecular techniques for detecting chromosomal abnormalities and for genetic testing.

- Provide examples of how, pharmacogenomics is changing the treatment of genetic conditions.

- Discuss how monoclonal antibodies may be used for treating disease.

- Understand the purpose of gene therapy, compare and contrast different gene therapy strategies, and recognize the limitations of gene therapy.

- Define regenerative medicine and provide examples of how cell and tissue transplantation and tissue engineering can be used.

- Understand what stem cells are, describe how they can be isolated, and provide examples of possible therapies that may be developed from stem cells.

- Compare and contrast therapeutic cloning and reproductive cloning.

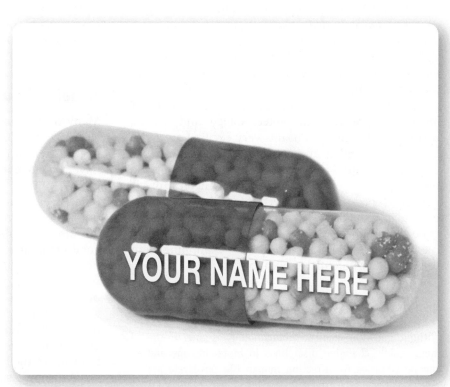

Customized medicine is a goal of medical biotechnology.

There is perhaps no topic in biotechnology that provokes greater optimism and debate than **medical biotechnology.** Applications of medical biotechnology have existed for decades. For instance, 100 years ago, leeches were commonly used to treat illness by so-called bloodletting. Some doctors believed that by using leeches to suck blood out of a patient, diseases were being removed from the body. Now, however, much better ways of treating disease have been found thanks to biotechnology. It is an exciting time to learn about medical biotechnology because advances in this field are occurring at a mind-boggling rate. In fact, even the leech is getting attention again—not for bloodletting but for enzymes found in its saliva that can dissolve blood clots and possibly be used to treat strokes and heart attacks.

Medical biotechnology incorporates many topics that we have already discussed. From developing new drugs to the prospects of using stem cells and cloning, the possibilities of medical biotechnology are incredible but also incredibly alarming to many people, including scientists.

In this chapter, we consider a range of different applications of medical biotechnology and discuss many potential impacts of this very exciting area. We begin by providing an overview of how molecular biology techniques can be used to detect and diagnose disease and by considering innovative medical products developed through biotechnology. We then present an introduction to applications and examples of gene therapy before going on to discuss regenerative medicine. The chapter concludes by examining stem cells and their potential applications.

FORECASTING THE FUTURE

Without question medical biotechnology is one of the most rapidly changing disciplines in biotechnology and the possibilities for affecting human life positively are amazing. Because of the range of different technologies being developed and the new applications in medical biotechnology, it is challenging to predict what the next major future discoveries will be. In this "forecasting," we pose several questions that are among the major challenges being addressed by medical biotechnologists.

- To what extent will personalized genomics and sequencing of individual genomes affect disease diagnosis and treatment in the future?
- What new medicines and treatments will result from our understanding of the genome and epigenome?
- What roles will nanotechnologies play for sensing and treating disease?
- What new treatments will emerge from regenerative medicine and the ability to engineer cells, tissues, and organs?

- Will gene therapy techniques advance to become more routine options for treating genetic diseases?
- Will stem cells become reliable, safe, and affordable options for treating disease?

11.1 The Power of Molecular Biology: Detecting and Diagnosing Human Disease Conditions

The year 2003 marked the 50-year anniversary since Nobel Prize winners James Watson and Francis Crick revealed the structure of DNA. Since then, molecular biology has advanced at an astonishing pace, providing molecular techniques that give scientists and medical doctors very powerful tools for combating human diseases.

Models of Human Disease

Many of the applications you will learn about in this chapter are possible because of important **model organisms.** In particular, mice, rats, worms, and flies have played critical roles in helping scientists study human disease conditions (Chapter 7). We think of ourselves as unmatched by other species not only for our ability to communicate through speech and writing but also for walking upright, creating music, making good pizza, and exploring distant planets. But we are not really unique at the genetic level. From yeasts and worms to mice, we share large numbers of genes with other organisms. A number of human genetic diseases also occur in model organisms. Therefore scientists can use model organisms to identify disease genes and test gene therapy and drug-based therapeutic approaches to evaluate their effectiveness and safety in preclinical studies before using them for **clinical trials** in humans.

Model organisms are critically important to scientists because we cannot manipulate human genetics for experimental purposes. It is, of course, unethical and illegal to force humans to breed or to remove their genes to learn how they function. However, these approaches are widely used to study genes in model organisms. Mice, rats, chicks, yeasts, fruit flies, worms, frogs, and even the zebrafish, a common fish in home aquariums, have all played important roles in widening our understanding of human genetics. Many important genes are highly conserved from species to species. If we identify important genes in model organisms, we can form hypotheses and make predictions about how these genes may function in humans.

Many genes identified in different model species have been shown to be related to human genes based

FIGURE 11.1 **Obese Mice** Model organisms are very valuable for helping scientists learn about human disease genes. The mouse on the left has been genetically engineered by gene knockout to lack the *obese* (*Ob*) gene, which produces a protein hormone called leptin, from the Greek word *leptos* meaning "thin." The *Ob* knockout mouse weighs almost five times as much as its normal sister (right).

on DNA sequence similarity. Such related genes are called **homologues**. A gene thought to play a role in human illness can be eliminated in model organisms through gene knockout. The effects on the organism can then be studied to learn about the functions of the gene. For instance, several years ago scientists discovered that mice can become obese if they lack a single gene called *Ob* (**Figure 11.1**). *Ob* codes for a protein hormone called leptin, which is produced by fat cells (adipocytes) and travels through the bloodstream to the brain to regulate hunger, essentially telling the brain when the body is full. The subsequent discovery of a human homologue for leptin has led to a new area of research with great promise for providing insight into fat metabolism in humans and the genetics that may influence weight disorders. Some childhood diseases of obesity are affected by mutation of the *Ob* gene. Extremely obese children in England have been treated with leptin and have responded very well in preliminary studies.

In developing embryos, some cells must die to make room for others. How does the body know where to develop certain organs and determine which cells must die to make room for others? Studies of the nematode *Caenorhabditis elegans,* an unsegmented roundworm, have greatly advanced the quest for answers to these important developmental questions. Maps of *C. elegans*, which has 959 cells, have been created, allowing scientists to determine the fate or lineage of all cells in the embryonic worm that develop to form the mature worm's nervous system, intestine, and other tissues. Of these cells, 131 are destined to die in a form of cell suicide known as programmed cell death, or

apoptosis. During the development of a human embryo, sheets of skin cells create webs between the fingers and toes; apoptosis is responsible for the degeneration of these webs prior to birth. We know that apoptosis is involved in neurodegenerative diseases such as Alzheimer disease, Huntington disease, amyotrophic lateral sclerosis (Lou Gehrig disease), and Parkinson disease as well as arthritis and some forms of infertility. Model organisms are helping scientists better understand the genes involved and thus to help find ways of slowing or stopping these degenerative processes.

The Human Genome Project and studies in comparative genomics have clearly demonstrated that we share a large number of genes with other organisms. Can you believe that you share approximately 50% of your genes with the pesky fruit flies you bring home on fruit from the grocery store? It may seem hard to believe that it takes only about twice as many genes to make a human as it does a fruit fly. And plants, such as rice, have even more genes than we do. We share nearly 40% of our genes with roundworms and 31% of our genes with yeast—the same yeast we use to help make bread rise and ferment alcoholic beverages. We share even more genes with mice; approximately 90% of our genes are similar in structure and function.

Many genes that determine our body plan, organ development, and eventually our aging and death are virtually identical to genes in fruit flies. Moreover, mutated genes that are known to give rise to disease in humans also cause disease in fruit flies. Approximately 61% of genes mutated in 289 human disease conditions are found in the fruit fly. This group includes the genes involved in prostate cancer, pancreatic cancer, cystic fibrosis, leukemia, and many other human genetic disorders. Heart disease is another example of a condition that scientists are studying in model organisms. Nearly 1 million people die each year in the United States from heart disease. Researchers are using gene knockouts to develop so-called heart attack mice that are deficient in the genes required for cholesterol metabolism. They hope that these mice will show elevated blood cholesterol levels similar to those that occur in atherosclerosis—hardening of the arteries—so that they can test therapies to combat heart disease.

Last, researchers have been hunting for a cure for AIDS for over 20 years. Creating a small animal model for AIDS is a high priority for many AIDS scientists. In addition to humans, HIV and related viruses cause disease in primates such as chimpanzees and rhesus macaque monkeys. But these animals are very expensive and ethical concerns have been raised about experimenting on fellow primates. Each animal can cost over $50,000, and they are available in only limited numbers. Researchers are making slow but steady progress toward developing rodent models infected with HIV. Even if such animals are created, there is no guarantee their disease will adequately mimic human AIDS. In humans, HIV infects and destroys human T lymphocytes (T cells). A main impediment to a rodent model is that HIV does not recognize and bind to receptor proteins on mouse T cells, the mechanism by which HIV infects human T cells. Nevertheless, scientists might be able to express human T-cell proteins on mouse T cells to trick the virus into infecting mice.

Biomarkers for Disease Detection

In theory, with the right diagnostic tools, it may be possible to detect almost every disease at an early stage. For many diseases, such as cancer, early detection is critical for providing the best treatment and improving the odds of survival. One detection approach is to look for **biomarkers** as indicators of disease. Biomarkers are typically proteins produced by diseased tissue or proteins whose production is increased when a tissue is diseased. Many biomarkers are released into body fluids such as blood and urine as a product of cell damage—released by dead and dying cells such as cells undergoing apoptosis. For example, a protein called **prostate-specific antigen (PSA)** is released into the bloodstream when the prostate gland is inflamed, and elevated levels can be a marker for prostate inflammation and even prostate cancer. Detecting individual genes or gene expression patterns also provides scientists with biomarkers for disease (see Figure 11.8 later in this chapter), and many biotechnology companies are actively involved in searching for better biomarkers that can be used for early detection and disease diagnosis.

The Human Genome Project Has Revealed Disease Genes on All Human Chromosomes

Many of the disease genes that can currently be tested for were discovered through the Human Genome Project. Prior to the genome project, only about 100 genetic diseases could be tested for; now there are over 2,000 diseases for which genetic tests are available. Through the Human Genome Project, scientists developed complex "maps" showing the locations of normal and diseased genes on all human chromosomes. Chromosome maps are available that pinpoint the locations of normal and disease genes of interest. **Figure 11.2** shows simplified maps for each chromosome, highlighting one or two prominent genes on each that are known to be involved in human genetic disease.

The Human Genome Project has led to the development of follow-up projects such as **The Cancer Genome Atlas Project (TCGA)**, a comprehensive effort to identify genomic changes involved in a variety of different cancers. Scientists are particularly interested in genetic changes that trigger normal cells to become cancerous cells in the brain, mammary glands, ovaries, pancreas, liver, and lungs because cancers of these organs affect large numbers of Americans. Eventually scientists expect that new information from TCGA will be used to better diagnose and treat cancer.

Detecting Genetic Diseases: Testing for Chromosomal Abnormalities and Defective Genes

Molecular biology techniques have proven to be extremely valuable for detecting many different genetic diseases.

Until relatively recently, most genetic testing occurred on fetuses for the purpose of identifying the sex of a child or to detect a small number of genetic diseases. Most of these procedures involved testing for genetic conditions that occur as a result of alterations in chromosome number or large structural abnormalities of chromosomes. If there are problems with chromosome separation during the formation of sperm or egg cells, a fetus may contain abnormal numbers of chromosomes. One of the best-understood examples of a disorder created by an alteration in chromosome number is **Down syndrome**. Most individuals with this condition have three copies of chromosome 21 (trisomy 21). Affected individuals show a number of symptoms, including cognitive impairments, short stature, and broadened facial features. Scientists have developed a strain of mice with almost a complete copy of human chromosome 21. These mice show characteristics of Down syndrome and may turn out to be a very valuable model for understanding the genetics of this condition.

Fetal testing for Down syndrome is fairly common, particularly in pregnant women older than 40 years, because the incidence of Down syndrome is related to the age of eggs produced by the woman. Trisomy 21 and other abnormalities in chromosome number can be tested in a fetus to provide parents with information that may be used to determine if they want the pregnancy to continue. If a defect is detected, genetic tests also provide information that can be used to treat fetuses during pregnancy and after the child is born.

So how is a developing fetus tested for Down syndrome? Two different techniques can be used: **amniocentesis** and **chorionic villus sampling**. Amniocentesis is performed when the developing fetus is around 16 weeks of age. A needle is inserted through

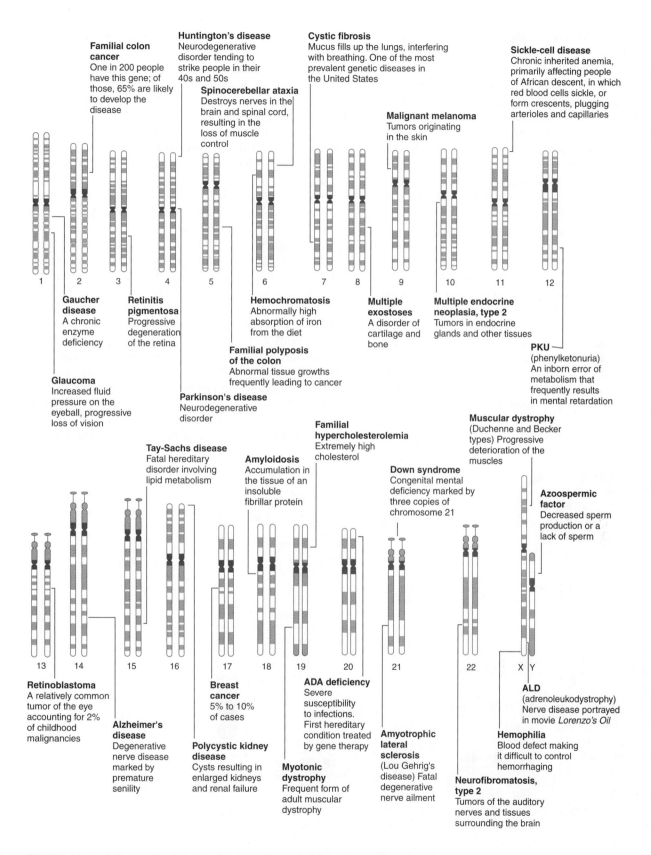

Familial colon cancer
One in 200 people have this gene; of those, 65% are likely to develop the disease

Huntington's disease
Neurodegenerative disorder tending to strike people in their 40s and 50s

Spinocerebellar ataxia
Destroys nerves in the brain and spinal cord, resulting in the loss of muscle control

Cystic fibrosis
Mucus fills up the lungs, interfering with breathing. One of the most prevalent genetic diseases in the United States

Malignant melanoma
Tumors originating in the skin

Sickle-cell disease
Chronic inherited anemia, primarily affecting people of African descent, in which red blood cells sickle, or form crescents, plugging arterioles and capillaries

Gaucher disease
A chronic enzyme deficiency

Retinitis pigmentosa
Progressive degeneration of the retina

Hemochromatosis
Abnormally high absorption of iron from the diet

Multiple exostoses
A disorder of cartilage and bone

Multiple endocrine neoplasia, type 2
Tumors in endocrine glands and other tissues

Glaucoma
Increased fluid pressure on the eyeball, progressive loss of vision

Familial polyposis of the colon
Abnormal tissue growths frequently leading to cancer

PKU
(phenylketonuria) An inborn error of metabolism that frequently results in mental retardation

Parkinson's disease
Neurodegenerative disorder

1 2 3 4 5 6 7 8 9 10 11 12

Tay-Sachs disease
Fatal hereditary disorder involving lipid metabolism

Amyloidosis
Accumulation in the tissue of an insoluble fibrillar protein

Familial hypercholesterolemia
Extremely high cholesterol

Down syndrome
Congenital mental deficiency marked by three copies of chromosome 21

Muscular dystrophy
(Duchenne and Becker types) Progressive deterioration of the muscles

Azoospermic factor
Decreased sperm production or a lack of sperm

Retinoblastoma
A relatively common tumor of the eye accounting for 2% of childhood malignancies

Alzheimer's disease
Degenerative nerve disease marked by premature senility

Breast cancer
5% to 10% of cases

Polycystic kidney disease
Cysts resulting in enlarged kidneys and renal failure

Myotonic dystrophy
Frequent form of adult muscular dystrophy

ADA deficiency
Severe susceptibility to infections. First hereditary condition treated by gene therapy

Amyotrophic lateral sclerosis
(Lou Gehrig's disease) Fatal degenerative nerve ailment

Neurofibromatosis, type 2
Tumors of the auditory nerves and tissues surrounding the brain

ALD
(adrenoleukodystrophy) Nerve disease portrayed in movie *Lorenzo's Oil*

Hemophilia
Blood defect making it difficult to control hemorrhaging

13 14 15 16 17 18 19 20 21 22 X Y

FIGURE 11.2 Disease Gene Maps of Human Chromosomes Maps show one or two representative genes on each human chromosome that are involved in a genetic condition. Many more genes than are shown in this figure are located on each chromosome. Note: Chromosomes are not drawn to scale.

the mother's abdomen into the pocket of amniotic fluid surrounding and cushioning the fetus (**Figure 11.3**). This fluid contains cells shed from the fetus, such as skin cells. Isolated cells are then cultured for a few days to increase cell numbers, after which the cells are treated to arrest them in mitosis to facilitate the viewing of mitotic chromosomes, which are spread onto a glass slide. The chromosomes are stained with different dyes that bind to proteins attached to the DNA, creating patterns of alternating light and dark bands on each chromosome. Based on the size of each chromosome and its banding pattern, chromosomes can be aligned into pairs. This procedure is called **karyotyping**; it can also be used to determine a child's sex based on the presence of the sex chromosomes (X and Y) (see Chapter 2).

During chorionic villus sampling (CVS) for fetal testing, a suction tube is used to remove a small portion of a layer of cells called the chorionic villus, fetal tissue that helps form the placenta (Figure 11.2). An advantage of CVS over amniocentesis is that enough cells are obtained so the sample can immediately be used for karyotyping. Another advantage of CVS is that the procedure can be done earlier in the pregnancy, around 8 to 10 weeks. Because the fetus is so small at that stage, CVS carries a higher risk than amniocentesis of disturbing it and causing a miscar-

riage (overall about 1% of amniocentesis and CVS procedures cause miscarriages).

Several researchers are close to producing noninvasive tests for fetal testing based on identifying and sequencing small amounts of fetal chromosomal fragments present in a pregnant woman's blood. Be on the lookout for further advances in these so-called **noninvasive prenatal genetic diagnosis (NIPD)** tests, which are predicted to be readily available as soon as 2013.

Karyotyping is easily carried out on adults to check for chromosomal abnormalities. Typically, blood is drawn from an adult and the white blood cells are used. A modern technique for karyotyping in both fetuses and adults is **fluorescence in situ hybridization (FISH)**. In FISH, a chromosome spread is prepared on a slide and then fluorescent probes are hybridized to each chromosome. Each probe is specific for certain "marker" sequences on each chromosome. In some cases, FISH can be performed with probes that fluoresce different colors—a procedure called **spectral karyotyping**. FISH is very useful for identifying missing chromosomes and extra chromosomes, but in particular FISH makes it much easier than conventional karyotyping to detect defective chromosomes. A number of human genetic diseases due to chromosomal

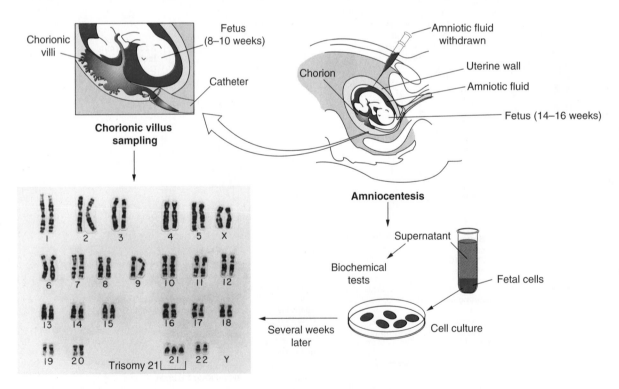

FIGURE 11.3   Amniocentesis and Chorionic Villus Sampling Fetal testing for chromosomal abnormalities is most commonly achieved through either amniocentesis or chorionic villus sampling. This karyotype from a person with Down syndrome shows three copies of chromosome 21 (trisomy 21).

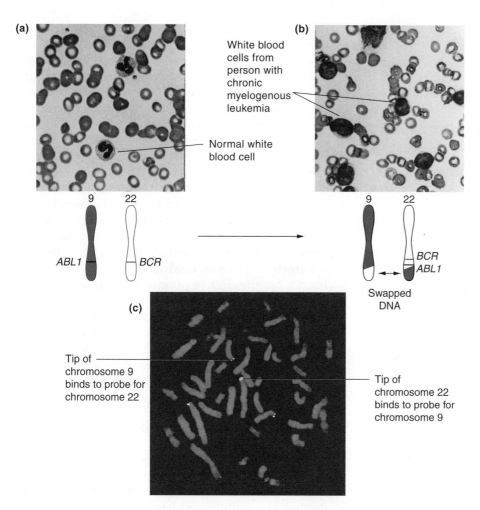

(a)

White blood cells from person with chronic myelogenous leukemia

Normal white blood cell

9   22

ABL1   BCR

(b)

White blood cells from person with chronic myelogenous leukemia

9   22

BCR
ABL1

Swapped DNA

(c)

Tip of chromosome 9 binds to probe for chromosome 22

Tip of chromosome 22 binds to probe for chromosome 9

FIGURE 11.4   FISH Can Be Used to Detect Chromosomal Defects Chronic myelogenous leukemia, a cancer of white blood cells created when genes on chromosome 9 and 22 are swapped, can be detected by FISH.

abnormalities occur when a portion of a chromosome is deleted or a piece of chromosome is swapped from one chromosome to another because of problems in chromosomal replication. For instance, in a type of leukemia (a cancer of the white blood cells) called chronic myelogenous leukemia, DNA is exchanged between chromosomes 9 and 22, so that genes from 9 are swapped onto 22 and vice versa (**Figure 11.4**). This exchange, called a *translocation*, can be detected by FISH using different-colored fluorescent probes for each chromosome.

More genetic diseases result from mutations in specific genes than abnormalities in the numbers or structures of chromosomes. As a result of the Human Genome Project, more sophisticated techniques have been developed to detect *individual* diseased genes in both fetuses and adults. Some genetic diseases can be detected in embryos and adults from either amniotic cells or blood cells, respectively, using **restriction fragment length polymorphism (RFLP) analysis** (pronounced "riff-lips"). The basic idea behind RFLP analysis is that defective gene sequences may be cut differently by restriction enzymes than their normal

complements because nucleotide changes in the mutant genes can affect restriction enzyme cutting sites to create more or fewer. Remember that RFLP analysis is used for DNA fingerprinting (Chapter 8). As an example, if DNA from a healthy individual and DNA from an individual with sickle-cell disease are both cut with restriction enzymes, they will be of different sizes because of the way restriction enzymes cut each gene. This can be clearly observed when the DNA fragments are subjected to Southern blot analysis with a probe for the ß-globin gene, the gene affected in sickle-cell disease (**Figure 11.5** on the next page). Hence we have the term *r*estriction *f*ragment *l*ength *p*olymorphisms—fragments of different lengths or forms ("poly" means many and "morphism" refers to the form or appearance of something) created by restriction enzymes.

Sickle-cell disease occurs when a person has two mutant versions of the ß-globin gene. The mutant copies of ß-globin protein produce an abnormal form of hemoglobin that affects the size and shape of red blood cells, giving them a characteristic "sickled" appearance (see also Figure 11.20).

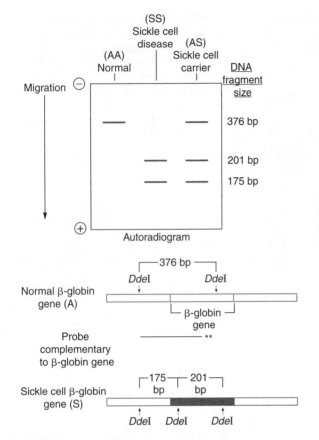

FIGURE 11.5   Using RFLP Analysis to Detect a Genetic Disease Human DNA cut with a restriction enzyme (*Dde*I) was subjected to agarose gel electrophoresis followed by Southern blotting to transfer the DNA to a nylon membrane and hybridization with a radioactive probe specific for the ß-globin gene. The normal ß-globin gene (A) contains two cutting sites for the enzyme *Dde*I; mutations in the sickle-cell (mutated) ß-globin gene (S) create three cutting sites for *Dde*I. Differences in DNA fragment sizes (polymorphisms) are detected by autoradiography, depending on where the probe binds to complementary sequences in the ß-globin gene. A healthy person with two copies of the normal ß-globin gene (homozygous, AA) shows a single band at 376 bp; a person with sickle-cell disease (homozygous, SS) would have two copies of the mutant ß-globin gene and show bands at 201 and 175 bp. A heterozygous person considered a "carrier" would have one normal and one defective ß-globin gene (AS) but would not have sickle-cell disease because there is one functioning copy of the ß-globin gene; bands show at 376, 201, and 175 bp.

The major disadvantage of RFLP analysis is that it can be used only to analyze gene defects in which a mutation changes a restriction site in a gene. **Allele-specific oligonucleotide (ASO) analysis** allows for the detection of a single nucleotide change in a gene even if the mutation does not change a restriction site. In this technique, DNA is isolated from human cells, usually white blood cells, and then amplified by the polymerase chain reaction (PCR) using primers that flank a disease gene of interest. Amplified DNA is then blotted onto nylon filters and hybridized separately to two different ASOs as probes. ASOs are small single-stranded oligonucleotide sequences, usually around 20 nucleotides in length. An ASO that will hybridize to a normal gene and another for the mutant gene are used. **Figure 11.6** shows an example of how ASO analysis can be used to test for the sickle cell gene. PCR-based tests such as this are becoming increasingly valuable for detecting diseased genes. One major advantage of PCR is its high sensitivity for detecting defects in small amounts of DNA. Consequently PCR and ASO analysis as well as FISH are being used to screen for gene defects in single cells from 8- to 32-cell embryos created by in vitro fertilization. Such **preimplantation genetic testing** allows individuals to select a healthy embryo prior to implantation.

Currently, several hundred defective genes can be tested for in adults or fetuses using many of the techniques we have described (Table 11.1). Refer to "Tools of the Trade" for several excellent websites where you can learn more about human disease genes.

Single nucleotide polymorphisms

If a segment of chromosomal DNA from two different people is compared by DNA sequencing, approximately 99.9% of the DNA sequence will be exactly the same. One of the many intriguing findings of the Human Genome Project was the discovery that **single nucleotide polymorphisms (SNPs**; pronounced "snips") represent one of the most common forms of genetic variation among humans. SNPs are single-nucleotide changes in DNA sequences that vary from individual to individual (see Figure 1.12).

SNPs have been found on all human chromosomes. It has been estimated that SNPs make up about 90% of human genetic variation and occur approximately every 100 to 300 base pairs (bp) in the human genome. Most SNPs are less likely to have an effect on a cell because they occur in non–protein coding regions (introns) of the genome. But when a SNP occurs in a gene sequence, it may cause a change in protein structure that produces disease or influences traits in a variety of ways, including conferring susceptibility for some types of disease conditions.

SNPs represent variations in DNA sequences that may ultimately influence how we respond to stress and disease. The first SNP discovered to be associated with a disease condition was for sickle cell disease. Because SNPs occur frequently throughout the genome, they serve as valuable genetic markers for identifying disease-related genes. Some SNPs are being used to predict susceptibilities to ailments such

DNA extracted from white blood cells and amplified by PCR

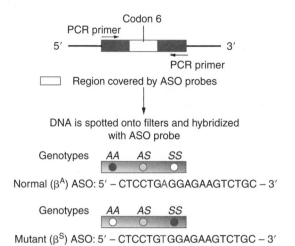

FIGURE 11.6 Using PCR and ASOs to Test for the Sickle Cell Gene ASO tests are very valuable for detecting single nucleotide mutations such as the one that causes sickle cell disease (see also Figure 2.20). In this example, DNA from white blood cells is amplified by PCR, blotted onto nylon membrane, and hybridized to fluorescently labeled ASO probes (probe binding shown as blue spots). The ASO for the normal hemoglobin gene (ß^A) will bind to DNA on the nylon only if the normal gene is present; the ASO for the mutant hemoglobin gene (ß^S) will bind to the defective gene only if it is present on the nylon. In this figure, probe binding is represented by blue DNA from individuals who are homozygous for the normal gene (AA) and have two copies of the ß^A allele; their DNA will bind only to the normal ß^A ASO. Individuals who are homozygous for the sickle cell hemoglobin gene (SS) have two copies of the ß^S allele and their DNA will bind only to the ß^S ASO and not the normal ß^A ASO. Heterozygous individuals (AS) have one normal copy of the gene and one mutant copy; as a result, their DNA will bind to both probes but produce a lighter hybridization signal than that from homozygous individuals.

as stroke, diabetes, cancer, heart disease, behavioral and emotional illnesses, and a host of other disorders that may have a genetic basis.

SNPs are thought to be so promising that pharmaceutical companies have invested millions of dollars in a collaborative partnership called the **HapMap Project**. Many SNPs on the same chromosome are clustered in groups called haplotypes. "Hap" is an abbreviation for haplotype. HapMap is an international effort among companies, academic institutions, and private foundations with an established goal of identifying and cataloguing the chromosomal locations (loci) of the more than 1.4 million SNPs that are present in the 3 billion bp of the human genome and to understand the roles of SNPs in disease diagnosis and treatment. As you just learned, ASO analysis is one way to detect SNPs.

Identifying sets of disease genes by microarray analysis

Another key technique for studying genetic diseases are **DNA microarrays**; also called gene chips (Chapter 3). A single microarray can contain probes for thousands of genes. Researchers can use microarrays to screen a patient for a pattern of genes that might be expressed in a particular disease condition. As shown in **Figure 11.7**, microarray data can then be used to predict the patient's risk of developing disease based on the patient's expressed genes for the disease.

For instance, microarrays created with probes for known disease genes or certain SNPs have become valuable for studying expressed genes in a patient. To do this, DNA or RNA is isolated from a patient's tissue sample—a blood sample or even a scraping of cells lining the cheeks. The patient's DNA is tagged with fluorescent dyes and then hybridized to the chip. Spots on the microarray where the patient's DNA bound are revealed by fluorescence (refer to Figure 3.18). Binding of a patient's DNA to a gene sequence on the chip indicates that his or her DNA has a particular mutation or SNP.

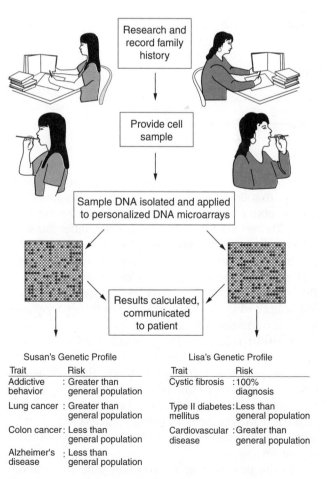

FIGURE 11.7 Using Gene Microarrays to Create a Genetic Profile

TABLE 11.1   GENETIC DISEASE TESTING

Genetic Disease Condition	Genetic Basis for Disease and Symptoms
Cancers (brain tumors; urinary bladder, prostate, ovarian, breast, brain, lung, and colorectal cancers)	A variety of different mutant genes can serve as markers for genetic testing.
Cystic fibrosis	Large number of mutations in the cystic fibrosis transmembrane conductance regulator (CFTR) gene on chromosome 7. Causes lung infections and problems with pancreatic, digestive, and pulmonary functions.
Duchenne muscular dystrophy	Defective gene (dystrophin) on the X chromosome causes muscle weakness and muscle degeneration.
Familial hypercholesterolemia	Mutant gene on chromosome 19 causes extremely high levels of blood cholesterol.
Hemophilia	Defective gene on the X chromosome makes it difficult for blood to clot when there is bleeding.
Huntington disease	Mutation in gene on chromosome 4 causes neurodegenerative disease in adults.
Phenylketonuria (PKU)	Mutation in gene required for converting the amino acid phenylalanine into the amino acid tyrosine. Causes severe neurological damage, including mental retardation.
Severe combined immunodeficiency (SCID)	Immune system disorder caused by mutation of the adenosine deaminase gene.
Sickle cell disease	Mutation in ß-globin gene on chromosome 11 affects hemoglobin structure and shape of red blood cells, which disrupts oxygen transport in blood and causes joint pain.
Tay-Sachs disease	Rare mutation of a gene on chromosome 5 causes certain types of lipids to accumulate in the brain. Causes paralysis, blindness, retardation, and respiratory infections.

There are even companies working on handheld chip devices that doctors can use to get nearly instant information about a patient's genetics. Microarrays are currently being used to identify genetic differences in patients with various types of cancer. Treatment strategies to combat cancer are being designed based on subtle differences in the expression of cancer-causing genes. But significant questions have been raised about the importance of quality control and consistency measures in using microarrays for genetic testing.

Protein microarrays are another relatively new option for disease diagnosis. They are used in much the same way as DNA chips. For instance, these chips can contain hundreds or thousands of antibodies spotted on a chip. By applying blood proteins from a patient, researchers have been able to detect illness indicated by the presence of proteins from disease-causing organisms.

In the next section we consider how biotechnology is creating new products that can be used to treat human disease.

11.2 Medical Products and Applications of Biotechnology

Identifying novel drugs and developing new ways to treat disease are major areas of medical biotechnology. We have discussed how different proteins—for example, insulin and human growth hormone—produced by recombinant DNA technology have been used to treat human disease conditions (see Chapter 4 and Chapter 5). Here we consider a few examples of important products of medical biotechnology that you will hear more about in the future.

The Search for New Medicines and Drugs

It is estimated that cancer may soon surpass cardiovascular disease as the leading cause of death in the United States. But many promising breakthroughs on the horizon may empower doctors by providing them with new strategies for treating different types of cancer.

Individuals respond differently to the anti-leukemia drug 6-mercaptopurine (6-MP).

The diversity in responses is due to variations (mutations, ■ or ✱) in the gene for an enzyme called TPMT, or thiopurine methyltransferase.

After a simple blood test, individuals can be given doses of medication that are tailored to their genetic profile.

Most people metabolize 6-MP quickly. Doses need to be high enough to treat leukemia and prevent relapses.

Others metabolize 6-MP slowly and need lower doses to avoid toxic side effects of 6-MP.

A small portion of people metabolize 6-MP so poorly that its effects can be fatal.

Normal dose

Dose for an extra slow metabolizer (TPMT-deficient)

FIGURE 11.8 **Pharmacogenomics** Different individuals with the same disease often respond differently to a drug treatment because of subtle differences in gene expression. The dose that works for one person may be toxic for another—a basic problem of conventional medicine. This example shows patients' responses to a chemotherapy compound called 6-mercaptopurine (6-MP), which has long been used to treat children with acute lymphocytic anemia (ALL), a form of blood cancer. Since it was found that genetic variations of the gene for the 6-MP-degrading enzyme thiopurine methyltransferase (TPMT) are important for determining how patients respond to 6-MP, physicians now routinely run genetic tests (or blood tests to measure the enzyme levels) on ALL patients before determining the proper dosage of 6-MP.

Scientists are investigating many of the genes involved in the growth of cancer cells, including genes called **oncogenes**. They produce proteins that may function as transcription factors and receptors for hormones and growth factors; they also serve as enzymes that in many ways help to change the growth properties of cancer-causing cells. Scientists are also actively studying tumor suppressor genes, which produce proteins that can keep cancer formation in check.

Oncogenes and tumor suppressor genes are getting so much attention because researchers are working on ways to make proteins encoded by these genes as targets for *small molecule inhibitors*—drugs that can bind to proteins and block their function. Similarly, researchers are working on drugs that can serve as "activators," binding to and stimulating important proteins that may be used to fight disease. In addition to small molecule drugs, there is a great deal of research designed to personalize medicine and improve drug delivery.

Pharmacogenomics for personalized medicine

The Human Genome Project and discovery of SNPs is partially responsible for a newly emerging field called

pharmacogenomics: it is customizing medicine by designing effective drug therapy and treatment strategies based on the specific genetic profile of a particular patient. Pharmacogenomics is based on the idea that individuals can react differently to the same drugs, which can have varying degrees of effectiveness and side effects in part because of genetic polymorphisms (**Figure 11.8**). Each year in the United States alone, over 100,000 deaths occur from the adverse effects of properly prescribed medications. It is unclear whether pharmacogenomics will be a cost-effective approach to medicine; nonetheless, this area of medical biotechnology holds great potential and has already demonstrated success in treating some conditions.

Many drugs currently used in **chemotherapy** can be effective against cancer cells because these drugs target rapidly dividing cells; however, such drugs also affect normal body cells that regularly reproduce, such as hair and skin cells and cells in the bone marrow, the last of which are responsible for making blood cells. As a result, hair loss, dry skin, changes in blood cell counts, and nausea are all related to the ways in which chemotherapy affects normal cells. Researchers have

TOOLS OF THE TRADE

Using the Internet to Learn about Human Chromosomes and Genes

We have discussed the importance of bioinformatics for analyzing genetic information and creating databases as tools that scientists around the world can use to compile, share, and compare DNA sequence information. The wealth of freely available databases that catalog information about human chromosomes and genes resulting in part from the Human Genome Project is a great example. An excellent way to learn about what the Human Genome Project has revealed is to review some of the chromosome maps available on the Web. For instance, if you are interested in the Y chromosome, you could review maps and descriptions of the genes found on it to find out why this chromosome is partially responsible for making male humans.

We encourage you to use the websites mentioned here and available through the Companion Website to follow a chromosome or gene of interest for a few months to see what kinds of information you can uncover. These sites are great resources and among the best sites for learning about human disease genes and chromosome maps.

You can, for example, use them to learn more about a rare disease gene related to a disease condition affecting someone close to you. These sites present up-to-date information that cannot be found in even the most recent books. If you cannot find a gene of interest at these sites, it probably has not been identified yet! The Department of Energy's Human Genome Program Information site provides excellent chromosome maps of identified genes and good historical information about the Human Genome Project. The Online Mendelian Inheritance in Man site (OMIM) is a great database of human genes and genetic disorders. In the keyword box, type in the name of a gene or disease that interests you. For example, type in "breast cancer" and then click the search button. When the next page appears, you will see a list of genes implicated in breast cancer, along with corresponding access numbers highlighted in blue as links. Clicking on one of the links will take you to a wealth of information about that gene, including background information, links to scientific papers about the gene, gene maps, and even nucleotide and protein sequence data (when available). You might also want to search this site to see if a gene has been found for a particular behavioral condition (for example, alcoholism or depression). The National Center for Biotechnology Information (NCBI) sponsors the Genomes and Maps website, which allows you to access detailed maps of chromosomes and disease genes using a feature called Map Viewer.

---

been looking for "magic bullet" drugs that destroy only cancer cells without harming normal cells. If such drugs were designed, patients might get well faster because the drugs would have little or no effects on normal cells in noncancerous tissues.

Consider the following example. Breast cancer is a disease that shows familial inheritance for some women. Women with defective copies of the genes called *BRCA1* or *BRCA2* have an increased risk of developing breast cancer, but many other cases of breast cancer do not exhibit a clear mode of inheritance. If a woman has a breast tumor thought to be cancerous, a small piece of the tissue could be used to isolate RNA or DNA for SNP and microarray analysis, which could then serve to determine which genes are involved in this particular woman's form of breast cancer. Armed with this genetic information, a physician could design a drug treatment strategy—based on the genes involved—that would be *specific* and *most effective* against this woman's cancer. A second woman with a different genetic profile for her breast cancer might undergo a different treatment.

Scientists at Genentech used this strategy to develop **Herceptin**, a type of monoclonal antibody (discussed in the next section) approved by the FDA in 1998. Herceptin binds to and inhibits HER-2, a protein produced by the human epidermal growth factor receptor 2 gene, which is overexpressed in about 25% to 30% of breast cancer cases. Women with HER-2–positive tumors (HER-2 overexpression) typically develop aggressive breast cancer with a greater likelihood of metastasis (spreading) and poorer prognosis for survival. Herceptin has proven to be effective in some women, but in others tumors become resistant to the antibody. A similar problem has occurred with other pharmacogenomics drugs developed for treating other cancers.

One of the first successful examples of pharmacogenomics involved a drug called **Gleevec,** introduced by Novartis in 2001 and used to treat chronic myelogenous leukemia (CML), the condition discussed in Figure 11.4. Gleevec targets the BCR-ABL fusion protein, which is created by an exchange of DNA between chromosomes 9 and 22 that occurs in CML; in doing so, Gleevec has proven to be a relatively effective way of treating the disease. Gleevec and related drugs have increased the survival rate of CML patients from 30% to nearly 90%. Visit the Howard Hughes Medical Institute website listed on the Companion Website and check out the animations link for a good way to observe the action of Gleevec.

Gene expression data from DNA microarrays are being used to diagnose patients based on the genes they express; then patients use pharmacogenomics treatment (if available) based on those genes. **Figure 11.9** shows an example of microarray data for individuals diagnosed with different forms of leukemia. Notice how the patients can be grouped into different genetic categories of leukemias based on the clusters of genes most actively expressed. Because each group of patients expressed large numbers of different genes and proteins, there is no reason to think that they would all respond well to the same chemotherapy. Knowing this, different chemotherapy approaches can be customized for each category of patients. As a result, patient survival rates have been greatly increased. A similar approach has been used for breast cancer patients, and it is expected that microarray data and pharmacogenomics approaches will increasingly become routine aspects of disease diagnosis and treatment. Several promising clinical trials are under

way for pharmacogenomic treatments of melanoma and many other cancers.

It will also be interesting to follow what happens with personalized genomics, the ability to have individual genomes sequenced, and the impact this will have on pharmacogenomics (Chapter 3). Similarly, the epigenome is being analyzed for its role in diseases and as a target for new treatment approaches (Chapter 2).

Nanotechnology and nanomedicine: Biotechnology at the nanoscale

Nanotechnology is an area of science involved in designing, building, and manipulating structures at the nanometer scale. A nanometer (nm) is one billionth of a meter. For reference, a human hair is approximately 200,000 nm in diameter, DNA is about 2 nm in diameter, and bonds between many atoms are around 0.15 nm long. Nanotechnology is a big business, with applications in materials manufacturing, energy, electronics, and engineering, but **nanomedicine**—applications of nanotechnology for improving human health—is of particular interest for medical biotechnology. Scientists envision tiny devices in the body carrying out a myriad of medical functions, including nanodevice sensors

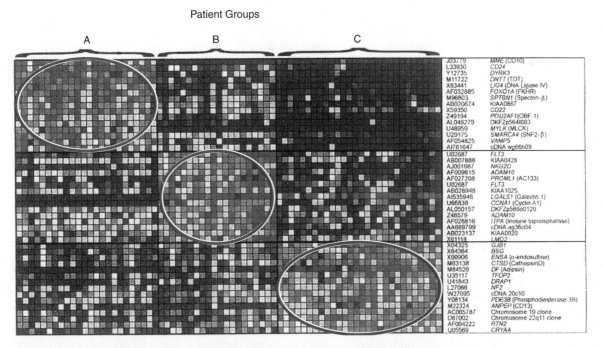

FIGURE 11.9 Microarrays for Gene Expression Analysis and Pharmacogenomics Shown here are microarray results indicating gene expression profiles in leukemia patients. Each column represents data for one patient, and each row is a different gene (gene symbols and names are on the far right side of the figure). Because this black-and-white image was reproduced from a color image, gray spots shown here represent highly expressed active genes. Red ovals highlight clusters of actively expressed genes (gray and white spots) that can be used to categorize patients into different groupings based on expressed and relatively inactive (dark spots) genes.

YOU DECIDE

Genetic Testing: Issues to Consider

Since genetic testing became available, a number of concerns have been raised about how genetic information could be used and what it may be used for beyond personal and private decisions. Consider some of these issues:

- Should we test unborn children or adults for genetic conditions for which there is currently no treatment or cure?

- What are acceptable consequences if parents learn their unborn child has a genetic defect?

- What are the psychological effects of a false result, which may indicate erroneously that a healthy person has a disease gene or a gene defect that goes undetected in a person with a genetic disorder?

- How do we ensure privacy and confidentiality of genetic information and avoid genetic discrimination? Who should have access to your genetic information? How could your genetic background be used to discriminate against you? How could your health or life insurance company's access to your genetic information affect your premiums? Could your premiums be raised based on "genetic" risk in the same way that premiums are raised based on other risks, such as how old you are and the car you drive?

- What are your obligations to inform others, such as a potential spouse or employer, of your knowledge about a possible genetic disorder?

- If genes are discovered for undesirable human behaviors, how would these genes be perceived in courts of law if accused criminals use genetics as their basis for a plea of not guilty by reason of genetics?

- Would society implement mechanisms to prevent or dissuade individuals with genetic defects from having children?

- Errors in genetic testing can have tragic consequences. Currently there are no federal standards for quality control of genetic testing in the United States. Should mandatory proficiency testing be a requirement to minimize errors?

As you can see, genetic testing is certainly not without its controversies and limitations, and there are few easy answers to these issues. Visit the "Your Genes, Your Choices" website listed on the Companion Website for a thought-provoking series of ethical dilemmas created by genetic testing and genetic technologies. What would you do if you had to face the scenarios presented at this site? You decide.

---

to monitor blood pressure, blood oxygen levels, and hormone concentrations as well as nanoparticles that can unclog blocked arteries and detect and eliminate cancer cells.

Many companies are working on nanotechnologies to develop innovative ways to improve drug delivery techniques and maximize their effectiveness. Sometimes even a well-designed drug is not as effective as desired because of delivery problems—getting the drug to where it must function. For instance, if a drug to treat knee arthritis is taken as an oral pill, only a small amount of the drug will be absorbed by the body and transported via blood to tissues of the knee joint. Other factors that influence drug effectiveness are a drug's solubility (its ability to dissolve in body fluids), drug breakdown by body organs, and drug elimination by the liver and kidneys.

Microspheres, nanoparticles between 1 to 100 nm in size that can be filled or coated with drugs, may be one way to improve drug delivery and effectiveness. These particles are often made of lipid materials that closely resemble the phospholipids in cell membranes. Delivery of microspheres as a mist sprayed into the airways through the nose and mouth has been used successfully for treating lung cancer and other respiratory illnesses such as asthma, emphysema, tuberculosis, and flu (**Figure 11.10**). Refer to Section 11.3 for a discussion of how microspheres called liposomes are used in gene therapy. Researchers are also investigating ways to package anticancer drugs into microspheres for implantation in the body adjacent to growing tumors; they are also working on anesthetics for pain management and adding microspheres to wafers and patches that can be used for drug delivery.

In 2006, the FDA approved Exubera, an inhalable version of insulin produced by Nektar Therapeutics of San Francisco and sold by the pharmaceutical giant Pfizer. Exubera is a recombinant form of insulin delivered as an inhalable powder; it offers diabetic patients the first alternative to needle-based delivery of insulin. But after barely a year, Pfizer stopped selling Exubera because of slow sales. Among the reasons that have been given for Exubera's failure are the unwillingness of physicians and patients to try something new, a bulky inhaler used to deliver the particles, and a poor marketing strategy.

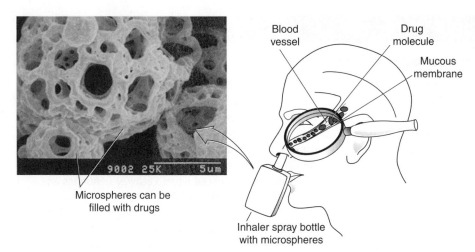

Microspheres can be filled with drugs

Inhaler spray bottle with microspheres

Blood vessel

Drug molecule

Mucous membrane

FIGURE 11.10 Microspheres for Drug Delivery Microspheres can deliver drugs to specific locations in the body or be distributed throughout the body depending on how they are used. In this example, drug-containing microspheres are sprayed into the nose, where they will enter blood vessels and rapidly enter the bloodstream to travel throughout the body.

Currently over a dozen nanoparticle-based drugs are in clinical trials in the United States, and most of these target cancers. Several nanotechnology-enabled drugs are making their way to the market, primarily in areas of cancer treatment, and over 150 nanotechnology cancer therapies are in development. Scientists have developed "smart drugs" using viruses or tiny nanoparticles such as gold particles that are introduced into the body to seek out and target viruses or specific cells, such as cancer cells; they then deliver a cargo intended to treat or destroy those cells rapidly, effectively, and silently, with few side effects (**Figure 11.11**). Some of these ideas have been tested in patients, with promising results, and nanotechnologists are optimistic about the powers of this technology.

Artificial Blood

Since the 1930s, blood transfusions have been performed routinely and successfully in the United States. Transfusions are often necessary for treating trauma victims, providing blood during surgeries, and treating people with blood-clotting disorders such as hemophilia. In the 1980s, the realization that HIV had contaminated many blood supplies led to new testing techniques that have made blood supplies much safer. Blood donated for transfusions is tested for pathogens such as HIV and hepatitis viruses B and C before it is stored, but donated blood has a shelf life of only a few months and must be refrigerated. Throughout many areas of the world, particularly in developing countries where screening procedures are not very good, there is a serious need for safe blood, free of infectious bacteria and viruses. These and other concerns have prompted scientists to seek ways to develop artificial blood or blood substitutes.

Major advantages of artificial blood could include a disease-free alternative to real blood, a constant supply of blood in the face of blood shortages and emergency situations, and a supply of blood that can be stored for long periods of time. Also, unlike donated blood, synthetic blood would not have to be matched to the recipient's blood type to avoid rejection by the immune system. A major limitation in the development of synthetic blood to date is that artificial bloods have been designed to serve the primary task of normal red blood cells—transporting oxygen to body tissues, a role carried out by the oxygen-carrying protein hemoglobin. Red blood cells are literally hemoglobin factories. But normal red blood cells perform other

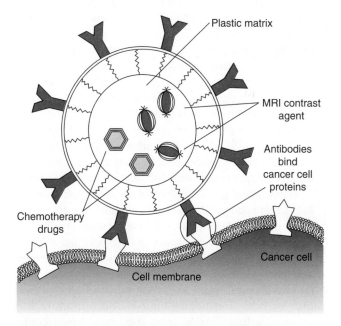

Plastic matrix

MRI contrast agent

Antibodies bind cancer cell proteins

Cancer cell

Cell membrane

Chemotherapy drugs

FIGURE 11.11 Tumor-Seeking and Tumor-Killing Nanoparticles Multifunctional nanoparticles have been developed that have shown promise for seeking out and destroying tumor cells. This example shows a plastic nanoparticle covered with antibodies against cancer cell proteins, which allow the particle to bind to cancer cells. Inside the particle are contrast agents, which can be used for magnetic resonance imaging or x-ray approaches to detect the tumor. The nanoparticle can also contain chemotherapy drugs that can diffuse out of it to kill tumor cells.

YOU DECIDE

Direct-to-Consumer Genetic Tests

The past decade has seen dramatic developments in **direct-to-consumer (DTC) genetic tests**. A simple web search will reveal many companies offering such tests, and there are approximately 1,900 diseases for which such tests are now available (in 1993 there were about 100 such tests). Most DTC tests require that a person mail a saliva sample, hair sample, or cheek cell swab to the company. For a range of pricing options, DTC companies largely use SNP-based tests such as ASO tests to screen for different mutations. For example, in 2007 Myriad Genetics, Inc. began a major DTC marketing campaign of its tests for *BRCA1* and *BRCA2*. Mutations in these genes increase risk of developing breast and ovarian cancer. DTC testing companies report absolute risk, the probability that an individual will develop a disease; but how such risks results are calculated is highly variable and subject to certain assumptions.

Such tests are controversial for many reasons. For example, the test is purchased online by individual consumers and requires no involvement of a physician or other health care professionals, such as a nurse or genetic counselor, to administer or to interpret results. There are significant questions about the quality, effectiveness, and accuracy of such products because the DTC industry is currently largely self-regulated. The U.S. Food and Drug Administration (FDA) does not regulate DTC genetic tests. There is currently no comprehensive way for patients to make comparisons and evaluations about the range of tests available and their relative quality.

Most companies make it clear they are not trying to diagnose or prevent disease, nor are they offering health advice. So what is the purpose of the information that these tests provide? Websites and online programs from DTC companies provide information on what advice a person should pursue if positive results are obtained. But is this enough? If the results are not understood, might negative tests provide a false sense of security? Just because a woman is negative for *BRCA1* and *BRCA2*

mutations *does not* mean that she cannot develop breast or ovarian cancer.

In June 2010, the FDA announced that five genetic test manufacturers (Illumina, Pathway Genomics, NaviGenics, 23andMe, deCODE Genetics) would need FDA approval before their tests could be sold to consumers. This action was prompted when Pathway Genomics announced plans to market a DTC kit for "comprehensive genotyping" in the pharmacy chain Walgreens. Pathway Genomics and the other companies have been selling their DTCs through company websites for several years. Pathway and others claim that because their DTC kits are approved by the Clinical Laboratory Improvement Amendments (CLIA), no further regulation is required. CLIA regulates certain laboratory tests but is not part of the FDA. This scenario in particular prompted discussion on how the FDA will oversee DTC genetic tests. However, at the time of publication of this edition, the FDA has not revealed any definitive plans to regulate or oversee DTC genetic tests. There are varying opinions on the regulatory issue. Some believe that the FDA has no business regulating DTC tests and that consumers should be free to purchase products according to their own needs or interests. Others insist that the FDA must regulate DTCs in the overall interest of consumers.

In 2010, the National Institutes of Health announced that it will create a **Genetic Testing Registry (GTR)** designed to increase transparency by publicly sharing information about the utility of their tests and research with the general public, patients, health care workers, genetic counselors, insurance companies, and others. The GTR is intended to allow individuals and families access to key resources to let make them better-informed decisions about their health and genetic tests. But participation in the GTR by DTC companies has not been made mandatory yet. Therefore will companies involved in genetic testing participate? Should DTC genetic tests be more carefully regulated by the FDA? You decide.

functions as well, such as providing the body with a source of iron, and hemoglobin is also important for removing carbon dioxide from the body. Researchers have yet to create blood substitutes that can perform all the functions of normal blood; nevertheless, many promising products are under development.

So how is artificial blood made? Artificial bloods are cell-free solutions containing molecules that can bind to and transport oxygen in much the same way as normal hemoglobin. Some blood substitutes are made from the hemoglobin of cattle; others are made

from human hemoglobin. Cow blood is collected from food cattle at slaughterhouses and then processed to purify the hemoglobin. Many other types of artificial blood being tested are produced using fluorocarbons, chemicals that can bind oxygen just as hemoglobin does and then release oxygen to the surrounding tissues. Ultimately, artificial blood products must provide safe alternatives to real blood transfusions. Much work remains to be done, but the potential benefit of these products has many companies investing large amounts of money and time to develop viable blood substitutes.

Vaccines and Therapeutic Antibodies

Vaccines can be used to stimulate the body's immune system to produce antibodies and provide a person with protection against infectious microbes (Chapter 5). Certainly vaccination has been very effective for protecting us from pathogens that cause polio, tetanus, typhoid, and dozens of others. Development of vaccines against some of the most deadly pathogens is a very active and important area of research.

Many scientists hope that vaccination may be useful against conditions such as Alzheimer disease and many different types of cancers, but vaccination for these purposes is still mostly unproven in humans. Cancer vaccines are being experimented with as therapeutic treatments that are not preventative but designed to treat a person who already has cancer. In this approach, a person is injected with cancer cell antigens in an effort to stimulate the patient's immune system to attack existing cancer cells. In fact, there is considerable excitement about new types of "naked DNA" vaccines in which plasmid DNA-encoding genes that produce antigens are injected directly into tissue, where cells take up the plasmid and express the antigens that stimulate antibody production by the body. In other types of DNA vaccines, an immune response is mounted against the DNA itself.

The primary purpose of vaccination is to stimulate antibody production by the immune system and thus to help ward off foreign materials (Chapter 5). However, antibodies themselves might be used to treat an existing condition as opposed to preventing infectious microbes from causing disease. Using antibodies in some types of therapy makes good sense because antibodies are very specific for the molecules or pathogens to which they are produced and can find and bind to their target with great affinity. Since their development in 1975, **monoclonal antibodies (MAbs)**, purified antibodies that are very specific for certain molecules, have been considered "magic bullets" for disease treatment. To make a MAb, a mouse or rat is injected with the purified antigen to which researchers are trying to make antibodies. **Figure 11.12** shows production of MAbs specific for proteins from human liver cancer cells. After the mouse makes antibodies to the antigen, a process that usually takes several weeks, the animal's spleen is removed. The spleen is a rich source of antibody-producing B lymphocytes, commonly called B cells. In a culture dish, B cells are mixed with cancerous cells, called myeloma cells, which can grow and divide indefinitely. Under the right conditions, a certain number of B cells and myeloma cells will fuse together to create hybrid cells called **hybridomas**.

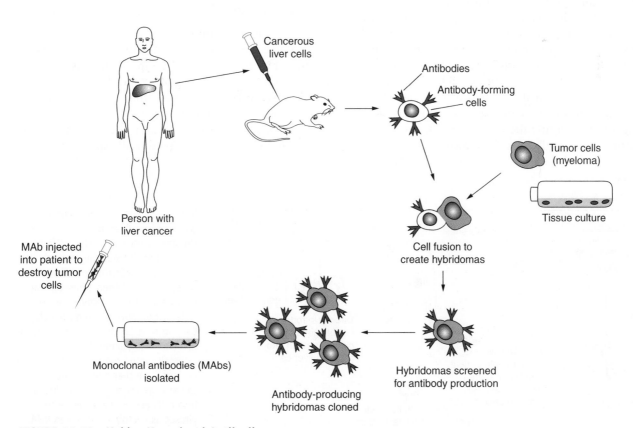

FIGURE 11.12 Making Monoclonal Antibodies

Hybridoma cells grow rapidly in liquid culture because they contain antibody-producing genes from B cells. These cells are literally factories for making antibodies. Hybridoma cells secrete antibodies into the liquid culture medium surrounding the cells. Chemical treatment is used to select for hybridomas and discard unfused mouse and myeloma cells, so that researchers have pure populations of fused, antibody-producing cells. Hybridomas can be transferred to other culture dishes and frozen at ultralow temperatures so that a permanent stock of cells is always available. Antibodies can be isolated from hybridoma cultures in large batches by growing hybridoma cells in batch culture using bioreactors.

Monoclonal antibodies can be injected into patients to seek out and target the antigens to which the MAbs were produced. The MAbs in Figure 11.12 would bind to liver cancer cells and work on destroying the tumor. In 1986, the FDA approved the first monoclonal antibody, OKT3, which was used to treat organ transplant rejection. In the 1990s, MAbs were developed to treat breast cancer (Herceptin) and lymphoma (Rituxan). There are currently over a dozen MAbs being used worldwide to treat cancer, cardiovascular disease, allergies, and other conditions. Scientists even envision attaching chemicals or radioactive molecules to MAbs in the hope that these might target damaged or cancerous cells and use their payloads to kill these cells. Therapeutic antibody strategies may also be of value for treating people addicted to harmful drugs, such as cocaine and nicotine. In the United States alone, more than 13 million people abuse drugs. Scientists believe that it may be possible to stimulate antibody production to drugs such as cocaine. These antibodies would then bind to the drug as the antigen, trapping and preventing the drug from affecting brain cells. Monoclonal antibodies have also been used for

several years in common tests for conditions such as strep throat, and most home pregnancy kits use MAbs to detect hormones produced during pregnancy. Monoclonals for disease treatment have still not lived up to their initial hype, and there have been some setbacks in the field. For example, MAb treatment of Alzheimer patients produced severe inflammation in several people due to a human antimouse antibody response. Humanizing antibodies can alleviate some of the problems with MAbs (Chapter 4). Increasingly it appears that MAbs will continue to be valuable tools for medicine in the twenty-first century. In the next section we consider gene therapy, a promising and controversial topic of medical biotechnology.

11.3 Gene Therapy

Gene therapy involves the delivery of therapeutic genes into the human body to correct disease conditions created by a faulty gene or genes. Think about the awesome power and potential of gene therapy—providing a person with normal genes to supplement defective genes and cure disease or even using normal genes to replace faulty ones. Here we provide an overview of gene therapy strategies for treating and attempting to cure disease.

How Is It Done?

The two primary strategies for gene delivery are **ex vivo gene therapy** and **in vivo gene therapy** (**Figure 11.13**). In ex vivo therapy (*ex* means "out of," *vivo* is Latin for "something alive"), cells from a person with a disease condition are removed from that person, treated in the laboratory using techniques similar to bacterial transformation, and then reintroduced into them into him or her. Technically speaking, intro-

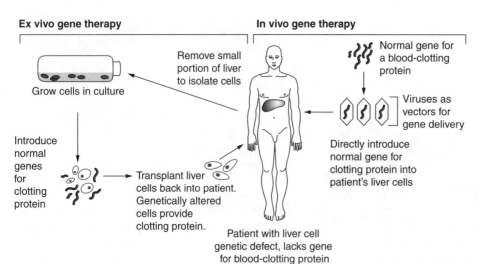

Ex vivo gene therapy

Remove small portion of liver to isolate cells

Grow cells in culture

Introduce normal genes for clotting protein

Transplant liver cells back into patient. Genetically altered cells provide clotting protein.

In vivo gene therapy

Normal gene for a blood-clotting protein

Viruses as vectors for gene delivery

Directly introduce normal gene for clotting protein into patient's liver cells

Patient with liver cell genetic defect, lacks gene for blood-clotting protein

FIGURE 11.13   Ex Vivo and in Vivo Gene Therapy in a Patient with a Liver Disorder Ex vivo gene therapy involves isolating cells from the patient, introducing normal genes for the clotting protein into these cells, and then transplanting cells back into the body where these cells will produce the required clotting protein. In vivo gene therapy involves introducing DNA directly into cells while they are in the patient. In either mode of gene therapy, genes may be introduced into cells as DNA packaged into viruses as vectors or as naked DNA.

ducing DNA into animal or plant cells is called **transfection**. For instance, liver cells from a patient suffering from a liver disorder would be surgically removed and cultured. Appropriate therapeutic genes would then be delivered into these cells using vectors and other approaches discussed in the next section. These genetically altered liver cells would then be transplanted back into the patient without fear of rejection of the tissue transplant because these cells came from the patient initially (Figure 11.13).

In vivo gene therapy does not entail removal of a patient's cells; DNA is introduced directly into cells and tissues in the body (Figure 11.13). One challenge of in vivo gene therapy is delivering genes only to the intended tissues and not to tissues throughout the body. Scientists have primarily relied on using viruses as vectors for gene delivery, but in some cases genes have been directly injected into some tissues. So far, ex vivo strategies have generally proven to be more effective than in vivo approaches.

Delivering the payload: Vectors for gene delivery

A major challenge that must be overcome if gene therapy is to become a reliable tool for treating disease is achieving a safe and effective delivery of therapeutic genes—the payload. Depending on the genetic condition to be treated, some therapeutic strategies may require long-term expression of a corrective gene, whereas others may require rapid expression for shorter periods of time. A majority of gene delivery strategies, both ex vivo and in vivo, rely on viruses as vectors to introduce therapeutic genes into cells.

A viral vector uses a viral genome to carry a therapeutic gene or genes to "infect" human body cells, thereby introducing the therapeutic gene. Scientists have considered various viruses such as **adenovirus**, which causes the common cold, a related virus called **adeno-associated virus (AAV)**; influenza viruses, which cause the flu; and herpes viruses, which can cause cold sores and some cause sexually transmitted diseases, as potential **vectors** for gene delivery. Even human immunodeficiency virus-1 has been considered as a gene therapy vector. For any viral vector to work, scientists must be sure that these vectors have been genetically engineered and inactivated so they neither produce disease nor spread throughout the body and infect other tissues.

Most viruses infect human body cells by binding to and entering cells and then releasing their genetic material into the nucleus or cytoplasm of the human cell. This is usually DNA, but some viruses contain an RNA genome. The infected human cell then serves as a host for reproducing the viral genome and producing viral RNA and proteins. Viral proteins ultimately assemble to create more viral particles that break out

of the host cells so they are free to infect other cells and repeat the life cycle.

It may seem strange that viruses would even be considered for carrying genes to cure human diseases. However, since viruses are very effective at introducing their genomes into cells, scientists reasoned that if viruses could be disabled so that they would not cause disease and genetically altered to deliver therapeutic genes safely, we could use viruses for beneficial purposes. In many ways viruses are perfectly designed as gene therapy vehicles or vectors for gene delivery. For instance, adenovirus—which approximately 80% to 90% of the population has been infected with in childhood because it causes the common cold—can infect many types of body cells fairly efficiently. Retroviruses such as **lentivirus,** including even HIV, are of interest as vectors because, on entering a host cell, they copy their RNA genome into DNA and then randomly insert their DNA into the genome of the host cell, where it remains permanently, a process called **integration**. A main reason why retroviruses are used for gene delivery is that they can integrate therapeutic genes into the DNA of human host cells, allowing permanent insertion of genes into the chromosomes of a patient's cells as a way of providing lasting gene therapy.

Viruses have also been widely studied as gene therapy vectors because some viruses infect only certain body cells. This might allow for *targeted* gene therapy—the ability to deliver genes only to the tissues infected by a certain virus. For instance, a strain of herpesvirus (HSV-1) primarily infects cells of the central nervous system. This strain is a candidate for targeted gene delivery to cells of the nervous system, which may be an effective way to treat genetic disorders of the brain such as Alzheimer disease and Parkinson disease. Researchers are also investigating ways that the genetically altered herpes viruses can be used to destroy brain tumors. Preliminary trials of this approach in humans have shown some promise.

Most human cells do not take up DNA easily. If they did, it would then be possible to transfect cells by simply mixing them with DNA in a tube in much the same way that transformation is achieved with bacterial cells. However, some success has been demonstrated for both in vivo and ex vivo strategies using "naked" DNA. Naked DNA is simply DNA by itself, without a viral vector, which is injected directly into body tissues. Small plasmids containing therapeutic genes are often used for this approach. Cells of certain tissues will take in some of the naked DNA and express genes delivered in this way.

Delivery techniques for naked DNA have been somewhat effective in the liver and in skeletal muscle. One of the major problems with transfecting human cells in vivo is that because a relatively small number of cells take up the injected DNA, there may not be enough cells expressing the therapeutic gene for gene therapy to have any effect on the tissue. Scientists are working on ways to overcome these problems and deliver naked DNA more effectively. For example, electroporation (recall that in Chapter 5 we discussed how electrical stimulation is used to move plasmids into bacterial cells) can be used to stimulate movement of DNA into cells.

One approach to deliver DNA without viral vectors involves **liposomes**, small-diameter hollow microspheres made of lipid molecules, similar to the fat molecules in cell membranes. Liposomes are packaged with genes and then injected into tissues or sprayed onto them. A similar technique involves coating tiny gold nanoparticles with DNA and then shooting these into cells using a DNA gun, a pressurized air gun that delivers gold or liposome particles through cell membranes without killing most cells. Biodegradable gelatin particles are also being studied as gene-carrying vectors. Short-term expression of genes through "gene pills" is being explored. In this approach, a pill delivers DNA to the intestines, where it is absorbed by intestinal cells; these then express therapeutic protein encoded by the DNA and secrete these proteins into the bloodstream.

Antisense RNA technology and RNA interference for gene therapy

We have discussed the use of **antisense RNA technology** as a way to block translation of mRNA molecules to silence gene expression (Chapter 3). This approach was used to create the Flavr Savr tomato (see Figure 6.7). The basic concept of antisense RNA technology is to design an RNA molecule that will serve as a complementary base pair to the mRNA you want to inhibit, thus blocking it from being translated into a protein (**Figure 11.14**). This approach for shutting off a gene is frequently called **RNA** or **gene silencing**. Since the development of antisense RNA technologies in the 1970s, scientists have thought that RNA silencing approaches would be promising ways to turn off disease genes as a gene therapy approach.

Antisense RNAs have been effectively used for gene silencing in cultured cells, but this technology has yet to live up to its promise as a treatment for disease. The recent emergence of **RNA interference (RNAi)** as a method to control mRNA stability and protein synthesis has reinvigorated gene therapy approaches by gene silencing (refer to Figure 3.19). With RNAi, double-stranded RNA molecules are delivered into cells where the enzyme Dicer chops them into 21-nt-long pieces called **small interfering RNAs (siRNAs;** Figure 11.14). The siRNAs then join with an enzyme complex called the **RNA-induced silencing complex (RISC)**, which shuttles the siRNAs to their target mRNA, where they bind by complementary base pairing.

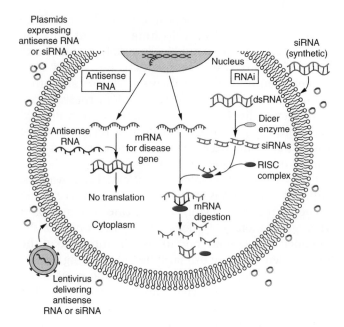

FIGURE 11.14 Antisense RNA and RNAi Approaches for Gene Therapy by Gene Silencing Antisense technology and RNAi are two ways to silence gene expression and turn off disease genes. In antisense technology (left) an antisense RNA molecule binds to the mRNA for the disease gene and prevents it from being translated into a protein. With RNAi technology, double-stranded RNA (dsRNA) molecules are delivered to the cell. The dsRNAs are then cleaved by Dicer into short interfering RNAs (siRNAs), which are escorted by the RISC protein complex to bind to their target mRNA causing its degradation and thus preventing translation.

The RISC complex degrades the siRNA-bound mRNAs so they cannot be translated into protein. Using gene therapy to stimulate naturally occurring **microRNAs (miRNAs)** that inhibit gene expression is another area of active investigation. A main challenge to RNAi-based therapeutics so far has been sustained in vivo delivery of the antisense RNA, dsRNA, miRNA, or siRNA to the target tissues. RNAs degrade quickly in the body. It is also hard to get them to penetrate cells and to target the right tissue. Two common delivery approaches are to inject the antisense RNA or siRNA directly or as a plasmid that is taken in by cells to be transcribed to make antisense or RNAi molecules (Figure 11.14). Liposome, lentivirus delivery mechanisms, and attachment of siRNAs to cholesterol and fatty acids are also used to deliver silencing RNAs. Another problem is that most complex diseases are not caused by just one gene. For example, as mentioned earlier in this chapter, the gene *BCL-2* is overexpressed in about 50% of breast cancer cases, and antisense RNA technology can silence this gene in vitro; but breast cancer is a multigene disease and it is not yet possible to silence all of the genes involved in it.

More than a dozen clinical trials using RNAi are under way but no RNAi drugs are on the market yet. Several antisense and RNAi clinical trials are under way in the United States for blindness. One study to combat macular degeneration, a form of blindness, attempts to minimize the expression of the *VEGF* gene. The protein encoded by *VEGF* promotes blood vessel growth. Overexpression of this gene leads to excessive production of blood vessels in the retina, which causes impaired vision and eventually blindness. In 2009 a promising RNAi trial for macular degeneration failed in late-stage clinical trials. Others diseases that have been targeted include several different cancers, influenza (scientists have had good success blocking flu infections in the lungs of mice), diabetes, multiple sclerosis, arthritis, and neurodegenerative diseases.

Curing Genetic Diseases: Targets for Gene Therapy

Most gene therapy researchers are focusing on genetic disorders created by single gene mutations or deficiencies—such as sickle cell disease—because in theory these conditions may be easier to cure by gene therapy than genetic diseases involving multiple genes that interact in complex ways. Current estimates indicate there may be more than 3,000 human genetic disease conditions caused by single genes. Table 11.1 includes some of the diseases that are potential candidates for treatment by gene therapy.

Recent trials for treating deafness, arthritis, melanoma, blindness, AIDS, malignant brain tumors, and other conditions have shown promise. For example, researchers from the University of Pennsylvania used gene therapy to restore retinal cone cell function and day vision in dogs with a condition called congenital **achromatopsia**. This is a rare autosomal recessive condition (1 in 30,000 to 50,000 humans) and affects the cone cells in the retina, which are essential for color vision and some aspects of visual acuity. The therapy cured both young and older canines and appears to be permanent. University of Pennsylvania and Children's Hospital of Philadelphia researchers also reported beneficial results for treatments of **Leber's congenital amaurosis (LCA)**, a degenerative disease of the retina that affects 1 in 50,000 to 100,000 infants each year and causes severe blindness. Young adult patients with defects in the *RPE65* gene were given injections of the normal gene. Complete vision was not restored to these patients, although four of the children who were treated gained enough vision to play sports. Several months after a single treatment with the gene, the patients are still legally blind but they can see more light, some of them can read the lines of an eye chart, and two who had stumbled through an obstacle course were able to navigate it.

Trials using antitumor genes to treat mice with **malignant melanoma**—a cancer of melanin-producing cells called melanocytes, which give the skin its pigmentation—have been successful and similar trials in humans are being planned.

Researchers at the University of Paris and Harvard Medical School have reported that 2 years after gene therapy treatment for ß-thalassemia, a blood disorder that involves a defect in the ß-globin chain of hemoglobin which reduces the production of hemoglobin, a young man no longer needs transfusions and appears to be healthy. A modified, disabled HIV-derived lentivirus vector was used to carry a copy of the normal gene and delivered via blood stem cells. There have also been reports of therapeutic gene integration near a growth factor gene called *HMGA2* that turned this gene on, reminiscent of what occurred in the French X-SCID trials described below. Long-term follow-up will be important to see whether this treatment is effective and safe. Here we consider a few well-studied, successful examples of gene therapy in action.

The first human gene therapy

The first human gene therapy was carried out in 1990 by a group of researchers and physicians at the National Institutes of Health in Bethesda, Maryland, led by W. French Anderson, R. Michael Blaese, and Kenneth Culver. The patient, 4-year-old Ashanti DaSilva, had a genetic disorder called **severe combined immunodeficiency (SCID)**. Patients with SCID lack a functional immune system because of a defect in a gene called adenosine deaminase (*ADA*). *ADA* produces an enzyme involved in metabolism of the nucleotide deoxyadenosine triphosphate (dATP). Mutation of the *ADA* gene results in the accumulation of dATP, which, at high concentration, is toxic to certain types of T cells, resulting in a near-complete loss of these cells in the patient with SCID. This condition is appropriately called "severe combined" immunodeficiency because mutations in the *ADA* gene deliver a knockout blow to the immune system's ability to make antibodies and fight off disease. Without functioning T cells, B cells cannot recognize antigen and make antibodies. Prior to gene therapy, most SCID patients did not live past their teens because their immune systems simply could not fight off infections.

To treat Ashanti, the normal gene for *ADA* was cloned into a vector that was then introduced into a retrovirus. An ex vivo gene therapy approach was used in which a small number of T cells were isolated from Ashanti's blood and cultured in the lab. Her T cells were then infected with the ADA-containing retrovirus, and the infected T cells were further cultured. Because retroviruses integrate their genome into the genome of host cells, the retrovirus was integrating

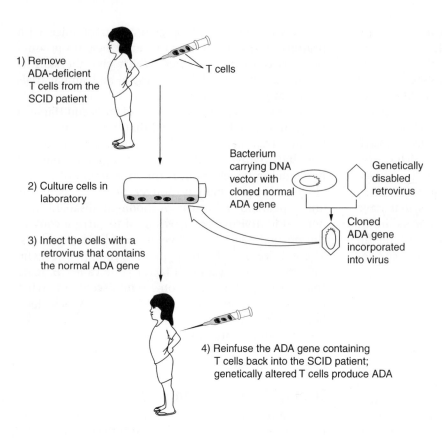

1) Remove ADA-deficient T cells from the SCID patient

T cells

2) Culture cells in laboratory

Bacterium carrying DNA vector with cloned normal ADA gene

Genetically disabled retrovirus

Cloned ADA gene incorporated into virus

3) Infect the cells with a retrovirus that contains the normal ADA gene

4) Reinfuse the ADA gene containing T cells back into the SCID patient; genetically altered T cells produce ADA

FIGURE 11.15 The First Human Gene Therapy An ex vivo gene therapy strategy was used in a 4-year-old SCID patient with a defective ADA gene.

the normal *ADA* gene into the chromosomes of Ashanti's T cells during this culturing. After a short period of culturing, these ADA-containing T cells were reintroduced back into Ashanti (**Figure 11.15**).

Ashanti received multiple treatments. Within a few months after gene therapy, the T-cell numbers in Ashanti began to increase. After 2 years, Ashanti's ADA enzyme activity was relatively high, and she was showing near-normal T-cell counts with about 20% to 25% of her T cells showing the added ADA gene. Ashanti is currently enjoying a healthy life. Since Ashanti's treatment, gene therapy has successfully restored the immune systems of over two dozen children with SCID.

Treating cystic fibrosis

Cystic fibrosis (CF) is one of the most common genetic diseases. Approximately 1,000 children with CF are born in the United States each year, and currently more than 30,000 people in the United States have been diagnosed with CF. This disease occurs when a person has two defective copies of a gene encoding a protein called the **cystic fibrosis transmembrane conductance regulator (CFTR)**. The normal CFTR protein serves as a pump at the cell membrane to move electrically charged chloride atoms (ions) out of cells. Chloride ions enter cells in a number of ways and are required for many cellular reactions. The CFTR is

important for maintaining the proper balance of chloride inside cells. Mutations in the *CFTR* gene, which may cause the total absence of the protein or result in defective protein, are responsible for CF.

The CFTR protein is made by cells in many areas of the body, including the skin, pancreas, liver, digestive tract, male reproductive tract, and respiratory tract (trachea and bronchi). An abnormally functioning or absent CFTR causes an imbalance in chloride ions inside the cells because the defective CFTR does not pump out these ions (**Figure 11.6**). In organs such as the trachea, an accumulation of chloride ions in these cells leads to the production of an extremely thick sticky mucus that clogs the airways. This occurs because water moves into chloride-rich cells in an effort to balance chloride concentrations inside the cells. Normally, mucus in the trachea helps sweep dust and particles out of the airways to keep these materials from reaching the lungs. But when water enters tracheal cells, the mucus becomes extremely thick. In addition to clogging the airways, the thick mucus provides an ideal environment for microbes to grow; as a result, patients with CF experience infections from bacteria such as *Pseudomonas*.

Infections of the airways and lungs can lead to pneumonia and respiratory failure, the leading cause of death among patients with CF. There are similar effects

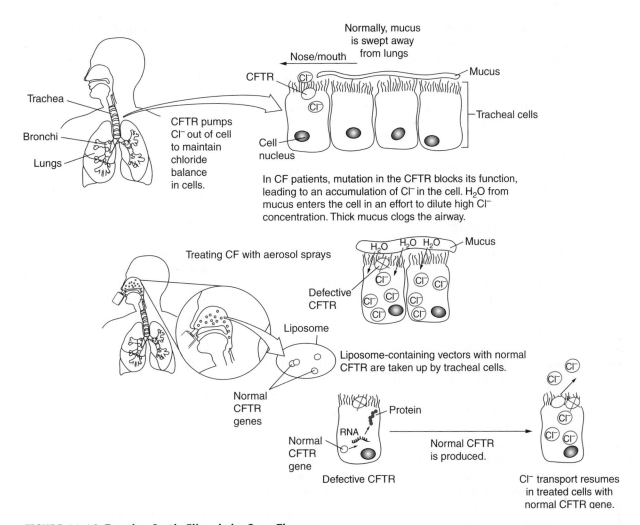

FIGURE 11.16 Treating Cystic Fibrosis by Gene Therapy

in many other organs of the body. Males experience infertility owing to problems related to ion transport in male reproductive organs, and both males and females have extremely salty sweat owing to abnormalities in ion transport in sweat glands. In fact, prior to the discovery of the *CFTR* gene, a sweat test was the standard for diagnosing CF and is still used today.

Treatments for patients with CF vary from back clapping—holding the patient almost upside down and slapping the back—and moving body positions to drain lung mucus, to using drugs to thin the mucus and antibiotic treatment to fight infections. However, there is no cure for CF. This disease often causes death due to lung malfunctions and respiratory infections early in life. Many new treatment strategies have enabled these patients to live to well into adulthood; as a result, the average life span of CF patients has increased from 19 years in the 1980s to over 37 years currently. In 1989, the *CFTR* gene was discovered, and by 1993 scientists had begun gene therapy trials by introducing

the normal *CFTR* gene into viral vectors. Recent studies are using liposomes and spraying these into the noses and mouths of CF patients as an aerosol or administering into the airways through a hose (Figure 11.16). Liposomes fuse with lipids in cell membranes of tracheal cells, releasing the normal *CFTR* gene into the cytoplasm of cells. The normal *CFTR* gene is copied into mRNA and the normal protein translated. The normal CFTR protein enters cell membranes and starts to transport chloride ions out of cells, thereby thinning the mucus and alleviating CF symptoms.

Gene therapy for CF is not yet a reliable cure. It is expensive, requiring multiple reapplications, because DNA delivered via liposomes does not integrate into chromosomes. Each time tracheal cells divide, and they divide rapidly, delivered genes are lost and more spraying is required. Also, *CFTR*-containing liposomes are taken up by a small percentage of tracheal cells. Even cells containing the delivered gene may not produce enough CFTR protein to allow for adequate transport of chloride ions. Furthermore, there have been problems with the expressed CFTR protein being toxic to cells. In addition, there are now over 1,500 known mutations of the *CFTR* gene, potentially resulting in variations of CF and complicating the administration of gene therapy. Although a gene therapy cure for CF is not yet available, scientists are aggressively moving forward on strategies that may eventually lead to the permanent introduction of the normal gene or correction of the defective *CFTR* in an effort to improve the lives of people with CF.

Challenges Facing Gene Therapy

Scientists have always been concerned about the potential risks associated with gene therapy and the safety of these procedures. Discussions about the safety of gene therapy greatly intensified after 18-year-old Jesse Gelsinger died during a gene therapy clinical trial at the University of Pennsylvania in 1999. Jesse's death was directly attributed to complications related to the adenovirus vector used to deliver therapeutic genes to treat him for a liver disorder (ornithine transcarbamylase deficiency) that affected his ability to break down dietary amino acids. Jesse's death was triggered by a massive inflammatory response to a modified adenovirus vector bearing the *ornithine transcarbamylase (OTC)* gene that had been injected into his hepatic artery. The vectors were intended to enter liver cells and result in the production of OTC protein in the hope that this treatment would cure him of his liver disease. Within hours of his first treatment, a massive immune reaction surged through Jesse's body. He developed a high fever, his lungs filled with fluid, multiple organs shut down, and he died 4 days later of acute respiratory failure.

During inquiries into this tragedy, it was learned that clinical trial scientists had not reported other adverse reactions to gene therapy and that some of the scientists involved in this trial were affiliated with private companies that could benefit financially from the trials. It was found that serious side effects seen in animal studies were not explained to patients during informed-consent discussions. The FDA then scrutinized gene therapy trials across the country, halted a number of them, and shut down several. Other research groups voluntarily suspended their gene therapy studies. Over 500 gene therapy clinical trials have been carried out around the world, a majority of these in the United States, and more than 600 trials are ongoing in 20 countries. Jesse Gelsinger was the first person in a gene therapy clinical trial to die as a result of his treatment. His death raised more questions about using viral vectors, placed greater emphasis on the development of nonviral vectors, and called for greater scrutiny of gene therapy and tighter restrictions on gene therapy trials.

In 2002, concerns about gene therapy involving retroviruses were further elevated as a result of trials in France for treating X-linked severe combined immunodeficiency syndrome (SCID-X). In this trial, 3 of 11 children treated developed leukemia because of the therapy, and one of them died in 2004. They had received injections of bone marrow cells that had been treated (ex vivo) with a retrovirus-delivered therapeutic gene. The 2 surviving leukemic children were 1 and 3 months old at the time of treatment and had returned home to a normal life, seemingly cured, until they developed a leukemia-like cancer about 2½ years later. Their cancer was caused by retrovirus vectors that randomly integrated the therapeutic gene into a critical location of the genome containing the promoter region for a gene called *LM02*, which encodes a transcription factor required for the normal formation of white blood cells. This integration led to aberrant transcription of the *LM02* gene and overexpression of *LM02*, which triggered the uncontrolled division of mature T cells.

This tragedy resulted in the temporary cessation of a large number of gene therapy trials, and the FDA completely stopped most retroviral studies. Trials eventually resumed, but with greater patient monitoring.

Currently there are more barriers to gene therapy than solutions to medical problems. These include the following:

- How can expression of the therapeutic gene be controlled in the patient? What happens if therapeutic genes are overexpressed or if a gene shuts off shortly after it has been introduced?

- How can scientists safely and efficiently target only the cells and tissues that require the therapeutic gene without affecting other cells in the body where the gene is not needed?

- How can gene therapy be targeted to specific regions of the genome to prevent the random integration problems encountered in the French trials or instability and movement of the inserted DNA? One active area of research showing promise involves targeted genome editing using proteins called **zinc finger nucleases (ZFNs)**. These nucleases act as DNA-cleaving genome scissors to cut out specific areas of DNA, such as a defective gene, and ZFNs can be used to cut out and replace a sequence. Plasmids encoding ZFNs to "edit" out and replace defective genes is an area of intense research in gene therapy. A ZFN-based strategy is being used in the first gene therapy trial under way for patients with HIV in an attempt to disrupt the *CCR5* gene, which encodes a protein that HIV uses to enter cells, in the hope that this will halt the spread of the virus.

- How can gene therapy provide lasting, permanent treatment without frequent administration of the therapeutic gene?

- How can rejection of the therapeutic gene be avoided? Whenever gene therapy is used, it is not always known if the recipient's immune system will reject the protein produced by therapeutic genes or reject genetically altered cells containing therapeutic genes.

- How many cells must express the therapeutic gene to treat the condition effectively? This will vary depending on the disease condition, but it remains to be determined if a majority of diseased cells must be affected by the therapeutic gene or if a disease can be treated by correcting only a small number of cells.

These and other barriers must be overcome before gene therapy becomes a safe and reliable treatment approach, but scientists are making excellent progress in this field. They are also making incredibly rapid advances in another hot area of medical biotechnology called regenerative medicine, the topic of the next section of this chapter.

11.4 The Potential of Regenerative Medicine

Currently, physicians treating most human illness are primarily limited to approaches such as surgical techniques, radiation treatment, and drug therapy. Although these approaches all have a place in medicine to treat certain human conditions, they do not offer the ability to regenerate tissue or restore the functions of damaged organs. For example, when a person has a heart attack or stroke, tissue damage often results. When an organ is damaged, the only way to restore its functions fully is to replace the damaged tissue with new ones—something that often does not occur naturally in heart or brain tissue.

When organ development occurs in the embryo, changes in the expression of many genes must occur in an ordered sequence. Unwanted changes that even drugs cannot fix may occur in cells, and no one drug can stimulate the growth and repair of new tissue when an organ is severely damaged. Even with new knowledge gained from the Human Genome Project, it is highly unlikely that any one drug or even a few drugs could be used to stimulate hundreds of changes in gene expression with the proper timing required for tissue regeneration and the restoration of organ function.

Regenerative medicine, growing cells and tissues that can be used to replace or repair defective tissues and organs, is an exciting field of biotechnology that holds the promise and potential for radically changing medicine and the delivery of health care as we know it. Most researchers in the field agree that the goal of regenerative medicine is not to extend the human life span and achieve immortality but to improve the quality of life by making it healthier.

Cell and Tissue Transplantation

Organ transplantation is not a new idea, but applications that involve transplanting specific cells and tissues to replace or repair damaged tissues are relatively new aspects of medical biotechnology research.

Fetal tissue grafts

Neurodegenerative diseases occur gradually, leading to progressive loss of brain functions over time. Alzheimer disease and Parkinson disease are perhaps the two best-known examples. These diseases rank first and second, respectively, as the most common neurodegenerative disorders. For Parkinson disease alone, approximately 50,000 cases are diagnosed yearly, and an estimated 500,000 Americans currently have the disease. Parkinson disease is due to the loss of cells in an area deep inside the brain called the *substantia nigra*. Neurons in this region produce a chemical called *dopamine*, a neurotransmitter or chemical used by neurons (nerve cells) to signal one another. Loss of these dopamine-producing cells causes tremors, weakness, poor balance, loss of dexterity, muscle rigidity, a reduced sense of smell, inability to swallow, and speech problems, among other effects. Most treatments involve drugs that increase the production or accumulation of dopamine in the brain; however, after about 4 to 10 years of drug treatment, the disease progresses and the effectiveness of these drugs diminishes, leading to a poor quality of life for the patient, who typically dies of complications related to the disease.

Unlike fetal neurons, which can divide, most adult neurons will not repair themselves when damaged, and most neurons do not undergo cell division. Scientists have long been interested in using fetal neuron transplants as a way of treating Parkinson disease and other neurological conditions. The basic idea is to introduce fetal neurons in the hope that these cells can establish connections with other neurons, replace the damaged brain cells, and restore brain function. After demonstrated success in rodents, fetal tissue transplants have been used since the late 1980s, and well over 100 patients have received such transplants. Most human fetal tissue comes from embryos or fetuses obtained from accident victims and legally aborted embryos. Patients receiving fetal transplants have shown varying degrees of improvement, including relief of parkinsonian symptoms in over 40% of patients and in some cases the almost complete elimination of most symptoms even several years later, but fetal transplants have not provided full recovery.

Over 250,000 individuals have been paralyzed by trauma to the spinal cord, and nearly 2 million people worldwide are living with spinal cord injuries. Each year approximately 85,000 people suffer spinal cord injuries, including roughly 10,000 in the United States. Damage can occur when the cord is crushed or the nerve fibers are severed. Incomplete or complete severing of the spinal cord may result in paraplegia, paralysis of the lower body, or quadriplegia, paralysis of the body from the neck down, depending on where the injury occurred. Many strategies have been used in attempts to repair spinal cord injuries. One approach has been to graft nerve fibers from fetal or adult neurons into the damaged area of the spinal cord so as to bridge the parts of the cord that were severed (**Figure 11.17**). Such bridge implants have shown promise in dogs and rats. As scientists learn more about the inflammatory chemicals that hinder nerve growth and the factors that stimulate it, it may be possible to use such molecules to minimize scar tissue formation, reduce damage caused by scar tissue to supporting cells called glial cells, block growth inhibitor molecules, and stimulate neuron regeneration at the same time.

Organ transplantation

Organ transplantation can and does save lives. Approximately 8 million surgeries related to tissue damage and organ failure are performed in the United States each year, but about 4,000 people also die each year while waiting for an organ transplant. At least 100,000 people die each year without ever qualifying to be on a waiting list. Well over $400 billion is spent on organ failure and tissue-related health care costs in the United States. This number represents nearly half of the nation's health care bill.

Autografting—that is, the transplantation of a patient's own tissue from one region of the body to another—can alleviate some transplantation problems. For example, coronary artery bypass operations involve removing segments of a vein from the patient's leg and connecting it surgically to arteries in the heart as a bypass around obstructed vessels. But if a patient needs a heart or a liver transplant, another person who can donate an organ for the recipient must be found. Even when a human donor who appears to be a match is found, organ rejection is a major problem. Rejection typically occurs when the recipient's immune system recognizes that the donor organ is foreign. Matching organs for transplantation involves tissue typing to check if a donor organ is compatible for a recipient.

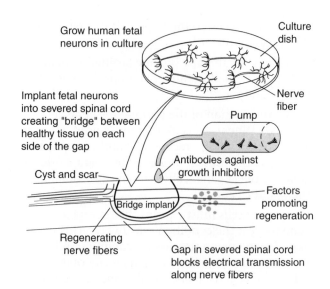

FIGURE 11.17 **Bridging the Gap** In many types of spinal cord injuries, a gap occurs when the spinal cord is severed. This blocks electrical transmission of nerve impulses and can result in paralysis and the loss of body sensations such as temperature, pain, and touch. One experimental strategy for repairing spinal cord injuries involves implanting fetal neurons into the severed area of the spinal cord to create a bridge between normal neurons. Scientists are using this strategy in combination with antibodies that may block chemicals that inhibit spinal cord repair and growth factors that can stimulate neuron regeneration in an effort to bridge the gap in spinal cord injuries.

Tissue typing is based on marker proteins found on the cell surface (membrane) of every cell in the body. Tissue typing proteins are part of a large group of over 70 genes called the **major histocompatibility complex (MHC)**, aptly named because *histo* means "tissues," and MHC molecules must be matched between donor and recipient to have a compatible organ for transplantation. There are many different types of MHC proteins. One common group, called the human leukocyte antigens (HLAs, named because they were first discovered on white blood cells, or leukocytes), are found on virtually all body cells. Immune system cells such as B and T cells recognize HLAs on all body cells present since birth as "self" (belonging to the same individual), whereas any other cells are "nonself," or foreign cells that may be attacked by the immune system and destroyed. Some common HLAs are found on most human tissues, and others are unique to a given individual. To have a successful transplantation of an organ from one human to another requires a close match of several types of HLAs between the donor organ and the recipient's cells; otherwise the recipient will reject the transplanted organ.

Since the first human liver transplant in 1963, transplant surgeons have been using immunosuppressive drugs to weaken the recipient's immune system and minimize organ rejection. Most transplant recipients must use immunosuppressive drugs for the rest of their lives. One obvious problem with this approach is that patients on immunosuppressive drugs can and do develop infections, which, because of their weakened immune systems, can be life-threatening. The lack of sufficient human organ donors and the problem of organ rejection are major reasons why scientists are looking at other ways to provide donor organs.

Xenotransplantation—the transfer of organs from different species—may one day become a viable alternative to human-to-human organ donation, thus helping relieve the tremendous need for human donor organs. Baboons were once considered the animal of choice for providing organs to human recipients. The first animal-to-human organ transplant in a child, carried out in 1984 by doctors at Loma Linda University Medical Center in California, involved transplanting a baboon heart into Baby Fae, a 12-day-old girl. Baby Fae lived with the baboon heart for 3 weeks before she died of complications related to organ rejection. Similar transplants have been performed without great success. Although baboons and other primates may still be candidates for providing organs, many groups are choosing to investigate the potential of using pigs as organ sources. Pigs may be a good choice because they are plentiful, easy to breed, and relatively inexpensive. Many pig organs are also similar in function and size to human organs. Progress on using pig organs for transplantation in humans has been slowed by concerns that viruses may be transmitted from pigs to humans, causing the transplanted organs to be rejected and creating other health problems.

Transplantation scientists have combined molecular techniques and transplantation technologies to produce cloned pigs that may help overcome current fears of organ rejection and viral disease transmission. Researchers at the University of Missouri have created cloned piglets that lack a key gene called *GGTA1* (ß-1, 3-galactosyltransferase). *GGTA1* produces a sugar on the surface of pig tissues, which, when transplanted into a human, would be recognized as a foreign antigen, leading to antibody production and rejection of the organ. The *GGTA1* knockout pigs were cloned using the nuclear transfer cloning techniques discussed in Chapter 7. Creating *GGTA1* knockout pigs may be a way to generate pigs that could produce organs for transplantation that the human immune system may not recognize as foreign (**Figure 11.18**).

Xenotransplantation does not always have to involve the transfer of a whole organ, as you will learn in the next section. Scientists are working hard to develop ways to deliver small clusters of cells as a technique for cell and tissue transplantation.

Cellular therapeutics

Cellular therapeutics involves using cells, instead of whole tissues or organs, to replace defective tissues or to deliver important biological molecules. One alternative for avoiding organ rejection of transplants is to use living cells that have been encapsulated into tiny plastic beads or tubes called **biocapsules** or microcapsules. Biocapsules may also contain genetically engineered cells designed to produce therapeutic molecules such as recombinant proteins.

Biocapsules have tiny holes in their walls, making them permeable to nutrient exchange and allowing molecules produced by the encapsulated cells to escape from the capsule and enter the bloodstream or surrounding tissues (**Figure 11.19** on the next page). For instance, capsules containing insulin-producing cells (beta cells) from the pancreas, implanted in patients with type I diabetes, would produce insulin that could travel out of the capsule into the bloodstream of the patient to all body organs requiring insulin. Another important feature of biocapsules is that they protect cells from being attacked by the recipient's immune system by hiding them within capsules, where immune cells and antibodies cannot reach and destroy them. Although not a permanent cure, biocapsules can provide lasting release of molecules into the body. This approach would likely require that biocapsules be changed every few months; however, in the case of diabetes, this might be a better alternative than daily injections of insulin.

FIGURE 11.18 Pigs Could Potentially Save the Lives of Patients Waiting for a Transplant These piglets have been engineered to lack a sugar-producing gene that causes human bodies to reject pig organs, potentially providing a source of rejection-free pig organs.

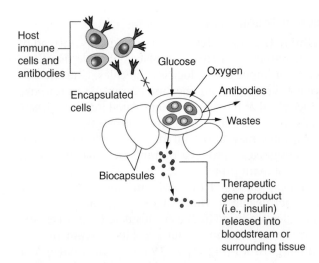

FIGURE 11.19 Biocapsules Encapsulated cells may provide valuable ways to deliver therapeutic molecules. Cells in the biocapsule are protected from attack by immune cells and antibodies of the host. At the same time, biocapsules allow molecules produced by the cells to leave the capsule and provide therapeutic benefits for the host. This figure illustrates how insulin-producing cells could be used to provide a patient with diabetes with a source of insulin.

Tissue Engineering

Whether you pay $30,000 or $300 for your first car, one thing is certain: over time parts will wear out and break. A trip to the local mechanic can repair and replace some car parts, but eventually the car wears out to the point of no return, and it is time to buy another car. Wear and tear on human body parts also take their toll. Over time, organs do not work as well as they should; in some cases, an organ may stop functioning altogether. Even if a person lives a relatively healthy life, the wear and tear of aging or a sudden event such as a stroke or heart attack will lead to a decline in organ function and perhaps organ failure. But our bodies are not like cars: we cannot go to a warehouse of body parts for replacements.

In the future, however, the emerging science of **tissue engineering** may provide tissues and organs that can be used to replace damaged or diseased tissues. This small but growing industry—there are over 60 biotechnology companies involved in tissue engineering in the United States alone and experts predict that the field will grow by 50% in the next decade—is actively involved in research to engineer human tissues and organs as replacements for worn out and damaged tissues. As we age, we outlive the functional abilities of our organs. It has been estimated that roughly 20% of people over the age of 65 in developing nations will benefit from some type of

tissue engineering application in their lifetimes. Tissue engineering and regenerative medicine approaches are also predicted to lead to substantial cost savings for care of chronic diseases such as congestive heart failure and diabetes. In the case of congestive heart failure, if regenerated heart tissues restored cardiac function for even a small percentage of individuals, cost savings could be several billion dollars annually worldwide.

Tissue engineering scientists often begin by designing and constructing a biomaterial framework or scaffold made of biological substances such as calcium, collagen, a polysaccharide called alginate, or biodegradable materials (**Figure 11.20a**). The scaffold is shaped as a mold of the tissue or organ to be made, and its purpose is to create a three-dimensional framework onto which cells are placed. Growing human cells on the scaffolding is called *seeding* because the cells literally act as "seeds" to create more cells that will grow over the scaffold. Scaffolds seeded with cells are bathed in a nutrient-rich medium and, over time, cell layers build up over the scaffolding material to assume the shape of the scaffold. Engineered sheets of human skin have already proven useful for treating severe burn victims who have lost substantial portions of their skin, a very painful and life-threatening condition. Sheets of skin grafts have proven to be successful organs grown by tissue engineering. Engineered bone structures for healing bone fractures have also worked fairly well. Scaffolding to engineer teeth and blood vessels has shown promise, and many products of tissue engineering are FDA-approved and being used to treat patients or in clinical trials.

In the 1990s, tissue engineering pioneer Charles Vacanti and colleagues made headlines around the world when they revealed a mouse with an engineered ear growing on its back. In this example, a biodegradable scaffold in the shape of an outer ear was attached to a mouse and then seeded with cartilage cells from cows. The cartilage cells infiltrated the scaffolding and produced cartilage as they grew; then, as the scaffolding degraded, the cartilage developed enough strength to support itself. The human ear–looking tissue was never transplanted onto a human. Because it was made of cow cells, it would have been rejected by the human immune system. Also, this tissue was just the outer ear without the inner ear structures that actually detect sound, but it provided strong evidence that tissue engineering could work.

Creating large and complex organs such as the liver, heart, and kidney has proven to be much more difficult, although fetal tissue has been used to grow a rudimentary kidney in rats that was able to produce a urine-like fluid. At least two phase II clinical trials are under way in which human bladders were created using tissue

FIGURE 11.20 Tissue Engineering (a) An example of collagen scaffolding used to engineer tissues. (b) A urinary bladder created from stem cells.

biopsies from patients' bladders to seed scaffolding and produce new bladders (Figure 11.20b). The field of tissue engineering is in its infancy. But progress is advancing at an incredibly rapid rate and, as discussed in the next section, applications involving stem cells are adding to this progress, and there is every reason to expect that engineered organs will become a reality.

The telomere story

In normal human cells, the ends of chromosomes contain sequences of DNA nucleotides called **telomeres**. Telomeres are usually 8,000 to 12,000 bp units of the repeating sequence 5'-TTAGGG-3'. Think of these as the plastic tabs at the ends of shoelaces that prevent the laces from unraveling, a sort of chromosome "cap." Normal cells have a limited ability to proliferate. Most human body cells can divide a maximum of 50 to 90 times before they show signs of aging—a process called

senescence—which eventually leads to cell death. A cell's life span is affected in part by telomeres.

Each time a cell divides, telomeres shorten slightly. This occurs because of a basic flaw that prevents DNA polymerase from completely copying the ends of both strands of a DNA molecule. In many ways, telomeres serve as a biological clock for counting down cell divisions leading to senescence and cell death. Telomeres shorten, and senescence occurs until the cell can no longer divide. If there are multiple copies of the TTAGGG repeats at the ends of chromosomes, cells can lose this DNA without the loss of precious gene sequences. Eventually loss of repeat sequences produces a critical loss of DNA, so cells no longer divide (**Figure 11.21** on the next page).

Scientists have long known that many cancer cells can divide indefinitely—a property called *immortality*. One way in which cancer cells achieve immortality is through the actions of an enzyme called **telomerase**. It repairs telomere length at the ends of chromosomes by adding DNA nucleotides to cap the telomere after each round of cell division. Telomerase is not active in normal cells but is active in over 90% of human cancers. By preventing telomere shortening, telomerase activity is a major reason why cancer cells can divide indefinitely. In fact, biologists call cancer cells immortal because of their ability to avoid senescence indefinitely.

Telomere shortening is involved in the aging process—in the aches and pains, wrinkles, arthritis, gray hair, and other symptoms humans experience as we age. Telomerase is not a "fountain of youth" cure for the effects of old age. Aging and cancer are far more complex processes that involve many proteins, not just telomerase. Although high levels of telomerase are found in almost every human cancer cell, telomerase itself does not *cause* cancer. Telomerase in combination with genetic mutations in genes that control cell division can create immortal cells that avoid senescence. Overproduction of telomerase often correlates with the aggressive growth of tumors. Researchers are working on cancer treatment strategies such as telomerase peptide vaccines to inhibit telomerase and stop cancer cells from dividing. Also, two companies have recently produced tests to measure the length of one's telomeres as an indicator of health status, although the diagnostic and predictive values of such tests are being debated.

From a tissue engineering perspective, scientists are investigating how introducing telomerase genes into cultured human cells can allow them to produce normal human cells that display immortality. If these efforts are successful, immortal human cells could be valuable in treating individuals with age-related disorders ranging from arthritis to neurodegenerative diseases. Such cells could also be used in many other ways, as for providing skin cells for healing bedsores and ulcers and

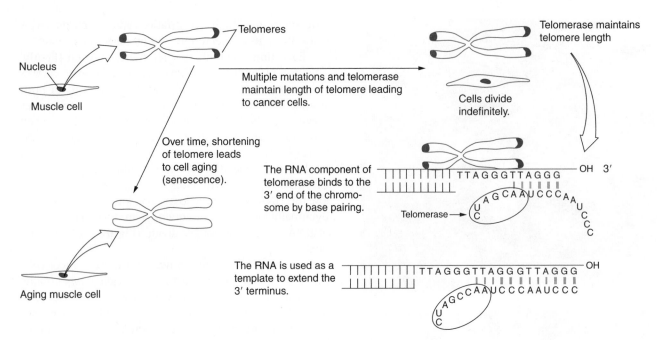

FIGURE 11.21   Telomere Shortening Leads to Cell Aging   Telomeres shorten with each round of DNA replication. Eventually telomere shortening leads to cell senescence and death. Through mutations that affect the functions of different genes and expression of the enzyme telomerase, cancer cells can divide indefinitely and avoid senescence. Telomerase plays a role in this process by continually filling in telomere sequences to prevent them from shortening during cell division.

treating patients with late-onset blindness, and muscular dystrophy. In the future, the technology might even be available to remove a patient's aging cells, introduce telomerase genes to these cells to extend the patient's life, and then return the cells to the patient. Some of this work could be done in combination with stem cells, which are perhaps the hottest and most controversial topic in medical biotechnology today.

Stem Cells

The tremendous promise and controversy surrounding stem cells has made stem cell research and related topics regular themes of front-page news items and TV headlines. Stem cells evoke emotional and controversial responses from scientists, clergy, politicians, and the general public. Among some people, the isolation and use of these cells engenders excitement, fear, anger, and a range of other emotions.

What are stem cells?

As you will soon learn, there are many different types of **stem cells**; but in general all stem cells have two basic characteristics that set them apart from other cell types: **self-renewal** and **differentiation** into specialized cells:

- Self-renewal: Stem cells grow and divide (proliferate) indefinitely by mitosis to create populations of identical stem cells.

- Differentiation: This is a complex process involving many genes that must be activated and silenced, and differentiating cells rely on chemical signals such as growth factors and hormones from other cells to help them change. Stem cells are special because they can eventually differentiate to form all of the more than 200 cell types that make up the human body. Stem cells are called **pluripotent** because they have the potential to develop into a variety of different types.

To understand what stem cells are, we must look briefly at the development of the human embryo. We do this by considering how **in vitro fertilization (IVF)** is carried out. IVF first gained public attention in 1978 when Louise Brown, the first test tube baby, was born. To create a child by IVF, sperm and egg from donor parents are mixed together in a culture dish to produce an embryo. After several days of division, the embryo is surgically implanted in the uterus of a woman, usually the egg donor, who has been treated with hormones to prepare her uterus for implantation. When a couple agrees to undergo IVF, several embryos are usually created, but often only one is implanted during each procedure. The remaining embryos are frozen for future use as needed. Potentially, the leftover embryos can be a source of **human embryonic stem cells**

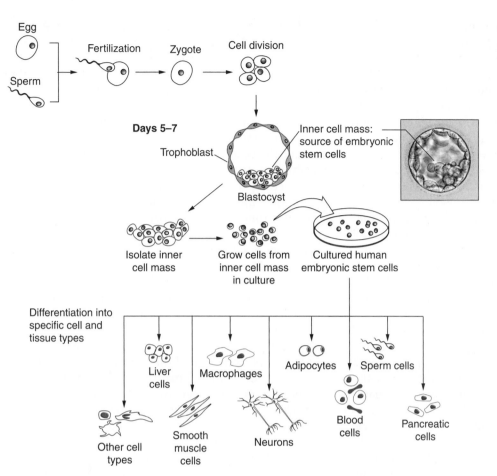

FIGURE 11.22 Isolating and Culturing Human Embryonic Stem Cells (hESCs) Cells isolated from the inner cell mass of human embryos can be grown in culture as a source of hESCs. Under the proper growth conditions, hESCs can be stimulated to differentiate into virtually any cell type in the body.

(hESCs), but they are also the source of a great deal of controversy.

An embryo goes through a predictable series of developmental stages (**Figure 11.22**). After an egg cell is fertilized, it is called a **zygote**. The zygote divides rapidly and, after 3 to 5 days, first forms a compact ball of about 12 cells called a **morula**, meaning "little mulberry." Around 5 to 7 days after fertilization, the embryo consists of a small hollow cluster of approximately 100 cells called a **blastocyst**. The blastocyst is approximately one-seventh of a millimeter in diameter. It contains an outer row of single cells called the trophoblast; this layer develops to form part of the placenta, which nourishes the developing embryo. The area of cells of primary interest to stem cell biologists is a small cluster of around 30 cells tucked inside the blastocyst, which form a structure known as the **inner cell mass**, the source of hESCs (Figure 11.22).

Cells of the inner cell mass develop to form the embryo itself. Stem cells in the inner cell mass have the ability to undergo differentiation. Successful isolation and culture of the first hESCs from a human blastocyst was reported in 1998 by James Thomson of the University of Wisconsin at Madison who had cultured

hESCs from rhesus monkeys 2 years earlier. Also in 1998, John Gearhart and colleagues at Johns Hopkins University isolated embryonic germ cells, primitive cells that form the gametes—sperm and egg cells—from human fetal tissue and demonstrated that these cells could develop into different cell types. These discoveries followed the work of other scientists who had isolated stem cells in species such as mice, pigs, cows, rabbits, and sheep. In fact, stem cell researchers credit much of what is now known about isolating hESCs from pioneering work initiated in mice in the 1980s—another outstanding example of how model organisms contribute greatly to the advancement of science.

Human ESCs avoid senescence and show no signs of aging, in part because they express high levels of telomerase. Several groups have maintained stem cells for over 3 years and over 600 rounds of division without apparent problems. Cultured cells such as these, which can be maintained and grown successively, are called **cell lines**. Stem cells also grow rapidly and can be frozen for long periods of time and still retain their properties. Under the right conditions, when they are stimulated with different molecules including hormones and substances called **growth factors**, stem cell lines can be coaxed to differentiate into different types of cells. This *directed differentiation* of stem cells into specific differentiated cells of interest is key to creating tissues for regenerative medicine applications. A major focus of stem cell research is experimentation to determine what controls the pluripotency of stem cells and to identify the factors that stimulate their differentiation into discrete cell types. These signals include substances called growth factors, hormones, and peptides, which stimulate differentiation in tissue-specific ways.

For example, signaling systems that involve transforming growth factor-ß (TGF-ß), bone morphogenetic proteins (BMPs), and other growth differentiation factors act on a gene for a transcription factor called *Nanog*. NANOG is one key protein that maintains hESCs in an undifferentiated pluripotent state. Once stem cells are produced and isolated, scientists use a number of tests to determine their pluripotency. In the laboratory under the proper culturing conditions, ESCs from humans, mice, rats, primates, and other species have been shown to differentiate into a myriad of cells including skin cells; brain cells (both neurons and glial cells, which support, nourish, and protect neurons); cartilage (chondrocytes); spermatozoa; osteoblasts (bone-forming cells); liver cells (hepatocytes); insulin-secreting pancreatic beta cells; muscle cells including smooth muscle, which forms the walls of blood vessels; skeletal muscle cells, which form the muscles that attach to and move the skeleton; and cardiac muscle cells (myocytes), which form the muscular walls of the heart. How do researchers obtain hESCs? hESCs for research purposes are derived from the blastocysts of embryos that are no longer needed for IVF or from human embryos created by IVF from sperm and egg cells donated in order to provide embryos for research purposes. Typically, leftover blastocysts would either be destroyed or frozen indefinitely, but with the consent of the couple who provided the sperm and egg, they can be used to derive hESCs, as described in Figure 11.22. In U.S. fertility clinics alone, an estimated 400,000 frozen unused embryos and several thousand eggs are discarded annually.

Other Sources of Stem Cells

Research on hESCs is very controversial because of their source—an early embryo. Scientists have discovered **adult-derived stem cells (ASCs)**, cells that reside in mature adult tissue and could be cultured and differentiated to produce other cell types. ASCs appear in very small numbers, and although they have been isolated from the heart, brain, intestine, hair, skin, pancreas, bone marrow, fat, mammary glands, teeth, muscle, and blood, they have not yet been discovered in all adult tissues.

Opponents of hESC research often claim that ASCs are a more acceptable alternative than using hESCs because isolating ASCs does not require the destruction of an embryo. ASCs can be harvested from people by fine-needle biopsy, through a thin diameter needle inserted into muscle or bone tissue. It may even be possible to isolate ASCs from cadavers. We also know that ASCs are present in fat (adipose) tissue, which could potentially be an outstanding source of stem cells, especially if you consider that over 500,000 L of fat tissue collected by liposuction and other cosmetic surgery techniques are discarded in the United States each year.

Experiments have shown that ASCs from one tissue can differentiate into another different specialized cell type. For instance, an ASC isolated from muscle tissue could be used to develop into a blood cell. But other studies have demonstrated that ASCs may not be as pluripotent as hESCs. Much more research is required to determine whether ASCs can be as valuable as hESCs might be.

Stem cells can be isolated from human amniotic fluid, the protective fluid that surrounds a developing fetus. In the lab, these **amniotic fluid–derived stem cells (AFSs)** have been coaxed to become neurons, muscle cells, adipocytes, bone, blood vessels, and liver cells. It is not entirely clear whether these cells are truly different from hESCs or ASCs; but if so, they may be a key breakthrough in stem cell technologies.

Cancer stem cells (CSCs), also called tumor-initiating cells, have been identified and implicated in the development of cancers, tumor progression, tumor metastasis, and the recurrence of cancers. Like normal stem cells, CSCs can self-renew and differentiate to form the tissues from which they were derived. Certain CSCs grow slowly in clusters or *niches* within a tissue. It is not clear what properties CSCs may have besides the ability to form a tumor. Researchers are also not sure whether CSCs are derived from normal cells or if they are involved in cancer tumor resistance to chemotherapies, but these cells are a focus of intense research and potential therapeutic treatments for the treatment of cancers.

Creating stem cells by nuclear reprogramming of somatic cells

Research on stem cell biology is an extremely active field. A primary focus continues to be alternative approaches for producing pluripotent stem cells without destroying an embryo. One of the most promising new approaches for doing this involves a technique called **nuclear reprogramming of somatic cells**. The basic concept of this approach is to use genes involved in cell development to push a somatic cell back to an earlier stage of development and affect gene expression and thus to reprogram the somatic cell

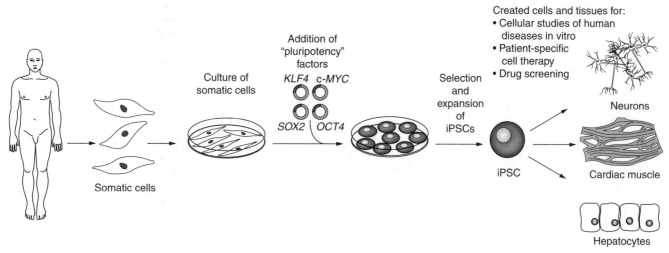

FIGURE 11.23   Nuclear Reprogramming of Somatic Cells to Produce Induced Pluripotent Stem Cells The introduction of four transcription factor genes (*Oct3/4, Sox2, c-myc, Klf4*) into mouse fibroblasts results in the formation of induced pluripotent stem cells (iPSCs). Invariably a number of these cells are defective in reprogramming and must be selected out during culture. They must also be selected for the expression of endogenous marker genes (such as *Nanog* and *Oct*) known to be expressed in pluripotent stem cells.

genetically to return to a pluripotent state characteristic of the stem cells from which it was derived. It was previously thought that once cells differentiated to become specific, specialized cell types, for example a skin cell, that their differentiation fate was irreversible. But we now know that this is not the case. One of the first successful reprogramming techniques involved fusing hESCs with skin cells called *fibroblasts*. The hybrid cells generated in this way displayed several properties of hESCs both in vitro and in vivo.

There are several different approaches to nuclear reprogramming, and these are being heralded as a revolution in stem cell research. One approach has involved using retroviruses to deliver four transgenes—*OCT3/4, SOX2, c-MYC*, and *KLF4*—into fibroblasts (**Figure 11.23**). Expression of these four genes, which encode transcription factors involved in cell development, "reprograms" the fibroblasts back to an earlier stage of differentiation. Such reprogrammed cells are called **induced pluripotent stem cells (iPSCs)**. iPSCs demonstrate many properties of hESCs, such as self-renewal and pluripotency, and appear to be indistinguishable from hESCs. iPSCs have been produced from human, mouse, rat, pig and monkey cells. Subsequently a cocktail of RNA molecules for the four genes described above has also been used to create iPSCs from somatic cells. Also, human neural stem cells have been reprogrammed to iPSCs by introducing only *OCT4*.

These iPSCs express genes such as *Nanog* and *Oct*, which are characteristic markers known to be expressed in undifferentiated hESCs. Other experiments have also demonstrated that iPSCs can differentiate into other cell types including neural cells and cardiac muscle cells. Nuclear reprogramming may be a way to generate patient-specific iPSCs without the need for an embryo and hESCs. iPSCs have been successfully derived by reprogramming human skin cells from patients. Reprogrammed skin cells taken from the face of a 36-year-old woman, connective tissue cells from the joints of a 69-year-old man, and skin cells from the foreskin of a newborn boy are examples. iPSCs have also been produced without using the *c-myc* gene. This is a potentially important advance because *c-myc* is a known oncogene.

Stem cells derived in this way could be used for patient-specific cell therapies. In addition, with iPSC technologies, it is theoretically possible to create disease-specific stem cells from individual patients. For example, one could take a tissue biopsy, such as skin, from a person with a particular disease and reprogram those cells into stem cells that could then be used to create cell types for combating the disease. Patient-specific iPSCs could be used for cell-based therapies without the risk of immune rejection. But even the most optimistic iPSC researchers believe that such cell therapies will not be ready for at least another decade or more.

Scientists are also very excited about how reprogrammed cells from patients can be used to study "diseases in a dish." Cultured reprogrammed cells from diseased patients are being studied to help scientists better understand human disease progression and disease processes. They are also being utilized for drug screening tests in order to determine the effectiveness of potential drug treatments for diseased cells. Labs around the world are making iPSCs from patients with a disease and differentiating these cells to become the tissues affected by a particular disease so that these diseases can be modeled in vitro and cells can be used for drug treatments.

On the surface, iPSCs circumvent some of the legal and ethical controversies associated with hESCs. But as promising as iPSCs are, there are challenges associated with them that will have to be addressed. For example, scientists still do not fully understand how pluripotent iPSCs may be and how to best control the potency of these cells. In addition, iPSCs:

- Are relatively inefficient to produce (only about one in 1000 somatic cells exposed to most reprogramming approaches becomes an iPSC)
- Require constant feeding to maintain viable cells lines
- Show low viability compared with other cell types once they have been stored frozen
- Can be prone to forming tumors
- Occasionally show spontaneous differentiation into mature cell types when in culture
- Can sometimes be difficult to use for directing differentiation into particular cell types

Research with iPSCs is progressing at an astonishing and exciting pace as scientists work to better understand the properties and capabilities of these cells. Promising results demonstrate that nuclear reprogramming may be a potentially viable way to generate person-specific stem cells without the need for an embryo. Watch for exciting new developments in the next few years involving nuclear reprogramming and iPSCs.

Potential applications of stem cells

The CDC's National Center for Human Statistics indicates that approximately 3,000 Americans die every day from diseases that may one day be treated by stem cell technologies. In the future, stem cell research may affect the lives of millions of people throughout the world.

There are many potential applications for stem cells—from growing healthy tissues, to studying them to understand and treat birth defects, to genetic manipulation for delivering genes in gene therapy approaches, to creating whole tissues in the laboratory using tissue engineering. Many scientists believe that stem cell technologies will play key roles in developing treatments for diseases such as stroke, heart disease, Parkinson disease, Alzheimer disease, Lou Gehrig disease, diabetes, and other conditions (refer to Table 11.2).

Potential and *promise* are frequently used words when stem cell applications are being discussed, but use of these cells for treating disease is still largely

TABLE 11.2	STEM CELL–BASED THERAPIES MAY POTENTIALLY BENEFIT MILLIONS OF PEOPLE
Disease Condition	**Number of Patients in the United States**
Cardiovascular disease	58 million
Autoimmune diseases	30 million
Diabetes	16 million
Osteoporosis	10 million
Cancers (urinary bladder, prostate, ovarian, breast, brain, lung, and colorectal cancers; brain tumors)	8.2 million
Degenerative retinal disease	5.5 million
Phenylketonuria (PKU)	5.5 million
Severe combined immunodeficiency (SCID)	0.3 million
Sickle cell disease	0.25 million
Neurodegenerative diseases (Alzheimer and Parkinson diseases)	0.15 million

Source: Adapted from Stem Cells and the Future of Regenerative Medicine, www.nap.edu/catalog/10195.html.

unproven. Here we consider some of the most promising examples of stem cell applications to date. Patients with leukemia frequently require chemotherapy or radiation treatment to destroy defective white blood cells. As a result, the patient's immune system is greatly weakened. Leukemia treatments may also involve blood transfusions to replace white blood cells and red blood cells damaged by chemotherapy. The use of stem cells to make white blood cells has already become an effective way to treat leukemia. Stem cells from umbilical cord blood have also been used to provide red blood cells for patients with sickle cell disease and those with other blood deficiencies. The isolation of stem cells from cord blood is becoming so popular that, in many U.S. states, parents can opt to pay to have cord blood stem cells frozen indefinitely should their child need them at some time in the future.

So far there have been a number of promising results in animal models as well as in human clinical trials using stem cells for tissue repair. For example, stem cells from fat have been used to form bone tissue in the human skull. The repair of heart tissue has shown strong potential. Stem cells might be used to replace dead and dying cells following trauma, such as a heart

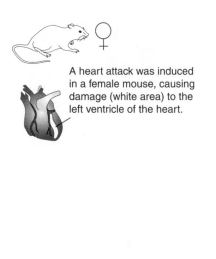

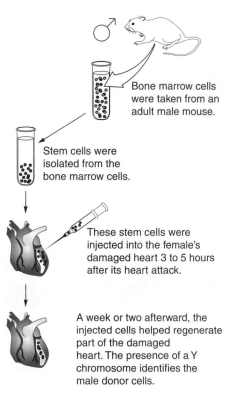

A heart attack was induced in a female mouse, causing damage (white area) to the left ventricle of the heart.

Bone marrow cells were taken from an adult male mouse.

Stem cells were isolated from the bone marrow cells.

These stem cells were injected into the female's damaged heart 3 to 5 hours after its heart attack.

A week or two afterward, the injected cells helped regenerate part of the damaged heart. The presence of a Y chromosome identifies the male donor cells.

FIGURE 11.24 Repairing a Damaged Heart with Adult Bone Marrow Stem Cells Mouse adult bone marrow stem cells can be used to repair areas of the mouse heart damaged by a heart attack.

attack. Heart attacks are a leading cause of death in the United States, killing nearly half a million people each year. The death of cardiac muscle cells weakens the heart and can prevent it from beating with the proper strength to maintain normal blood flow. Adult cardiac muscle cells do not repair themselves well. Several groups of researchers have injected adult stem cells from different sources into damaged areas of the mouse heart (**Figure 11.24**). These stem cells can develop into cardiac muscle cells, form electrical connections with healthy muscle cells, and improve heart function by over 35%. Researchers have reported improved pumping efficiency in mice at least a year after treatment. A similar approach was used to transplant hESCs into the ventricular walls of damaged hearts in pigs and in humans involved in clinical trials. In both examples transplanted stem cells differentiated to form cardiac muscle cells, which restored a significant percentage of electrical activity and contractility to the damaged areas. One study used stem cells from human umbilical cord blood to improve cardiac function in rats.

Scientists are optimistic that this approach may someday work in humans. Consider this: in the future, a surgeon may order a few grams of cardiac muscle cells from a regenerative medicine lab to transplant into a heart attack patient in much the same way that surgeons routinely order blood from a blood bank for a transfusion during a surgical procedure.

A potentially promising and innovative treatment using iPSCs to correct sickle cell anemia in mice has been demonstrated (**Figure 11.25** on the next page). In this work, iPSCs were produced from skin cells of transgenic mice that express a mutated version of the human sickle cell hemoglobin gene and display sickle cell disease. These iPSCs were genetically engineered to correct the hemoglobin gene mutation. The corrected iPSCs were then induced to form blood stem cells and transferred into donor sickle cell mice, which produced functional red blood cells that, in turn, corrected the disease condition. This is an incredibly exciting result, combining aspects of both stem cell technologies and gene therapy.

Earlier in this chapter, we discussed problems faced by patients who suffer from spinal cord injuries. In the last few years, researchers have disproved a long-standing belief that the human brain and spinal cord cannot grow new neurons. Adult stem cells have been isolated from the human brain and used to make neurons in culture, and scientists have already demonstrated that ESCs can be differentiated to form neurons that can be injected into mice and rats to improve neural function in those with spinal cord injuries. Researchers at Johns Hopkins University have demonstrated that human stem cell transplants can enable mice with paralyzed hind limbs to walk. These studies were carried out on mice that were paralyzed after

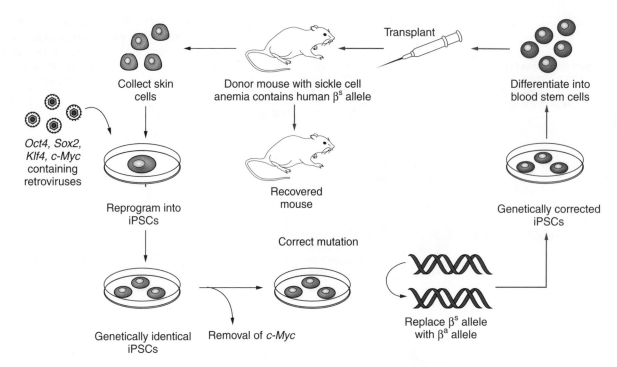

FIGURE 11.25   A Mouse Model for Correcting Sickle Cell Anemia Using iPSCs

they were infected with a virus similar to the polio-myelitis virus that causes polio. Recently, the California-based biotechnology company Geron Corporation received approval from the FDA for the first U.S. clinical trial using hESCs to treat individuals with spinal cord injuries. There is still much work to be done in the field of spinal cord repair and regeneration, but researchers are optimistic that adult neural stem cell transplants may be ready for human clinical trials in the next 3 to 5 years, offering hope to the many individuals who are affected by spinal cord injuries.

Many fundamental questions about stem cells, however, must be answered before stem cell technologies can become viable treatment strategies. Currently studies that have used pluripotent stem cells of any source and introduced those into animals or patients have largely been plagued by problems. What do you think is the main issue here? The main problem is how to control the differentiation of pluripotent stem cells into desired tissues of interest. For example, if the stem cells are injected into a tissue, say skeletal muscle, it is difficult to control how they will respond to differentiation cues in vivo. As a result, such cells often develop into many different and undesirable cell and tissue types other than the types that were intended.

When stem cells instead of differentiated adult tissues are injected, scientists cannot fully control the spread of the cells to other places in the body, nor can they control the differentiation of stem cells into tissues other than those that were intended. Injected hESCs have formed tumors, including types of tumors called *teratomas*, which contain mixtures of differentiated tissues such as teeth, bone, and hair, all in one tumor. Another problem is avoiding chromosomal abnormalities that are known to occur when stem cells differentiate. For example, alterations in chromosome number (trisomy 12, trisomy 17, and others) frequently occur when stem cells differentiate. Patients in unregulated stem cell clinics in China, Thailand, Korea, Romania, and other countries have died as a direct result of complications after having received injections of stem cells.

It is important to recognize that the most effective and safe stem cell treatments in the future are likely to involve differentiating stem cells in vitro into desired cell and tissue types and then introducing those cells into a patient as appropriate instead of injecting them directly.

Other important questions to be answered are these:

- Why do stem cells self-renew and maintain an undifferentiated state?

- What factors trigger the division of stem cells?

- What are the growth signals (chemical, genetic, environmental) that influence the differentiation of stem cells?

■ What factors affect the integration of new tissues and cells into existing organs?

■ Can nuclear reprogramming of somatic cells or other approaches that do not require an embryo become reliable techniques for producing pluripotent stem cells with the properties of hESCs?

■ Which diseases can be most effectively treated by stem cell technologies?

■ What strategies will be most effective for delivering stem cell treatments?

Answers to these and many other questions will help scientists and physicians in their quest to develop stem cell–based applications for treating human disease.

Cloning

We have discussed many types of cloning in this book. Remember that *cloning* refers to making a copy of something—a gene, a cell, or an entire organism. Recombinant DNA technology is used for gene cloning. When bacteria or cultured cells divide in a petri dish or bioreactor, clones of cells are being produced. But clearly no aspect of cloning is as controversial as animal cloning. With the announcement of the cloning of Dolly the sheep in 1997, the world was immediately faced with the prospect that advances in biotechnology could lead to human cloning. Dolly's creation generated a great increase in public awareness and additional discussion about cloning. In this section, we consider the scientific implications of human cloning applications.

Therapeutic Cloning and Reproductive Cloning

There are two main approaches to cloning: **reproductive cloning** and **therapeutic cloning** (Table 11.3). The intent of reproductive cloning is to create a baby. Dolly was the first of many mammals to be produced by reproductive cloning. Unlike reproductive cloning, therapeutic cloning provides stem cells that are a genetic match to a patient who requires a transplant. In therapeutic cloning, the chromosomes from a patient's cell (for instance, skin cells) are injected into an enucleated egg—an egg that has had its nucleus removed—that is then stimulated to divide in culture to create an embryo (**Figure 11.26** on the next page). The embryo produced will not be used to produce a child; instead, it will be grown for several days until it reaches the blastocyst stage so it can be used to harvest stem cells. Stem cells isolated from this embryo can be grown in culture and then introduced into the donor patient (Figure 11.26).

Prior to the development of iPSCs, therapeutic cloning was seen as a potential way to provide patient-specific stem cells that could be used to treat disease without fear of immune rejection by the recipient because he or she was the original source of the cells. In theory, stem cells from a patient can also be used to create cell lines from humans with genetic diseases to provide scientists with unprecedented potential to study and learn more about human disease conditions. We say "in theory" because it has not been proven that therapeutic cloning in humans can work.

Many scientists do not like the term *therapeutic cloning* because it implies creating a human clone. Creating stem cells to treat human diseases is not the same as cloning a human being.

Most stem cell researchers prefer the term **somatic cell nuclear transfer (SCNT)** because *nuclear transfer* truly describes the biological processes taking place. Recall that we discussed the details of SCNT earlier (Chapter 7). For therapeutic cloning, the blastocyst would be used as a source of stem cells and then destroyed. For reproductive cloning, the blastocyst would be implanted in a woman's uterus to allow it to grow and develop to form a baby, which would take the normal 9 months. Reproductive cloning by nuclear transfer is a very inefficient process at best. In the case of Dolly the sheep, it took 277 nuclear fusion attempts to produce only one successfully implanted embryo that developed completely and formed Dolly (Chapter 7). We will briefly discuss controversies surrounding the work of Dr. Woo Suk Hwang of Seoul National University (see "You Decide" on page 331; Chapter 13). Recall that Dr. Hwang had claimed to have cloned a human embryo by SCNT, but this work was proven to be fraudulent.

Many scientists think that reproductive cloning of humans is unethical, immoral, and scientifically unsafe; others think that it may be inevitable at least somewhere in the world. There have been several well-publicized announcements from a few private groups about their ongoing plans to produce humans by reproductive cloning. These claims have generated much skepticism and have been widely disapproved and strongly denounced by most of the scientific community. At the time this book was printed, no definitive proof existed that a human has ever been cloned.

Regulations Governing Embryonic Stem Cell and Therapeutic Cloning in the United States

Although there are international guidelines on the ethical use of stem cells, there is currently no international policy governing stem cell use. In the United States, the FDA has rules regarding the purity of stem

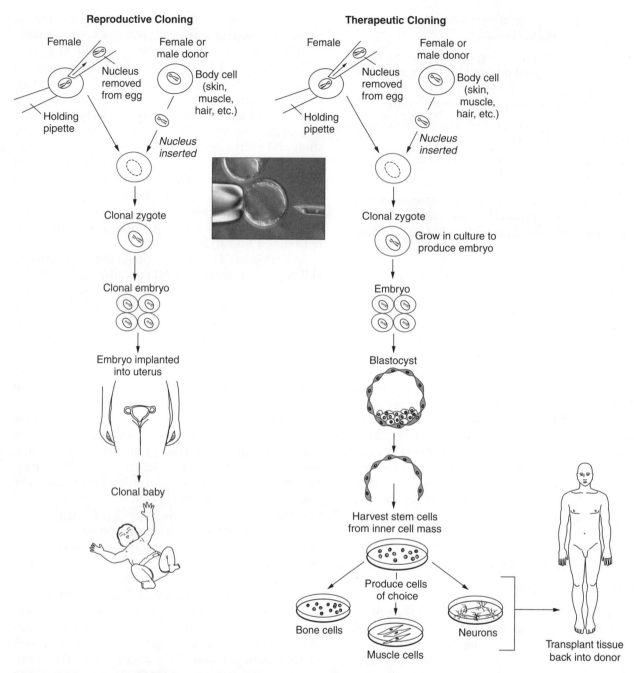

FIGURE 11.26   Reproductive Cloning and Therapeutic Cloning In reproductive cloning, the goal is to produce a cloned baby. In therapeutic cloning, stem cells that are genetically identical to the cells taken from a patient are produced to provide patient-specific stem cell therapy. This photo shows a holding pipette (left side of egg) holding an egg while the nucleus is being removed with a glass micropipette (right).

cells and their applications in clinical trials and medical products. Shortly after hESCs were first isolated in the United States, the National Bioethics Advisory Committee and the National Institutes of Health (NIH) began working on guidelines for hESCs research. On August 9, 2001, President George W. Bush announced a ban on the use of federal funds to create an embryo

for the purpose of isolating ESCs. This ban did provide for the use of federal funds for research on 78 cell lines that had already been established and were available through the NIH prior to this date.

However, only 21 of these lines became available for research, and many of the lines turned out to be far less valuable than initially believed. Some failed to

TABLE 11.3 COMPARISON OF STEM CELLS, THERAPEUTIC CLONING, AND REPRODUCTIVE CLONING

	Embryonic Stem Cells (ESCs)	Adult Stem Cells (ASCs)	Induced Pluripotent Stem Cells (iPSCs)	Therapeutic Cloning (Somatic Cell Nuclear Transfer)	Reproductive Cloning
Final or "end" product	Undifferentiated stem cells (isolated from fetal or embryonic tissue such as an embryo at the blastocyst stage) growing in culture	Undifferentiated stem cells (isolated from adult tissue such as bone marrow cells) growing in a culture dish	Undifferentiated stem cells created from somatic cells; may be patient-specific	Undifferentiated stem cells growing in a culture dish (obtained from the person who will also serve as the recipient of these cells)	"Cloned" human
Purpose/ application	Source of stem cells for research and for treating human disease conditions such as replacing diseased or injured tissue	Source of stem cells for research and for treating human disease conditions such as replacing diseased or injured tissue	Potential source of patient-specific stem cells for studying disease and for treating human disease conditions	Source of stem cells that are genetically matched to recipient for treating human disease conditions such as replacing diseased or injured tissue	Create, duplicate, or replace a human by producing an embryo for implantation, leading to the birth of a child
Embryo Required	Yes	No	No	Yes	Yes
Surrogate mother required	No	No	No	No	Yes
Human created	Depends on how one defines and embryo	No	No	Depends on how one defines and embryo	Yes
Time frame	A few weeks of growth in culture	A few weeks of growth in culture	Weeks to months	A few weeks of growth in culture	9 months, the duration of a normal biological pregnancy (after growth of the embryo in culture)

grow without differentiating. Others showed genetic instability with abnormalities in chromosome number. Some lines were contaminated by mouse feeder cells (which were commonly used to provide key nutrients to growing stem cells in the early days of ESC work). Many scientists and stem cell advocacy groups fought hard to lift the ban on federal funding for the creation of new hESC lines.

In 2006 a bill to lift the ban on federal funding passed the Senate but was subsequently vetoed by President Bush. Congress tried again with a bill in 2007 which was also quickly vetoed by President Bush. In March 2009 President Barack Obama issued an executive order to have the ban lifted and charged the NIH to develop guidelines for governing federal funding of ESC work within 4 months. By April 2009

the NIH had released draft guidelines to allow federal funding for research on hESCs (but not for the use of federal funds to derive hESCs).

In late 2010, a District of Columbia federal judge ruling temporarily blocked President Obama's 2009 order to expand ESC funding based on a lawsuit ruling that the Obama order violated a previous law banning federal funding for research to destroy an embryo. This ruling created major turmoil for stem cell researchers relying on federal funding for their work and resulted in an injunction prohibiting the use of federal funds for hESC work, although it did allow work from funds previously awarded to continue. To demonstrate the impact of this ruling on U.S.-based companies in stem cell research, stock share prices for such companies dropped over

YOU DECIDE

Stem Cell and Cloning Debates

Frequently, when science and medicine produce innovative discoveries, society is not prepared for the consequences of new technology. In 1850, the development of anesthesia was considered very controversial. Many worried about unanticipated adverse reactions from anesthesia, and religious groups protested over "painless" childbirth based on scripture, suggesting that Eve was to go forth from the Garden of Eden to deliver children in pain. Over 150 years later, few people dispute anesthesia as an important tool for complex surgeries and even routine procedures such as having a wisdom tooth extracted.

When recombinant DNA technology was first developed, there was great fear and speculation about what would result from such experiments (recall from Chapter 3 the Asilomar Conference to discuss the dangers of this work). Recombinant DNA technology has resulted in many innovative and safe products that have been used to treat more than 250 million people worldwide. Not unlike the anesthesia and recombinant DNA controversies, clergy, politicians, researchers, and the general public currently debate the merits of stem cells and cloning. At the root of these debates is the source of human stem cells, in particular hESCs and their potential uses and abuses. In large part, hESCs are controversial because of their source—the early human embryo. Knowing that hESCs may have enormous potential for treating and curing many devastating diseases and providing people with an opportunity for healthier, longer lives, what do you think about their use? ESCs are perhaps the most controversial scientific issue ever debated by the public and by politicians. The range of questions surrounding stem cells and cloning is seemingly endless.

- Is it acceptable to produce a human embryo for the sole purpose of destroying it for other uses?
- Some fear that stem cells and cloning technologies will cause a great need for human eggs to support research. Is it acceptable to pay women to collect their eggs surgically?
- What is the moral status of early embryos created by therapeutic cloning?
- What rights does a cell donor have to stem cell lines or technologies created from cells they have donated? Should tissue donors share in the commercial potential and monetary awards of stem cell line created from their cells?

Some people believe that a person is formed at the moment an egg is fertilized, so they consider therapeutic cloning the equivalent of killing a child deliberately for the benefit of another person. Others believe that the early embryo is a cluster of living cells with the *potential* for forming a person but the early embryo itself is not a human being.

- Should we justify destroying embryos that are developed through in vitro fertilization approaches?
- Why not use these embryos in an attempt to reduce pain and suffering in other humans?

Scientists define life in many ways. Biologists agree that the cluster of cells called the blastocyst is alive at the cellular level. Although all life forms deserve respect, the blastocyst is not a person because it does not have limbs, a nervous system, organs, or other physical features of a human individual. So a major source of debate continues about whether we should assign moral status to human embryos and if so, at what stage.

- Does the moral value of an embryo increase as it develops? Or is its moral value equal to that of a baby or adult? If an early embryo is deemed a living person, then it has all the rights of other living persons. Consequently to destroy an embryo intentionally is immoral. Taxpaying citizens must decide how their money will be spent and what they believe is ethical, responsible, and safe research. The scientific establishment must share its knowledge to ensure that citizens make well-informed decisions on such topics as ESCs and cloning.

Human cloning is banned in the United States, but many countries have less restrictive policies. There is concern that severe restrictions on stem cell and cloning research in the United States could create a "brain drain" in which top scientists in these fields will move to countries where cloning is legal. The biotechnology industry in the United States could suffer as a result. Many private companies took the lead in stem cell research, using private funding, and in many cases these companies moved research operations to states and countries that were most supportive of stem cell research.

Even if therapeutic cloning is eventually approved in the United States, will its acceptance make it more likely that people will accept reproductive cloning? Probably not. Therapeutic cloning, which is intended to be used to treat illness, is a very different issue than creating a new human. If creating iPSCs turns out to be a viable

Continued

way to produce stem cells, scientific and ethical debates about therapeutic cloning and the use of hESCs may become irrelevant because stem cells could be generated without an egg or embryo.

Various public opinion polls have reported mixed feedback from Americans about the use of hESCs. Recent polls indicate that the percentage of Americans who consider the use of hESCs morally acceptable has gradually increased from about 50% to 65% over the past 8 years. Generally, Americans expect the highest level of health care in the world. If scientists in another country use therapeutic cloning to produce treatments for Parkinson disease, Alzheimer disease, and others, how will the American public feel about not having access to such technologies? Some have even argued that reproductive cloning is a fundamental right of people living in the United States. How would you feel if you were part of an infertile couple who could not have a biologically related child any way other than through reproductive cloning? Are there proper and ethically acceptable applications of using early embryos and their stem cells? Can the same be said for cloning?
You decide.

8% when the ruling was first announced. The NIH was forced to order an immediate shutdown of hESC research by its investigators. By spring 2011, the U.S. Court of Appeals lifted the 2010 injunction, but it is likely this case will eventually go to the U.S. Supreme Court. Check the NIH website on stem cells, which is listed in the Companion Website, for updates on current regulations and other resources.

Many private foundations are providing hundreds of millions of dollars for stem cell researchers, and several states have enacted legislation to create stem cell research institutes and provide state-funded support for ESC research. Some of the more active states include California (which in 2004 passed Proposition 71 approving a budget of nearly $300 million in bonds and over $3 billion overall), New Jersey, Connecticut, Illinois, Maryland, and Wisconsin. Around the world, regulations vary on the production of new hESC lines and policies on therapeutic cloning. For instance, production of new lines and therapeutic cloning is legal in the United Kingdom, Israel, South Korea, China, and Singapore. Therapeutic cloning is banned in Brazil, Australia, and the European Union (although in member nations that allow it, hESCs can be derived from unused embryos from in vitro fertilization procedures).

Legitimate stem cell research centers have been established around the world in countries considered major powers as well as in many relatively small countries (Belgium, Sweden, Turkey, Israel, Switzerland, etc.). But as mentioned earlier, many stem cell treatment clinics have also sprouted up around the world and patients desperate for cures have taken to traveling to other countries seeking stem cell–based cures, which are often overhyped and promoted as successful despite baseless information and adverse side effects. Often called "stem cell tourists," patients desperate for a cure have been known to travel around the world for unproven stem cell treatments, and the number of individuals involved in this "tourism" continues to grow at an alarming rate. The European Union has laws similar to FDA regulations in the United States that govern stem cell usage. But it is clear that international regulations must be established to govern safe applications of stem cells and avoid the fraudulent and tragic cases of their inappropriate and unethical use. Such abuses and the resulting tragedies will only impede progress and erode confidence in the legitimate treatments being developed.

MAKING A DIFFERENCE

In 1968, about 40 years before hESCs were isolated, physicians performed the first successful bone marrow transplant. Bone marrow contains ASCs, specifically hematopoietic stem cells, which can produce all of the different cell types present in blood. Since this first transplantation, bone marrow transplantation has become routine; it is commonly used to treat a variety of blood and bone marrow diseases, blood cancers such as leukemia, blood clotting disorders, and immune disorders of the bloodstream. Many of these diseases were once thought to be incurable and, in the early days of bone marrow transplants, there were many skeptics and many challenges to be overcome. In some ways research with hESCs and iPSCs is facing some of these same challenges and the future success of these technologies remains to be seen. In the United States alone, some 10,000 patients are treated by bone marrow transplants each year, and the success of such treatments is "making a difference" for many patients and their families.

CAREER PROFILE

Career Options in Medical Biotechnology

Medical biotechnology offers an exciting range of potential career choices, primarily in biotechnology and pharmaceutical companies. Courses in chemistry, cell biology, molecular biology, biochemistry, and bioinformatics are essential, but industry experts recommend that students interested in a career in medical biotechnology identify an area of interest and then gain life and work experience that will helps them fit in with a company's needs. In particular, an internship or summer research experience at a local chemical, pharmaceutical, or biotechnology company is highly recommended.

If possible, get involved in a research project with a professor while you are an undergraduate student. Undergraduate research can provide invaluable opportunities for learning how to plan research projects, execute and troubleshoot experiments, and interpret data. Such research may even lead to a presentation at regional or national meetings—a great way to network and interact with other scientists—or even a publication. Your research professor can also provide an essential reference to help you land your first job.

Hiring experts strongly recommend that students study companies of different sizes to get a feel for the type of work and size of company that appeals to them. Consider what you want to do, and then identify companies that intrigue you and whose needs match your interests and skills. Are you interested in drug discovery and development, regenerative medicine and stem cell technologies, gene therapy, cancer research, aging research, surgical materials, genomics, or curing childhood diseases? Would you prefer a large company to a smaller, more personalized company or even a biotechnology start-up company?

With the wealth of information available online, this is easier than ever. Most biotechnology companies have websites, and excellent links to these sites are presented in Chapter 1. On an interview, it can help to demonstrate a little background knowledge and a true interest in the company with which you would like to land a job.

There are many entry-level job opportunities in medical biotechnology for people with associate's or bachelor's degrees. Many start as laboratory technicians. In some companies, this position involves routine procedures such as preparing solutions and setting up materials for experiments, but individuals who show initiative are often given latitude to do more and to assume decision-making responsibilities in research projects. Application scientists develop new products and procedures, often working directly at the lab bench conducting research. In some companies, however, application scientists may give on-site product demonstrations and present technical seminars to potential customers. Application scientists are also frequently involved in customer relations issues, teaching customers how to troubleshoot a product, interpret data, and the like. Clinical scientists work together with physicians and research scientists to help carry out clinical trials for testing a new drug or medical product. Medical biotechnology companies also employ people who want to combine an interest in science with business skills through marketing and product sales positions.

The interview process at most biotechnology companies is very rigorous. Companies want people who are highly organized, systematic, and attentive to detail. Medical biotechnology research is a team effort. Good writing and communication skills are essential because you will be working with other people routinely and interacting in verbal and written form. In addition, virtually every biotechnology and pharmaceutical company today seeks people with computer skills who can work together with teams of software engineers and information technology professionals to create new software with which to manage and store data, test hypotheses, and model molecular structures.

Industry insiders consistently point out that enthusiasm, an ability to take novel approaches to problem solving, and a commitment to professional growth are valuable tools that companies look for in potential employees. People who are eager and excited about the challenge of applying their skills in teamwork approaches to problem solving are highly desired. Even if a person's educational background is not an exact match for a particular position, as is often the case, enthusiasm and a willingness to learn can determine whether the applicant is hired. Biotechnology companies seek people who want to make a difference—if you have a burning desire to use science to help improve human health, a career in medical biotechnology might be for you.

QUESTIONS & ACTIVITIES

Answers can be found in Appendix 1.

1. How can molecular biology techniques be used to identify genetic disease conditions in fetal or adult humans? Provide two examples.

2. What is gene therapy? Explain the differences between ex vivo and in vivo gene therapy, and give an example of a human genetic disease treated by each approach. Provide two examples of how therapeutic genes can be delivered into cells, and discuss the challenges scientists face in making gene therapy an effective technique for treating human genetic disease conditions.

3. Compare and contrast different types of stem cells, including embryonic stem cells, adult stem cells, amniotic stem cells, and induced pluripotent stem cells. Include an explanation of where each type of stem cell comes from and how each type can be isolated. Give two examples of how stem cells may be used to help treat human disease conditions.

4. Compare and contrast therapeutic cloning and reproductive cloning by preparing a table listing the pros and cons of each technology.

5. Define pharmacogenomics and explain how it may change health care in the future.

6. Briefly describe how the Human Genome Project will lead to advances in medical biotechnology.

7. Visit Clinical Trials.gov and search for MAb treatments currently in clinical trials. Visit the American Society for Gene Therapy site to search for gene therapy clinical trials currently in progress.

8. The use of federal funds for embryonic stem cell research has been highly controversial. Should taxpayers' dollars be used for this research? Should there be restrictions on the kinds of research that may be supported by federal funds? Do you know if your state representatives in the U.S. House of Representatives support the use of federal funding for embryonic stem cell research? (see http://www.house.gov/internet)

9. Describe two gene-silencing techniques and how they may be used for gene therapy.

10. What is regenerative medicine? Describe potential ways in which tissue engineering may be used to treat disease?

11. What are SNPs? How can SNPs be used to diagnose human genetic disease conditions?

12. What genetic testing techniques would you use to detect single-gene defects? What techniques would you use to detect chromosomal defects such as abnormalities in chromosome number or structure?

Visit www.pearsonhighered.com/biotechnology

To download learning objectives, chapter summary, "Keeping Current" web links, glossary, flashcards, and jpegs of figures from this chapter.

Biotechnology Regulations

After completing this chapter, you should be able to:

- Describe the Animal and Plant Health Inspection Service (APHIS) and the U.S. Department of Agriculture's permitting process, including the precautions that must be taken to prevent accidental release of bioengineered plants into the environment.

- List the six criteria that must be met before a plant can be eligible for "notification" under APHIS guidelines.

- Describe the role of the Environmental Protection Agency (EPA) in regulating biotechnology products.

- Describe the role of the U.S. Food and Drug Administration (FDA) in regulating food and food additives produced using biotechnology.

- Describe the FDA's role in regulating pharmaceutical products, including phase testing.

- Cite examples of the regulatory agencies' ability to respond expeditiously to emerging situations.

- Describe the function of patents in science and explain how patents encourage discovery.

- Describe the circumstances in which DNA sequences are patentable.

FDA Inspector at work. The FDA is one of the government agencies responsible for evaluating the safety of biotech products.

The use of a mislabeled drug or consumption of a contaminated food can cause death. Medicines and food are regulated by the government. These regulations have profound effects on the biotechnology industry, especially in the United States. It can take years to test a new drug to confirm its safety and effectiveness; in the meantime, lives may be lost that the new drug might have saved. There is a cost if food must be discarded, just as there is a cost if agricultural productivity is diminished because potentially toxic herbicides are not used. Clearly, there is a conflict between ensuring safety and reducing costs. New and innovative biotechnology products in medicine and agriculture promise benefits but also pose risks. When former President Bill Clinton and U.K. Prime Minister Tony Blair pledged to limit gene sequence patents by (among other things) sharing information gathered by researchers working on the Human Genome Project, panic spread in the biotech industry. Many applaud a restriction on the patenting of gene sequences because they argue that such patenting is morally unacceptable. At the heart of this debate is what it means for something to be "patentable." Your opinion about that may change once you have read about protections and patents in this chapter.

FORECASTING THE FUTURE

Biotechnology pioneer Craig Venter (a primary contributor to the original human genome sequencing effort) has played a key role in establishing bioethical regulations for synthetic genomes in the industry. When the first gene transfer experiments occurred in the 1970s, society's fear that "a genetic monster" would be created resulted in a voluntary moratorium of experiments until safeguards were established—safeguards that Venter requested in the 1990s before his team transplanted a synthetic genome into a bacterium. Venter and other biotechnologists knew that without such safeguards, their experiments might also be stalled owing to society's fears of genetic engineering. Because of Venter's initial efforts to establish safeguards and regulations, the results of his original experiments are being used as the basis for future work crucial to biotechnology: a designed alga that can make new hydrocarbons from CO_2 and be used in refineries, the speeding up of vaccine production, the cleanup of polluted water, and the production of new food chemicals, for example. These experiments will undoubtedly lead to even more new federal regulations and protections, but the groundwork for experimentation coupled with safety regulations was laid in the 1970s.

12.1 The Regulatory Framework

During the nineteenth century in America, peddlers freely promoted their "patented" medicines, but often their main concern was to skip town with their profits before local authorities could catch on to the fact that most of these so-called medicines were scams. At the same time, many large companies were poisoning the environment—often without any real understanding of the long-term consequences of their actions. Eventually the government could no longer ignore threats to public safety from either small or large businesses; as a result, it established protective agencies. The new federal agencies were charged with overseeing the safety of food and medicine and protecting the environment from polluters, and these agencies regulate those same industries today.

By the time biotechnology emerged as a field of research, a regulatory system to oversee the industry was already in place. The 1974 Asilomar Conference was a voluntary meeting to establish guidelines initiated by researchers themselves when some of those involved in gene transfer saw the potential for spread of disease genes if safeguards were not put in place The National Institutes of Health (NIH) was the first federal agency to assume regulatory responsibility over biotechnology. Among other things, the NIH published research guidelines for recombinant DNA techniques. Recall that the Recombinant DNA Advisory Committee of the NIH sets the guidelines for working with recombinant DNA and recombinant organisms in the laboratory (Chapter 3). The NIH continued to monitor and review all DNA research until 1984. At that point, it became clear that a different set of regulations was needed.

The government published the "Coordinated Framework for Regulation of Biotechnology," in 1986, which made biotechnology research a joint responsibility of the NIH, the USDA, and the EPA; it describes the federal system for evaluating products developed using modern biotechnology. This groundbreaking report proposed that biotechnology products should be regulated much as traditional products were. For example, bioengineered plants would be monitored by the same agencies that regulated other plants; bioengineered drugs would fall under the same rules that applied to all other drugs. In essence, this report proposed that biotechnology products would not pose regulatory or scientific issues that were substantially different from those posed by traditional products. This concept is central to understanding the regulation of biotechnology in the United States. The framework is based on health and safety laws developed to address specific product classes.

Three agencies are responsible for regulating most of the products that are the result of biotechnology: the U.S. Department of Agriculture (USDA), the Environmental Protection Agency (EPA), and the U.S. Food and Drug Administration (FDA). The USDA makes sure that a given organism is safe to grow, the

TABLE 12.1 PRIMARY FEDERAL REGULATORY AGENCIES IN THE UNITED STATES

Regulatory Oversight of Biotechnology Products Agency	Product Regulated
U.S. Department of Agriculture	Plants, plant pests (including microorganisms), animal vaccines
Environmental Protection Agency	Microbial/plant pesticides, other toxic substances, microorganisms, animals producing toxic substances
U.S. Food and Drug Administration	Food, animal feeds, food additives, human and animal drugs, human vaccines, medical devices, transgenic animals, cosmetics

Major Laws that Empower Federal Agencies to Regulate Biotechnology	
Law	**Agency**
The Plant Protection Act	USDA
The Meat Inspection Act	USDA
The Poultry Products Inspection Act	USDA
The Eggs Products Inspection Act	USDA
The Virus Serum Toxin Act	USDA
The Federal Insecticide, Fungicide, and Rodenticide Act	EPA
The Toxic Substances Control Act	EPA
The Food, Drug, and Cosmetics Act	FDA, EPA
The Public Health Service Act	FDA
The Dietary Supplement Health and Education Act	FDA
The National Environmental Protection Act	USDA, EPA, FDA

Source: www.fda.gov.

EPA makes certain that it is safe for the environment, and the FDA determines whether it is safe to consume (see Table 12.1). These three agencies cooperate to achieve their goals, and all are often involved in the oversight and regulation of a single biotechnology product. Even though bureaucracies are often depicted as slow to change and act, it is important to understand that the regulatory process is constantly being refined to ensure that products, both bioengineered and traditional, are safe.

12.2 U.S. Department of Agriculture

Created in 1862, the USDA has many functions related to the advancement and regulation of agriculture, including regulating plant pests, plants, and veterinary biologics. A **biologic** is broadly defined as any medical preparation made from living organisms or their products. Insulin and vaccines are examples of biologics.

Animal and Plant Health Inspection Service

The **Animal and Plant Health Inspection Service (APHIS)** is the branch of the USDA responsible for protecting U.S. agriculture from pests and diseases. Because genetically engineered plants are potentially invasive, they are treated as plant pests and regulated by the agency under the requirements of the Federal Plant Pest Act. APHIS provides permits for developing and field testing genetically engineered plants. If an experimental organism poses a potential threat to pre-existing agriculture, the agency makes certain that safeguards are in place.

APHIS Regulatory Process

After a new plant has been engineered, APHIS requires several years of **field trials** to investigate issues such as disease resistance, drought tolerance, and reproductive rates. APHIS will not approve a field trial application unless precautions are taken to prevent accidental cross-pollination. For example, trial planting sites for transgenic corn must be at least 1 mile away from fields where seed corn is grown and harvested. A 25-foot perimeter of fallow (unplanted, tilled) land must surround the entire trial site. In some cases, the tassels at the ends of the ears of corn must be bagged to confine the pollen (as seen in **Figure 12.1**), and the tassels themselves must later be removed., Plants must not escape the trial site, which means that plants that naturally reseed must be destroyed and that no plants can be removed unless they are to be destroyed. The entire trial site is usually kept fallow for the season following the trial harvest. APHIS employees inspect the sites before, during, and after trials to make sure that all requirements are followed.

An example of this protection can be illustrated by the fact that in 2002, two growers in Iowa and Nebraska were required to destroy all the seed corn produced from fields where they had previously grown transgenic soybeans for protein (human antibody) extraction but did not remove a few "volunteer" plants that could have contaminated the subsequent seed corn crop (as required by the USDA). They were also fined $250,000 each for this infraction. The case illustrates that the regulators are watching and the regulations are being enforced.

The APHIS approval process

Of course the ultimate objective is to harvest a marketable product. The first step in that process is to petition APHIS for **deregulated status**. Genetically engineered plants with this status are monitored in the same way as traditional plants. APHIS reviews field-test reports, scientific literature, and any other pertinent records before it determines whether the plant is as safe to grow as a traditional variety.

APHIS considers three broad areas while evaluating the petition for deregulation:

1. *Plant pest consequences.* APHIS examines the biology of the plant to evaluate the possible threat to other plants. This is known as the "plant pest" risks. The agency investigates the possibility that the new genetic material might lead to a new plant disease. It also considers whether crossbreeding with native plants could create new "superweeds."

2. *Risks to other organisms.* The agency investigates the risk to wildlife and desirable insects that might feed on the crop or be exposed to the pollen.

3. *Weed consequences.* Finally, the agency considers the possibility that the new plant will be unwelcome and invasive—in other words, a weed. The plant's reproductive strategies are an especially important consideration. A plant that reseeds itself easily, has a great mechanism for spreading its seeds, and is resistant to cold and drought can

FIGURE 12.1 APHIS Regulations in Action Corn growing in a field with its tassels bagged to prevent pollination of unintended plants in nearby fields in compliance with regulations from the Animal and Plant Health Inspection Service of the USDA.

present a real problem in the future. For instance, if the dandelion were presented as a new biotechnology plant, it would be rejected on the basis of its amazing success in propagation.

The APHIS notification process

An alternative system called **notification** is in place to fast-track some new agricultural products. Six criteria must be met before notification becomes an option.

1. The new agricultural product must be one of only a limited number of eligible plant species. These species—including corn, cotton, potatoes, soybeans, tobacco, and tomatoes—have all been thoroughly studied in past trials, and most of their characteristics are well understood.

2. The new genetic material must be confined to the nucleus of the new plant; it cannot be floating

in the cell on plasmids or viral vectors. (Nuclear genes tend to remain with the new engineered plant; see Chapter 3 for a refresher on vectors used in the genetic engineering of plants.)

3. The function of the genes being introduced must be known, and *no* plant disease can be caused by the protein being expressed.

4. If the new agricultural product will be used for food, the new genes cannot cause the production of a toxin, an infectious disease, or any substance used medically

5. If the gene is derived from a plant virus, it cannot have the potential to create a new virus.

6. The new genetic material must not be derived from animal or human viruses (to prevent spread as a new human disease).

If a bioengineered plant meets all six of these criteria, the agency will approve the field trial within 30 days. Under notification, the plant developer is still required to meet the standards demanded under the permitting process. The same precautions must be taken to prevent accidental spread of the new plant, and detailed records of the trial must be maintained.

After the safety of the new plant has been determined, a grower can cultivate, test, or use the plant for cross-breeding purposes without monitoring or approval by APHIS. The new biotechnology product can then be brought to market. But these requirements may not be all that is needed, and the EPA regulates all new products introduced into the environment.

12.3 The Environmental Protection Agency

Established in 1970, the EPA has responsibilities ranging from protecting endangered species to establishing emission standards for cars. Another major duty is regulating pesticides and herbicides. As a natural extension of that responsibility, the EPA regulates any plant that is genetically engineered to express proteins that provide pest control (Chapter 6).

The EPA also supervises the use of herbicide-tolerant plants (**Figure 12.2**). In this case, the EPA is interested not only in the plant but also in the herbicides that will be used in the fields where the herbicide-tolerant plants are grown. Of particular importance is the pesticide residue remaining on the crop. The EPA monitors these levels to make certain that the crop itself is safe for human consumption. Both USDA (APHIS) and EPA approvals are needed.

FIGURE 12.2 "Ice-Minus" Sprayers Protected by the EPA "Moon suits," like the ones illustrated here, are used to protect the sprayers against unknown biohazards while they apply ice-minus bacteria to strawberry plants to reduce freezing (full description of ice-minus bacteria is in Chapter 5).

Experimental Use Permits

Plant developers who intend to create a plant that expresses pesticidal proteins must first contact the EPA. Field experiments involving 10 acres or more of land or 1 acre or more of water cannot be conducted without an **experimental use permit (EUP)** issued by the agency. EUPs require careful record keeping because the plant cannot be registered for use as a pesticide without clear evidence that there will be no "unreasonable adverse effects" as defined by U.S. pesticide law. In other words, the plants themselves must meet the same standards as chemical pesticides used in the fields.

Deregulation and Commercialization

The next phase of the process is commercialization, during which the product is approved for market. However, before anything is available for sale, the EPA spends about a year reviewing the data collected during experiments. This review concentrates on four general areas of concern. First, the EPA considers the source of the gene, how it is expressed, and the nature of the pesticide-protein produced. Second, the health effects of the bioengineered plant are studied in depth. Experimenters use animal models to measure the impacts of the protein on living organisms, including how easily the protein is digested and whether it can trigger allergies. Third, because the EPA is responsible for protecting the environment at large, the agency also must investigate the "environmental fate" of the pesticide protein. One concern is the rate of degradation of the pesticide and whether it lingers in the soil or water. Another issue is the possibility that the pesticide protein might escape through cross-pollination with weed plants to create a superweed. Fourth, the EPA is interested in the effects on nontarget species, which can include helpful insects such as honeybees and ladybugs as well as other animals including fish, birds, and rodents (Chapter 6).

Like the USDA, the EPA can grant deregulated status to any plant that meets its requirements. Any engineered plant with deregulated status is registered with the EPA and can be sold or distributed like any other plant. But even in these cases, the mission of the EPA is not complete. The agency must be prepared to respond to new information, and it has the power to amend or revoke existing regulations whenever required (Chapter 6).

The bureaucratic regulatory framework in the United States is responsive to emerging public concerns and proceeds with caution. Perhaps most importantly, however, the regulatory bodies depend on good scientific evidence to come to their conclusions. The public is similarly protected by regulations relating to the release of foods and drugs.

12.4 The U.S. Food and Drug Administration

The FDA is charged with making certain that the foods we eat and the medicines we use are safe and effective. Because both food and drug products are hot sectors of biotechnology development, the FDA is busy monitoring new developments in the industry. They make sure that biotechnology products are as safe as their conventional counterparts.

Food and Food Additives

When a new food product or additive is being developed, the FDA serves as a consultant to biotechnology developers (companies, university researchers, etc.) and advises them on testing practices. Studies focus on whether the new product has any unexpected, undesirable effects. The expressed protein is evaluated to see if it is substantially the same as naturally occurring protein in food. Because many common foods—including milk, shellfish, peanuts, and eggs—are known to contain allergy triggers, any protein derived from such sources must be considered a potential allergen and labeled accordingly. Companies are not required to consult with the FDA in developing biotechnology food products that are already on the market. However, most companies in the United States have taken advantage of the FDA's expertise before bringing a food product to market, because if a food product or additive poses no foreseeable threat, the FDA can grant

generally-recognized-as-safe (GRAS) status, which may help to give the product a positive reception when it reaches the market.

An example of a product with GRAS status is genetically engineered chymosin, which is used in 80% of the cheese manufactured in the United States (Chapter 4). Sometimes a product that has been deemed safe by the FDA will still have a tough time in the marketplace despite this. An example is the use of bovine somatotropin (BST), which is injected into dairy cattle to increase their production of milk. Although BST occurs naturally in cows, consumers have concerns that it will show up in milk sold in stores. The debate continues as to whether there are negative effects on cattle or humans who consume the milk so produced. No hard data suggesting that BST presents any danger have been uncovered, even after an 8-year study by the FDA. However, if BST proves to be unsafe, the FDA has the responsibility and power to prevent its use.

The FDA Drug Approval Process

The FDA is also responsible for regulating new pharmaceutical products. Biotechnology drugs must meet the same exacting standards as any other new drug. Imagine for a moment that a new biotechnological treatment for cystic fibrosis has shown promising results during animal trials. The first step to getting the treatment to patients would be to contact the FDA and file an **investigational new drug (IND) application** with the **Center for Drug Evaluation and Research.** Before approving the substance for testing in human patients, the FDA would consider the results of previous experiments, the nature of the substance itself, and the plans for additional testing. To provide effective regulatory review of biological products, the Center for Drug Evaluation and Research conducts active mission-related research programs. This research greatly expands knowledge of fundamental biological processes and provides a strong scientific base for regulatory review.

Phase testing of drugs

If the FDA is satisfied that a new drug warrants further investigation, does not present obvious risks, and will be tested using scientifically sound methods, it attains IND status. The next step is phase testing, which is more stringent than the testing that foods undergo. First, during phase I (safety), between 20 and 80 healthy volunteers take the medicine to see if there are any unexpected side effects and to establish the dosage levels. Phase II (efficacy) begins the testing of the new treatment on 100 to 300 patients who actually have the illness the drug is designed to treat. If no detrimental side effects are noted and the drug seems to have some positive effect, it is ready to go to phase III testing to assess its benefit compared with other drugs. This phase involves between 1,000 and 3,000 patients in double-blind (where neither the patient nor the clinician knows whether placebo or drug is being administered, so as to remove potential bias from affecting the results) tests and lasts for 3½ years. Generally a drug can be marketed only after its benefits and long-term safety have been established in phase III studies. Most drugs never make it this far. Only 20% of drugs that go through phase I testing make it all the way to phase III and on to FDA approval as an **NDA (new drug authorization).** No therapeutic drug product can be marketed in the United States until an NDA has been issued by the FDA. Even after a new drug enters the market, it continues to be monitored (phase IV) by the FDA indefinitely. The number of new drugs that the FDA has approved has fallen by half since 1996, with only 20 approved in 2005. Phase III trials consume 70% of clinical development costs, yet 40% of compounds now fail at this final hurdle largely owing to the inability to demonstrate superiority over placebo (see **Figure 12.3**). A biotech company will file for a **Biological License Agreement (BLA)**, Similar to an NDA, if they are seeking approval of a biologically derived product such as a viral therapy, blood compound, vaccine, or protein derived from animals. The FDA reviews information that goes on a drug's professional labeling and, as part of the approval process, the FDA inspects the facilities where the drug will be manufactured.

Exceptions to phase testing

An important exception to this testing procedure must be noted. The FDA does permit the approval of drugs and vaccines intended to counter biological, chemical, and nuclear terrorism without first proving their worth in phase II and III trials. This is also true for "orphan drugs," as designated by the FDA Office of Orphan Products Development (OOPD); these are drugs with small numbers of beneficiaries but with great benefit, such as human botulism immune globulin (BIG), an antibody for botulism poisoning. Because it would be unethical to expose human beings deliberately to smallpox or nerve gas to test the value of such treatments, there are no pools of patients on which these treatments could be tested. The OOPD administers the major provisions of the Orphan Drug Act (ODA), which provide incentives for sponsors to develop products for rare diseases. The ODA has been very successful—more than 200 drugs and biological products for rare diseases have been brought to market since 1983. In contrast, the decade prior to 1983 saw fewer than 10 such products come to market. New drugs of this kind will still undergo phase I testing because that part of the process requires exposure to the drug only, not to the dangerous substance it is meant to counteract; therefore the FDA can determine that the drug itself will not cause unwanted side effects. For example, the FDA expedited approval of the drug Cipro (ciprofloxacin), an antibiotic used to treat anthrax.

The Development of Good Laboratory (GLP), Clinical (GCP), and Manufacturing (GMP) Practices

In 1975, FDA inspection of a few pharmaceutical testing laboratories revealed, among other problems, poorly conceived, carelessly executed experiments; inaccurate record keeping; and poorly maintained animal facilities. These deficiencies led the FDA to institute the **good laboratory practice (GLP)** and **good manufacturing practice (GMP)** regulations to

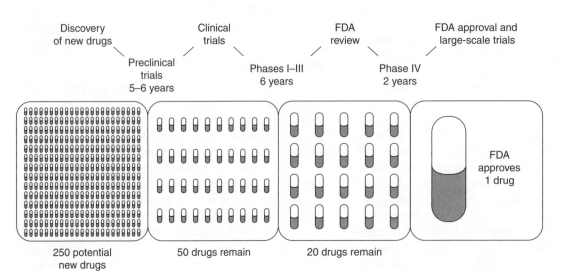

Discovery of new drugs Clinical trials FDA review FDA approval and large-scale trials

Preclinical trials 5–6 years Phases I–III 6 years Phase IV 2 years

250 potential new drugs 50 drugs remain 20 drugs remain FDA approves 1 drug

FIGURE 12.3 The Drug Pipeline Only 20% of the drugs that enter phase I will make it to phase III, and 40% of these will fail due to the inability to show superiority over placebo.

govern animal studies of pharmaceutical products. GLPs require that testing laboratories and manufacturers follow written protocols (standard operation procedures, or SOPs), have adequate facilities, provide proper animal care, record data properly, and conduct valid toxicity tests. A similar set of standards, **good clinical practices (GCPs)**, protect the rights and safety of human research participants and ensure the scientific quality of clinical experiments.

Faster Drug Approval versus Public Safety

Biotechnology companies spend about $1.2 billion to bring a biological product to market, and it takes 8 to 15 years to receive FDA approval for marketing. Because of financial concerns, biotechnology companies would like to speed up this process; however, the FDA would prefer to keep the process at its present pace to guarantee adequate testing. A compromise may be needed to provide products that the public needs (such as gene therapies to treat fatal diseases) while maintaining safety that is acceptable to regulatory agencies like the FDA.

For instance, there are currently five suppliers of human growth hormone. Each company was required to get individual approval from the FDA, each doing its own phase testing. The U.S. Pharmacopeia Convention (USP) has proposed an alternative method whereby the five companies could share early portions of the testing; this change would mean that each of these companies could accelerate the testing and proceed to the human safety trials a little earlier. From the point of view of the pharmaceutical companies, this change in methodology would address the backlog of products being submitted for approval by using standardized methods without compromising safety. Whether this change will be approved can only be determined by time.

Adaptive phase testing for faster results

In an attempt to get quicker results with smaller numbers of individuals in clinical drug trials, the FDA is encouraging drug companies to design clinical trials with flexible enrollment and dosing. Not everyone agrees that this is a good idea.

In a typical clinical trial, drug dosage and the number of patients in the control and experimental groups are predetermined and unchangeable. In contrast, adaptive trials allow changes in drug dosage, patient pool sizes, and other changes in response to incoming data. In other words, the parameters are changed "on the fly" as new results are received. Considering that phase II and III trials can easily drag on for 5 or more years, everyone is supportive of changes that will possibly speed results.

Bayer Healthcare used an adaptive approach in its phase II trials for a new cancer drug. They did not know ahead of time which cancer would respond best to the new drug, so they enrolled a number of patients with different advanced cancers. Within a short time they knew that the drug responded best to kidney cancer, so they "adapted" their trials to enroll and test only patients suffering from advanced kidney cancer.

The major concern is that the committee to decide on adaptive responses should be completely objective. It is quite possible that drugs in trials will fail faster and cost companies more, and results could be manipulated with the right "adaptations." The FDA is releasing guidelines for evaluating multiple possible outcomes and methods to enrich trials with patients who are most likely to benefit. Will adaptive design provide faster results? Will the cost of drugs go up or down? Who will benefit the most, the patient or the drug company?

FDA Regulations to Protect Laboratory Workers

Laboratory workers who handle gene vectors, recombinant DNA, and biological organisms containing recombinant DNA, bacteria, and fungi risk infections as a result of their work. Biotechnology companies are aware of the potential liabilities and financial implications related to employee infection and have made implementation of the appropriate biosafety practices a high priority. There are three levels of protection against biological hazards: A class 1 cabinet protects the operator from airborne material generated at the work surface, where the air flows over the work surface and is filtered before being vented back into the room. A class IIB cabinet draws high-efficiency particulate (HEPA)-filtered air across the work surface, refilters the air before venting into the room, and is suitable for bacterial and tissue cultures. A class III cabinet isolates the material completely in a gas-tight area, which is required only in working with the most hazardous agents (**Figure 12.4** on the next page).

12.5 Legislation and Regulation: The Ongoing Role of Government

The regulation of biotechnology is a matter of politics as well as science. Regulatory agencies are government organizations, and they exist to enforce laws. Ideally, those laws rest on good scientific practice, but they also reflect philosophical values. It is not uncommon

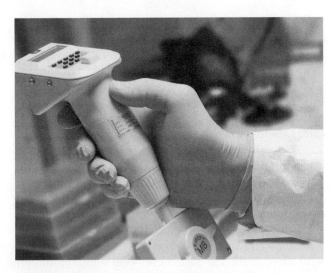

FIGURE 12.4 Lab Technician Uses a Micropipette Under Biosafety Level 1 Conditions

for value systems to clash, and when they do, a long process of negotiation and compromise may be needed to reach a consensus.

Labeling Biotechnology Products

A controversial issue is the labeling of foods that contain genetically modified organisms (GMOs). From the perspective of the FDA and the current White House administration food labels must include information about the ingredients and any claims made or suggested by the manufacturer. The FDA also requires special labeling of foods that present known safety, usage, or allergen issues. If a biotechnology food product included a protein that was not usually found in the food and was a known allergen, the FDA would require special labeling to notify allergic consumers of the risk. Many chocolate candies include a warning that they may contain traces of peanuts. However, because biotechnology foods are not inherently more or less dangerous than their traditional counterparts, the FDA does not require labeling to indicate the method of production. If the product is to be sold in Europe, however, the European Union does require labeling when genetic modification has occurred.

Some consumer groups are pursuing a change in labeling laws that would require information about the use of genetic engineering in food production. They argue that this disclosure is essential to their right to know and to make informed choices based on their own values and beliefs. An interesting example of this real-world debate focuses on the use of BST to increase milk production, as discussed earlier. Over the past decade, a series of comprehensive studies have indicated that milk produced using BST presents no significant health risks. For this reason, the FDA

continues to hold the position that no label is required. Furthermore, when dairy producers began labeling their products as "BST Free," the FDA intervened, stating that such labels are misleading because milk production in dairy cows depends on naturally occurring BST. This has spurred sales of milk labeled "from cows not treated with BST." Some people fear that the introduction of proteins into foods will stimulate allergies or otherwise add antibiotic resistance substances to our foods (see Chapter 7). Safeguarding against potential hazards is what the regulatory agencies have been doing for decades.

The future is likely to present many more debates as society adjusts to the role of biotechnology in our world. For this reason, one of the primary responsibilities of all regulatory agencies is to provide an avenue for public comment. This aspect of regulation is an important element of democracy in action and is shared by more than one agency, as you can see in Table 12.2. You can learn more about the role of public comment in shaping policy, or even participate yourself, when you visit the regulatory agency websites listed on the Companion Website.

12.6 Introduction to Patents

A **patent** gives an inventor or researcher exclusive rights to a product and prohibits others from making, using, or selling the product in these broad categories:

- **Utility patents** protect useful processes, machines and manufactured goods. Examples are computer hardware and medicines.

- **Design patents** protect against the unauthorized use of new and original designs for manufactured goods. Examples are the "look" of an athletic shoe or a bicycle helmet.

- **Plant patents** protect invented or discovered asexually reproduced plant varieties.

To be granted a patent, a discovery must meet three basic requirements: it must be novel, it must be nonobvious, and it must have some utility. (The concept of "utility" is discussed in more detail later in this chapter.) Traditionally, patents have been awarded for inventions, but many natural products clearly meet all three requirements and fall into one of the categories described above. For example, a newly discovered DNA sequence can be novel, nonobvious, and potentially very useful, and the **U.S. Patent and Trademark Office (USPTO)**, which oversees patents, generally does not hesitate to issue patents on sequences for proteins of known function.

TABLE 12.2 EXAMPLES OF SHARED RESPONSIBILITIES BY FEDERAL REGULATORY AGENCIES

New Trait/ Organism	Regulatory Review Conducted by	Reviewed for
Viral resistance in food crop	USDA	Safe to grow
	EPA	Safe for the environment
	FDA	Safe to eat
Herbicide tolerance in food crop	USDA	Safe to grow
	EPA	New use of companion herbicide
	FDA	Safe to eat
Herbicide tolerance in ornamental crop	USDA	Safe to grow
	EPA	New use of companion herbicide
Modified oil content in food crop	USDA	Safe to grow
	FDA	Safe to eat
Modified flower color in ornamental crop	USDA	Safe to grow
Modified soil bacteria that degrade pollutants	EPA	Safe for the environment

Source: www.fda.gov.

When patent laws were first conceived, no one could have imagined patenting DNA or proteins. It was expected that inventors would be engineering better mousetraps, not better mice. The USPTO issued the first patent on a bacterium with a unique gene sequence (genetically engineered organism) in 1980. Since that time, the office has granted more than 2,000 patents for plant, animal, and human genes. Patents have also been granted for transgenic animals and plants, monoclonal antibodies and hybridomas, isolated antigens and vaccine compositions, and methods for cloning or producing proteins.

These patents have translated into big profits for biotechnology companies, but they have also stirred considerable controversy. Many people believe that genes—especially human genes—should not be treated like mousetraps. To their way of thinking, it should not be possible to patent any component of human life such as DNA and proteins.

The process of patenting genes fully fits the letter and spirit of patent law. Still, many concerns and uncertainties remain. The challenge for the government is to find a way to protect the rights of individual researchers while promoting science as a whole.

Obtaining a Patent

In the United States, patents are enforced for up to 20 years from the earliest date of filing. That means other companies can market their own versions of the product after the original developers have had exclusive rights for two decades. This plan both encourages new discoveries and prevents long-term monopolies.

YOU DECIDE

Clinical Trials for Investigational Drugs

If you were diagnosed with a serious life-threatening disease, would you want to be part of a clinical trial for a new drug? New FDA regulations announced in September 2010 improve your chances of benefit. The FDA has issued new rules to clarify the safety information that must be reported during clinical investigations of drugs and biologics. The agency now requires that certain safety information that previously was not required to be reported to FDA be handed over within 15 days of a safety issue turning up. Investigators will have to divulge findings from clinical or experimental studies that suggest a significant risk to study participants; serious suspected adverse reactions that occur at a rate higher than expected; as well as serious adverse events from studies that determine what percentage and at what rate a drug is absorbed by the bloodstream; they must also provide studies determining whether a generic drug has the same effectiveness as the brand name drug. These new rules are designed to protect the patient without slowing the rate of drug discovery. Are the risks of a new drug worth trying with new regulations in place? You decide.

Source: FDA Doles Out New Rules on Clinical Trial Safety Data Reporting, *GEN*, September 29, 2010.

To obtain a U.S. patent, a company must file an application that adequately describes the product. The company must also disclose what it considers to be the best use of the product at the time that the application is filed. USPTO experts with specialty technical backgrounds examine all patent applications. Patents are awarded on a first-come, first-served basis. If two companies file applications for the same discovery at around the same time, the two sides could be headed for a legal battle. In such cases, a little preparation goes a long way. The following guidelines protect a product that is heading for patenting:

1. *Record Keeping.* Every researcher should keep detailed notes within a bound notebook or in a secure electronic format as specified in FDA regulations. Each entry should be signed, dated, and witnessed by an individual who is not directly involved in the research. The notebook should contain all conceptual ideas and supportive data. This evidence may be used to support the company's patent or to invalidate rival patents.

2. *Preparation.* Researchers should continue to monitor the activities of potential competitors and others in their field through trade literature, published patent applications, and issued patents. The USPTO website is a valuable source of information. You can also check commercial databases such as Derwent, Medline, and Biosis.

TOOLS OF THE TRADE

A Safe and Efficient Regulatory Work Environment

What can a group of well-trained scientists with excellent government funding, limited regulations, and a large supply of germ plasma produce? Chinese scientists have created the largest plant biotechnology production capacity outside of North America. They have shown that government support is essential if new products are to be developed quickly in developing countries. As a result, farmers in China are cultivating more acres of genetically modified plants than any other developing country, and they are doing this with the support and encouragement of their government. These crops are monitored and regulated by the Chinese government under regulations developed within the country.

Although China has spent the past 50 years building the most successful agricultural research system in the developing world (employing more than 70,000 scientists), research in modern plant biotechnology did not begin until the mid-1980s. In response to rising pesticide use and the emergence of a pesticide-resistant bollworm population in the late 1980s, Chinese scientists began research on genetically modified cotton. In 1997, after devastating pest-related crop losses, the commercial use of genetically modified cotton was approved. The response by China's poor farmers to the introduction of Bt cotton eliminated any doubt that genetically modified crops can play a role in developing countries. By 2000, farmers planted Bt plant varieties on 20% of China's cotton acreage and reduced an average of 13 sprayings (and a savings of $762 per hectare per season). Although yields and the price of Bt cotton and non-Bt varieties were the same, the cost savings and reduction in labor enjoyed by Bt cotton users reduced the cost of producing a kilogram of cotton by 28%.

A simplified approval process is not the only reason for the Chinese success. Chinese scientists have identified more than 50 plant species and more than 120 functional genes that scientists are using in plant genetic engineering. Of these genetic technologies, 353 were developed between 1996 and 2000. China's Office of Genetic Engineering and Safety Administration approved 251 cases of genetically modified plants for field trials, environmental releases, or commercialization. Of these, regulators approved 45 genetically modified plant applications for field trials, 65 for environmental release, and 31 for commercialization. Transgenic rice, resistant to three of China's major rice pests—stem borer, plant hopper, and bacterial leaf blight—have passed at least 2 years of environmental release trials.

Unlike the rest of the world, in which most plant biotechnology research is financed privately, China's government funds almost all of its plant biotechnology research and planned to raise these research budgets by 400% before 2005 (although they fell somewhat short of their goal). Some U.S. companies have found this favorable regulatory environment an enticement. Monsanto (based in St. Louis, Missouri) is pursuing a joint venture with the Hebei Provincial Seed Company in China to develop other plant varieties. China is vigorously pursuing opportunities for contract research, and it has become an exporter of biotechnological research methods and commodities. The sales of genes, markers, and other tools, in addition to genetically modified varieties, may cause other countries with more restrictive policies regarding genetic modification to look at this world leader.

The Value of Patents in the Biotechnology Industry

For the biotechnology industry, strong patents mean strong businesses. Patents are the primary method by which a successful biotechnology company will be valued during all stages of its development. Suppose a team of scientists discovered a vaccine against a deadly virus. Without a patent, no pharmaceutical company would go forward with the daunting tasks of conducting costly clinical trials, obtaining FDA approval, and developing a comprehensive marketing plan. Without patent protection, a rival company could buy the drug off the shelf, duplicate it, and sell it at a discount without spending a dime on research.

Patenting DNA sequences

Although the vast majority of DNA sequences may never have medical or agricultural value, some could play an important role in producing food or fighting disease. Naturally, organizations that identify those sequences want to enjoy the fruits of their efforts by obtaining protection under the patent laws. The genomic frontier has been mapped, and today's pioneers are eagerly and aggressively staking out what they regard as their territory.

Under patent guidelines, applicants must assert a utility for the claimed invention that is specific, substantial, and credible. To claim a **specific utility** in the case of a DNA sequence, a researcher must know exactly what the DNA sequence does. It is not enough to say that it is a probe or a marker (see Chapter 3). The researcher must disclose what, precisely, it probes or marks (for instance, a probe for a specific human disease gene). A **substantial utility** defines a real-world use—for example, a cloned DNA fragment encoding a protein with a known function or use. To meet the **credible utility** requirement, the researcher must convince the patent office that the application is backed by sound science.

Even if a gene sequence is proved to have utility, it can still be denied a patent if it is too similar to an existing patented sequence. Remember that the function of a gene can be dramatically altered by switching a single nucleotide (Chapter 3). For this reason, patents on gene sequences are usually very specific. Put another way, a segment of DNA usually must match a patented sequence exactly in order to violate—or "infringe" on—that patent. If a company claims patent infringement on two similar sequences, a court must decide if the sequences have essentially the same function. This is known as the **doctrine of equivalents**. The requirements for protecting intellectual property rights and the sale of products with patents are not the same outside the United States and require separate applications in these countries.

According to several studies, approximately 20% of all human genes have been patented, primarily by private biotechnology companies. Not surprisingly, the most heavily patented genes are typically those implicated in human health and disease. Almost 50% of known cancer genes have been patented. In some cases, genes such as *BRCA1*, one of the genes involved in breast cancer, have patent rights issued to multiple companies, resulting in disputes over commercial rights and ownership of such sequences. Some genes have up to 20 patents covering rights to commercial uses of the genes, as for diagnostic tests for genetic testing. See **Figure 12.5** for a map of patent hot spots on chromosome 20.

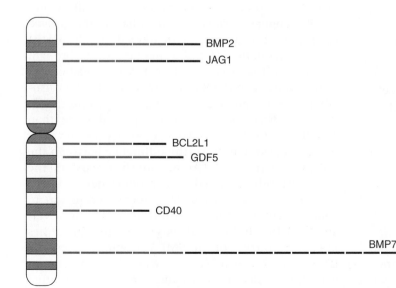

FIGURE 12.5 An Illustration of Patent Activity on One Human Chromosome Chromosome 20 is shown, with each horizontal bar representing a patent claiming a gene sequence located in that region. The labels indicate the loci of highly patented genes.

YOU DECIDE

Would We See Drug Development without Patents?

Gene patents are here to stay, but questions remain. For instance, what happens when a single company "owns" a biotechnology product with potentially lifesaving applications? As Table 12.4 illustrates, many companies have applied for patents on gene therapies for cancer.

History shows that medical practice can be affected by patent law. For example, Myriad Genetics holds exclusive patents on *BRCA1* and *BRCA2*, the genes that largely determine a woman's risk for breast cancer. The company demands fees from any lab that runs blood tests for the genes, a move that threatened to make the test unavailable to many women. In January 2001, the National Cancer Institute (NCI) struck a deal with Myriad Genetics that allows all NCI and NIH labs to use the test at a discount. Still, some other labs are no longer able to afford the test. A landmark decision brought

by the American Civil Liberties Union against Myriad Genetics in March 2010 brought into question this ability to patent genes.

The U.S. Patent Office has issued about 35,000 patents for gene sequences. It is very likely that this decision will be reversed or limited by the appeals process, but it will undoubtedly lead to a better definition of "products of nature." Many in biotech see this decision as a stimulus to innovation, since "manipulated by the hand of man" will be better defined and lead to specifically directed products. The court held that diagnostic and screening methods that utilize gene sequences are not patentable.

Will this lead to more expensive biotech products or less screening tests? Were the courts correct in their decision? You decide.

12.7 International Biotechnology Regulation

Biotechnology is a global enterprise. At the present time the world community is still involved in preliminary negotiations about the regulation of biotechnology products. There have been some promising beginnings. The United Nations (UN), for example, has produced a voluntary code for the deliberate release of GMOs. Public comment has been solicited since it was proposed in 2000, and no final binding regulations have been agreed on. An agency of the UN, the Food and Agriculture Organization (FAO), is focusing its attention on biotechnology in developing nations, providing guidance on safety and risk assessment. In general, no consensus or clear set of commonly held standards for the import and export of biotechnology products exists. There is reason to be optimistic, however. Recent developments in Europe can serve as a model for international cooperation in the trade and development of biotechnology products.

The European Union

In recent years, the European Union (EU) has begun to stand out as both a hotbed of innovation and an attractive market for biotechnology products. The formation

of the EU in 1993 dramatically streamlined the process for introducing new biotechnology products to the marketplace. Today, the EU stands as a textbook example of a government body committed to promoting scientific progress.

A key part of the new system is the **European Agency for the Evaluation of Medicinal Products (EMEA)**, the equivalent of the FDA. The EMEA authorizes medicinal products for human and veterinary use. It also helps spread scientific knowledge by providing information on new products in all 11 official EU languages. Once a product has been approved by the EMEA, it can be marketed in all of the 15 countries in the EU. New drugs can now be brought to market throughout the EU in 1 year instead of 4 or 5, compared with the 8 to 11 years that the FDA usually takes to approve a new drug. Since its inception, the centralized procedure has approved more than 100 human medicines and 25 veterinary products.

There is one noteworthy difference between the EMEA and the FDA. As previously discussed, the FDA is closely involved with product development from the moment a company seeks permission to conduct clinical trials. Within the EU, however, approval for clinical trials is granted by authorities in each country, not the EMEA. In many cases, the EMEA has no prior contact with the company or even knowledge of the product until it is submitted

CAREER PROFILE

Quality Control Inspector

Quality control inspectors in biotechnology companies are responsible for inspecting and checking products to make sure they meet the specified levels of quality described on the package inserts (and that the products' specifications are filed with the appropriate regulatory agencies). This is a good career for a detail-oriented person who likes to work in the lab. It is an area that has a continued high rate of employment. Quality control is involved at every stage of production. Some inspectors examine raw materials to verify specified requirements, others test equipment and verify accuracy, and others test the standard operating procedures established by the company (based on FDA, USDA, EPA, and related regulations).

Although a high school diploma and experience in quality control is adequate in some companies, most employers prefer an associate's degree in a technical area and a few years of experience. Quality control inspectors are often required to determine error levels in measurements, making a good math background a necessity. Good hand-eye coordination and mechanical aptitude are also required. A strong background in the regulations that govern the products of the industry is a must. Quality control inspectors with a bachelor's degree and experience can be promoted to manager. Many transfers occur within biotechnology companies from quality assurance to quality control. The quality assurance department is responsible for maintaining all of the paperwork that is reviewed when any regulatory agencies show up for an inspection. A good career ladder exists in many companies from quality assurance to quality control.

---

for review. According to the most recent polls (2000), 75% of pharmaceutical companies in the EU are satisfied with the process as it stands.

International regulations can affect U.S. companies

Some 48 million potential doses of flu vaccine were lost by the Emeryville, California company Chiron when British regulators from the Medicines and Healthcare Products Regulatory Agency (MHRA) found bacterial contamination in some lots of the vaccine. Companies that operate in the United Kingdom must follow GMP regulations that are as strict as or stricter than those in the United States. After the MHRA closed the plant in 2004, the FDA performed its own investigation and listed 20 violations in a warning letter sent to Chiron in December 2004. The company found *Serratia* bacteria in nine of its flu vaccine lots. Because it could not trace where the problem started, it was forced to destroy 91 lots of vaccine. According to Current Federal Regulations section 211, standard operating procedures are required for the traceability of problems like these.

How can other companies avoid lapses in GMP like the Chiron incident? FDA regulators recommend that firms should evaluate every aspect of production, including people, procedures, equipment, materials, record keeping, audits, and training. Current GMPs require continual updating, and companies are required to provide their employees with this training.

MAKING A DIFFERENCE

Federal regulations for drug manufacture (FDA Section 211) require that all employees be trained in current Good Manufacturing Procedures (cGMP). Biotechnology companies routinely provide training and time to make these regulations clear to their employees. The FDA monitors compliance and fines those who are out of compliance. In May 2010, the company Genzyme was fined $175 million for not fixing practices at its Allston filling plant near Boston. The FDA found tiny particles of steel, rubber, and fiber in drugs products at the plant. The company was required to shut down the plant until the problems were fixed. Compliance is part of working in the biotechnology industry and requires vigilance, but reporting problems is also a cGMP requirement. In most cases, companies are happy to change procedures that are likely to lead to the contamination of products because they want to produce sterile, lifesaving products (and can't risk the disgrace of a regulatory noncompliance report).

QUESTIONS & ACTIVITIES

Answers can be found in Appendix 1.

1. Which U.S. agency regulates food additives?

2. Which U.S. agency regulates GM crops?

3. Which U.S. agency regulates water quality?

4. Which U.S. agency regulates environmental release of genetically engineered organisms?

5. How do patents protect drugs and devices?

6. How have new streamlined approval processes affected the European drug market?

7. Where are the greatest number of patent applications filed in cancer therapy?

8. In your opinion, what should have happened according to FDA regulations to prevent the Fluvirin incident?

9. In your opinion, what is wrong with drugs being advertised for purposes other than what was specified in their FDA approval?

10. Why are drugs often available in Europe before they become available in the United States?

Visit www.pearsonhighered.com/biotechnology

To download learning objectives, chapter summary, "Keeping Current" web links, glossary, flashcards, and jpegs of figures from this chapter.

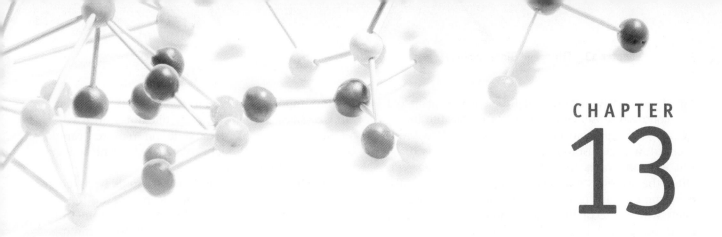

Ethics and Biotechnology

After completing this chapter, you should be able to:

- Define bioethics and explain how it relates to biotechnology.
- Understand and apply different approaches to ethical thought.
- Identify potential ethical dilemmas associated with biotechnology.
- Pose questions and approaches that address the ethical problems identified in this chapter.
- Identify outcomes and pitfalls associated with different ethical approaches.
- Discuss interactions among science, economics, communication, and public policy.
- Understand and explain controversies and ethical issues surrounding genetic testing, stem cells, genetic modified organisms, cloning, and other bioethical topics.
- Describe possible pathways to a career in bioethics.

Bioethics considers topics such as human life. John and Lucinda Borden testified at a U.S. House hearing on embryo research. John holds their twins Luke and Mark; Lucinda holds a picture of the twins as embryos. Luke and Mark were adopted as frozen embryos.

In some ways, **ethics** might seem like plain old common sense. Of course it is unethical to intentionally cause cancer in a group of people without their knowledge just to test your new anticancer drug. But in many cases the choices are not so clear-cut. Often the decisions must deal with potential trade-offs, compromises, or possibly even sacrifices. Bioethics deals with some of the most fundamental questions confronting our society. And in many respects the decisions made now can affect the future of science, humanity, and the world in which we live. The intent of this chapter is not to tell you what to think about bioethics. Instead, the intent is to get you to understand *how* to think about bioethics—to encourage you to ask questions, to think about how to ask the right questions, acquire all the facts, and make decisions based on information rather than emotional reactions.

FORECASTING THE FUTURE

Ethics and related issues in all areas of science are increasingly gaining attention among the public, scientists, politicians and others. In the future, as new biotechnology applications emerge, discussions and increased awareness of ethical concerns associated with these applications will almost always precede the actual development of the application. Using stem cells as an example, well before any stem cell applications were proven and ready to be implemented, ethicists were discussing *potential* ethical issues of stem cell applications. With the way information is rapidly communicated now, there will be few situations in which a biotechnology application or discovery is not preceded by ethical debate. So rather than trying to do the nearly impossible—that is, to forecast the most immediately pressing ethical issues that biotechnology will encounter over the next few years—we want to emphasis that, increasingly, ethical thinking and decision making will be important skills for the future. Almost every application you have learned about in this book, including those we forecast as hot areas in the future, will have significant ethical components to them.

13.1 What Is Ethics?

Ethics identifies a code of values for our actions, especially toward other humans. In simple terms, ethics could be considered a guide to separate right from wrong and good from evil. The area of ethics that deals with the implications of biological research and biotechnological applications, especially regarding medicine is called **bioethics**. It considers social and moral s and potential outcomes of the use of biological edical technologies.

It is important for you to appreciate that ethics is a *dilemma-based discipline*. Ethical dilemmas arise when an important problem or situation requires careful consideration and thought to make what one believes to be sound ethical decisions.

One fundamental question that should be asked in dealing with bioethical issues is not "Can this be done?" but "Should this be done?" And if something should be done, the question becomes "How can it be done in the right way?" Such questions are important for everyone to consider, especially in areas of biotechnology, where discoveries and their applications can have a great impact on human health and the environment. These questions get to the heart of not only society but also of science and its role in society.

For example, consider the photo shown in **Figure 13.1**. Scientists have implanted mouse cells into chicken embryos to demonstrate that the mouse cells could receive signals from the chicken embryo to induce development of teeth—something that does not normally occur in chickens. Chick embryos started to form teeth, but they were not allowed to develop into adults. But some people are outraged at experiments such as this because it creates images of "Frankenstein-like" mutant chickens with teeth that aren't normal. The rooster image shown in Figure 13.1 was created by photographers for shock value and headlines;

FIGURE 13.1 Growing Teeth in Chickens. Should This Be Done? The image shown here is an artificial depiction of a rooster with teeth; no such animal actually exists. This is a deliberately inaccurate image presented for its shock value and to stimulate your thinking about bioethics.

it is not real. So just because technology is available to create a chicken embryo with teeth, should it be done in the first place? Should the embryo be allowed to grow into an adult chicken to see if it will develop teeth? Is this ethical? What do you think? For more information on the research we refer to here, see the paper by Mitsiadis et al. in the reference list on the Companion Website.

Approaches to Ethical Decision Making

Before we explore bioethical issues, we first examine some of the more important methods of ethical thought. The study of ethics is as old as humanity itself. Questions of our duties and responsibilities to other members of the human community have been with us for ages. Hippocrates (c. 460–361 B.C.) might be considered the first **bioethicist**. He emphasized the patient rather than the disease in his practice of medicine, viewing the worth of the individual and the sanctity of human life as of primary importance. For years, physicians have pledged to follow the central tenets of the **Hippocratic Oath**—"do not kill," "to help, or at least do no harm"—in their duty to patients and to their profession.

Ethical thought and methods to approach problems can often be divided between two main viewpoints (although there are certainly other approaches as well). The **utilitarian approach**, or consequential ethics, started with the Scottish philosopher Jeremy Bentham (1748–1832) and the English philosopher John Stuart Mill (1806–1873). This approach states that something is good if it is useful, and an action is moral if it produces the "greatest good for the greatest number." The second main approach—the **deontological (Kantian) approach**, or duty ethics—comes primarily from the German philosopher Immanuel Kant (1724–1804) and focuses on certain imperatives, or absolute principles, which we should follow out of a sense of duty and should dictate our actions.

Modern bioethics can be traced to much more recent times and is primarily the work of two ethicists in the 1970s: Joseph Fletcher and Paul Ramsey. These men refined the two primary approaches to ethical thought mentioned—Fletcher for utilitarianism (also termed "situational ethics") and Ramsey for deontology (or "objectivism").

Utilitarianism emphasizes consequences, not intentions. Another way to phrase this emphasis is that "the ends justify the means." The idea is to calculate what the consequences of an action would be and to weigh different consequences against one another. If we can analyze various courses of action to determine which will have the greatest positive effect on the greatest number of people, then we can provide an answer to the question of what we ought to do. In some ways, this is an elegant method for making decisions. All avenues can be assessed for potential benefit, and the decision becomes more quantitative. The disadvantage of this calculation is that we must assign a value to all of the things considered. How do we assign these values? Some would say that some of the most important things in life (love and family) are not easily quantified, whereas other things (material goods and life span) could be emphasized in the calculations because they are quantifiable. Another source of concern could also be the individual doing the calculating and assigning values. For example, the utilitarian calculation would be less than ideal if done by someone who believed that males are worth more than females or that the primary consideration should be the profit margin.

Deontology, or objectivism, starts from the point of view that there are at least some absolutes (definitive rules that cannot be broken) and that we have a moral obligation or commitment to adhere to these absolutes. One absolute usually mentioned is for the value of human life, expressed by Kant in this way: "Act in such a way that you always treat humanity, whether in your own person or in the person of any other, never simply as a means, but always at the same time as an end." Another way to say this is in terms of respect for others—that we ought to treat others as ends in themselves rather than as means to an end.

This approach has often been associated with religious traditions, but it is a mistaken notion to assume that an objective ethical approach is solely a religious approach or even that all religious approaches to ethical issues begin with the same absolutes. Deeply held convictions are just that—personal points of reference to which an individual adheres. An advantage of objectivism is that it gives firm guidelines in many situations in which ethical decisions are required, providing a clear-cut ethical formula for decision making. A disadvantage of this approach is that it may be, or at least may be perceived to be, too rigid in its decision-making process, not taking what may be important factors into account or considering possible changes in values. And there may be situations or issues where no clear conviction or absolute exists, requiring further examination of the issue to define the moral imperatives and discern the course of action.

To illustrate the difference between utilitarianism and objectivism, consider a situation in which a person is desperately hungry and has no money to buy food. Wandering by a store, the person sees a loaf of bread left out on a table. A utilitarian calculation might take into account the person's need, the available food, and the minuscule loss in value to the store and consider that it would be okay to take the bread. An objectivi approach might have the absolute "it is wrong to steal" and consider it unethical to take the bread.

Note that there can be other ethical approaches and methods beyond the two main approaches mentioned here, and even these two approaches to ethical decision making can sometimes be blended. In approaching ethical decisions, a key objective is to gather information, consider the facts, and make a thoughtful, informed decision. And in debates on ethically contentious issues, it is neither wise nor polite to deride or belittle another person's decision. Be sensitive to the effects of your own conduct. Aim to understand and defend your own ethical decisions rationally, and strive to consider the decisions of others.

The idea of the likelihood of an event, the **statistical probability**, is another important concept to keep in mind in making ethical decisions. It is crucial to determine accurately what chance exists for a "bad" event to happen. Another consideration might be just how negative the effect of the possible event would be. Because many consequences take time to develop, the probability of an outcome occurring must be part of the consideration.

Bioethicists and scientists often use statistical measures to determine **risk assessments**, the likelihood that something harmful or unintended will happen. Risk assessment is part of your decision-making process every day. For instance, you know that each time you drive your car there is a risk that you might be in an accident. Assuming that you are not afraid of driving, your risk assessment probably tells you that the likelihood of an accident does not outweigh the need

to get where you want to go. Cell phone use, air travel, drinking alcohol, the risk-benefit of taking antibiotics, cold medicines, vaccines or other drugs (versus the side effects) or making poor dietary choices are other good examples of risk assessments in your regular life.

Ethical Exercise Warm-up

What follows is a typical scenario regarding tough ethical decision making. Although the situation may seem a bit harsh because it deals with a life-or-death scenario, some of the biotechnologies being developed also deal with issues of life and death and have far-reaching consequences.

A family pulls up to the Grand Canyon in their car. The parents get out to check out a refreshment stand and lock the car doors, leaving three young children asleep in the back seat. But they do not set the brake well enough. After they've moved some distance away, the car slowly begins to roll toward the edge of the cliff. You are the only one who notices. A large man is standing close to where the car is headed. You could push him in front of the car, which would stop its rolling, although he would either be crushed or pushed over the edge himself. What would you do?

First, do you have all the facts in this situation? So far we have painted the scenario so you have only been given two choices: to let the car go over the edge or to sacrifice the one man for the three children. The question seems to be simply "Can you trade his life for theirs?" A utilitarian approach might consider this is an ethical trade, one life for three, and also consider

---

TOOLS OF THE TRADE

Careful Thought and an Open Mind Are Powerful Tools for Bioethicists

In contrast to the various laboratory and industrial applications of biotechnology, bioethics requires no specialized equipment. The main tools of the trade are logic and an open, inquisitive mind. Bioethicists must be able to assess a situation carefully, considering the possibilities from many different angles. They must include medical, religious, philosophical, legal, scientific, and social concerns and outcomes in their assessments. Bioethicists must be able and willing to consider many differ-ent ___ints, which are often conflicting. Bioethicists ___ inquisitive, able to ask probing questions, ___ what the many factors that underlie the ___ may also find it necessary to do exten-___ searches, delving into the background of

a topic from numerous perspectives. Language skills can be an additional asset, because not all perspectives or literature may be accessible in a single language. Sorting through possible outcomes and pointing out potential pathways for the resolution of concerns requires reason and logic as well as patience. In the modern world, knowledge of regulatory agencies and rules is also extremely important. This means that bioethicists must not only be able to understand the regulations and laws currently in place but be able to interact with policy makers to facilitate the creation of sound regulations and laws. Finally, good interpersonal and communication skills (both oral and written) are essential for making the concerns, questions, and outcomes clear to all involved.

YOU DECIDE

Right or Wrong?

The chief executive officer (CEO) of a small startup biotech company is involved in negotiating a multimillion-dollar merger of his company with a larger biotech company because of a promising stem cell cloning technology his company has developed. However, the CEO is also aware of a significant problem with this technology that has recently turned up. Does the CEO reveal the problem and risk jeopardizing the merger, or does the CEO allow the merger to move ahead without discussing the problem? You decide.

that the three are young lives; an objectivist approach might consider that it is wrong to endanger or kill any human life, no matter the eventual consequences. Are there other possible choices or alternatives?

13.2 Ethics and Biotechnology

Nature has been doing biotechnology experiments for much longer than humans have. Bacteria routinely swap plasmids and recombination and mutation occur, allowing the expression of new genes or new combinations of genes that were not present before. However, the game changes when humans get involved and create new genetic combinations. As we discussed in Chapter 3, because of concerns with recombinant DNA technology as a new method, scientists met at a conference in Asilomar, California, in 1975 and called for a **moratorium** (a temporary but complete stoppage of any research) until the safety of the technique and possible consequences could be assessed. Recall that a primary concern was that genetically engineered bacteria would escape from the laboratory into the environment, possibly creating new diseases, spreading old diseases in a more virulent manner, or creating ecosystem changes that might lead to decimation of some species.

In the end, scientists determined that recombinant DNA technology could be controlled in a way that would preserve safety for humans and for the environment while allowing the science to continue. In particular, guidelines were developed for different levels of biosafety containment depending on the inherent dangers of the experimental system used. For example, experiments with nonpathogenic bacteria and nonpathogenic gene sequences require only minimal safety equipment, whereas experiments with known pathogens, human cells, or potentially pathogenic genes require more stringent containment procedures.

Cells and Products

As we have discussed throughout this book, both bacteria and eukaryotic cells can be genetically engineered to express foreign genes and proteins. This has proven to be an indispensable approach for producing medically valuable products. Think about the ethical challenges involved with the genetic modification of individual cells and the products that have emerged as a result of these changes.

When a product concerns human application, such as a therapeutic recombinant protein, there is not only the obvious issue of safety but also the issue of **efficacy** or effectiveness in its intended use. This is obviously important for an intended patient because the patient wants an effective treatment. But efficacy is also important for the manufacturer; if the product is not effective, there can be tremendous economic waste. For any drug, an important consideration will be the dose at which the drug is effective, with minimal side effects and toxicity. Consequently it will also be important to establish whether the drug poses any **carcinogenic** or **teratogenic** hazards. These considerations prevent future problems—such as finding out that the drug cures the disease at one dose but kills the patient or causes cancer or birth defects at another dose—because they raise the ethical concern of harming rather than helping the patient and the potential problems involved in using the patients themselves as guinea pigs to test drug effectiveness.

We must be mindful of the **humane treatment** of animals in preclinical studies. We need to determine how many experimental animals will be the minimum needed to test the drug, what types of treatments will be necessary for the tests, and whether rodents or primates must be used. The choice of species can affect the action of the drug. One of the best examples of species difference in drug action occurred with the drug **thalidomide**, which was originally designed as a mild sedative. It was tested in standard laboratory rodents and found to be safe. However, many of the pregnant women who took this sedative gave birth to babies with severe birth defects. When the drug was tested on marmosets (a type of monkey), birth defects similar to those seen in humans resulted. It turns out that drug metabolism can vary among species. We consider the question of human experimentation in detail later, but the possible differences in effects on humans versus various animal species must always be kept in mind.

GM Crops: Are You What You Eat?

A key advance in genetic engineering, and also one of the biggest controversies, has come from the production of **genetically modified (GM) organisms**, particularly GM crops. The aim is to produce plants that can resist pests, disease, or harsh climates, facilitating the production of crops. Yet many people are opposed to GM crops and have an aversion to ingesting genetically modified food.

GM crops and other genetically modified plants present several areas of concern. The first area involves the plant itself. Scientists must determine whether the alterations in the plant's genetics provide a benefit to the plant or at least do not produce a less vigorous plant. One question that might be worth answering is whether the integrity of the species (maintaining the original genetic composition of a species without major change, so that it is still essentially the same species) is somehow preserved along with the alteration. You should seek to define for yourself whether such **species integrity** is important or whether creating a "better" plant species is more desirable than trying to maintain an "old" species. In doing so, determine for yourself whether the genetic modification of organisms, in this case plants, violates any ethical codes.

Another question, on a broader scale from the first, is the possible effect of altered plants on the ecosystem and on overall **biodiversity** (the range of different species present in an ecosystem). We must determine the effect of the introduction of a genetically

YOU DECIDE

Buyer Beware?

Because of public concern over the possible safety of GM foods, there have been proposals for conspicuous labeling of all GM foods, even if there is only a minute concentration of any GM product in the food. The cost of testing and labeling could add significantly to the price of these foods, and the actual need for labeling in terms of safety is disputed. Should public fears over GM foods require that all such foods be labeled for consumers? Is this a good marketing scheme? You decide.

modified plant on the local environment. Because we are focusing on crop plants, the desired effect will likely be not only to increase the growth and production of the GM crop plant but also to reduce the harm that might be caused by potential pests and diseases.

One example is Roundup-Ready soybeans. This soybean is genetically modified to resist Roundup herbicide, allowing a farmer to spray the crop and kill noxious weeds that would interfere with growth of the crop without harming the soybeans. Another example is Bt corn (**Figure 13.2**). Recall from Chapter 6 that this GM crop is engineered to produce a toxin from the bacterium *Bacillus thuringiensis,* which kills corn borer larvae and other insect pests that feed on the plant.

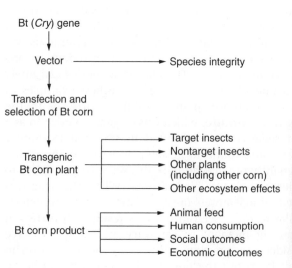

FIGURE 13.2   Possible Interactions and Considerations for a GM Crop, Bt Corn Bt-modified crops were designed to protect plants against pests such as the cotton bollworm (top photo) and corn borer; but soon after their use, concerns were raised about possible effects on nontarget insects such as monarch butterflies (bottom photo).

Some research has suggested a possible toxic effect of Bt crops on monarch butterflies, even though they were not a target insect and do not feed on corn. It would thus be important to know if the toxin affects specific species or groups of insects and whether non-target insects can also be affected. One question that was considered was whether the toxin could be spread or was confined solely to the corn plant. Because corn is wind-pollinated, the pollen might be carried to other plants and be toxic to some insects at a distance. Researchers had to determine the likelihood of this happening. For the monarch butterflies, the study indicated that corn pollen could be spread by wind to milkweed plants (which are a food source for monarchs) located next to the GM cornfield. Monarchs feeding on the milkweed could then ingest the corn pollen (and the toxin). Scientists had to ask whether this was a likely occurrence and how much pollen and toxin it would take to kill a monarch butterfly. Think about the types of experiments you might design to test these questions. In recent years, several long-term studies have shown no adverse effects of monarch exposure to Bt crops.

There might be possible effects of cross-pollination between Bt corn and the transfer of engineered genes to other noncrop species (for example, other plants that might acquire the toxin gene and kill desirable insects, such as monarch butterflies). You would have to consider the likelihood of such an occurrence. The whole question of introducing GM plants into the natural environment needs careful scrutiny to consider possible long-term changes to the ecosystem and the effects on biodiversity. Another consideration is the spread of GM plants to other areas beyond where they are cultivated.

Another question to consider regards the product of the GM crop: we must consider how it will be used, whether it is safe to feed to animals, and whether it is safe for humans. We must also gather all the facts and make informed evaluations. Because the toxin is directed against insects, it seems unlikely that it would affect animals or humans. But making this assumption is not enough; evidence and tests are needed that could verify its safety. Think of some experiments you could do to test the safety of a GM product.

We should not focus only on the particular gene in question but also ask whether other genes or products present in the GM crop might have to be considered. For example, in many cases antibiotic-resistance genes are used as selection markers for genetically engineered cells. Are the genes still present in the GM crop? And if so, can these genes be transferred from ingested food to gut bacteria? You should again ask what the likelihood is that DNA or proteins would survive digestion. Figure 13.2 shows a some of the possible

YOU DECIDE

Should the Public Be Consulted About the Release of Genetically Modified Insects?

In December 2010, some 6,000 GM mosquitoes were deliberately released into an uninhabited forest in Malaysia—to the surprise of locals and international followers. The Malaysian trial was an effort to control dengue fever, a virus-based disease transmitted by the mosquito *Aedes aegypti*. The dengue virus has spread to over 100 countries, and the World Health Organization (WHO) considers the at-risk population for dengue infection to comprise nearly 2.5 billion people. In some parts of the world, dengue fever is a major cause of death in children. Dengue fever causes severe and painful flu-like symptoms in adults, which can sometimes lead to lethal complications. British biotech company Oxitec had previously released 3.3 million GM insects in trials in the Cayman Islands in 2009 and 2010 in the world's first GM mosquito field trial designed to control dengue-carrying carrier insect populations. The company claims that their Cayman Island trials reduced that insect population by about 80%. In both cases the public and even researchers in the field were generally unaware of the releases until after results of the trials were announced. There are no policies on informing the public about such experiments. Should there be? Is it sufficient to simply inform the public that GM insects have been released? You decide.

interactions and considerations. We must also determine whether the product from a GM crop should be quarantined after the crop has been cultivated. This again comes back to our question of safety, especially regarding human exposure to or consumption of the product.

One consideration to keep in mind is whether the Bt toxin would survive food preparation and still stimulate an allergic reaction. Some countries strictly limit the importation of GM crops, and there is a movement to label foods to designate their origin in terms of potential genetically modified plants. Some people believe that, based on the data, these restrictions are not valid and may unduly frighten consumers. What would be your reaction to finding a "GM" label on your pizza or cornflakes?

Social and economic questions also arise from the potential use of GM crops. The ability to modify plants for better, less costly production could drastically change

YOU DECIDE

How Much Return on the Investment?

Monsanto, a company based in St. Louis, Missouri, created Roundup-Ready soybeans, which are genetically engineered to withstand Monsanto's Roundup weed killer, a commonly used herbicide that kills almost all plants. Roundup-Ready seed costs several dollars per bushel more than conventional soybean seeds. Monsanto has used private investigators to check out reports that farmers in New Jersey had been recycling seed harvested from Roundup-Ready soybeans planted the previous year. Using seed harvested from a previous crop is a common practice for many farmers and allows them to save money on seed for planting a new crop. Because Monsanto invested so much money in their technology, they wanted farmers to buy new Roundup-Ready seeds each year, but farmers claim they were never told their seed could not be replanted. Monsanto claims farmers are using their high-tech expensive product for free.

Should the farmers be allowed to continue their traditional practice of recycling seed? Or should biotechnology companies be able to enforce restricted uses of their products? You decide.

the agricultural industry. Potentially, more abundant food could be available at a reduced cost both to the farmer and to the consumer. These advantages may be offset by potential disadvantages, however, such as the safety concerns described previously. Other possible uses for GM crops include the production of medically useful compounds. Safety and efficacy again become key considerations in the ethical assessment for use of these compounds. Current U.S. policy requires not only the usual range of tests for safety and efficacy of medical compounds but also that growth of the GM plants be restricted. Fields must be surrounded only by other plants that should not cross-pollinate with the GM plant, all plant material must be removed at harvest, and the field cannot be used for a number of years after harvest.

Animal Husbandry or Animal Tinkering?

Genetic modification of animals raises many of the same questions posed by the genetic modification of plants. Early applications of biotechnology to animal husbandry have included antibiotic supplements in feeds and injections of growth hormone or steroids to increase growth of the animals. Consider the application of these supplements and injections from an ethical standpoint. First, because these are agricultural animals, the effects of genetic modification on the products from the animals (milk, meat, and so on) and the safety of these products for human consumption were the main concern. One consideration was the length of time the supplements (especially hormones) would persist within the animal—that is, whether they would still be present when the animal products were consumed by humans. If so, scientists must also determine whether there would be any effects on the consumer and whether the hormone, if present, would survive any cooking or the digestive process.

Little concern has been expressed about the effect of GM agricultural animals on the environment, but there are still questions about species integrity and the health of the animal as well as the safety of animal products for human consumers. Similar considerations are in order regarding animals that have been genetically engineered to produce medically useful products. These could include transgenic animals modified to produce a clotting factor in their milk, transgenic animals such as pigs engineered with human genes so their organs could be transplanted into humans without being rejected, or cloned cows to be used for meat. Chinese scientists have created transgenic cows that express human milk. At what point would you consider such alteration of an animal to be unethical?

Be sure to consider whether there is a point at which the animal might acquire enough human genes, cells, or attributes that you would consider it human and whether this would change your ethical viewpoint on using the animal for research. Refer back to Figure 13.1. Recently, English and French scientists implanted mouse cells into chicken embryos to demonstrate that mouse cells could recognize developmental signals from chicken cells to stimulate tooth formation. Biologists are interested in this in part because although the genes involved in these developmental signals are not active in birds (teeth were lost in birds about 70 to 80 million years ago), it indicates that these genes can be activated and can stimulate tooth development when the proper cells are present. So what? Who wants to make chickens with teeth? These are good questions, and you may certainly ask whether these types of experiments are ethical uses of animals. But also consider that information from these experiments can be fundamentally valuable to clarify tooth development and potentially in the future to develop novel treatments for tooth formation, replacement, and regeneration in humans—the main reasons for this research.

As described in Chapter 10, genetically modified wild species of animals such as salmon present another set of ethical questions, including environmental concerns.

Synthetic Genomes and Synthetic Biology

Research on synthetic genomes has resulted in transplantation of a synthetic genome into a bacterial strain to change the recipient microbe into the organism of the donor synthetic genome (Chapter 5). These and other future applications of synthetic biology are raising many ethical questions about what should and should not be done with synthetic genomes and their potential uses to create novel life forms.

The Human Question

Many of the thorniest questions regarding biotechnology, and some of the most contentious debates, revolve around humans. Even simple scientific procedures can evoke strong emotions and stir profound controversy when humans are the subjects. Why do you think the use or potential use of humans experimentally causes such strong reactions? Later we discuss how the mere definition of what is human has become an area for debate.

For now, we look at a simple example based on a potential anticancer drug generated through biotechnology. Suppose the drug has moved along to the point where it was ready for clinical trials. We must decide to whom we will administer the drug as a test. Because it is an anticancer drug, we will naturally give it to patients who have the type of cancer targeted by this drug. We must decide, however, whether to give it to all patients with that type of cancer or only to those in the most advanced stages of disease—those who may have exhausted all other means of treatment and for whom the new experimental drug is a last resort. The rationale is that, for this subset of patients, there is no other possible treatment; therefore the experimental treatment presents at least some hope. However, the drug may work better with patients who are earlier in the progression of the cancer and thus might be more effective. Nonetheless, the patients who are in earlier stages of the disease have other alternatives that have already shown safety and efficacy (at least to some extent).

After you have picked the patient group for the clinical trial, another dilemma arises. Patients have a right to be informed fully of the potential effects of the experimental treatment, both good and bad. Only when so informed can they be willing participants in the trial. This is termed **informed consent**. Patients give their consent to proceed with the experiment, fully informed of the potential benefits and risks that lie ahead. The concept of informed consent is vital to any procedure involving a human. If a patient is unable to give consent on her or his own (because the individual is too young or in a coma, for example), a family member or guardian can give proxy consent. Determine for yourself why informed consent is so vital.

Placebos present another problem as far as experimental procedures on humans are concerned. In a placebo-controlled study standard scientific practice involves using an experimental group (in this case, patients who would receive the drug) and a placebo-controlled group (in this case, patients who would receive a placebo, a safe but noneffective treatment

---

YOU DECIDE

Should Clinical Trial Data Be Public or Do They Belong to the Drug Company?

In 2002, the National Institutes of Health (NIH) launched Clinicaltrials.gov to help patients and physicians find information on nearby clinical trials. The current registry contains information on over 108,400 clinical trials in more than 174 countries. This registry has been voluntarily utilized by most biotech and drug companies because it excludes early-stage trials, where sensitive business information could be lost. For several years a congressional bill (S.467—the Fair Access to Clinical Trial Act; FACT Act) has been debated; it was to call for mandatory disclosure of the results of clinical trials in a public database. Some claim that the information could help patients determine whether to participate, but some in the drug industry say it is unnecessary and will stifle innovation.

The FACT Act was never passed as a law. The bill proposed to require open access to information from phase 1 trials, which enroll healthy patients in small numbers. According to some industry spokespersons, however, such information would be of little value to patients. The advocates for the bill claim that companies could benefit from sharing early-stage trials to prevent new drugs from repeating the same mistakes that had already produced negative results. In defense of the industry, biotech companies rely on venture capital funding, and if early data are shared with other companies, it is claimed that funding for innovation would dry up. Is the benefit worth the risk that it may reduce funding for drug research? Is there a better solution? You decide.

such as a sugar pill or saline injection). In a completely randomized **double-blind trial**, neither the patients nor the doctors administering the treatment would know who received the real drug and who received the placebo. Informed consent should play a role in this type of experiment. Using placebos is part of objective science, but we must look at whether it is ethical. Objective science may not always be the best approach or the ethical approach.

What Does It Mean to Be Human?

Many of the current ethical debates about stem cells and cloning revolve around the moral status of the human embryo. As we discussed in Chapter 11, stem cells may hold the potential for tremendous breakthroughs in **regenerative medicine**, making possible the repair or replacement of damaged and diseased tissue in many diseases such as heart disease, stroke, Parkinson disease, and diabetes. As you know, there are many possible sources of stem cells, including embryos, fetuses, umbilical cord blood, adult tissues, and even "tamed" tumor cells (**Figure 13.3**). These sources of stem cells are in addition to nuclear reprogramming options such as making induced pluriopotent stem cells from human somatic cells. Much of the stem cell debate has centered on the scientific question of the abilities of different stem cells to transform into other tissue types for the treatment of diseases; however,

key elements in the debate have been ethical questions about embryos as a source of stem cells.

As a society, we have not been able to agree whether it is ethical to destroy early-stage human embryos for research that may potentially be of benefit to thousands of patients. Unlike organ donation, in which an individual may donate one of two paired organs while alive or after his or her death, the process of harvesting embryonic stem cells destroys the donor (the embryo; Figure 13.3). The use of human embryos as a source of stem cells is a very controversial topic, in part because it raises questions such as, "When is the embryo considered to be a person? Is it ethical to destroy a human embryo to harvest stem cells that may benefit many other humans?

We must resolve the focal question, which examines the moral or ethical status of a human embryo. Biologically the embryo is a human being, species *Homo sapiens*, just starting out on the developmental journey. We must weigh the relevance of different developmental milestones for being considered human against the simple biological fact that the embryo is a member of our own species. Biologically, an embryo is a member of the human species, but the question of the moral status of a human embryo goes beyond the biological and also revolves around what some have termed the status of **personhood**. This term has been used to define an entity that qualifies for protection based not on an intrinsic value but rather on certain attributes,

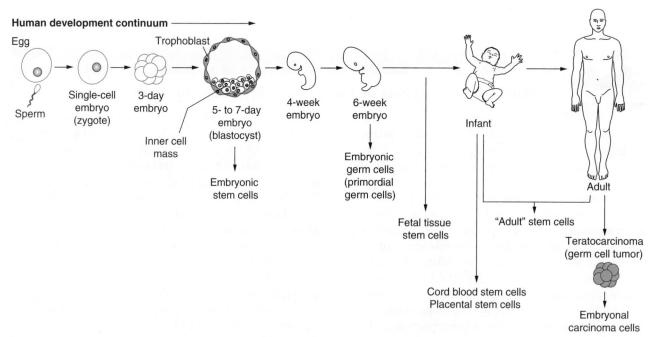

FIGURE 13.3   **Sources of Stem Cells** Human embryos, fetal tissue, infants, and adults are all potential sources of stem cells.

YOU DECIDE

Celebrity, Shame, and Stem Cells

Dr. Woo Suk Hwang of Seoul National University in South Korea was thought to be a pioneer in stem cell research. Dr. Hwang was treated to free airfare to travel around the world, high fees for speaking engagements, high-society parties, regular appearances on television, and other perks not typically associated with the life of a scientist. Hwang became a source of national pride in Korea primarily for claims that his research group had created the first cloned human embryo and stem cells derived from human patients. This work was published in 2004 and 2005 in highly heralded papers in the prestigious journal *Science*. But by December 2005, Woo Suk Hwang's celebrity and his career were quickly falling apart in disgrace and he resigned from his position. Current and former members of his own research team as well as a collaborating lab revealed evidence suggesting the falsification of data and allegations of unethical means for obtaining donated eggs. It was subsequently determined that data and figures had indeed been falsified and manipulated; the lab had not cloned an embryo or derived patient-specific stem cells as claimed. Among these issues of unethical actions was concern about how eggs had been collected for the work (initially Hwang reported that anonymous donors had provided the eggs) and lies about the number of eggs used. It turned out that junior researchers in Hwang's lab were among the

donors and that donors were paid for their eggs. Women were also paid to take fertility drugs to provide eggs. Hwang's fall from grace was highly publicized around the world, and it served to further support the claims of groups speaking out against stem cell work—groups that were already skeptical regarding the highly lauded potential of stem cells.

Bioethicists claimed that Hwang failed to adhere to the West's ethical standards when he paid women for their eggs and used junior researchers who worked for him to provide eggs. But the South Korean Health Ministry deemed there was no ethical problem because the eggs were donated voluntarily, with no evidence that junior researchers were coerced to provide them. The ministry, in other words, determined that Hwang had not violated existing laws. The ministry further explained its conclusions by saying that Western-trained physicians and scientists considered donations from vulnerable junior researchers as unethical because they might feel pressured to satisfy their supervisors and donate involuntarily. The ministry went on to say that although researchers in South Korea abide by ethical procedures in conducting research, they generally do not focus on ethical issues as much as is common in the West. Was it ethical for Dr. Hwang to collect eggs from junior researchers? You decide.

---

such as self-awareness. List for yourself the advantages and disadvantages of this concept of personhood. One concern with it might be the question of who decides which attributes count in evaluating whether a particular human being can be valued as a person.

Regarding embryo research, some have taken this attitude: "Not a person, not a problem." They reason that a human embryo is microscopic, not yet possessing a beating heart, brainwaves, arms, or legs. Of course those things develop later, but at a very early stage the embryo lacks what we usually associate with our concept of humanity. There are obviously various views of the status of the human embryo, and no consensus. Some say it is simply a clump of cells, just like a chunk of skin. Others believe it is a form of human life deserving of profound respect—a potential person. Still others maintain that an embryo has the same moral value as any other member of the human species. Consider the question of whether *any* human cell deserves respect as a potential person. When combined, an egg cell and a

sperm cell form an embryo, and any somatic cell can contribute a nucleus to form a cloned embryo. Consider, then, whether there is something special about having a complete human genome in an egg cell.

Considered in the context of using human embryos to isolate embryonic stem cells, the debate regarding the necessary destruction of the embryo is contentious. Certainly, embryonic stem cells can theoretically be used to form any tissue, with the potential for transplants to repair or replace damaged or diseased tissue. This might suggest that it would be ethical to destroy human embryos for research if, from that destruction, it meant that such research could lead to treatments for disease. However, based on current published evidence, one question that first must be asked is whether embryonic stem cells are as good for potential treatments as claimed. Another consideration that has both ethical and scientific components is whether other viable alternatives, such as adult stem cells or induced pluripotent stem cells, are just as good.

The ethical questions then take on a new form, asking, on the one hand, whether research on embryos is necessary if science is to explore all possible avenues for medical breakthroughs or, on the other, whether the calculation now indicates that the alternative makes ethically contentious research unnecessary. Of course one ethical position would be that even if there are no alternatives, the ethical cost is too high to justify the destruction of embryos.

Interestingly, even some people who view a human embryo as a potential person and not a realized individual oppose destruction of human embryos on ethical grounds. The concern is not directly with the embryo and its status but rather with how society views any human life. From their point of view, we are embarking on a slippery slope in which the destruction of human beings for medical use or experimentation might move from the use of embryos to the use of born individuals. Once again, we return to the question of personhood, especially as a societal construct that might rank different humans based on their quality of life and on their usefulness to society.

Spare Embryos for Research versus Creating Embryos for Research

As we consider the question regarding the moral status of the embryo, there are varying views based on the original purpose for which the embryo was created or on the method by which the embryo was created. Contrary to a common misconception that aborted fetuses are used for stem cell research, the primary source of embryos for this research is "excess" embryos from in vitro fertilization, or IVF (Chapter 11). These embryos, left in frozen storage after a couple has used other embryos for implantation and ideally pregnancy and birth, may be donated (with the couple's consent) for research. Depending on the IVF clinic or the country, frozen embryos may be discarded after a certain period of time. List for yourself the necessary ethical considerations in this instance. Some say that it is ethically valid to use these embryos for research if they would otherwise be discarded—that some ethical good could be salvaged from their existence if they were to contribute to a potential therapy or new in scientific knowledge. Others say their destruction for research crosses an ethical line inconsistent with the purpose for which they were originally created.

Another potential source of human embryos is the specific creation of embryos for research purposes. Some have argued that the use of spare embryos for research is ethically valid (reasoning that this could be salvaging a potential good out of certain destruction) but that the specific creation of embryos for research is not. To determine whether this argument is ethically consistent, consider whether there is any difference in the embryos biologically or only in a view of the intent or use of the embryos.

Should Humans and Other Animals Be Cloned for Any Reason?

Creation of embryos by somatic cell nuclear transfer cloning raises many of the same questions encountered with stem cell research, with the added complexity of the technique and the potential "identity" of the clone (**Figure 13.4**). A cloned embryo created for transfer into a uterus and implantation faces many risks and

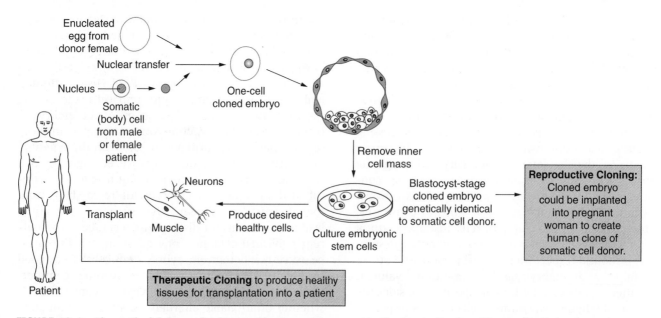

FIGURE 13.4 Theoretical Scheme for Human Cloning or Therapeutic Cloning to Produce Tissues for Transplantation

safety factors for its own development and growth, both before and after birth. The success rate for live births and the subsequent survival of the clones is extremely low. As we discussed in Chapters 7 and 11, we know that there can be a number of problems, some subtle and some very severe, with the health of clones.

One consideration is whether creating a cloned human embryo with the intent of initiating a pregnancy and live birth should be considered another type of assisted reproductive technology similar to IVF or whether (because of the safety risks and low success rates) it should be considered unethical human research. Societal questions regarding the identity of a born human clone must also be taken into consideration. For example, if a couple decided to create a cloned child using a donor cell from the wife, the clone would not be a genetic daughter but rather the wife's sister, a late-born twin, and would not be related to the husband at all. Ethical considerations related to the clone include how the lack of relatedness to one parent might change kinship and family relationships. Given that the clone is a "copy" of one of the "parents," the clone, once born, might be expected to "live a better life" than the person who was cloned. Another potential concern arises if a previously existing person, now deceased, had been cloned. The genetic makeup of the clone would already be known, already dictated, because the process of cloning reproduces a previously existent individual. A clone may be expected to live up to that genetic legacy, with heightened expectations

by the parents and others based on what was achieved by the donor of the clone's genetic material.

The cloned cat "cc" (**Figure 13.5**) looks very similar to the cat that donated her genetic material, but her coat pattern differs slightly from the donor's. These differences occur because of subtle epigenetic changes during development. Our genes determine many of our physical characteristics and predispose us to various diseases or behaviors, but after we are born there are many experiences and environments that make us who we are and who we will become. Those experiences cannot be duplicated, so the clone will grow up differently than the one who was cloned and may behave quite differently. A clone of Einstein might become an artist instead of a scientist. A great deal more affects us and our makeup than just our genetics, including our environment and experiences.

So far we have considered the creation of a cloned human embryo with the intent of producing a live-born child. However, this is not the only proposal for the creation of cloned human embryos. Creating human embryos for **therapeutic cloning** could lead to matched embryonic cells for patients (Figure 13.4) and to valuable human research models for the study of genetic diseases and cancer. Although this might sound like a potentially valuable and ethically valid reason for creating cloned human embryos, others argue that such embryos should not be produced.

FIGURE 13.5 The First Cloned Cat, "cc" (the Carbon Copy Cat)

One argument against the creation of cloned human embryos is not based on the embryo's inherent value as a human or person but on the argument that the creation of human embryos for such purposes could lead to human commercialization, making any human life a commodity to be bought, sold, and used, thus cheapening life in the process. Still others have argued that we should not be creating human embryos in a manner that manufactures human life, the so-called designer embryos.

One argument in favor of creating cloned human embryos relies on the assumption that if an embryo is not created by normal means (in this case meaning by fertilization), it is not human. Each of these arguments is based on different definitions of what it means to be human and the value placed on human life. Interestingly, nearly 30 years ago, similar ethical debates about what it means to be human arose when Louise Brown, the so-called test tube baby, was conceived by in vitro fertilization. And now even the most conservative groups, including many religious sects, consider in vitro fertilization to be a completely ethical way in which to conceive an embryo.

Patient Rights and Biological Materials

Consider how you would feel if the following scenario happened to you. You are being treated for a disease such as leukemia and you donate blood, bone marrow, and spleen cell samples for analysis as part of your treatment. You later learn that the physicians treating you developed cell lines from your tissues, which they patented and then used to bring substantial financial rewards. Such situations have in fact occurred and resulted in a number of patients' lawsuits. In these cases, patients have claimed that prior to providing their **informed consent** to extract the cells, not told that their physicians would conduct research on their cells and that they would benefit financially. Patients and lawyers have claimed that these examples represent a breach of the doctor–patient relationship and an example of a "principle of unjust enrichment."

In most of these cases, patients have sought a share of financial compensation for their cells. In several high-profile cases, courts have ruled that physicians do have a duty to disclose their personal interest in research and potential economic matters unrelated to patient treatment. But courts have also ruled that donors of cells and other biological materials do not have ownership rights of their biological materials, and physicians and scientists who develop patents and receive financial benefits from these materials are not liable for patient compensation.

Many donors are unaware that they do not own their own cells and they relinquish a large degree of control over their tissues when they donate them. Related to this, there have even been situations where donated sperm samples were used to fertilize eggs and produce children without the consent of the individual who donated the sperm, thus making men fathers without their consent. Overall, U.S. regulations are increasingly favoring written, informed consent without coercion for any donor subject. As stem cells and regenerative medicine technologies become more common, undoubtedly the issue of patients' rights regarding biological materials will become even more complex and require greater scrutiny and ethical consideration.

Genetic Information

The Human Genome Project has led to the identification of genes responsible for or contributing to many disease states. This knowledge has also led to new strategies for genetic testing and treating genetic disease. But because the story told in our DNA could become so easily read, there is a growing concern over the privacy of that information. One's DNA sequence can be a truly unique identifier. Researchers especially must take care to guard the confidentiality of those who have donated DNA for sequencing and testing. Because a free flow of scientific information ensures rapid dissemination of ideas and facilitates scientific advances, we next look at some of the ways in which scientists can safeguard genetic information to assure individual research subjects or groups of subjects that their privacy will be maintained. This is important not only to reassure research subjects but also for scientists to maintain the continued trust of people and their willingness to participate in research projects. A sampling of some of ethical questions about genetic information include the following (refer back to Chapter 11 for more a extensive discussion and further questions about genetic testing):

- Should we test people for genetic disorders for which there are no effective treatments?
- What ethical obligation do physicians and scientists have to divulge genetic testing results if analysis of a person's DNA reveals mutations unrelated to the original reasons for the test?
- A negative result from a genetic test does not always rule out the future development of certain diseases, nor does a positive result always mean that an individual will develop a disease. How do we effectively communicate the results of genetic tests and actual risks to the person being tested?
- Should it be possible for someone to be tested for non-disease-related genes affecting such traits as intelligence, skin color, height, or weight?

■ *Identifiability*, the potential for disseminated genetic data to be associated (or reveal the confidential identify) of specific individuals, is a major concern. How can electronic medical record keeping prevent identifiability even when patients agree to share or release certain aspects of their medical records?

We have discussed controversies associated with direct-to-consumer genetic tests (Chapter 11). As the identification of genetic traits becomes more routine in clinical settings, physicians will have to ensure their patients' privacy. There are significant concerns as to how genetic information could be used negatively by employers, insurance companies, government agencies, or through perceptions acquired by the general public. Genetic privacy and the prevention of genetic discrimination will be increasingly important in the coming years. In 2008, the U.S. House of Representatives passed bill H.R. 483, the **Genetic Information Nondiscrimination Act (GINA)**. The Senate also passed an identical bill (S. 358), and shortly thereafter President George W. Bush signed the GINA into law. This act prohibits discrimination based on genetics or the improper use of genetic information in health insurance and employment.

More or Less Human?

The first successful example of gene therapy involved children suffering from severe combined immunodeficiency syndrome (SCID; Chapter 11). Individuals with SCID are born without an effective immune system, primarily because of one defective gene in their immune cells. Successful treatments have used adult stem cells from the children's bone marrow, using genetic engineering to add the correct gene and replacing the engineered stem cells back into the patients (refer to Figure 11.15).

Think about some of the ethical considerations associated with gene therapy technology. Gene therapy treatments such as those used to correct SCID seem wildly successful, so one might think there should be no concerns. However, because these are experimental medical procedures, there are certainly issues of informed consent, safety, and efficacy. One safety issue that has arisen is the potential for cancer formation. As discussed in Chapter 11, an adenovirus vector used for gene therapy has led to leukemia in several children. This consequence has aroused considerable concern about future genetic therapies, especially using this particular viral vector. So one ethical consideration might be the risks associated with altering an individual's genetics, especially the specificity of the targeting for insertion of the replacement gene.

Current and proposed somatic gene therapies involve the treatment of existing genetic diseases. However, we should consider the possibility of genetic treatments for conditions where there is only a genetic *tendency* toward a particular disease and there is no certainty that the disease will occur. One possibility is breast cancer, in which genes (such as mutations in *BRCA1* and *BRCA2*) have been identified that can increase the risk for development of the cancer. We must consider the difference between a genetic disease and a genetic attribute. For a genetic disease such as SCID, it is known that an individual with such a genetic problem will definitely develop the disease. However, a genetic *attribute* does not mean that the disease condition already exists or that it will inevitably develop. If we consider attributes that may not necessarily lead to a disease or may not even be associated with an increased risk for disease, we extend our considerations of genetic engineering beyond the therapeutic and begin to contemplate whether we will consider other potential genetic modifications as ethically acceptable, going beyond treatment for an existing condition to alterations and even enhancements of the human genetic composition.

Consider whether enhancing our individual genetics, perhaps to increase muscle mass or the oxygen-carrying capacity of red blood cells, might also be considered medically necessary for the health of the individual, even if it were really just a personal preference. The prohibition of such genetic alterations could be regarded as an infringement of individual rights. Because this individual's genome would be altered to something other than the normal, we must reconsider the question of whether he or she would still be considered human as well as just what we might include in that definition. Many questions surround the possibility of **gene doping** in athletes; that is, using gene therapy for genetic modification to enhance human traits important for sports. Genetic techniques applied to animals have demonstrated enhanced muscle function, increased blood cell production and oxygen content, and enhanced metabolism and endurance performance.

The question of what makes an organism human becomes more pressing when we look at the potential of **germline genetic engineering**. In this case, the genetic alteration is done in the sperm, egg, or early embryo. This can actually make the manipulation more effective than somatic genetic engineering because the genetic modification can affect every cell in the individual's body, making the genetic alteration inheritable. Potentially this could mean the elimination of some genetic diseases, because those genes could be removed from the human gene pool. Of course, any problems resulting from the genetic alteration could also be inheritable, as could any genetic enhancements. Consider whether this type of genetic modification, which would affect not only the treated individual but also future generations, might be ethically acceptable. Some of the potential outcomes of removing or adding genes in the human gene pool could include the elimination of genetic diseases or disease susceptibility, enhanced human performance, increased or decreased human life span, and increased cancer risks.

One Nobel laureate has been an advocate of this type of research because it could lead to the development of a "better human being"; but we might pause to consider just what constitutes "better" and whose definition should be utilized in these decisions. Another potential outcome sometimes mentioned for germline genetic alteration could be the creation of a new, superior species of human—the basic concept behind **eugenics**. One potential negative associated with this outcome could be splitting humans into different genetic social classes.

13.3 Economics, the Role of Science, and Communication

We cannot escape the fact that money plays a major role in research decisions. Private investments as well as government funds fuel research and development, and individuals and companies alike seek to use biotechnology for discoveries that will be profitable. Biotechnology is a business, after all.

Not only is scientific research expensive, but projects, companies, and careers may live or die depending on the funding for and profitability of the research. There is clearly not enough money to fund every scientific proposal, whether the proposal is from a university faculty member doing research or a company working to develop a new product. Research funding is often evaluated based on whether it will increase our knowledge about fundamental aspects of science or produce a discovery or product that improves our lives.

Many college and university researchers seeking funding submit research proposals to funding agencies. These proposals are then reviewed by panels of scientists who make recommendations to the funding agencies about which projects should be funded. But the amount of money and even some of the decisions on funding direction also come from those providing the money, especially the U.S. Congress. Most of the decisions are based on the potential success of the science, its novelty, and its potential application to health and general knowledge. However, some of the decisions may also be affected by how well the scientists

YOU DECIDE

What to Treat Using Gene Therapy

In Chapter 11, we discussed some of the unresolved scientific questions surrounding gene therapy. There are also many ethical concerns about gene therapy and the risks associated with it. Do the patient and his or her family members understand the risks associated with gene therapy trials? For instance, Jesse Gelsinger (recall from Chapter 11 that Jesse died as a result of his gene therapy) had a relatively mild form of a disease that was being successfully treated with medication. Should he have been a gene therapy trial participant? Gene therapy is currently very expensive. Should everyone have access to gene therapy treatments regardless of the cost? What types of disease and conditions should be subject to gene therapy (for example, only deadly diseases)? What about treatable diseases for which gene therapy would be used so patients would not have to take daily but effective medications? What about cosmetic conditions such as baldness? You decide.

are known by the reviewers or by political pressure from members of Congress.

We should consider not only how research proposals are evaluated and their chances of success but also the possibility of their contributing to the breadth of scientific knowledge (because in science, it is difficult to know where the next major breakthroughs might come from and what background knowledge might lead to such breakthroughs). Additional considerations are the costs of and access to any treatments that might be produced. Funding for research might be so costly that only the rich would be able to benefit from it, or it might prove to be a colossal waste of money and time; it might simply suggest that other avenues of research would have a better chance of producing useful results. Because the possible sources of funding for research are limited, funding decisions may have to be weighed in terms of whether funding one type of research might decrease the funding of another, potentially more successful line.

A biotechnology company with a potentially profitable idea typically has to seek funding from venture capitalists and other investors willing to fund the R&D necessary for product development. If such investors are risk-averse but seeking a profit, who will fund risky experimental research that is expensive but has great potential for saving lives or leading to some new and useful technology?

We could even extend the discussion of economics to ethical issues such as whether it is acceptable to pay women to donate eggs for stem cell research. The lack of quality eggs left over from IVF coupled with the increased demand for eggs as stem cell research becomes more common is driving a need for human eggs. In June 2009, New York became the first state to allow public money to be used in this way—up to $10,000! Some of the ethical issues here revolve around the risk associated with hormonal stimulation, potential complications, and the invasive approaches required to harvest eggs. Many economic and noneconomic questions revolve around this issue:

- Should women be paid as egg donors?

- If so, should there be limits on such payments?

- Should there be a limit on how many times a woman can donate eggs over a particular period of time or should she be allowed to serve as a "serial" donor, giving dozens of eggs over time?

- Will paying for eggs create a market wherein women use egg donation as a form of employment?

Intellectual property rights are also important as we consider the ethical implications of research and its funding. The **patenting** of intellectual property—whether it be isolated genes, new cell types, GMOs, or even embryos—can be lucrative for the discoverers but may also pose ethical and scientific dilemmas. For example, consider the possibilities for a human gene that has been isolated and characterized—and then patented by the discoverer. The person or company holding the patent could require that anyone attempting to do research with the patented gene pay a licensing fee for its use. Should a diagnostic test or therapy result from the research, more fees and royalties might be required. This could make it difficult or expensive to carry out research on some genes or limit the clinical use of the patented gene. In Chapter 12 we discussed some of the controversies surrounding the patenting of genes. In 2011, a federal appeals court ruled that human genes can be patented, reversing a lower court's ruling involving a test for breast cancer (i.e., the *BRCA1* and *BRCA2* genes). It is expected than an appeal will be filed by the American Civil Liberties Union to challenge this ruling.

Some physicians have already complained that they cannot afford to pass on the charges of certain genetic tests to their patients owing to licensing restrictions. However, limiting or preventing patents for genes or genetic tools could remove the incentive for pursuing such research, especially for companies that hope to profit from their research. Compile a list of potential ethical problems associated with patenting of genes or cells and consider possible alternatives or compromises that might be reached to allow research to continue.

Ethical questions regarding biotechnology also touch on science's role in society (**Figure 13.6**). Science and technology have provided discoveries and inventions that make our lives happier and healthier. But it is important to consider whether scientists should have unlimited freedom for research. It is often difficult to know what the source of the next major breakthrough will be, and sometimes, new discoveries arise from unexpected sources and paths that many scientists may not even consider worth following. We must determine how science can best serve society. The whole concept of regulating something as unpredictable and free-flowing as scientific discovery is difficult. It is also hard to know who should decide these questions—scientists because they know how science works, policy makers because they must set the rules, or society because it is most affected by the decisions. Consider what types of regulations might best serve the needs both of science in furthering discovery and of society in furthering health care and ethical interests. Bioethics becomes critically important to the biotechnology industry because ethical discussions and debate are often the driving force for making laws in general (laws that

"I FIND IT HARDER AND HARDER TO GET ANY WORK DONE WITH ALL THE ETHICISTS HANGING AROUND."

FIGURE 13.6 Ethical Decision Making Is an Essential Part of Science

govern the population's behavior and certainly laws that will regulate the biotechnology industry) once it becomes a general consensus of a particular population that a law is necessary.

Accurate, honest communication is vital to the success of science in general and biotechnology in particular. Scientists must be willing and able to communicate openly and candidly with other scientists but also and more importantly with the general community. A public that cannot understand and appreciate the importance of the contributions that science makes to their daily lives will not support its endeavors. Straightforward communication is necessary, without overstating the potential of the research or the imminence of the results and without using confusing terminology. If the public and the policy makers believe they have been misled, the

outcomes might be disastrous for both science and society. If the public believes scientists are insensitive to ethical questions in their research, scientists will have a hard time earning the public's trust. Consideration of all the facts is important. Integrity in research is essential, but so is integrity in the communication of science.

Many of these ethical questions involve difficult decisions affecting not only yourself but other lives as well, both now and into the future. It would always be easier if the decisions, especially the tough ones, could be made for you. But the latter approach assumes that the decision maker has all the facts or has objectives that match your own. It also involves giving up your individual freedom in making those decisions, as well as your individual responsibility. With freedom comes responsibility—if you make

---

CAREER PROFILE

Career Paths in Bioethics

There are over 100 academic bioethics research centers around the world. Public interest in bioethics is at an all-time high, and there are many exciting career opportunities in this field. Traditionally, medical ethics has been a concern from the time Hippocrates swore to "First, do no harm"; but bioethics became a discipline only in the 1980s. As concerns over issues such as abortion and euthanasia were raised, doctors consulted with religious scholars, priests, rabbis, and philosophers to develop the field of bioethics. In the late 1980s, there was increasing consultation and involvement from lawyers and scientists. All these pathways (medical, religious, philosophical, legal, scientific) have contributed to the development of the discipline of bioethics, and there is still no very clear path to follow for a bioethics career.

Many people working in bioethics train in science or medicine initially and then pursue a master's degree in bioethics. Some begin in law school, some start with a Ph.D. in philosophy, and now there are Ph.D. programs in bioethics. Other starting points may be in medical anthropology, medical sociology, the history of medicine, or nursing. There is still much controversy over the best training to develop proper credentials for bioethics. One nearly universal agreement is that training should be interdisciplinary: It should include a broad knowledge base, language skills, and the ability to communicate across such multiple disciplines as religion, science, and medicine.

Choosing the right path of training depends to a great extent on a person's interests. Start by assessing your own interests and skills as they relate to bioethics. You should also investigate what possible jobs are available so you can anticipate the necessary training. A good place to start is the NIH "Careers in Bioethics" website (see the Keeping Current list of websites on the Companion Website).

Skills in this area can convince employers of your potential. When biotechnology employers look at potential employees, they do not simply look at academic credentials and work experience. Meeting the specifications in the job description is essential, but possessing some of the intangible skills that can make prospective employers call you for an interview is often more important. Employers often look for well-rounded employees who not only will work well in teams but also are diplomatic, resourceful, and able to build networks among their peers. Let employers know that you possess these skills, even if you have acquired them over the years in other work experiences than those called for in the application. Employers also want to know that their workers can tackle complex issues with integrity and appropriate ethical consideration.

Be prepared to defend any so-called soft skill you possess with clear examples, but don't be afraid to sell yourself or your abilities.

Adapted from P. Watson (2003), Transferable Skills for a Competitive Edge, *Nature Biotechnology*, 21: 211.

choices, you are responsible for those choices, and your choices affect not only you but others as well. A person who is clever knows the best means to achieve an end; a person who is wise knows which ends are worth striving for.

MAKING A DIFFERENCE

For this "Making a Difference" we chose to highlight a policy mentioned earlier in the chapter, the Genetic Information and Nondiscrimination Act (GINA), signed into law in 2008 after 13 years of debate in Congress. The ethical significance of this law is that it is intended to prohibit discrimination against individuals based on genetics. While the overall effectiveness of this law remains to be seen it is significant that at time when genetic testing is becoming more common, there will be a greater need to protect citizens from insurance and employment discrimination based on genetic data. Thus GINA has the potential to make a difference in preventing abuse based on genetic information.

QUESTIONS & ACTIVITIES

Answers can be found in Appendix 1.

1. What are the two leading approaches to ethical thought, and how do they differ?

2. "Last September a little girl from California named Molly received a lifesaving transplant of umbilical cord blood from her newborn brother, Adam. Molly, who was then 8 years old, suffered from a potentially fatal genetic blood disorder known as Fanconi anemia. But what made the procedure particularly unusual was that Adam might not have been born had his sister not been sick. He was conceived through in vitro fertilization, and physicians specifically selected his embryo from a group of others for implantation into his mother's womb after tests showed that he would not have the disease and he would be the best tissue match for Molly." Was this ethical?

   Adapted from R. M. Kline (2001), Whose Blood Is It, Anyway? *Scientific American*, 284(4): 42–49.

3. Make a chart with a list in one column of all the possibilities you can think of for interactions a GM plant might have with the ecosystem. Then, in the second column, list the possible outcomes, good or bad, for each of the listed interactions.

4. If a GM crop was engineered to produce growth hormone, what are some of the possible ethical outcomes and questions that must be anticipated?

5. Is there a difference between a human being and a person? Why or why not?

6. There is never enough money available, even from the government, to support every scientific research proposal that is submitted to funding agencies. How should decisions be made about which of these proposals are a waste of money and which will yield discoveries that may change human life forever? When competing scientists are applying for research grants, what principles should guide decisions on how the available tax money is to be spent? Should ethical concerns be a part of such funding decisions?

7. Suppose that gene therapy could be used to lower levels of cholesterol and saturated fat in the blood, thus allowing people to consume fat-rich fried foods with little risk of heart disease. However, this treatment might not work for everyone, and there would be no way of knowing in advance who would benefit. As a result, some people would inevitably die prematurely from heart disease. Should such gene therapy be approved for use in humans? Answer this question by briefly describing how a bioethicist who believes in a utilitarian approach might view this scenario, and contrast this perspective with a deontological viewpoint.

8. When drug companies find out that a drug target is not responding to their new drug, do they have a legal obligation to inform other drug companies?

9. If a food allergen can be removed from food products by using a recombinant plant food source and a consumer is harmed by the unwillingness of a producer to utilize the GM source, could the company be found liable for the harm created?

10. If a single cell is removed from an embryo in vitro and kept to be used for future organ regeneration but the embryo is not harmed and can continue to develop normally, is this an ethical use of human embryonic tissue?

11. Michelle, a 21-year-old doctoral candidate in sciences at one of the best universities in the United States, is an attractive woman. To earn a little extra cash, she responds to an ad in local newspaper from a fertility clinic seeking women eager to donate eggs for infertile couples seeking to have children. Michelle will be paid for these eggs, about $5,000 each. Over a period of several years she donates eggs for infertile couples and eggs for somatic-cell nuclear transfer procedures used to create embryos for deriving embryonic stem cells.

During this time, Michelle earns around $45,000 for her donated eggs. She will never know if any her eggs actually led to a successful pregnancy. Is this ethically OK with you? Is it acceptable to pay someone for their eggs? What information do you think potential egg donors should be told about before or after making a donation?

12. If a company develops a vaccine for a deadly disease without a cure but early studies show that the vaccine is effective in only about 40% of patients, should the company wait to bring the vaccine to market while they work to improve its efficacy (which may not be possible) and people suffering from the disease die, or should it be available immediately?

13. If you support releasing GM organisms such as plants or insects into the environment, develop a plan for controlling the spread of these organisms.

Visit www.pearsonhighered.com/biotechnology

To download learning objectives, chapter summary, "Keeping Current" web links, glossary, flashcards, and jpegs of figures from this chapter.

1 Answers to Questions & Activities

Chapter 1

1–2. Open-ended; answers variable.

3. Bioremediation

4. Pharmacogenomics involves prescribing a treatment strategy based on the genetic profile of a patient—fitting the drug to the patient's genetics. An RNA or DNA sample from a patient with a medical condition can be analyzed to determine whether the genes that person expresses match the genetic profile of a particular disease process. If they do, a specific treatment approach can be designed according to the best-known treatment strategy based on the patient's genetic profile.

5. No product can leave a biotechnology company until it passes rigorous tests by the internal unit called quality control (QC). Quality assurance (QA) is responsible for monitoring everything that enters and leaves a company to ensure product safety and quality and tracing sources of problems identified by complaints as specified by regulatory agencies, such as the U.S. Food and Drug Administration (FDA). QA and QC measures are important for every step of research and development (R&D) as well as for product manufacturing, including packaging and marketing.

6–10. Open ended; answers variable.

11. Biotechnology companies often begin with a discovery by a scientist working in academia. For example, a gene or protein is found to be important in a disease process and may have potential applications as a recombinant DNA product. Starting a biotechnology company, a "start-up" company, to develop this protein as a product then requires significant initial funding to do the R&D, such as preclinical research in animals and then ultimately FDA clinical trials (if the protein is to be used in humans). Sometimes a company starts with initial "seed money," which may come from a few individual investors. Then if the product shows potential, the person or persons starting the company will seek venture capital (from corporations or groups wishing to provide money for early-stage, high-risk, potentially high-reward-earning companies) or "angel" investors (individuals who provide capital for a start-up in exchange for ownership and profit making in the company).

12. The pharmaceutical industry is primarily involved in producing, selling, and marketing drugs for treating humans. This is one similarity to many (medical) biotechnology companies; however, the way in which pharma as opposed to biotech makes such drugs differ. Pharma companies primarily use synthetic chemical processes to make drug compounds, whereas biotech companies use living cells to produce a drug. Pharma companies also often produce medical devices. In recent years many large pharma companies have acquired smaller biotech companies; as a result, the many traditional pharma companies are also doing biotechnology R&D. For example, the pharma company Roche purchased the biotech company Genentech.

13. Open ended. No specific answer. Students sign up for biotech e-news.

Chapter 2

1. Genes are sequences of DNA nucleotides—usually from 1,000 to around 4,000 nucleotides long—that provide the instructions (code) for the synthesis of RNA. Most genes produce mRNA molecules encoding proteins, but some genes produce RNAs that do not code for proteins. Genes are contained in chromosomes, which are tightly packed coils of DNA and protein. Chromosomes enable cells to separate DNA evenly during cell division. Chromosomes contain multiple genes, and the number of genes in a chromosome can vary depending on its size.

2. The complementary strand will be an antiparallel strand with the sequence 3'-TCGGGGCTGAGATAAG-5'.

3. Biologists use the phrase "gene expression" to talk about the production of mRNA from a particular gene. From mRNA, cells translate proteins that are responsible for many aspects of cell structure and function.

4. Chargaff's rules state that the percentage of adenines is approximately equal to the percentage of thymines in a genome; the percentages of guanines and cytosines are also roughly equal. This is true because DNA consists of complementary base pairs. Adenines in one strand form hydrogen bonds with thymines in the opposite strand; guanines form hydrogen bonds with cytosines. This knowledge could be applied as follows: if DNA contains approximately 13% adenines, then it would also contain approximately 13% thymines. Combined, these bases account for approximately 26% of the DNA in the bacterium; therefore, the rest of the DNA (74%) consists of equal amounts of guanines and cytosines. Therefore guanine would comprise approximately 37% of the genome and cytosine approximately 37%.

5. DNA is a double-stranded molecule located in the nuclei of cells. Each strand of DNA is made of building blocks

called *nucleotides,* which consist of a pentose sugar, phosphate group, and base. DNA is the inherited genetic material of cells; its genes contain instructions for the synthesis of proteins. RNA is copied from DNA through a process called *transcription.* RNA is a single-stranded molecule, which is an important structural difference compared with DNA. RNA contains not only nucleotides—including the base uracil, which replaces the thymine present in DNA—but also a different pentose sugar (ribose) than that in DNA (deoxyribose). Three major types of RNA are transcribed: mRNA, tRNA, and rRNA; other forms of RNA (snRNA and siRNA) are involved in mRNA splicing and the regulation of gene expression respectively. After transcription, RNA molecules move into the cytoplasm, where they are required for protein synthesis (translation).

6. Seven codons, six amino acids coded.

    Start   Stop

    5'-AGCACCAUGCCCCGAACCUCAAAGUGAAA
    CAAAAA-3'

    The amino acid sequence is methionine-proline-arginine-threonine-serine-lysine. Remember that actual mRNAs and their encoded proteins are much longer than shown in this example and that stop codons do not code for an amino acid.

7. Only the bottom strand produces a functional mRNA with a start codon. This sequence is 5'-UUUAUG-GGUUGGCCCGGGUCAUGAUU-3'. The amino sequence of a polypeptide produced from this sequence is:

    Methionine–glycine–tryptophan–proline–glycine–serine

    Messenger RNA copied from the bottom strand of DNA with a T inserted between bases 10 and 11 is 5'-UUUAUG-GGAUAGGCCCGGGUCAUGAUU-3'. This will result in an A (bold) inserted in the mRNA sequence transcribed, which now creates a stop codon (UAG) in the mRNA, which will produce a truncated (shortened) peptide of only two amino acids, methionine–glycine.

8. Messenger RNA (mRNA) is an exact copy of a gene. It acts like a "messenger" of sorts by carrying the genetic code in the form of codons, encoded by DNA, from the nucleus to the cytoplasm, where this information can be interpreted to produce a protein. Ribosomal RNA (rRNA) molecules are important components of ribosomes. Ribosomes recognize and bind to mRNA and "read" along the mRNA during translation. Transfer RNA (tRNA) molecules transport amino acids to the ribosome during protein synthesis. Each tRNA contains an amino acid and an anticodon sequence that base pairs with codon sequences during translation.

9. Operons are clusters of genes that typically regulated simultaneously under the control of a single promoter. Operons are most prevalent in bacterial genomes. Bacteria can use operons to carefully control the expression of gene products (proteins) involved in similar processes, such as nutrient metabolism (the lactose operon is a classic example), depending on the demands of the cell.

10. *Gene regulation* is a broad phrase used to describe ways in which cells can control gene expression. Gene regulation

is an essential aspect of a cell's functions. Cells use many complex mechanisms to regulate gene expression. In this chapter, we highlighted transcriptional control as a regulatory process. Gene regulation allows cells to tightly control the amount of RNA and protein they produce in response to particular needs of the cell. Refer to Figure 2.14 for an overview of gene regulatory mechanisms.

11. Refer to Figure 2.9.

12. Mutations that affect protein structure and function include nonsense mutations (which affect a codon by creating a stop codon), missense mutations (which change a codon to code for a different amino acid with different properties than the original amino acid coded for), and frameshifts created by insertions or deletions. Silent mutations have no effect on protein structure and function because these mutations of a codon do not affect the amino acid coded for by the mutated codon.

13. 1' carbon.

14. The epigenome comprises structural aspects of DNA that do not involve direct mutation of a nucleotide or changes in DNA sequences. The epigenome includes methylation of histone proteins on chromosomes, acetylation, noncoding RNAs that bind to the genome, and other modifications. Epigenetic modifications of chromosomes affect gene expression. In recent years we have learned that epigenetic modifications play a role in disease; thus these modifications are being targeted for new therapeutic approaches to treating disease.

Chapter 3

1. Gene cloning is the copying (cloning) of a gene or a piece of a gene. The terms *recombinant DNA technology* and *genetic engineering* are often used interchangeably. Technically, recombinant DNA technology involves combining DNA from different sources, whereas genetic engineering involves manipulating or altering the genetic composition of an organism. For instance, ligating a piece of human DNA into a bacterial chromosome is an example of recombinant DNA technology, and placing this piece of recombinant DNA into a bacterial cell to create a bacterium is considered genetic engineering.

2. DNA ligase is used to form phosphodiester bonds between DNA fragments during recombinant DNA experiments. This is an important step in a cloning experiment because hydrogen bonds between sticky ends of DNA fragments are not strong enough to hold a recombinant DNA molecule together permanently. Restriction enzymes cut DNA at specific nucleotide sequences (recognition sequences). Frequently DNA is digested into fragments with restriction enzymes as an important step in many cloning experiments prior to using DNA ligase to join fragments together.

3. The gene sequence cloned from rats could be used to create primers that might be used to amplify DNA from human cells in an effort to amplify the complementary gene in humans. If one were successful in obtaining PCR products from this experiment, the PCR products could be sequenced and compared with the rat gene to search

for similar nucleotide sequences (suggesting that these genes are related). Also, PCR products could be used as probes in library screening experiments to find the full-length human gene or as probes for Northern blot analysis to determine if mRNA for this gene is expressed in human tissues.

4. Refer to Section 3.2.

5. Approximately 32,768 (2^{n-1}, where n = number of cycles).

6. Genomic libraries contain both introns and exons, whereas cDNA libraries contain DNA reverse-transcribed copies of mRNA expressed in a given tissue. cDNA libraries are the libraries of choice when expressed genes are being cloned. In searching for a gene involved in obesity, one could make and screen a cDNA library from adipocytes. Cloning gene regulatory regions, for instance, promoter and enhancer sequences described in Chapter 2, can be accomplished with genomic DNA libraries because they contain both exons and introns.

7. Searching for *diabetes* reveals a list of sequences for genes related to insulin-dependent and noninsulin-dependent diabetes mellitus. Clicking on any of the highlighted numbered links will take you to informative pages providing great detail about each gene. Searching with the accession number 114480 reveals information on genes for familial breast cancer.

8. This piece of DNA can be cut once by *Bam*HI at the beginning of the sequence and once by *Eco*RI at the end of the sequence. There are no cutting sites for *Sma*I in this sequence. A search for all enzymes in the database reveals approximately 50 restriction enzymes that can cut this sequence. This demonstration should provide you with an appreciation of how powerful computer programs can be for helping molecular biologists analyze DNA sequences.

9. Corrected answers are available on the website as students work on the questions.

10. Open ended; answers variable.

11. The ddNTPs are missing an oxygen at the 3' carbon.

12. The human genome consists of approximately 3.1 billion base pairs. The genome is approximately 99.9% the same between individuals of all nationalities and backgrounds. Less than 2% of the human genome codes for genes. The vast majority of our DNA is non–protein coding. The genome contains approximately 20,000 to 25,000 protein-coding genes. Many human genes are capable of making more than one protein. Chromosome 1 contains the highest number of genes. The Y chromosome contains the fewest.

13. The "omics" revolution of modern biology refers to the rapid expansion of new disciplines of research that have resulted from genomics studies, as reflected by new terms using the suffix *omics* or *ome*. Generally such studies involve a large-scale comprehensive analysis. For example, proteomics involves the study of all the proteins in a cell or tissue; metabolomics involves the study of all the proteins and metabolic products involved in a metabolic process, such as carbohydrate (sugar) metabolism.

Chapter 4

1. When you access www.ncbi.nlm.nih.gov, you will find a tab labeled Entrez. This is the public database for proteins, which is updated by researchers and peer-reviewed regularly. The amino acid sequences of proteins common to multiple organisms can be compared and the structures of some of these can be seen (although special software may be needed for viewing). Most researchers use this database to orient their research and analyze the work of colleagues.

2. Directed molecular evolution technology focuses on the mutations of a specific gene, selecting the best-functioning protein from that gene irrespective of the benefits or hazards it may have for the organism.

3. PSI has helped researchers to discover the functions of proteins, design experiments, and solve key biomedical problems. It has also enabled the more rapid identification of promising new structure-based medicines, helped to produce better therapeutics for treating both genetic and infectious diseases, and facilitated the development of technology and methodology for protein production.

4. Only the active genes in a cell make mRNA, the source of cDNA.

5. The process of protein production is being patented, and its steps must be repeatable. The product must be efficacious if it is to receive a patent. If organisms produce proteins on their own, why should companies be allowed to patent proteins? The patenting process (described in Chapter 12) requires knowledge of the DNA sequence, the amino acid sequence, and the function of the protein. If the "utility" of the protein has not changed (i.e., it has not been transferred to a different organism), it is not patentable. Gene transfer usually results in protein expression in a different organism (e.g., as in the case of an insect-resistant protein that has been transferred into a susceptible plant) for a specific purpose. If a patent is approved, it will require this new verifiable evidence.

6. The earliest fractions.

7. Last from the column.

8. More selective.

9. MS, HPLC, x-ray crystallography.

10. The procedure lost your protein.

Chapter 5

1. There are several important structural differences between prokaryotic and eukaryotic cells. Bacteria and Archaea are prokaryotes; eukaryotic cells include plant cells, animal cells, fungi, and protozoa, which are structurally more complex than bacteria. Prokaryotic cells are smaller than eukaryotic cells, do not contain a nucleus, have relatively few organelles, and contain a cell wall. Prokaryotes also have smaller genomes than eukaryotic cells; the prokaryotic genome usually consists of a single circular chromosome. Prokaryotic cells have served many important roles in biotechnology. Bacteria are used as hosts in gene cloning experiments, recombinant proteins

produced by bacteria are important for research, some proteins are used in medical applications, microbes are used for manufacturing many foods and beverages, and bacteria are used as sources of antibiotics and as hosts for the production of subunit vaccines.

2. Yeasts are unicellular fungal eukaryotes and bacteria are prokaryotic cells. Most yeasts have larger genomes than bacteria. Through fermentation, yeasts are used to make breads and dough as well as brewing beers and wines. Yeasts serve many important roles in research. The yeast two-hybrid system is a valuable technique for studying protein interactions. Because the yeast genome contains many genes similar to human genes, geneticists study it for clues about the functions of many human genes.

3. DNA binds to cells when cells and DNA are chilled on ice. A brief heat-shock step at about 37°C to 42°C causes DNA to enter cells. Electroporation has the advantages of being a very quick procedure, enabling many samples to be processed quickly. Moreover, much smaller amounts of DNA can be used than in calcium chloride transformation, and electroporation can also be used to introduce DNA into plant and animal cells.

4. All vaccines are designed to stimulate the immune system of the recipient to make antibodies or activated T cells to a pathogen—for example, a bacterium or virus. During antibody production, immune memory cells, such as antibody-producing B cells and memory B cells, are also produced. By creating antibodies and memory cells in the recipient, the goal of vaccination is to provide the recipient with protection against a particular pathogen should he or she be exposed to it. Three major types of vaccines are subunit vaccines, which consist of molecules from the pathogen (usually proteins expressed by recombinant DNA technology); attenuated vaccines, which are made from live pathogens that have been weakened to prevent replication; and inactivated vaccines, which are prepared by killing the pathogen and injecting dead or inactivated microbes into the vaccine recipient.

5. Anaerobic microbes are those microorganisms that do not require oxygen to convert sugars into energy (in the form of ATP). Under anaerobic conditions, some microbes use lactic acid fermentation; others use alcohol (ethanol) fermentation to derive energy. Lactic acid and ethanol are waste products of anaerobic fermentation that are important components of many foods. Lactic acid is found in cheeses and yogurts; ethanol is found in alcoholic beverages such as wine and beer.

6. One way to control alcohol content is to carefully regulate the amount of oxygen added to a fermentation bioreactor while making wine. When facultative anaerobes (microorganisms that can use cellular respiration or fermentation) are used for fermentation, they will, in the presence of oxygen, degrade sugars by aerobic cellular respiration and produce less alcohol. In the absence of oxygen, such microbes will degrade sugars by fermentation, and thus the alcohol content of the wine will be higher.

7. Open-ended questions; answers variable.

8. The study of microbial genomes can help scientists in many ways. The sequencing of genomes can reveal new genes, including genes that may be used in biotechnology applications (e.g., genes that encode novel enzymes with important properties for various commercial uses, including bioremediation). Scientists can study genes of disease-causing pathogens (including potential agents of bioterrorism) and then use this knowledge to help fight disease. The study of genomes can help scientists learn more about bacterial metabolism, the relatedness of bacterial strains, and lateral gene transfer (the sharing of genes between species).

9. Open-ended activity; answers variable.

10. Open-ended questions, but some people believe that requiring Gardasil vaccination for teens presumes sexual activity in teens. This may be common for some teens, but it should not be assumed that all teens are sexually active.

11. Fusion proteins are generated by inserting a cloned gene for a protein of interest into an expression vector that allows production of a protein fused to a tag protein such as green fluorescent protein or maltose-binding protein. This fused protein can then be passed over an affinity column, which will bind the tag portion of the fusion protein and allow the fusion protein to be isolated from a bacterial extract of proteins.

12. Refer to Table 5.1.

13. A preventative or prophylactic vaccine such as Gardasil is designed to provide immune protection before you are exposed to a particular pathogen but will typically not protect or cure you of a preexisting condition.

14. Metagenomics involves analyzing genomes from environmental samples. Important potential applications include identifying previously unknown microbes, genes and proteins with commercially valuable properties; learning about the impact of environmental changes on microbial community; and studying the genetic relatedness of communities of microbes in various environmental samples.

Chapter 6

1. Approval of food crops produced in this manner must be obtained separately from regulatory agencies. GM crops that are destined for animal feeds or are not ingested have been approved, but no human consumption GM crops can be approved without extensive testing to demonstrate lack of allergic response in humans.

2. Agricultural biotechnology innovations can reduce the need for pesticide application because plants have the ability to protect themselves from certain pests and diseases. It not only decreases water usage, soil erosion, and greenhouse gas emissions through more sustainable farming practices but also improves productivity of marginal cropland, especially where acres for planting are decreasing around the world.

3. The benefits of agricultural biotechnology include reduction of undesirable qualities such as saturated fats in

cooking oils, elimination of allergens, an increase in nutrients that help reduce the risk of chronic disease, and better delivery of proper nutrients such as vitamin A in commonly consumed crops.

4. With biotechnology, researchers can identify specific genetic characteristics, isolate them, and then transfer them to valuable crop plants. This technique is more precise and efficient than traditional crossbreeding and can increase food production through higher yields.

5. Foods produced from our current biotechnology methods do not require special labeling in the United States (at the time of this printing).

6. Although some evidence of transfer to other plants and insects has been recognized, no long-term damage to beneficial insects has been demonstrated. In fact, genetic engineering has produced more examples of "integrated pest management" in a shorter time than ever before.

7. Biotechnology-enhanced products have been available for only about 5 years. Long-term studies are not required if the gene product already exists in the environment in another organism. If found, adverse conditions will be appropriately regulated.

8. Monocot plants do not accept the Ti plasmid, requiring other methods to be used to transfect genes.

9. The germplasm was developed and donated for free, although intellectual property rights legal battles have held up its introduction.

10. Regulating multiple genes will make the development of insect resistance much more difficult.

Chapter 7

1. They provide the continuous dividing ability needed for the culture to produce antibodies.

2. The antibodies produced by the mouse are not exactly the same as those in the human, producing reaction to them as if they were antigens.

3. They have grafted human spleen tissue into mice that have no functioning immune systems.

4. Pain, suffering, and what is deemed as unnecessary testing. All experimentation must be approved by an independent committee that seeks to reduce, replace, and refine testing and finds that the animal is the best choice for the testing.

5. A human gene replaces the knocked out animal gene, and potential drugs targeting that human gene are tested for their effectiveness.

6. The heat acts as stimulant to the inserted genes that target the cancer cells.

7. Because, using these fish, it takes only 5 days to determine toxicity to inserted human transgenes.

8. Usually the animal most comparable to humans is chosen based on genetic similarity (either naturally or by gene transfer).

9. Researchers must prove which animal (and how many such animals) are needed for the proposed research. They must also prove that other alternatives cannot be used and that the most humane treatment is used.

10. One of the two nuclei is removed and other DNA is used to replace it.

Chapter 8

1. A polymorphic DNA locus is a sequence of DNA that has many possible alternatives (for example, different numbers of multiple repeats of tandem nucleotide sets, such as GC, GC, GC, and so on).

2. The polymorphic DNA sequences tested in fingerprinting have no known effect. They usually occur in regions of DNA that are not translated into protein.

3. Half should occur in the father and half should occur in the mother.

4. Degraded DNA in smaller quantities (as few as three cells with DNA) can be amplified using PCR primers and used for comparison. RFLP requires much larger quantities of undegraded complete DNA.

5. DNA transferred to an evidence sample from any source can be amplified by PCR primers, resulting in confusing unrepeatable results that break the chain of evidence.

6. Eyewitness testimony does not have the statistical strength of DNA profile comparison.

7. The DNA fingerprint of the cabernet sauvignon plant contained bands from both parents.

8. As a cross check against contamination, twice the band strength will appear for the original result.

9. Errors do occur in dramatizations for public viewing.

10. Disputed paternity or identification of victims that have close relatives, as in major disasters like a tsunami or the 9/11 attack, rely on mitochondrial DNA comparison.

Chapter 9

1. Open-ended activity; answers depend on the processes described.

2. The addition of fertilizers such as carbon, nitrogen, phosphorus, and potassium is often an important step in many bioremediation processes. Fertilization, also called nutrient enrichment, stimulates the growth and activity (metabolism) of microorganisms in the environment. These microorganisms, usually bacteria, also divide more rapidly to create more bacteria. By stimulating the metabolism of bacteria and increasing their number at a contaminated site, pollutants are usually degraded more rapidly. Adding oxygen to a contaminated site is effective when the microorganisms involved in the cleanup are relying on aerobic biodegradation.

3. Many formerly polluted sites throughout the United States are being redeveloped for industrial and residential use. Sometimes houses developed in these areas are sold at lower prices than other comparable homes; however,

not all states require builders to disclose the history of the land to potential homeowners. One obvious problem is the concern that the site may not be fully cleaned up. If the housing development relies on groundwater for its drinking water, the water is typically tested regularly to check for chemical pollutants. However, this can be problematic because even trace amounts of some chemicals may go undetected, and the health effects of these chemicals may not be fully known. Similarly, remaining chemicals in the soil can also affect residents' activities, such as playing in the yards and growing gardens in the soil. A major problem with redeveloping bioremediation sites for residential use is that it may take many years (or generations) to determine if residents are experiencing health effects from chemicals remaining at the site. Even if residents experience some health problems, it is often very difficult to determine if these effects are caused by pollutants at the site.

4. One approach might involve studying lead-containing structures to find out whether bacteria are growing on them. For instance, one could study lead pipes left in the environment and, over time, determine if bacteria are growing on these pipes. Bacterial growth on a lead surface could be an indication that the bacteria have developed a way to avoid the toxic effects of lead. This could be a sign that these cells may be capable of degrading lead. Then such bacteria could then be isolated and experiments using microcosms could be designed to test whether the bacteria could be used to degrade lead.

5–8. Open-ended activity; answers variable.

9. Open-ended activity; answers variable, although to date no significant bioremediation activities other than bioremediation by indigenous microbes has been reported.

10. It was reasonable to expect that zones of hypoxia might develop because, as aerobic microbes degraded oil , they would consume oxygen in local areas of the ocean. Declining oxygen levels in the water could cause hypoxia, resulting in fish kills. However, given the great depth of the spill, wave and wind action were likely responsible for causing sufficient mixing of the water, thus preventing large zones of hypoxia and resulting fish kills from occurring.

11. Open-ended activity; answers variable.

12. Open-ended activity; answers variable but a primary criticism has been to question whether these microbes actually metabolize arsenic and incorporate arsenic into its DNA, as claimed by the scientists publishing this work.

Chapter 10

1. Open-ended activity; answers variable. Potential answers might include initiatives to clean up habitat areas where clams and oysters could be grown; incentives for farmers to produce clams and oysters by aquaculture; and development of novel technologies to combat oyster pathogens that destroy oysters.

2. As discussed in Chapter 8, transgenic animals contain genes from another source. The transgenes may be from related species (for example, introducing the salmon gene for growth hormone into trout) or from very different species (for example, introducing the *luciferase* [*lux*] gene from bioluminescent marine bacteria or fireflies into salmon). Transgenic fish can be created using a number of different techniques, but one prominent technique includes microinjecting the transgene into blastocyst-stage embryos to allow incorporation of the transgene into embryonic tissues. Polyploid fish such as triploids, which contain three complete sets of chromosomes, are usually created by either chemical or electrical treatment of sperm or egg cells to produce diploid gametes that can be used to fertilize haploid gametes.

3. Open-ended activity; answers variable.

4. Examples of benefits include providing food sources, improving populations of fish and shellfish, creating farming industries in noncoastal that do not have fishing or seafood industries, and providing fish for recreational fishing. Examples of problems include some species that are not suitable for aquaculture or too expensive to raise through farming, diseases and illnesses that might decimate fish stocks because most farmed fish are genetically similar, wastes from fish farms that could create pollution problems, and threats to the genetic potential of native species when escaped farmed fish enter the population.

5. Open-ended activity; answers variable.

6. Open-ended activity; answers variable. Answers should, however, mention that organisms that grow under unique or extreme environmental conditions (such as pressure, heat, and ocean depths) are among the most highly studied for the identification of unique molecules that might be potentially valuable. For instance, mussels, which adhere tightly to structures to withstand the constant pounding of waves, produce unique molecules in their byssal fibers that provide adhesive strength.

7. Refer to Table 10.1.

8. Many aquatic biotechnologists believe that biotechnology will play an important role in improving finfish and shellfish populations in the future. One obvious way in which this may be done is to use bioremediation approaches to detect and clean up environmental pollution. Another approach may involve using biotechnology to learn about the pathogens that cause disease in aquatic organisms and develop ways to prevent or treat such diseases. Transgenic and polyploidy approaches may be used to continue to produce aquatic organisms with improved resistance to disease. In addition, aquaculture may be used to grow species that can be stocked, so as to increase dwindling populations of aquatic organisms.

9. Federal agencies such as the FDA and the U.S. Department of Agriculture (USDA) believed that there was no need to regulate the GloFish because it posed no threat to public health, since tropical fish are not used for food. The U.S. Fish and Wildlife service determined that the fish posed little threat to the environment because unmodified zebrafish have been sold in the United States and released into the wild with no known consequences. Many animal rights groups and others have strongly protested development of the GloFish primarily because

they perceive it to be an abuse of genetic engineering to create a transgenic species solely for the purpose of its enjoyment as a pet. Concerns have also been raised about the potential release of this genetically modified (GM) pet into the wild. Some people have also raised the point that the GloFish has provided a precedent for the unregulated development of other GM pets and GM animals that may pose environmental threats. Antibiotech groups also cite this as an example to raise public skepticism about biotechnology because if the GloFish can go unregulated, what other areas of biotechnology may go unregulated?

10. Biofilms are accumulations of livings organisms such as bacteria that coat a surface. For example, biofilms occur on teeth, heart implants, and intravenous lines. In marine environments, algae and mollusks growing on the hulls of ships exemplify biofilming. Many aquatic organisms combat biofilming by producing compounds that kill or inhibit the growth of biofilming microbes. Therefore scientists are interested in identifying these compounds so that they can be used in commercial applications to combat biofilming.

11. The Limulus amebocyte lysate (LAL) test relies on enzymes from the blood of the horseshoe crab (*Limulus polyphemus*) to detect endotoxins produced by toxic microbes. It is used to check the sterility of instruments such as medical devices.

Chapter 11

1. Amniocentesis and chorionic villus sampling are ways to sample tissue from developing fetuses. Fetal cells or adult tissue (usually white blood cells, skin, cheek, or hair cells) can be tested by karyotype analysis to check for abnormalities in chromosome number and structure. Molecular techniques such as restriction fragment length polymorphism (RFLP) analysis and the antistreptolysin-O (ASO) test can be used to detect defective genes in human cells. In the future, DNA microarrays (gene chips) will play a greater role in genetic testing, allowing for the analysis of thousands of genes at the same time.

2. The purpose of gene therapy is to deliver therapeutic genes into humans to treat or cure disease conditions. Ex vivo gene therapy involves removing cells from a patient, inserting a gene or genes into these cells, and then injecting or implanting them into the patient. Treatment of SCID is an example of ex vivo gene therapy. In vivo gene therapy occurs within the body by delivering genes directly into the body (for example, treating cystic fibrosis by nasal sprays). Some techniques for delivering therapeutic genes include using viruses as vectors for gene therapy, injecting naked DNA, and using liposomes to deliver genes. Gene therapy scientists face many challenges, including ensuring the safe delivery of genes (especially when viral vectors are used), targeting the therapeutic gene to the correct cells and tissues, and finding ways to get sufficient expression of the therapeutic gene to cure the given condition.

3. Embryonic stem cells (ESCs) are isolated from early embryos at the blastocyst stage. Blastocysts are usually derived from embryos left over from in vitro fertilization procedures. To isolate and culture ESCs, the inner cell mass is dissected out of the blastocyst, and these cells are then grown in a tissue culture dish. In contrast, adult-derived stem cells (ASCs) are isolated from mature adult tissues. These cells are found in small numbers in many tissues of the body, such as muscle and bone. A small sample (biopsy) of adult tissue is removed (for instance, a needle can be used to biopsy a sample of adult bone marrow stem cells), and ASCs can then be grown in culture. By treating ESCs or ASCs with different growth factors, scientists can stimulate stem cells to differentiate into different types of body cells. Amniotic stem cells are harvested from the amniotic fluid surrounding a developing embryo. Induced pluripotent stem cells (iPSCs) are produced by reprogramming somatic cells to become stem cells. Stem cell biologists envision many ways in which stem cells can be used to treat disease. Stem cells could be implanted in the body to replace damaged tissue, they might be good vectors for the delivery of therapeutic genes, and they could be used to grow tissues in organs in culture for subsequent transplantation into humans.

4. Refer to Table 11.3 for a comparison of therapeutic and reproductive cloning.

5. Pharmacogenomics is customized or personalized medicine created by analyzing a person's genetics and designing a drug or treatment strategy that is specific for that particular person based on the genes involved in that person's medical condition.

6. The Human Genome Project has revealed the location of all human genes including those involved in normal conditions as well as in disease processes. Identifying these genes is an important first step toward understanding how genes function and how certain diseases develop. The identification of disease genes will enable scientists to develop tests that can be used to screen individuals for the likelihood that they will inherit a particular genetic condition. As we learn more about different genes and the proteins they encode, it is expected that more specific forms of medical treatment will be available in the form of drugs designed to affect the specific genes and proteins involved in genetic diseases and conditions that have a genetic basis. Such treatments may also include gene therapy strategies. Understanding how human genes are regulated and affected by conditions such as stress and environmental factors will lead to a greater understanding of preventive measures that may be used to minimize genetic disease.

7. Open-ended; answers variable.

8. Open-ended; answers variable.

9. Antisense RNA technologies and RNAi are two common gene silencing techniques that could be used to inhibit a gene involved in a disease process. Refer to Chapter 11 for a description of each technique.

10. Regenerative medicine involves creating cells, tissues, and organs that can be used to repair or replace dead or damaged tissues in a person.

11. SNPs, or single-nucleotide polymorphisms, are single-base changes in a DNA sequence that are responsible for the subtle genetic differences between individual humans. When they occur in protein-coding gene sequences, SNPs can represent mutations that affect gene function and cause disease. Thus SNPs can be detected and used as a way to identify genetic changes involved in disease.

12. Single-gene defects can be detected by techniques such as RFLP analysis and ASO testing. Changes in chromosomal number and structural abnormalities can be detected by various karyotyping and techniques such as FISH.

Chapter 12

1. FDA.

2. EPA.

3. EPA.

4. FDA, EPA, USDA.

5. For the U.S. PTO to issue a patent, the product must have utility, be nonobvious, and be functional.

6. Approvals that took 4 years can now occur in 1.

7. Europe, where every step of the purification process is usually patented.

8. Firms should evaluate every aspect of production, including people, procedures, equipment, materials, record keeping, audits, and training.

9. The FDA approval is only for the purposes demonstrated to be effective by the trial process. A company may not advertise "off-label" uses under the FDA approval.

10. The EMEA has no knowledge of a drug product until it is submitted for approval; therefore a drug will take a shorter time for approval.

Chapter 13

1. Utilitarian approach and deontological (Kantian) approach. Utilitarianism tries to weigh all possible outcomes and produce the greatest good for the greatest number, so that the end result justifies the means. The deontological approach starts with certain absolutes that cannot be crossed in order to maintain an ethically correct outcome, no matter how much good might be achieved.

2. Most bioethicists concluded that this was ethically appropriate because Adam was not physically harmed by providing cord blood for his sister Molly. In addition, no evidence indicated that Adam's parents treated Adam poorly or as if he had been conceived only to save his sister. However, what do you think about parents selecting the offspring they want based on genetics (a process that is not uncommon when doing in vitro fertilization)?

3–6. Open-ended activity; answers variable.

7. A bioethicist following a utilitarian thought process would propose a "greatest good for the greatest number" answer. If this form of gene therapy can help lower cholesterol and saturated fats levels in humans and allow most people to enjoy fatty foods, then it is ethically all right to assume the risk that some smaller number of people may die prematurely because of this therapy.

8. From an ethics perspective one could argue that for the overall benefit of improving human health, drug companies should share such information to help facilitate the development of drugs that cure. However, there is no legal requirement for companies to do so.

9–13. Open-ended question; answers variable.

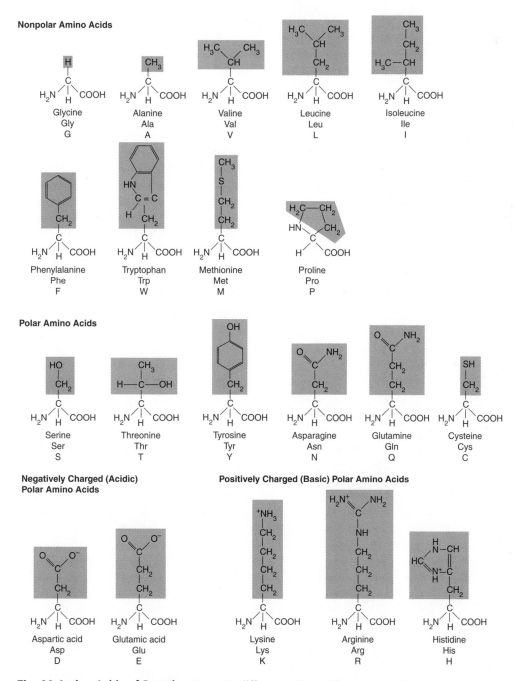

Nonpolar Amino Acids

Glycine
Gly
G

Alanine
Ala
A

Valine
Val
V

Leucine
Leu
L

Isoleucine
Ile
I

Phenylalanine
Phe
F

Tryptophan
Trp
W

Methionine
Met
M

Proline
Pro
P

Polar Amino Acids

Serine
Ser
S

Threonine
Thr
T

Tyrosine
Tyr
Y

Asparagine
Asn
N

Glutamine
Gln
Q

Cysteine
Cys
C

Negatively Charged (Acidic)
Polar Amino Acids

Positively Charged (Basic) Polar Amino Acids

Aspartic acid
Asp
D

Glutamic acid
Glu
E

Lysine
Lys
K

Arginine
Arg
R

Histidine
His
H

The 20 Amino Acids of Proteins Up to 20 different amino acids can be used to synthesize a protein. Each amino acid has a common structure but side chains (shaded) vary from amino acid to amino acid. These side chains determine the chemical properties of amino acids. Notice that some amino acids are nonpolar while others are polar and there are also negatively charged and positively charged amino acids. Both the three letter code and single letter code of each amino acid is also shown.

Credits

Photo

Chapter 1: Opener: Dr. Yorgas Nikas/Photo Researchers, Inc. **1.1(a):** Professor John Doebley. **1.1(b):** Max Gibbs/PhotoLibrary. **1.1(c):** Richard White. **1.4(b):** SIU BioMed/Custom Medical Stock Photo. **1.6:** Noah Berger/AP Images. **1.7:** Derek Bromhall/PhotoLibrary. **1.8:** Orchid Cellmark, Inc. **1.9:** Accent Alaska/Alamy. **1.10:** AP Images. **1.15:** Jim Reme.

Chapter 2: Opener: alengo/iStockphoto. **2.6:** A. L. Olins, University of Tennessee/Biological Photo Service. **2.7:** Dr. David Adler. **2.19(a):** Cheryl Powers/Photo Researchers, Inc. **2.19(b):** National Institutes of Health. **2.20:** David Cooney.

Chapter 3: Opener: Michael A. Palladino. **3.3(a):** Eye of Science/Photo Researchers. **3.3(b):** Michael A. Palladino. **3.10:** Michael A. Palladino. **3.12:** Pfizer, Inc. **3.14:** Peter Lansdorp. **3.16(a):** Michael A. Palladino. **3.16(b):** Michael A. Palladino. **3.18(1):** Sam Ogden/Photo Researchers, Inc. **3.18(3):** Volker Steger/Photo Researchers, Inc.

Chapter 4: Opener: Michael A. Palladino. **4.3:** The Protein Data Bank/RCSB. **4.15:** Amersham Biosciences.

Chapter 5: Opener: Dr. Gopal Murti/Photo Researchers, Inc. **5.1(a):** S. H. Pincus and S. F. Hayer. **5.1(b):** Manfred Kage/Photolibrary/Peter Arnold, Inc. **5.1(c):** CNRI/Photo Researchers, Inc. **5.2(a):** John Durham/Photo Researchers, Inc. **5.2(b):** Maximilian Stock Ltd/Photo Researchers, Inc. **5.3:** Charlotte K. Mulvihill and Catherine Pongratz. **5.6(a):** Jurgen Freund/bluegreenpictures. **5.6(b):** Kenneth Lucus/Biological Photo Services. **5.7(1):** SciMAT/Photo Researchers, Inc. **5.7(2):** Ian O'Leary/Dorling Kindersley. **5.7(3):** Michael A. Palladino. **5.16(a):** Eye of Science/Photo Researchers, Inc. **5.16(b):** C. D. Humphrey and T. G. Ksiaze/CDC. **5.16(c):** Janice Carr/CDC/National Escherichia, Shigella, Vibrio Reference Unit at CDC. **5.17(a):** A Dowsett/Photo Researchers, Inc. **5.17(b):** T. H. Chen and S. S. Elberg, Inf. Imm 15:972-977. **5.19:** AP Images.

Chapter 6: Opener: Bill Thieman. **6.5(a):** Bio-Rad Laboratories Diagnostics Group. **6.9(a):** Mediscan/Corbis. **6.9(b):** Nature's Images, Inc./Photo Researchers, Inc. **6.11:** Fancy/Alamy. **6.13:** Bill Thieman.

Chapter 7: Opener: Keith Weller/USDA/APHIS Animal and Plant Health Inspection Service. **7.2(a):** Lila Solnica-Krezel. **7.2(b):** Dorling Kindersley Media Library. **7.5:** Ben Margot/AP Images. **7.8:** Geoff Tomkinson/Photo Researchers, Inc.

Chapter 8: Opener: Southern Illinois University/Photo Researchers, Inc. **8.12:** Courtesy of Seminis Inc.

Chapter 9: Opener: Maciej Dakowicz/Alamy. **9.4:** Jennifer Bower and Ralph Mitchell. **9.9:** RJH/Alamy. **9.10:** Michael A. Palladino. **9.12:** ASM Publications. **9.15:** Kelly P. Nevin. **9.17(a):** Jack Smith/AP Images. **9.17(b):** Courtesy of the United States Navy. **9.18:** © Accent Alaska.com/Alamy. **9.19(a):** Wild Wonders of Europe/Carwardine/bluegreenpictures.com. **9.19(b):** REUTERS/Hoi-Ying Holman Group/Handout. **9.20:** TEM of D. radiodurans acquired in the laboratory of Michael Daly, Uniformed Services University, Bethesda, MD.

Chapter 10: Opener: Jim Zipp/Photo Researchers, Inc. **10.1:** Peggy Greb/USDA/APHIS Animal and Plant Health Inspection Service. **10.2:** Michael A. Palladino. **10.3(a):** Dr. Kevin M. Fitzsimmons. **10.3(b):** PhotoStock-Israel/Photo Researchers,

Inc. **10.3(c):** Mark Burnett/Photo Researchers, Inc. **10.3(d):** Steven David Miller/bluegreenpictures. **10.4(a):** Dr. Kevin M. Fitzsimmons. **10.4(b):** Ian Nolan/Alamy. **10.7(a):** Dwight Smith/Shutterstock. **10.7(b):** American Association for the Advancement of Science. **10.7(c):** USDA/ARS/Agricultural Research Service. **10.8:** Michael A. Palladino. **10.9:** Garth Fletcher. **10.12:** Herbert Waite. **10.13:** Chris Wilson/Alamy. **10.14(a):** U.S. Environmental Protection Agency Headquarters. **10.14(b):** Ted Kinsman/Photo Researchers, Inc. **10.15:** Deborah L. Santavy.

Chapter 11: Opener: NatUlrich/Shutterstock. **11.1:** John Sholtis/Amgen Inc. **11.4(a):** Centers for Disease Control and Prevention (CDC). **11.4(c):** Lisa G. Shaffer. **11.9:** DeRisi, J. (1996): Nat. Genet., 14:457–60 adapted for Hollon, T. (2003): The Scientist, Suppl 2, p. 30, 9/22/03 with modification. **11.10:** Robert S. Langer and Kenneth J. Germeshausen. **11.18:** Bill Ling/Dorling Kindersley Media Library. **11.20(a):** Becton Dickinson and Company. **11.21(b):** Wake Forest University School of Medicine. **11.26:** Anne Bower.

Chapter 12: Opener: Alliance Images/Alamy. **12.1:** U.S. Dept. of Agriculture. **12.2:** Steven E. Lindow. **12.4:** Corbis Bridge/Alamy.

Chapter 13: Opener: Manny Ceneta/AFP/Newscom. **13.1:** Eric Carlson. **13.2(a):** USDA Forest Service. **13.2(b):** USDA Forest Service. **13.5:** Richard Olsenius/National Geographic Image Collection. **13.6:** Sidney Harris.

Special Features: All chapter openers: Denis Vrublevski/Shutterstock. **Career Profile:** mrt/Shutterstock. **Tools of the Trade:** Shutterstock. **You Decide:** Tom Grundy/Shutterstock.

Text/Art Credits

Figure 1.3: John E. Smith, *Biotechnology*, Third Edition, 1996, © J. E. Smith 1981, 1988, © Cambridge University Press 1996, reproduced with permission.

Figure 1.3: John E. Smith, *Biotechnology*, Third Edition, 1996, © J. E. Smith 1981, 1988, © Cambridge University Press 1996, reproduced with permission of Cambridge University Press and the author.

Figure 1.5: *Nature Biotechnology*, Volume 22, December, 2004 by Nature Publishing Company. Reproduced with permission of MacMillan Publishers Ltd. and Nature Publishing Group in the format Journal via Copyright Clearance Center.

Figure 1.17: Ernst & Young, Beyond Borders: The Global Biotechnology Report, 2011. Reprinted by permission.

Table 1.1: Ernst & Young, Beyond Borders: The Global Biotechnology Report, 2011. Reprinted by permission.

Figure 2.21: From "Brain function and chromatic plasticity" by Catherine Dulac, Volume 465, 2010, *Nature* by Nature Publishing Group. Reproduced with permission of Macmillan Publishers Ltd. Nature Publishing Group in the format Journal via Copyright Clearance Center.

Figure 3.13: Adapted from "The development and Impact of 454 sequencing" Volume 26, Issue 10, 2008, *Nature Biotechnology* by Nature Publishing Company. Reproduced with permission of Macmillan Publishers Ltd. and Nature Publishing Group in the format Journal via Copyright Clearance Center.

Figure 3.22: Courtesy of Mikael Häggström.

Figure 4.1: John E. Smith, *Biotechnology*, Third Edition, 1996, © J. E. Smith 1981, 1988, © Cambridge University Press 1996, reproduced with permission.

Figure 4.1: John E. Smith, *Biotechnology*, Third Edition, 1996, © J. E. Smith 1981, 1988, © Cambridge University Press 1996, reproduced with permission of Cambridge University Press and the author.

Figure 4.7: From *Protein Biotechnology*, by Gary Walsh and Dennis Headon, Figure 3.1, p. 42. Copyright © 1994, John Wiley & Sons, Limited. Reproduced with permission.

Figure 4.10: From *Proteins to PCR* by David W. Burden, Figure 5.4, p. 99. Copyright © 1995 by Springer-Verlag. Reprinted with permission.

Figure 4.11: From *Proteins to PCR* by David W. Burden, Figure 554, p. 102. Copyright © 1995 by Springer-Verlag. Reprinted with permission.

Figure 4.13: From *Proteins to PCR* by David W. Burden, Figure 5.6, p. 104. Copyright © 1995 by Springer-Verlag. Reprinted with permission.

Figure 4.16: from "Antibodies Detect Protein Building" figure 4.17, p. 44 from *Genomics & Proteomics* November/December 2001. Copyright © 2003 *Genomics & Proteomic*, a publication of Reed Business Information, a division of Reed Elsevier Inc. All rights reserved. Reprinted by permission.

Figure 5.14: From "Single-shot readout of an electron spin in silicon" by Andrea Morello, et al., Volume 467, Issue 7316, September 2010, *Nature* by Nature Publishing Group. Reproduced with permission of Macmillan Publishers Ltd. and Nature Publishing Group in the format Journal via Copyright Clearance Center.

Figure 5.16: From *American Biotechnology Laboratory*, Volume 24, Number 1 © 2006. Reprinted by permission.

Figure 5.18: Reprinted by permission of Jared Schneidman Design.

Table 5.6: From "U.S. food safety under siege?" Volume 22, Number 12, 2004, *Nature Biotechnology* by Nature Publishing Company. Reproduced with permission of Macmillan Publishers Ltd. and Nature Publishing Group in the format Journal via Copyright Clearance Center.

Figure 6.1: Courtesy of International Service for the Acquisition of Agri-biotech Applications.

Figures 6.2, 6.3, 6.4, 6.6, and 6.10: from *Recombinant DNA*, Second Edition by James D. Watson, et al. © 1992 by James D. Watson, Michael Gilman, Jan Witkowski and Mark Zollar. Used with permission of W. H. Freeman and Company.

Figure 6.12: From "Bonkers about biofuels" by Stephan Herrera, Volume 24, Issue 7, 2006, *Nature Biotechnology* by Nature Publishing Company. Reproduced with permission of Macmillan Publishers Ltd. and Nature Publishing Group in the format Journal via Copyright Clearance Center.

Figure 7.1: From "A mightier mouse with human adaptive immunity" by Marie Kosco-Vilbois, Volume 22, Issue 6, 2004, *Nature Biotechnology* by Nature Publishing Company. Reproduced with permission of Macmillan Publishers Ltd. and Nature Publishing Group in the format Journal via Copyright Clearance Center.

Figure 7.10: From "Knockout rats via embryo microinjection of zinc-finger nucleases" by A.M. Geurts, Volume 325, Issue

5939, 2009, *Science* by American Association for the Advancement of Science. Reproduced with permission of American Association for the Advancement of Science in the format Journal via Copyright Clearance Center.

Figure 7.11: from *Recombinant DNA*, Second Edition by James D. Watson, et al. © 1992 by James D. Watson, Michael Gilman, Jan Witkowski and Mark Zollar. Used with permission of W. H. Freeman and Company.

Figure 8.2: Reprinted by permission of Dr. Kenneth C. Jones.

Figure 8.3: Reprinted by permission of Dr. Kenneth C. Jones.

Figure 8.4: Reprinted by permission of Dr. Kenneth C. Jones.

Figure 9.5: reprinted with permission from "Anaerobes to the Rescue" by Derek R. Lovley, figure on p. 1445, *Science* 293: 1444–1446 (2001) August 24, 2001. Copyright © 2001 by the American Association for the Advancement of Science.

Figure 9.16: From "Plants tackle explosive contamination" by Richard B Meagher, Volume 24, Issue 2, 2006, *Nature Biotechnology* by Nature Publishing Company. Reproduced with permission of Macmillan Publishers Ltd. and Nature Publishing Group in the format Journal via Copyright Clearance Center.

Figures 10.5 & 10.6: reprinted with permission from "Antifreeze proteins and their genes" by Garth L. Fletcher, Sally V. Goddary and Yaling Wu, figures 3 and 10, *Chemtech* 1999, 30(6), 17–28. Copyright © 1999 by the American Chemical Society. Used with permission.

Table 10.1: Reprinted by permission of Environmental Defense Fund.

Table 10.2: Reprinted by permission of Environmental Defense Fund.

Figure 11.5: from *Principles of Cell & Molecular Biology*, Second Edition by Lewis J. Kleinsmith and Valerie M. Kish, figure 3-53, p. 110. Copyright © 1995 by HarperCollins College Publishers. Reprinted by permission of Pearson Education, Inc.

Figure 11.8: Reprinted by permission of National Institute of General Medical Sciences.

Figure 11.11: From "Nanotechnology takes aim at cancer" by R.F.Service. From *Science* 310: 1132–1134, November 2005. Reprinted with permission from American Association for the Advancement of Science.

Figure 11.23: From "Brain function and chromatin plasticity" by Catherine Dulac, Volume 465, Is. 7299, 2010, *Nature* by Nature Publishing Group. Reproduced with permission of Macmillan Publishers Ltd. and Nature Publishing Group in the format Journal via Copyright Clearance Center.

Figure 11.24: Courtesy of Dr. Donald Orlic.

Figure 11.25: Reprinted from *Cell*, Volume 132, Issue 1, Copyright 2008, with permission from Elsevier.

Table 11.2: Reprinted with permission from the National Academies Press, Copyright 2002, National Academy of Sciences.

Figure 12.5: From Intellectual Property Landscape of the Human Genome, *Science* 310: 239. Reprinted with permission from American Association for the Advancement of Science.

Glossary

A (aminoacyl) site (of a ribosome): Portion of a ribosome into which aminoacyl tRNA molecules bind during translation.

accession number: Unique identifying letter and number code assigned to every cloned DNA sequence that is catalogued in databases such as GenBank (for example, BC009971 is the accession number for one of the human keratin genes, which produces a protein that is a major component of skin and hair cells). The accession number for any sequence can be used by scientists around the world to retrieve database information on that particular sequence.

acetylase: Enzyme produced by the lacA gene of the lactose (lac) operon in bacteria.

achromatopsia: A rare autosomal recessive genetic disease that affects cone cells in the retina, which are essential for color vision and some aspects of visual acuity.

acquired mutations: These occur in the genome of somatic cells and are not passed along to offspring; can cause abnormalities in cell growth leading to cancerous tumor formation, metabolic disorders, and other conditions.

adenine: Abbreviated A; purine base present in DNA and RNA nucleotides.

adenosine triphosphate (ATP): A nucleoside-triphosphate that contains the nitrogenous base adenine. ATP is the primary form of energy used by living cells.

adenovirus: Virus that causes the common cold.

adult-derived stem cells (ASCs): Stem cells derived from tissues of an adult, as opposed to embryonic stem cells, which are derived from a blastocyst; can differentiate to produce other cell types.

aerobes: Organisms that use oxygen for their metabolism.

aerobic conditions: Conditions in which oxygen is present.

aerobic metabolism: Metabolism that requires oxygen.

affinity chromatography: A separation technique, based on the unique match between a molecule and its column-bound chemical counterpart (like antigen/antibody), that involves passing proteins or other substances in solution over a medium that will bind to ("have an affinity for") specific components of the solution. It is used to isolate fusion proteins from a mixture of bacterial cell proteins.

agarose gel electrophoresis: See *gel electrophoresis*.

agricultural biotechnology: A discipline of biotechnology that involves plants and their applications, including genetic engineering of plants for agricultural purposes.

Agrobacter: Soil bacteria that invade injured plant tissue.

alcohol (ethanol) fermentation: Enzymatic breakdown of carbohydrates (sugars) in the absence of oxygen; products include ATP, CO_2, and ethanol (alcohol) as waste products; important type of microbial metabolism used for the production of certain types of alcohol-containing beverages.

allele-specific oligonucleotide (ASO) analysis: Genetic testing technique that involves using PCR with oligonucleotides specific to a disease gene to analyze a person's DNA.

alternative splicing: Splicing sometimes can join certain exons and cut out other exons, essentially treating them as introns. This process creates multiple mRNAs of different sizes from the same gene. Each mRNA can then be used to produce different proteins with different, sometimes unique, functions. Alternative splicing allows several different protein products to be produced from the same gene sequence.

amino acids: Building blocks of protein structure; combinations of 20 different amino acids can join together by covalent bonds in varying order and length to make a polypeptide.

aminoacyl transfer RNA (tRNA): Transfer RNA (tRNA) molecule with an amino acid attached.

amniocentesis: A technique for obtaining fetal cells from a pregnant woman to analyze fetal cells to determine the genetic composition of the cells, such as the number of chromosomes or the sex of the fetus.

amniotic fluid–derived stem cells (AFSs): Stem cells in the amniotic fluid of pregnant women that have been found to have many of the same traits as embryonic stem cells.

amylase: An enzyme that digests starch.

anaerobes: Organisms that do not require oxygen for their metabolism.

anaerobic metabolism: Lacking oxygen; an organism, environment, or cellular process that does not require oxygen.

angel investors: Investors that save your company from financial ruin at the last minute (like an angel).

angiogenesis: The growth of new blood vessels.

Animal and Plant Health Inspection Service (APHIS): The branch of the USDA responsible for protecting U.S. agriculture from pests and diseases.

animal biotechnology: A diverse discipline of biotechnology that involves the use of animals to make valuable products such as recombinant proteins and organs for human transplantation; also includes organism cloning.

annotation: Bioinformatics approach that involves searching databases to determine whether the sequence and function of a gene sequence has already been determined.

antibiotic: A substance produced by microorganisms that inhibit the growth of other microorganisms; commonly used to treat bacterial infections in humans, pets, and farm animals.

antibiotic selection: See *selection*.

antibodies: Proteins produced in response to a non-self molecule by the immune system; antigen-binding immunoglobulins, produced by B cells, that function as the effector in an immune response.

antibody-mediated immunity: Portion of the immune system dedicated to producing antibodies that combat foreign materials; also known as humoral immunity.

anticodon: Three-nucleotide sequence at the end of a tRNA molecule. During translation, the anticodon binds to a specific codon in an mRNA molecule by complementary base pairing.

antifreeze proteins (AFPs): Category of proteins isolated from aquatic organisms that live in cold environments; these proteins have the unique property of lowering the freezing temperature of body fluids and tissues.

antigens: Molecules unique to specific surfaces that can stimulate antibody response; substances that trigger antibody production when introduced into the body.

antimicrobial drugs: Chemicals that inhibit or destroy microorganisms.

antiparallel: Refers to the 5′ and 3′ directionality of strands of DNA in which the two strands joined together by hydrogen bonds between complementary base pairs are oriented in opposing directions with respect to their polarity.

antisense molecule: See *antisense RNA*.

antisense RNA: An RNA molecule that is complementary to a native RNA.

antisense RNA technology: See *antisense RNA*.

apoptosis: Involves a complex cascade of cellular proteins that cause cell death; controlled apoptosis is an important biological process that remodels tissues during development; uncontrolled apoptosis is involved in degenerative disease conditions such as arthritis; also known as "programmed cell death."

aquaculture: Farming finfish, shellfish, or plants for commercial or recreational uses.

aquatic biotechnology: The use of aquatic organisms such as finfish, shellfish, marine bacteria, and aquatic plants for biotechnology applications.

Archaea: See *domain Archaea*.

astaxanthin: A pigment that gives shrimp their pink color. Recombinant astaxanthin is used to change the color of aquaculture species such as salmon.

attenuated vaccine: Vaccine consisting of weakened, live microorganisms.

autografting: The transplantation of a patient's own tissues from one region of the body to another; for example, skin or hair grafting, coronary artery bypass surgery.

autoradiograph: An image created by exposing photographic film to a radioactively labeled compound.

autoradiography: Technique that uses film to detect radioactive or light-releasing compounds (such as DNA probe) in cells, tissues, or blots; produces photographic film image called an autoradiogram or autoradiograph.

autosomes: Chromosomes whose genes are not primarily involved in determining an organism's sex; chromosomes 1–22 in humans.

avian flu: "Avian influenza virus" refers to influenza A viruses found chiefly in birds, but infections with these viruses can occur in humans.

avidin: A protein in egg whites that binds biotin and inhibits bacterial growth.

B lymphocytes (B cells): White blood cells (leukocytes) that develop in the bone marrow and can mature into antibody-producing cells called plasma cells.

bacilli: Rod-shaped bacteria.

Bacillus thuringiensis (Bt): A bacterium that produces a crystalline protein that dissolves the cementing substance between certain insect midgut cells.

bacteria: See *domain Bacteria*.

bacterial artificial chromosomes (BACs): Large circular vectors that can replicate very large pieces of DNA; used to clone pieces of human chromosomes for the Human Genome Project.

bacteriophages: Viruses that infect bacterial cells; often simply called phages.

baculoviruses: Viruses that attack insect cells.

band shifting: Equal molecular-weight fragments of DNA that do not migrate uniformly because of variations in the gel matrix.

Basic Local Alignment Search Tool (BLAST): Internet-based program from the U.S. National Center for Biotechnology Information (NCBI) that can be used for DNA nucleotide-sequence-comparison (alignment) studies and for sequence searches in GenBank and other databases.

basic sciences: Research disciplines that examine fundamental aspects of biological processes, often without obvious direct applications (for curing disease or making a product).

batch (large-scale or scale-up) processes: Growing microorganisms such as bacteria or yeast and other living cells such as mammalian cells in large quantities for the purpose of isolating useful products in a batch.

ß-galactosidase: Enzyme produced by the lacZ gene of the lactose (lac) operon in bacteria; ß-galactosidase degrades the disaccharide lactose.

ß-sheet: One of two secondary structures of proteins.

bioaccumulation: Progressive concentration or accumulation of a substance, such as a chemical pollutant, as it moves up the "food chain" from organism to organism.

bioaugmentation (seeding): Adding bacteria or other microorganisms to a contaminated environment to assist native (indigenous) microbes in the bioremediation processes.

biocapsules: Tiny spheres or tubes filled with therapeutic cells or a chemical substance such as a drug; can be implanted for therapeutic purposes; also known as microcapsules.

biodegradation: The use of microorganisms to break down chemicals (not necessarily pollutants).

biodiversity: The range of different species present in an ecosystem.

bioethicist: A person who studies bioethics.

bioethics: The area of ethics (a code of values for our actions) concerning the implications of biologic and biomedical research and biotechnological applications (particularly with respect to medicine).

biofilming (biofouling): The attachment of organisms to surfaces; examples include the attachment of shellfish to the hull of a ship and the attachment of microorganisms to teeth.

biofuel: A product of biological organisms that can substitute or enhance existing fuels.

bioimpedance: A technique involving the use of low-voltage electricity to measure muscle mass and the fat content of tissues; used, for example, to measure the fat content of fish to be sold for consumption.

bioinformaticists: Scientists specializing in the bioinformatics.

bioinformatics: Interdisciplinary science that involves developing and applying information technology (computer hardware and software) for analyzing biological data such as DNA and protein sequences; also includes the use of computers for the analysis of molecular structures and creating databases for storing and sharing biological data.

biologic: Any medical preparation made from living organisms or their products.

biological: Pertaining to biology.

Biological License Agreement (BLA): Filed by a biotech company seeking approval of a biologically derived product such as a viral therapy, blood compound, vaccine, or protein derived from animals.

bioluminescence: The release of light by living organisms.

biomarker proteins: A protein biomarker (see *biomarkers*).

biomarkers: Substances used as indicators of a biologic state. A characteristic that is objectively measured and evaluated as an indicator of normal biologic processes, pathogenic processes, or pharmacologic responses to a therapeutic intervention.

biomass: The dry weight of living material in part of an organism, a whole organism, or a population of organisms.

biopiles: Large piles of contaminated soil that have been removed from the original site. Air and vacuum systems are used to help dry these piles and release chemicals by evaporation. Vacuums are sometimes used to collect chemical vapors and trap them in filters.

bioprocessing engineers: Work at the frontiers of biologic and engineering sciences to "bring engineering to life" through the conversion of biologic materials into other forms needed by mankind.

bioprocessing: The use of biologic systems to manufacture (process) a product.

bioprospecting: Endeavors to capitalize on indigenous knowledge of natural resources. However, bioprospecting may also describe the search for previously unknown compounds in organisms that have never been used in traditional medicine.

bioreactors: Cell systems that produce biologic molecules (which may include fermenters).

bioremediation: The use of living organisms to process, degrade, and clean up naturally occurring or man-made pollutants in the environment.

biosensors: Living organisms used to detect or measure biological effects of some factor, such as a chemical pollutant, or condition.

biosimilar drugs: Also called "follow-on" biologics. Subsequent version of a recombinant protein product after the original patent has expired. The biosimilar is produced by a different company than the innovator holding the initial patent. As a result, when a biosimilar is made, the exact production processes are not the same as the innovator's hence the product or its manufacturing is "similar" but not identical to the original protein. The equivalent of a generic for a pharmaceutical drug.

biotechnology: A broad area of science involving many different disciplines designed to use living organisms or their products to perform valuable industrial or manufacturing processes or applications that will solve problems.

bioterrorism: The use of biological materials (live organisms or their toxins) as weapons to cause fear in or harm against civilians.

bioventing: Pumping air or oxygen-releasing chemicals such as hydrogen peroxide into contaminated soil or water to stimulate aerobic degradation by microorganisms.

bioweapons: Biological materials used as weapons.

blastocyst: Hollow cluster of approximately 100 cells formed about 1 week after fertilization of an egg.

blunt ends: Double-stranded ends of a DNA molecule created by the action of certain restriction enzymes.

BRCA1 gene: A gene found in about 65% of human breast tumors. Commonly part of breast cancer diagnosis for treatment.

byssal fibers: Ultrastrong, protein-rich threads created by mussels and other shellfish. Byssal fibers have unique adhesive properties and can withstand great stress from stretching and shearing forces; attach shellfish to substrates such as rocks.

CAAT box: Short nucleotide sequence (CAAT) usually located approximately 80 to 90 base pairs "upstream" (in the 5′ direction) of the start site of many eukaryotic genes; part of promoter sequence bound by transcription factors used to stimulate RNA polymerase.

calcitonin: A thyroid hormone that stimulates calcium uptake by digestive organs and promotes bone-hardening (calcification).

callus: A loose collection of de-differentiated plant tissue.

Cancer Genome Atlas Project (TCGA): NIH sponsored project to identify and map important genes and genetic changes involved in cancer.

cancer stem cells (CSCs): Stem cells which develop to form cancerous tumors.

carbohydrases: Enzymes that digest carbohydrates.

carcinogen: A cancer-causing chemical.

carcinogenic: Cancer causing agents as chemicals and X-rays.

carrageenan: Polysaccharide (sugar) derived from seaweeds; wide range of uses in everyday products such as syrups, sauces, and adhesives as a "thickening" or bulking agent.

cDNA: DNA synthesized as an exact copy of mRNA called complementary DNA (cDNA). The mRNA is degraded by treatment with an alkaline solution or enzymatically digested; then DNA polymerase is used to synthesize a second strand to create double-stranded cDNA.

cell culture: Growing cells in laboratory conditions outside of a whole organism (in vitro); usually a term applied to growing mammalian cells.

cell line: An established or immortalized cell line that has acquired the ability to proliferate indefinitely, either through random mutation or deliberate modification. Numerous well-established cell lines are representative of particular cell types.

cell lysis: See *lysis*.

cellular therapeutics: Cell-based therapies. Including, for example, stem cells and implanted cells.

cellulase: Bacterial enzyme that degrades the polysaccharide cellulose (a primary component of the plant cell wall).

Center for Drug Evaluation and Research: A branch of the FDA that tracks total drug submissions routed to the drug review process comprised primarily of New Drug Application (NDA) related submissions, Abbreviated New Drug Application (ANDA) related submissions and Investigational New Drug (IND) related submissions.

centrifugation: Involves using an instrument called a centrifuge to apply a rotational force to samples to separate components based on their weight. Rotational forces measured in revolutions per minute (rpm) or times gravity (g). Centrifugation has a number of important applications related to separating components of a mixture (for instance, separating proteins from DNA, separating different cell types).

centromere: Constricted region of a chromosome formed by intertwined DNA and proteins that hold two sister chromatids together.

chemotherapy: The treatment of cancer and other diseases with specific chemical agents or drugs that have a toxic effect on diseased cells or disease-causing microorganisms.

chimera: An organism with more than one type of cell; some cells are normal and some are genetically modified.

chitin: A complex polysaccharide polymer composed of repeating units of a sugar called N-acetylglucosamine. Chitin forms the hard outer shell (exoskeleton) of crabs, lobsters, crawfish, shrimp, and other crustaceans. Also found in insects and other organisms.

chitosan: Polysaccharide polymer derived from chitin. Used in many applications from health care to agriculture to dyes for fabrics and in dietary supplements for weight loss.

chloroplast: The photosynthetic organelle of complex plants.

cholera: Acute intestinal infection of humans caused by the bacterium Vibrio cholerae that causes severe diarrhea, which can result in dehydration and death, particularly in young children, if not treated. Spreads by contaminated water and food; bacterium often found in water supplies of developing countries with poor sanitary conditions.

chorionic villus sampling: Technique for sampling fetal cells from the placenta to determine the genetic composition of these cells, such as the number of chromosomes or the sex of the fetus.

chromatin: Strings of DNA wrapped around proteins; present in the nucleus of eukaryotes and the cytoplasm of prokaryotes; coils tightly together to form chromosomes during cell division.

chromatography: A columnar separation method that uses resin beads with special separation capabilities.

chromosomes: Highly folded arrangements of DNA and proteins; they package DNA to allow even separation of genetic material during cell division.

chymosin: See *renin*.

clinical trials: Experimental process of testing products before approval of a drug, or treatment plan for widespread use in humans; clinical trials involve several "phases" of carefully planned experiments in different numbers of human participants to test drug effectiveness and safety.

clone: A genetically identical copy of a cell or whole organism; also describes the process of making copies of a gene, cell, or organism.

cocci: Bacteria with a spherical shape.

CODIS, Combined DNA Index System: CODIS enables federal, state, and local crime labs to exchange and compare DNA profiles electronically, thereby linking crimes to each other and to convicted offenders.

codons: A codon comprises a combination of three nucleotide sequences in an mRNA molecule. With few exceptions, each codon codes for a single amino acid; 64 possible codons compose the genetic code of mRNA molecules.

cohesive (sticky) ends: Overhanging single-stranded ends of a DNA molecule created by the action of certain restriction enzymes.

colchicine: Substance derived from crocus flowers that blocks microtubule formation; used to stop cell division (for instance, when creating polyploid organisms).

collagenase: Protease (protein-digesting enzyme) used in a number of biotechnology applications.

colony hybridization: A procedure for binding a single-stranded DNA probe to DNA molecules from bacterial colonies; used for library screening to identify a gene of interest.

comparative genomics: Studies that allow researchers to investigate gene structure and function in organisms in ways designed to understand gene structure and function in other species including humans.

competent cells: Bacterial cells that have been chemically treated to be able to take in DNA (transformation) from their surrounding environment; to be "competent" to accept DNA.

complementary base pairs: Refers to the nucleotides adenine (A) and thymine (T) [or uracil (U) when referring to RNA], and guanine (G) and cytosine (C); in a DNA molecule, A&T and G&C nucleotides join by hydrogen bonds to "complement" each other.

complementary DNA (cDNA): DNA copy of an mRNA molecule; mRNA can be copied into cDNA by the enzyme reverse transcriptase (for instance, when preparing a cDNA library).

complementary DNA (cDNA) library: DNA copies of all mRNA molecules expressed in an organism's cells; can be "screened" to isolate genes of interest.

composting: Mixing soil with hay, straw, grass clippings, wood chips, or other similar "bulking" materials to stimulate biodegradation by soil microbes; also used to degrade everyday materials such as vegetable scraps, grass clippings, weeds, and leaves.

computer-automated DNA sequencing methods: Computer assisted methodologies for DNA sequencing. Allow for sequencing and analysis of large amounts (high-throughput) of sequence data compared to older manual sequencing methods.

contigs (contiguous sequences): Restriction enzyme–digested pieces of entire chromosomes are sequenced separately, and then computer programs are used to align the fragments based on overlapping sequence pieces called contigs (contiguous sequences).

Coomassie stain: A sensitive stain that reacts with proteins.

Coordinated Framework for Regulation of Biotechnology: A 1984 governmental report that proposed that biotechnology products be regulated much like traditional products.

copy number: The number of copies of a particular DNA molecule or gene sequence such as a plasmid in a bacterial cell.

copy number variations (CNVs): DNA segments larger than 1 kb such as deletions, insertions and duplications in the genome that vary between individuals; account for much of the genome diversity identified between humans.

cosmid vectors (cosmids): Large circular double-stranded DNA vectors that are used for gene-cloning experiment in bacteriophages.

credible utility: Requirement of the USPTO that the researcher must convince the patent office that the application is backed by sound science.

crown gall: A hardened mass of protruding dedifferentiated plant tissue usually resulting from an attack from Agrobacter.

cryopreservation: Storage of tissue samples at ultra low temperatures (−150 to −20°C).

customer relations specialists: Sometimes work in QA divisions of a company. One function of customer relations is to investigate consumer complaints about a problem with a product and follow up with the consumer to provide an appropriate response or solution to the problem encountered.

cystic fibrosis transmembrane conductance regulator (CFTR): Protein that serves as a "pump" to regulate the amount of chloride ions in cells; mutations in the CFTR are the cause of the genetic disorder cystic fibrosis.

cytokines: Stimulating factors that enhance an animal's immune system function.

cytoplasm: The inner contents of a cell, consisting of fluid (cytosol) and organelles; its boundaries are defined by the plasma membrane.

cytosine: Abbreviated C; pyrimidine base present in DNA and RNA nucleotides.

cytosol: Gel-like, nutrient-rich aqueous (water-based) fluid of the cytoplasm.

dideoxyribonucleotide (ddNTP): Modified nucleotide. A ddNTP differs from a normal deoxyribonucleotide (dNTP) because it has a hydrogen group attached to the 3′ carbon of the deoxyribose sugar instead of a hydroxyl group-OH.

denaturation: A process that involves using high heat or chemical treatment to break hydrogen bonds in DNA or RNA molecules to separate complementary base pairs and create single-stranded molecules; also refers to the act of changing protein structure (through heat or chemical treatment).

deontological (Kantian) approach: Developed by German philosopher Immanuel Kant, this approach suggests that absolute principles (of moral or ethical behavior) should be followed to dictate our actions.

deoxyribonucleic acid (DNA): Double-stranded nucleic acid consisting of bases (adenines, guanines, cytosines, thymines), sugars (deoxyribose), and phosphate groups arranged in a helical molecule; inherited genetic material that contains "genes," which direct the production of proteins in a cell.

depolymerization: A process of reducing multiunit structures to single units.

deregulated status: First step in the APHIS (USDA) review of a genetically altered product in which field-test reports, scientific literature, and any other pertinent records are checked before APHIS determines whether the plant is as safe to grow as traditional varieties.

design patent: A design for an article of manufacture submitted to the USPTO that is dictated primarily by the function of the article (such as a 3D shape).

diabetes mellitus: A condition of elevated blood sugar. There are several different forms of diabetes. In some people (type 1 or insulin-dependent diabetes mellitus), the hormone insulin is missing.

diafiltration: See *dialysis.*

dialysis: A separation of water and a solute involving a semipermeable membrane.

dicer: An enzyme that cuts microRNA transcripts into short single-stranded pieces of 21 to 22 nucleotides.

dicotyledonous: A term describing a seed with two embryonic seed leaves (like peanuts).

dideoxyribonucleotide (ddNTP): Nucleotides used for certain types of DNA sequencing methods. Structurally different than nucleotides because they lack oxygen atoms at positions 2 and 3 of the deoxyribose sugar. As a result ddNTPs are called "chain-terminating nucleotides" because they cannot form phosphodiester bonds.

differential display PCR: A PCR technique that uses random primers to amplify DNA in two or more different samples with the purpose of identifying randomly amplified genes that are different (differentially expressed) in one sample compared with another sample.

differentiation: Cellular maturation process; involves changes in patterns of gene expression that affect the structure and functions of cells. For instance, embryonic stem cells differentiate to produce mature cells such as neurons, liver cells, skin cells, and all other cell types in the body.

diploid: Refers to a cell or an organism consisting of two sets of chromosomes: usually, one set from the mother and another set from the father. In a diploid state, the haploid number is doubled; thus, this condition is also known as 2n.

diploid number: Two sets of chromosomes (2n); often used to describe cells with two sets of chromosomes, typically one set from each parent.

directed molecular evolution: A process that selects for proteins after extensive mutations.

direct-to-consumer (DTC) genetic tests: DNA sequence analysis or genetic testing of human DNA for comparison to known sequences that have been associated with abnormal human conditions; DTCs can be purchased directly without prescription.

DNA-binding domains: Regions with folded structural arrangements of amino acids called motifs that interact directly with DNA.

DNA fingerprinting: An analysis of an organism's unique DNA composition as a characteristic marker or fingerprint for identification purposes, such as forensic analysis, remains identification, and paternity. DNA fingerprinting is also used in biologic research (for example, to compare related species based on their DNA sequences).

DNA helicase: Enzyme that separates two strands of a DNA molecule during DNA replication.

DNA library: See *genomic DNA library* or *complementary (cDNA) library.*

DNA ligase: Enzyme that forms covalent bonds between incomplete (Okazaki) fragments of DNA during DNA replication; routinely used for joining DNA fragments in recombinant DNA experiments.

DNA microarray (gene chip): A chip consisting of a glass microscope slide containing thousands of pieces of single-stranded DNA molecules attached to specific spots on the slide; each spot of DNA is a unique sequence.

DNA polymerase III: One form of DNA polymerase; the primary DNA polymerase which copies DNA in prokaryotes.

DNA polymerases: Key enzymes that copy strands of DNA during DNA replication to create new strands of nucleotides; these have important applications for synthesizing DNA in molecular biology experiments.

DNAse: DNA-degrading enzyme.

DNA sequencing: Laboratory technique for determining the nucleotide "sequence" or arrangement of A, G, T, and C nucleotides in a segment of DNA.

doctrine of equivalents: If a patent holder claims patent infringement on two similar DNA sequences, a court must decide whether the sequences have essentially the same function.

domain Archaea: A classification (domain) of prokaryotic cells; a common feature of these microbes is that they live in extreme environments, for example, boiling hot springs.

domain Bacteria: Taxonomic category of living organisms that includes bacteria; one of three domains (Bacteria, Archaea, Eukarya), domain Bacteria is the most diverse domain consisting of primarily unicellular prokaryotes.

domains: Taxonomic categories above the kingdom level: Archaea, Bacteria, and Eukarya.

dot blot: A detection strip coated with DNA sequences that reflect sites in human DNA where antibody-forming alleles (gene) differ.

double-blind trial: Study, such as a clinical trial, in which neither the patients nor the researchers administering the treatment know which patients receive a drug treatment and which patients receive a placebo; researchers learn the treatment of each patient after the study is completed.

Down syndrome: Human genetic disorder first described by John Langdon Down. Most commonly results from individuals having three copies of chromosome 21 (trisomy 21). Characteristics of affected individuals include a flat face; shortness; short, broad hands and fingers; protruding tongue; and impaired physical, motor, and mental development.

downstream processing: Purification processes involved in producing a pure product such as a recombinant protein.

DPT vaccine: Childhood vaccine designed to provide immune protection against bacterial toxins diphtheria, pertussis, and tetanus (contains killed Bordetella pertussis cells). Microbes producing these toxins cause upper respiratory infections, whooping cough, and tetanus (muscle spasm and paralysis), respectively.

E site: A groove through which tRNA molecules leave the ribosome.

efficacy: The effectiveness of a particular product such as a drug or medical procedure.

effluent: Water that has been treated to remove chemical pollutants or sewage; treated water as it leaves a treatment facility, such as water leaving a sewage-treatment plant.

electroporation: A process for transforming bacteria with DNA that uses electrical shock to move DNA into cells; can also be used to introduce DNA into animal and plant cells.

embryo twinning: Splitting embryos in half at the two-cell stage of development to produce two embryos.

embryonic stem cells (ESCs): Cells typically derived from the inner cell mass of a blastocyst; cells can undergo differentiation to form all cell types in the body.

Encyclopedia of DNA Elements (ENCODE) Project: Genome project to identify DNA regulatory sequences.

endotoxins (lipopolysaccharides): Molecules that are toxic to cells; endotoxins are part of the cell wall of certain bacteria; endotoxins cause cell death.

enhancers: Specific DNA sequences that bind proteins called transcription factors to stimulate ("enhance") transcription of a gene.

enucleation: Removal of the DNA from the nucleus of a cell.

Environmental Genome Project: NIH program designed to study the impacts of environmental chemicals on human genetics and disease.

Environmental Protection Agency: See *U.S. Environmental Protection Agency*.

environmental genomics (metagenomics): Involves sequencing genomes for entire communities of microbes in environmental samples of water, air, and soils from oceans throughout the world, glaciers, mines—virtually every corner of the globe.

EnviroPig: A transgenic pig expressing the enzyme phytase in its saliva. It is deemed environmentally friendly because it produces substantially less phosphorus in its urine and feces than do nontransgenic pigs.

enzymes: Catalytic proteins.

ES (embryonic stem) cell transfer: Embryonic stem cells (ES cells) are collected from the inner cell mass of blastocysts. Transformed ES cells are usually injected into the inner cell mass of a host blastocyst.

Epigenome: Modifications in an organisms genome that are not due to mutations in DNA sequence.

ethidium bromide: Tracking dye that penetrates (intercalates) between the base pairs of DNA. Commonly used to stain DNA in gels because ethidium bromide fluoresces when exposed to ultraviolet light.

eugenics: The creation of a new superior species of human.

eukaryotic cells: Cells that contain a nucleus, including plant and animal cells, protists, and fungi.

European Agency for the Evaluation of Medicinal Products (EMEA): Agency that authorizes medicinal products for human and veterinary use. It also helps spread scientific knowledge by providing information on new products in all 11 official EU languages.

exons: Protein-coding sequences in eukaryotic genes and mRNA molecules.

ex situ bioremediation: Removing contaminated soils or water from the site of contamination for clean-up at another location (as opposed to cleaning up pollutants at the contaminated site—a process called in situ bioremediation).

ex vivo gene therapy: Gene-therapy procedure that involves removing a person's cells (such as blood cells) from the body, introducing therapeutic genes into these cells, and then reintroducing the cells back into a person.

experimental use permit (EUP): Permit issued by the EPA for field experiments with new plant varieties with pesticidal capabilities that involve at least 10 acres of land or 1 acre of water; these experiments cannot be conducted without an experimental use permit.

expressed sequence tags (ESTs): Small incomplete pieces (tags) of gene sequence such as those derived from a cDNA library; represent mRNAs expressed in a given tissue.

expression vector: DNA vector such as a plasmid that can be used to produce (express) proteins in a cell.

extrachromosomal DNA: DNA that is not part of an organism's nuclear (chromosomal) DNA; examples include plasmid and mitochondrial DNA.

fermentation: A metabolic process that produces small amounts of ATP from glucose in the absence of oxygen and also creates byproducts such as ethyl alcohol (ethanol) or lactic acid (lactate). Fermenting microbes (bacteria and yeast) are important for producing a variety of beverages and foods, including beer, wine, breads, yogurts, and cheeses.

fermenters (bioreactors): Containers for growing cultures of microorganisms or mammalian cells in a batch process. Fermenting vessels allow scientists to control and monitor growth conditions such as temperature, pH, nutrient concentration, and cell density. Fermenters do not require oxygen when anaerobic organisms are fermenting nutrients. Bioreactors use non-fermenting organisms that require oxygen.

fertilization (nutrient enrichment): Addition of nutrients (fertilizer such as nitrogen and phosphorus) to a contaminated environment to stimulate growth and activity of naturally occurring soil microorganisms that will aid bioremediation.

field trials: Tests outside of the laboratory required by APHIS (USDA) after a new plant has been engineered in a laboratory; may require several years to investigate everything about the plant, including disease resistance, drought tolerance, and reproductive rates.

5′ end: The end of a strand of DNA or RNA in which the last nucleotide is not attached to another nucleotide by a phosphodiester bond involving the 5′ carbon of the pentose sugar.

finance divisions: Unit responsible for overseeing company finances.

fluorescence in situ hybridization (FISH): Laboratory technique that uses single-stranded DNA or RNA probes labeled with fluorescent nucleotides to identify gene sequences in a chromosome or cell in situ (Latin for "in its original place").

Food and Drug Administration (FDA): An agency of the federal government that regulates food and drug safety.

forensic biotechnology: The analysis and application of biologic evidence such as DNA and protein-sequence data to help solve or recreate crimes.

forensic science: Application of a broad spectrum of sciences to answer questions of interest to a legal system.

frameshift mutation: Mutation resulting from the addition or deletion of nucleotides that causes a shift in the genetic code-reading frame of a gene.

functional genomics: Genetic-sequencing analysis of genes active during a cellular event.

fungi: Eukaryotic organisms that belong to kingdom Fungi.

fusion proteins: A "hybrid" recombinant protein consisting of a protein from a gene of interest connected (fused) to another, well-known protein that serves as a tag for isolating recombinant proteins.

gametes: Haploid cells often called "sex" cells include human sperm and egg cells (ova). Gametes join together during fertilization to form a zygote (see *meiosis*).

gel electrophoresis: Laboratory procedure that involves using an electrical charge to move and separate biomolecules of different sizes, such as DNA, RNA, and proteins, through a semisolid separating gel matrix. Examples include agarose gel electrophoresis and polyacrylamide gel electrophoresis (PAGE).

GenBank: Renowned public database of DNA sequences provided by researchers throughout the world; resources for sharing and analyzing DNA sequence information.

gene: A specific sequence of DNA nucleotides that serves as a unit of inheritance. Genes govern visible and invisible characteristics (traits) of living organisms in large part by directing the synthesis of proteins in a cell.

gene chip: See *DNA microarray*.

gene cloning: The process of producing multiple copies of a gene.

gene doping: Defined by the World Anti-Doping Agency as "the non-therapeutic use of cells, genes, genetic elements, or of the modulation of gene expression, having the capacity to improve athletic performance."

gene expression: Term generally used to describe the synthesis ("expression") of RNA by a particular cell or tissue. For instance, if mRNA for the fictitious gene korn is detected in a tissue by PCR or Northern blot analysis, that tissue is said to express the korn gene.

gene gun: A microprojectile device used to propel genes on the surface of metal particles into cells.

Genentech: California company. Name derived from genetic-engineering technology. Founded in 1976 and widely recognized as the world's first biotechnology company.

gene regulation: General term used to describe processes by which cells can control gene activity or "expression" (RNA and protein synthesis).

gene therapy: The use of therapeutic genes to treat or cure a disease process; also refers to the delivery of genes to improve a person's health.

generally-recognized-as-safe (GRAS) status: Status granted by the FDA if a food product or additive poses no foreseeable threat.

generic drugs: Copies of brand-name products that generally have the same effectiveness, safety, and quality but are produced at a cheaper cost to the consumer than the brand-name drugs.

genetic code: Information contained in bases of DNA and RNA that provide cells with instructions for synthesizing proteins; consist of three-nucleotide combinations called codons.

genetic engineering: The process of altering an organism's DNA. This is usually by design.

Genetic Information Nondiscrimination Act: This act prohibits discrimination based on genetics and the improper use of genetic information in health insurance and employment.

Genetic Testing Registry (GTR): A branch of the NCBI projected to be developed by late 2011, it will provide access to information about genetic tests for inherited and somatic genetic variations, including newer types of tests such as arrays and multiplex panels. GTR information about tests primarily will be based on voluntary data submissions by test developers and manufacturers.

genetically modified (GM) foods: Foods derived from sources that have been genetically modified (and must be approved) for consumption.

genetically modified (GM) organisms: GM organisms included selectively bred and transgenic organisms. For example, GM crops have been produced to plants that can resist pests, disease, or harsh climates, allowing better production of crops.

genome: All of the genes in an organism's DNA.

Genome 10K plan: Project to create a genomic "zoo" containing the sequences of 10,000 vertebrate species.

genomic DNA library: Collection of DNA fragments containing all DNA sequences in an organism's genome; can be "screened" to isolate genes of interest.

genomics: The study of genomes.

germline genetic engineering: Genetic alteration of sperm, eggs, or embryos to create inheritable genetic changes in offspring.

Gleevec: A drug used to treat certain types of leukemia (cancer that begins in the white blood cells) and other cancers of the blood cells. Imatinib (Gleevec) is also used to treat gastrointestinal stromal tumors.

glycosylation: A natural process of adding sugar units to proteins by complex cells.

glyphosphate: A small molecule that interacts and blocks 5-enolpyruvylshikimate-3-phosphate synthase (EPSPS), a key enzyme in plant photosynthesis. Resistant plants have a substitute pathway obtained by gene transfer.

Golden Rice: Genetically engineered rice with high vitamin A content.

good clinical practice (GCP): An FDA requirement governing clinical trials that protects the rights and safety of human subjects and ensures the scientific quality of experiments.

good laboratory practice (GLP): An FDA requirement that testing laboratories follow written protocols, have adequate facilities, provide proper animal care, record data properly, and conduct valid toxicity tests.

good manufacturing practice (GMP): Regulations instituted by the FDA to govern animal studies of pharmaceutical products.

green fluorescent protein (GFP): Protein produced by the bioluminescent jellyfish *Aequorea victoria*. Protein fluoresces when exposed to ultraviolet light; the GFP gene is used as a reporter gene.

Gram stain: Technique for staining the bacterial cell wall that can be used to divide bacteria into different categories, gram-positive or gram-negative bacteria.

growth factors: Molecules that stimulate cell growth and division.

growth hormone (GH): A peptide hormone produced by the pituitary gland; accelerates the growth of bone, muscle, and other tissues.

guanine: Abbreviated G; purine base present in DNA and RNA nucleotides.

HAMA response: When mouse hybridoma-produced human antibodies produce enough "mouse" antigens, they evoke this undesirable immune response in humans.

haploid: A single set of 23 chromosomes.

haploid number: A single set of chromosomes (n); often used to describe cells with a single set of chromosomes.

HapMap Project: See *International HapMap Project.*

helix: One of two secondary structures of proteins.

hemoglobin: Oxygen-binding and oxygen-transporting protein of red blood cells in vertebrates. Mutation of globin genes that encode hemoglobin results in a number of different human blood disorders with a genetic basis, including sickle-cell disease.

Herceptin: Herceptin is approved for the treatment of early-stage breast cancer that is **H**uman **E**pidermal growth factor **R**eceptor **2**-positive (HER2+) and has spread into the lymph nodes, or is HER2+ and has not spread into the lymph nodes. If it has not spread into the lymph nodes, the cancer needs to be estrogen receptor/progesterone receptor (ER/PR)-negative or have one high risk feature.

high-performance liquid chromatography (HPLC): High-pressure separation of similar proteins by using incompressible beads with special properties.

Hippocratic Oath: An oath that embodies the obligations and duties of medical doctors; one expectation of the oath is that physicians treat patients with the intent of not worsening human health ("first, do no harm").

histones: DNA-binding proteins that are important for chromosomal structure.

homologous pairs: Pairs of chromosomes or genes that occur in organisms with multiple sets of chromosomes.

homologous recombination: A gene recombines with its corresponding existing gene on a chromosome (with which it has a high degree of sequence similarity), replacing part or all of the gene.

homologues: Related genes in different species; genes share sequence similarity because of a common evolutionary origin.

hormones: Molecules involved in chemical signaling between cells and organs; transported in body fluids, hormones interact with and control the activity of many cells.

human antimouse antibody (HAMA) response: Mouse hybridomas produce enough "mouse" antigens that they still evoke an undesirable immune response in humans.

human embryonic stem cells: Stem cells are immature cells that can grow and divide to produce different types of cells such as skin, muscle, liver, kidney, and blood cells. Most stem cells are obtained from embryos (embryonic stem cells or ESCs). Some ESCs can be isolated from the cord blood of newborn infants.

Human Epigenome Project: Project designed to reveal epigenetic changes in different cell and tissue types and to evaluate potential roles of epigenetics in diseases,

humane treatment: Compassionate and sympathetic approach, as applied to treating humans or animals.

Human Genome Project: An international effort with overall scientific goals of identifying all human genes and determining (mapping) their locations to each human chromosome.

Human Microbiome Project: Project designed to identify the genomes of all microorganisms (bacterial, yeast, viral) living in or on the human body.

human papillomavirus (HPV): Causes about 70% of cervical cancers (HPV strains 16 and 18) and a large percentage of genital warts (caused by HPV strains 6 and 11). Cervical cancer affects 1 in 130 women, nearly half a million women worldwide, and approximately 70% of sexually active women will become infected with HPV.

Human Proteome Project: A federal initiative proposed to study the structures and functions of all human proteins (the proteome).

Humulin: The recombinant form of human insulin, became the first recombinant DNA product to be approved for human applications by the U.S. Food and Drug Administration.

hybridization: The joining of two DNA strands by complementary base pairing; for instance, binding of a single-stranded DNA probe to another DNA molecule.

hybridomas: Hybrid cells used to create monoclonal antibodies; created by fusing B cells with cancerous cells called myeloma cells that no longer produce an antibody of their own.

hydrophilic: Water-loving (portion of protein molecule).

hydrophobic interaction chromatography (HIC): Chromatographic separation based on binding of hydrophobic parts of proteins to the column.

hydrophobic: Water-hating (portion of protein molecule).

hydroponic systems: A form of aquaculture in which tanks of flowing water are used to grow plants; in some cases, water from fish aquaculture tanks is used to provide nutrients for plant growth in a polyculture approach.

hydroxyapatite (HA): Structural component of bone and cartilage.

in situ bioremediation: Cleaning up pollutants at the actual site of contamination; "in place" clean-up as opposed to removing contaminated soils or water from the clean-up site for remediation at another location—a process called ex situ bioremediation.

in situ hybridization: Laboratory technique that uses single-stranded DNA or RNA probes, usually labeled with radioactive or color-producing nucleotides, to identify gene sequences in a chromosome or cell in situ (Latin for "in its original place").

in vitro: Occurring within an artificial environment such as a test tube; from Latin "in glass."

in vitro fertilization (IVF): Assisted-reproduction technology in which sperm and egg cells are removed from patients and cultured in a dish (in vitro) to achieve fertilization.

in vivo: Occurring within a living organism; from Latin "in something alive."

in vivo gene therapy: Gene therapy procedure that involves introducing therapeutic genes directly into a person's tissue or organs without removing them from the body.

inactivated vaccines: Vaccine consisting of killed microorganisms.

inclusion bodies: Foreign proteins that concentrate in transformed cell.

indigenous microbes: Naturally occurring microorganisms living in the environment.

induced-pluripotent stem cells (iPSCs): Nuclear reprogramming of mouse and human cells, heralded as a revolution in stem-cell biology research. One approach has involved using retroviruses to deliver four transgenes Oct3/4, Sox2, c-myc, and Klf4 into fibroblasts. Expression of these four genes, which encode transcription factors involved in cell develop-

ment, "reprograms" the fibroblasts back to an earlier stage of differentiation. iPS cells demonstrate many properties of hESCs, such as self-renewal and pluripotency, and appear to be indistinguishable from hESCs.

influenza: Caused by a large number of viruses that belong to the influenza family of viruses. Influenza kills approximately 500,000 to 1 million people worldwide each year. Because flu viruses mutate so rapidly, no one-size-fits-all vaccine protects against all strains.

informed consent: Applies to clinical trials in humans. Patients are made aware of potential beneficial and harmful effects of a particular treatment so that they are informed of the risks of a procedure before they agree (consent) to participate.

inherited mutations: A change in DNA structure or sequence of a gene passed to offspring through gametes; can be a cause of birth defects and genetic disease.

initial public offering (IPO): First sale of stock by a formerly private company. It can be used by either small or large companies to raise expansion capital and become publicly traded enterprises.

inner cell mass: Layer of cells in the blastocyst that develop to form body tissues; a source of embryonic stem cells.

Innocence Protection Act: A law that gives the convicted access to DNA testing, prohibits states from destroying biological evidence as long as a convicted offender is imprisoned, prohibits denial of DNA tests to death-row inmates, and encourages compensation for convicted innocents.

inorganic compounds: See *inorganic molecules.*

inorganic molecules: Molecules that do not contain carbon.

insulin: Protein hormone produced by cells of the pancreas; involved in glucose metabolism by cells; deficiencies in insulin production or insulin-receptor production can cause different forms of diabetes.

insulin-dependent (type I) diabetes mellitus: Caused by an inadequate production of insulin by beta cells. The decreased production of insulin results in an elevated blood glucose concentration that can cause a number of health problems such as high blood pressure, poor circulation, cataracts, and nerve damage. People with type I diabetes require regular injections of insulin to control their blood sugar levels.

integration: Inserting DNA into a genome. A process by which retroviruses such as lentivirus and HIV can enter host cells, copy their RNA genome into DNA and then randomly insert their DNA into the genome of the host cell, where it remains permanently.

International HapMap Project: International effort to develop haplotype maps of human genetic variation; includes SNP analysis of the genome.

International Laboratory for Tropical Agricultural Biotechnology (ITLAB): A not-for-profit research laboratory funded by Monsanto and others to research plant varieties that could benefit countries by improving their food quality.

introns: Non–protein-coding sequences in eukaryotic genes and primary transcripts that are removed during RNA splicing.

investigational new drug (IND) application: A formal request to the FDA to consider the results of previous animal experiments, the nature of the substance itself, and the plans for further testing.

ion-exchange chromatography (IonX): The attachment of protein to a column based on its charged side groups.

isoelectric focusing: Migration of a protein until its charge matches the pH of the medium.

isoelectric point (IEP): The pH at which the charge on proteins matches that of the surrounding medium.

Kantian approach: See *deontological (Kantian) approach*.

karyotyping: A laboratory procedure for analyzing the number and structure of chromosomes in a cell.

knock-in animals: Animals can have a human gene inserted to replace their own counterpart by homologous recombination to become knock-ins.

knock-outs: An active gene is replaced with DNA that has no functional information.

laboratory technicians: Entry-level laboratory jobs with a range of responsibilities such as preparing solutions and mediums, ordering laboratory supplies, cleaning and maintaining equipment; may sometimes involve bench research.

lac operon: A well-characterized bacterial operon that contains a series of genes (lacZ,Y, Z) responsible for metabolism of the sugar lactose.

lac repressor: Inhibitory protein encoded by the lacI gene of the lactose (lac) operon in bacteria; in the absence of lactose, repressor blocks transcription of the lac operon.

lactic acid fermentation: See *fermentation*.

lagging strand: During DNA replication, the strand of newly synthesized DNA that is copied by DNA polymerase in a discontinuous (interrupted) fashion, 5′ to 3′ away from the replication fork as a series of short DNA pieces called Okazaki fragments.

landfarming: Process by which contaminated soil is removed from a site and spread into thin layers on a pad that allows polluted liquids (usually water) to leach from the soil. Chemicals also vaporize from spread-out soil as the soil dries.

leachate: Water or other liquids that move (leach) through the ground from the surface or near the surface to deeper layers.

leading strand: During DNA replication, the strand of newly synthesized DNA that is copied by DNA polymerase in a continuous fashion, 5′ to 3′ into the replication fork.

leaf-fragment technique: A method of plant cloning from asexual plant tissue.

Leber's congenital amaurosis (LCA): Degenerative disease of the retina that affects 1 in 50,000 to 100,000 infants each year and causes severe blindness; caused by defects in the *RPE65* gene.

legal specialists: In biotechnology companies, typically work on legal issues associated with product development and marketing, such as copyrights, naming rights, and obtaining patents. Staff in this area also address legal circumstances that may arise if problems are found with a product or litigation from a user of a product.

leukocytes: White blood cells; important cells of the immune system; include B and T lymphocytes and macrophages.

ligands: Binding components.

limulus amoebocyte lysate (LAL) test: A procedure involving blood cells (amoebocytes) from the horseshoe crab (Limulus polyphemus); an important test used to detect endotoxin and bacterial contamination of foods, medical instruments, and other applications.

lipases: Fat-digesting enzymes.

liposomes: Small hollow spheres or particles made of lipids; can be packaged to contain molecules such as DNA and medicines for use in therapeutic procedures (for example, gene therapy).

luciferase: A light-releasing enzyme present in bioluminescent organisms.

lyophilization: The process of freeze-drying.

lysis: The dissolution or destruction of cells.

lytic cycle: A process of bacteriophage replication that involves phages infecting bacterial cells and then replicating and rupturing (lysing) the bacterial host cells.

macrophage: Term literally means big eaters; macrophages are phagocytic white blood cells that engulf and destroy dead cells and foreign materials such as bacteria.

major histocompatibility complex (MHC): Tissue-typing proteins present on all cells and tissues; recognized by a person's immune system to determine whether cells are normal body cells or foreign; MHCs must be "matched" for successful organ transplantation.

malaria: Caused by the protozoan parasite Plasmodium falciparum and transmitted by insects. Worldwide, Plasmodium strains are developing resistance to the most commonly used antimalarial drugs.

manufacturing assistant: Entry-level job includes material handlers, manufacturing assistants, and manufacturing associates. Supervisory and management-level jobs usually require a bachelor's or master's degree in biology or chemistry and several years of experience in manufacturing the products or type of product being produced by a company.

manufacturing technicians: Must strictly comply with Food and Drug Administration (FDA) regulations at all stages of producing biotech drugs.

mariculture: The cultivation or "farming" of aquatic organisms (animals or plants).

marker gene: See *reporter gene*.

marketing specialist: Marketing and sales position in biotechnology companies; marketing specialists are often involved in designing ad campaigns and promotional materials to market effectively a company's products.

Mass Fatality Identification System (M-FISys): M-FISys was essentially built in response to the 9/11 tragedy. Gene Codes did not have to write entirely new software, and they were able to customize M-FISys as necessary. In addition to analyzing mitochondrial DNA, M-FISys incorporated male-specific variations in the Y chromosome called Y-STRs to aid in the identification of individuals.

mass spectrometry (mass spec): Separation for identification of chemicals (e.g., proteins) based on charge-to-mass ratio.

maternal chromosomes: The 23 chromosomes inherited from your mother.

medical biotechnology: A diverse discipline of biotechnology dedicated to improving human health; includes a spectrum of topics in human medicine from disease diagnosis to drug discovery, disease treatment, and tissue engineering.

meiosis: A nuclear division process that occurs during formation of gametes. Meiosis involves a series of steps that reduce the amount of DNA in newly divided cells by half compared with those in the original cell.

membrane filtration: Using a thin membrane sheet with small holes (pores) to filter materials (such as large proteins or whole cells) out of a solution.

messenger RNA (mRNA): A template for protein synthesis, is an exact copy of a gene, contains nucleotide sequences copied from DNA that serve as a genetic code for synthesizing a protein, and is then bound and "read" by ribosomes to produce proteins.

metabolic engineering: Modifying an energy-generating or energy-requiring chemical process (usually to improve energy generation or to reduce energy use) through a biotechnical or chemical process.

metabolomics: A biochemical snapshot of the small molecules produced during cellular metabolism, such as glucose, cholesterol, ATP, and signaling molecules that result from a cellular change.

metagenomics: The sequencing of genomes for entire communities of microbes in environmental samples of water, air, and soils from oceans throughout the world, glaciers, mines—virtually every corner of the globe.

metallothioneins: Metal-binding proteins.

microbes: Tiny organisms that are too small to be seen individually by the naked eye and must be viewed with the help of a microscope. Although the most abundant microorganisms are bacteria, microbes also include viruses, fungi such as yeast and mold, algae, and single-celled organisms called protozoa.

microbial biotechnology: Discipline of biotechnology that involves the use of organisms (microorganisms such as bacteria and yeast) that cannot be seen individually by the naked eye to make valuable products and applications.

microbial diagnostics: Methods for identifying microbes for different purposes such as clinical diagnosis of bacteria infections, detecting microbes causing food outbreaks, and detecting bioweapons; many of these techniques are based on molecular analysis of DNA.

Microbial Genome Program (MGP): U.S. Department of Energy program to map and sequence genomes of a broad range of microorganisms.

microcosms: Test environments designed to mimic polluted environmental conditions; may consist of bioreactors or simply a bucket of polluted soil; allows scientists to test clean-up strategies on a small scale under controlled conditions before trying a particular clean-up approach in the environment.

microfiltration: Removal of particles 40 µm or smaller.

microorganisms (microbes): Organisms that cannot individually be seen with the naked eye; include bacteria, yeast, fungi, protozoans, and viruses.

microRNAs (miRNAs): A new class of non-protein coding RNA molecules. MicroRNAs are part of a rapidly growing family of small RNA molecules about 20 to 25 nucleotides in size that play novel roles in regulating gene expression.

microsatellites: Also known as short tandem repeats (STRs), microsatellites are short repeating sequences of DNA usually consisting of one-, two-, or three-base sequences (for example, CACACACACA); microsatellites are important markers for forensic DNA analysis.

microspheres: Tiny particles that can be filled with drugs or other substances and used in therapeutic applications.

missense mutation: A mutation changing a codon to another codon that codes for a different amino acid.

mitochondrial DNA: Small circular DNA found in mitochondria responsible for proteins unique to mitochondria. Mutations in this DNA can be followed from mother to offspring, because the egg is the predominant source of mitochondria.

mitosis: A nuclear-division process that occurs during cell division in eukaryotic somatic cells and prokaryotes; divided into four major stages (prophase, metaphase, anaphase, and telophase). Mitosis results in the even separation of DNA into dividing cells.

MMR vaccine: Measles, mumps, and rubella (German measles) vaccine designed to provide immune protection against common childhood diseases.

model organisms: Nonhuman organisms that scientists use to study biologic processes in experimental laboratory conditions; common examples include mice, rats, fruit flies, worms, and bacteria.

molecular pharming: The use of plants as sources of pharmaceutical products.

monoclonal antibodies (MABs): Antibody proteins produced from clones of a single ("mono") cell; these proteins are highly specific for a particular antigen.

monocotyledonous: Describes a plant with a single embryonic seed leaf (like corn).

moratorium: As it relates to science, a temporary but complete stoppage of any research.

morula: Latin term meaning "little mulberry;" solid ball of cells that forms during embryonic development in animals, created by repeated cell division of the zygote.

motifs: These regions, called DNA-binding domains, have folded structural arrangements of amino acids called motifs that interact directly with DNA.

multilocus probes: The probe will bind only to complementary sequences of DNA, located at more than one site in the genome.

mutagens: Physical or chemical agents that cause mutations.

mutation: A change in the DNA structure or sequence of a gene.

myelomas: Antibody-secreting tumors.

nanomedicine: Applying nanotechnology to improve health; for instance, microsensors that can be implanted into humans to monitor vital values such as blood pressure.

nanotechnology: Engineering and structures and technologies at the nanometer scale.

National Institutes of Health (NIH): Government agency that is the focal point for medical research in the United States; houses world-renowned research centers and agencies that are an essential source of funding for biomedical research in the United States.

New Drug Authorization (NDA): Approval of a new drug by the Food and Drug Administration (FDA) after the drug has successfully passed through Phase III clinical trials.

nitrogenous base: Important component of DNA and RNA nucleotides; often simply called a "base." Nitrogen-containing structures include double-ring purines (which include the bases adenine and guanine) and single-ring pyrimidines (which include the bases cytosine, thymine, and uracil).

noninvasive prenatal genetic diagnosis (NIPD): tests based on fragments of fetal DNA from cells that have died and been digested by enzymes. Free-floating DNA fragments end up in the mother's bloodstream and can be tested using a blood sample from the mother.

nonsense mutations: Mutations that change a codon into a stop codon; these usually produce a shortened, poorly functioning, or nonfunctional protein.

Northern blot analysis: Laboratory technique for separating RNA molecules by gel electrophoresis and transferring (blotting) RNA onto a filter paper blot for use in hybridization studies.

Northern blotting: See *Northern blot analysis.*

notification: A declaration to the USDA that can be used to "fast-track" some new agricultural products, in which the plant species are well characterized, introduce no new disease, are contained within the nucleus, do not produce a toxin, and are not produced from a plant, animal, or human virus.

nuclear envelope: A double-layered membrane, is typically the largest structure in an animal cell.

nuclear reprogramming of somatic cells: Used for isolating stem cells without creating an embryo. The basic concept of this approach is to use genes involved in cell development to push a somatic cell back to an earlier stage of development and affect gene expression, to reprogram the somatic cell genetically to return to a pluripotent state characteristic of the stem cells from which it was derived.

nucleic acids: Molecules composed of nucleotide building blocks; two major types are DNA and RNA.

nucleotide: Building block of nucleic acids; consists of a five-carbon (pentose) sugar molecule (ribose or deoxyribose), a phosphate group, and the bases adenine (A), guanine (G), cytosine (C), thymine (T), or uracil (U).

nucleus: Membrane-enclosed organelle that contains the DNA of a eukaryotic cell.

nutrient enrichment: See *fertilization.*

nutrigenomics: A new field of nutritional science focused on understanding interactions between diet and genes.

oligonucleotides (oligos): Short, single-stranded synthetic DNA sequences; used in PCR reactions and as DNA probes.

oncogenes: Cancer-causing genes; they are usually part of the genome and have normal functions; when mutated, oncogenes contribute to the development of cancer.

operator: Region of DNA, such as that located in an operon, that binds to a specific repressor protein to control expression of a gene or group of genes such as an operon.

operons: Gene units common in bacteria; typically consisting of a series of genes, located on adjacent regions of a chromosome, and regulated in a coordinated fashion. Many operons are involved in bacterial cell metabolism of nutrients such as sugars. Refer to the lac operon as a well-characterized bacterial operon.

organelles: Small structures in the cytoplasm of eukaryotic cells that perform specific functions.

organic molecules: Molecules that contain carbon and hydrogen.

origins of replication: Specific locations in a DNA molecule where DNA replication begins.

osteoporosis: Category of bone disorders that generally involve a progressive loss of bone mass.

oxidation: The removal of one or more electrons from an atom or molecule.

oxidizing agents: Atoms or molecules that accept electrons during a redox reaction and cause the oxidation of other atoms or molecules; known as electron acceptors, oxidizing agents become reduced when they accept an electron.

P (peptidyl) site (of a ribosome): Portion of a ribosome into which peptidyl tRNA molecules bind during translation.

p arm: "Petit" or small/short arm of a chromosome.

paleogenomics: The analysis of ancient DNA such as DNA from fossils; also called "stone-age genomics."

palindrome: A word or phrase that reads the same forward and backward (for example, "a toyota"). In the context of biotechnology, a DNA sequence with complementary strands that reads the same forward and backward. Most restriction enzyme recognition sequences are palindromes.

papain: Protein-digesting enzyme.

parthenogenesis: Creating an embryo without fertilization by pausing DNA division in an egg and allowing the egg to develop with an increased number of chromosomes; for instance, preventing chromosomes from separating in a human egg creates a diploid cell that may develop to form an embryo.

patent: Legal recognition that gives an inventor or researcher exclusive rights to a product and prohibits others from making, using, or selling the product for a certain number of years (20 years from the date of filing).

patenting: The process of receiving a patent (see *patent*).

paternal chromosomes: Copies of chromosomes inherited from the father.

pathogens: Disease-causing organisms.

pentose sugar: A five-carbon sugar; important components of DNA and RNA nucleotide structure. The pentose sugar deoxyribose is contained in DNA nucleotides; the pentose sugar ribose is contained in RNA nucleotides.

peptidoglycan: Structure in bacterial cell walls consisting of specialized sugars and short, interconnected polypeptides.

peptidyl transferase: The rRNA enzyme that is part of the large ribosomal subunit; catalyzes the formation of peptide bonds between amino acids during translation.

peptidyl tRNA: The tRNA molecule attached to a growing polypeptide chain; located in the P site of a ribosome during translation.

permease: Enzyme produced by the lacY gene of the lactose (lac) operon in bacteria; permease transports the disaccharide lactose into bacterial cells.

Personal Genome Project (PGP): An international project to develop personal genomics technologies and practices with several goals such as effective, informative, and responsible management of personal genome data, analysis and open sharing of personal genome sequences, and the establishment of how personal genomes may be used for improving human health and fighting disease.

personalized genomes: Sequencing a genome for an individual.

personalized genomics: A branch of genomics where individual genomes are sequenced and analyzed using bioinformatics tools that may lead to personalized medicines.

personhood: A term popular in bioethics that defines an entity that qualifies for protection based on certain attributes not intrinsic (built-in) or automatic values.

pharmaceutical companies: Companies creating drugs for the treatment of human health conditions.

pharmacogenomics: A form of customized medicine in which disease-treatment strategies are designed based on a person's genetic information (for a particular health condition).

phase I: The first clinical phase of FDA testing in which 20 to 80 healthy volunteers take the medicine to see whether any unexpected side effects are present and to establish the dosage levels.

phase II: The second clinical phase of FDA testing in which the testing of the new treatment on 100 to 300 patients who actually have the illness occurs.

phase III: The third clinical phase of FDA testing in which between 1,000 and 3,000 patients in double-blinded tests are tested after phase II has shown no adverse side effects; usually lasts for 3.5 years.

phase testing: A statistically significant number of trials are required on cell cultures, in live animals, and on human subjects in the three-phase testing processes specified by the Food and Drug Administration.

phosphodiester bonds: Covalent bonds between the sugars of one nucleotide and the phosphate group of an adjacent nucleotide; join nucleotides within strands of DNA and RNA.

phytoremediation: Using plants for bioremediation.

placebo: A blank or ineffective treatment. Used in the scientific practice of having a control group in an experiment, such as a drug trial, that receives an ineffective pill or treatment such as a sugar pill or injection of water instead of the actual medication being tested.

plant patent: A patent received for a genetically unique living plant.

plant transgenesis: Gene transfer to a plant from another species.

plaques: Small clear spots of dead bacteria appearing on a culture plate, caused by bacterial cell lysis by bacteriophage.

plasma (cell) membrane: A double-layered structure, consisting of lipids, proteins, and carbohydrates, that defines the boundaries of a cell; performs important roles in cell shape and regulating transport of molecules in and out of a cell.

plasma cells: Antibody-producing cells that develop from B lymphocytes after B cells are exposed to foreign materials (antigens).

plasmids (plasmid DNA): Small, circular, self-replicating double-stranded DNA molecules found primarily in bacterial cells. Plasmids often contain genes coding for antibiotic resistance proteins and are routinely used for DNA-cloning experiments.

pluripotent: Term used to describe cells, such as embryonic stem cells, with the potential to develop into other cell types.

point mutations: A single base change in DNA sequence.

polarity: Refers to the 5′ and 3′ ends of DNA and RNA molecules.

polyadenylation: Addition of a short sequence or "tail" of adenine (A) nucleotides to the 3′ end of an mRNA molecule; occurs during RNA splicing in eukaryotes; poly(A) tail is important for mRNA stability in the cytoplasm.

polyculture (integrated aquaculture): Raising more than one aquatic species in the same environment. For instance, cultivating fish together with aquatic vegetation.

polygalacturonase: An enzyme naturally produced by plants that digests tissue, causing rotting.

polymerase chain reaction (PCR): Laboratory technique for amplifying and cloning DNA; involves multiple cycles of denaturation, primer hybridization, and DNA polymerase synthesis of new strands.

polypeptide: A chain of amino acids joined by covalent (peptide) bonds; usually greater than 50 amino acids in length.

polyploids: Organisms with an increased number of complete sets of chromosomes.

posttranslational modifications: Protein modifications that occur naturally after initial synthesis.

precipitates: Combines because of mutual attraction.

preimplantation genetic testing: PCR and ASO analysis as well as FISH are being used to screen for gene defects in single cells from eight- to 32-cell–stage embryos created by in vitro fertilization. Allows individuals to select a healthy embryo before implantation.

Prialt (ziconotide): FDA-approved peptide conotoxin purified from the marine cone snail *Conus magus* by the Elan Corporation of Ireland. Conotoxins peptides are natural neurotoxins that block neural pathways that relay pain messages to the brain. Prialt is a painkiller used to treat chronic, severe forms of pain, such as back pain.

primary sequence: Amino acid sequence of a protein.

primary transcript (pre-mRNA): Initial mRNA molecule copied from a gene in the nucleus of eukaryotic cells; undergoes modifications (processing) to produce mature mRNA molecules that enter the cytoplasm.

primase: Enzyme that adds small RNA segments to a single strand of DNA as an early and necessary step for DNA replication.

primers: Oligonucleotides complementary to specific sequences of interest; used in PCR reactions to amplify DNA and DNA-sequencing reactions.

principal/senior scientists: Science leadership position in biotechnology companies; senior scientists are usually Ph.D. or M.D.-trained individuals who plan and direct the research priorities of a company.

prions: Infections protein particles. They attract normal cell proteins and induce changes in their structure, usually leading to the accumulation of useless proteins that damage cells. Prion diseases can occur in sheep and goats (scrapie) and cows (bovine spongiform encephalitis, or "mad cow" disease). Human forms of these brain-destroying diseases include kuru and transformable spongiform encephalitis (TSE). All these diseases involve changes in the conformation of prion precursor protein, a protein normally found in mammalian neurons as a membrane glycoprotein.

probe: Single-stranded DNA or RNA molecule (labeled, for instance, with radioactive or fluorescing nucleotides) that can bind other DNA or RNA sequences by complementary base pairing and be detected by a process such as autoradiography; important laboratory technique for such applications as identifying genes and studying gene activity. Protein probes are usually antibodies that bind to structures with high affinity (e.g., antigens) and can be detected by fluorescent tags.

product complaint specialists: A Quality Control (QC) specialist that certifies that a product meets the criteria specified in its package insert or other specifications submitted to regulatory agencies.

prokaryotic cells: Cells that lack a nucleus and membrane-enclosed organelles; only examples are bacteria and Archae.

promoter: Specific DNA sequences adjacent to a gene that direct transcription (RNA synthesis); binding site of RNA polymerase to begin transcription.

pronuclear microinjection: This method introduces the transgene DNA at the earliest possible stage of development of the zygote (fertilized egg).

prostate-specific antigen (PSA): A protein released into the bloodstream when the prostate gland is inflamed, and elevated levels can be a marker for prostate inflammation and even prostate cancer.

proteases: Protein-digesting enzymes.

proteins: Macromolecules consisting of amino acids joined by peptide bonds; major structural and functional molecules of cells.

protein chip: See *protein microarray*.

protein microarray: A "chip" similar to a DNA microarray; consists of a glass slide containing thousands of individual proteins attached to specific spots on the slide; each "spot" contains a unique protein.

protein sequencing: Identification of amino acid sequence by cleavage of each amino acid, one at a time.

Protein-specific antigen (PSA): Protein produced only by the prostate; biomarker for prostate inflammation and prostate cancer.

proteolytic: Protein–lysing characteristic.

proteome: The entire complement of proteins in an organism.

proteomes: Families of proteins.

proteomics: Study of protein families.

protoplast fusion: Fusing of plant cells lacking cell walls to produce a fused cell that can become a clone.

protoplast: A naked plant cell (without cell wall).

PSI (Protein Structure Initiative): An ongoing effort begun in 2000 to accelerate discovery in structural genomics and contribute to understanding biological function. Funded by the U.S. National Institute of General Medical Sciences (NIGMS), its aim is to reduce the cost and time required to determine three-dimensional protein structures and to develop techniques for solving challenging problems in structural biology.

PulseNet: Partnership of bacterial DNA fingerprinting laboratories, developed by the U.S. Centers for Disease Control and Prevention (CDC) and the U.S. Department of Agriculture, designed to provide rapid analysis of contaminated food, with the purpose of identifying contaminating microbes and preventing outbreaks of food-borne disease.

pyrosequencing: Reaction used in next-generation sequencing methods; involves synthesizing DNA from primers bound to template DNA of interest. Reactions release pyrophosphate (PPi) which can be detected and quantified to determine the number of nucleotides incorporated into a new sequence.

q arm: Long arm of a chromosome.

quality assurance (QA): All activities involved in regulating the final quality of a product, including quality-control measures.

quality control (QC): Procedures that are part of the QA process involving laboratory testing and monitoring of production processes to ensure consistent product standards (of purity, performance, and the like).

quasi-automated x-ray crystallography: Fast x-ray analysis of protein structure by using algorithms and data analysis.

quaternary structure: Proteins containing more than one amino acid chain that is folded into a tertiary structure.

real-time or quantitative PCR (qPCR): New applications in PCR technology make it possible to determine the amount of PCR product made during an experiment. Uses primers made with fluorescent dyes and specialized thermal cyclers that enable researchers to quantify amplification reactions as they occur.

recognition sequence: See *restriction site*.

Recombinant DNA Advisory Committee (RAC): NIH panel responsible for establishing and overseeing guidelines for recombinant DNA research and related topics.

recombinant DNA technology: Technique that allows DNA to be combined from different sources; also called gene or DNA splicing. Recombinant DNA is an important technique for many gene-cloning applications.

recombinant proteins: Commercially valuable proteins created by recombinant DNA technology and gene-cloning techniques; examples include insulin and growth hormone.

redox reaction: Combination of oxidation and reduction reactions.

reduction: The addition of one or more electrons to a molecule.

regenerative medicine: A discipline of medical biotechnology that involves repairing or replacing damaged tissues and organs by using tissues and organs grown through biotechnological approaches.

renin (chymosin): Protein-degrading enzyme derived from the stomach of milk-producing animals such as cows and goats; used in cheese production; recombinant form called chymosin.

replica plating: Technique in which bacteria cells from one plate can be transferred to another plate to produce a replica plate with bacterial cells growing in the same location as on the original plate; can also be carried out with other cells such as yeast, or viruses.

reporter genes: Genes (such as the lux genes) that can be used to track or monitor (report on) expression of other genes.

reproductive cloning: Cloning process that creates a new individual; process typically involves using the genetic material of a single cell to create an individual with the single genetic composition of its creator cell.

research and development (R&D): All of the processes involved in basic research (for example, pre-clinical research) and development of a potential product. The lifeblood of a biotechnology company, R&D is how companies identify new technologies, drugs etc. for commercialization.

research assistants/associates: Laboratory positions in which individuals are primarily involved in carrying out experiments under the supervision of other scientists such as principal or senior scientists.

restriction enzymes (endonucleases): DNA-cutting proteins found primarily in bacteria. Enzymes cleave (cut) the phosphodiester backbone of double-stranded DNA at specific nucleotide sequences (restriction sites). Commercially available restriction enzymes are essential for molecular biology experiments.

restriction fragment length polymorphism (RFLP): Unique patterns created when DNA with different numbers of restriction enzyme digestion sites (due to DNA variation) is separated on an electrophoresis gel and detected with dyes or other means.

restriction fragment length polymorphism (RFLP) analysis: DNA fingerprinting technique that involves digesting DNA into fragments of different lengths by using restriction enzymes; patterns of DNA fragments are analyzed to create a "fingerprint."

restriction map: An arrangement or "map" of the number, order, and types of restriction-enzyme cutting sites in a DNA molecule.

restriction sites: Specific sequences of DNA nucleotides recognized and cut by restriction enzymes.

retroviruses: Viruses that contain an RNA genome and use reverse transcriptase to copy RNA into DNA during the replication cycle in host cells.

retrovirus-mediated transgenics: Infecting embryos with a retrovirus (usually genetically engineered) before the embryos are implanted. The retrovirus acts as a vector for the new DNA.

reverse transcriptase: Viral polymerase enzyme that copies RNA into single-stranded DNA. This commercially available enzyme is used for many molecular biology experiments, such as creating cDNA.

reverse transcription PCR (RT-PCR): Laboratory technique that involves using the enzyme reverse transcriptase to copy RNA from a cell into cDNA and then amplifying cDNA by the polymerase chain reaction (PCR); a valuable technique for studying gene expression.

ribonucleic acid (RNA): Single strands of nucleotides produced from DNA. Different types of RNA have important functions in protein synthesis.

ribosomal RNA (rRNA): Small RNA molecules that are essential components of ribosomes.

ribosomes: Organelles composed of ribosomal ribonucleic acid (rRNA) and proteins assembled into packages called subunits. Ribosomes bind to mRNA and tRNA molecules and are the site of protein synthesis in prokaryotes and eukaryotes.

risk assessments: Analysis of potential risks and benefits of a procedure, drug, technology etc.

RNA-induced silencing complex (RISC): RISC unwinds the double-stranded siRNAs, releasing single-stranded siRNAs that bind to complementary sequences in mRNA molecules. Binding of siRNAs to mRNA results in degradation of the mRNA (by the enzyme slicer) or blocks translation by interfering with ribosome binding.

RNA interference (RNAi): RNA-based mechanisms of gene silencing. These siRNAs are bound by a protein–RNA complex called the RNA-induced silencing complex (RISC).

RNA polymerase: Copies RNA from a DNA template; different forms of RNA polymerase synthesize different types of RNA.

RNA primer: Short, single-strand RNA sequence that binds DNA and is used to start DNA replication by DNA polymerase.

RNAse: RNA-degrading enzyme.

RNA or gene silencing: General term for techniques that are used to inhibit a gene or RNA molecule from expressing the protein it encodes.

RNA splicing: The removal of nonprotein coding sequences (introns) from primary transcript (pre-mRNA) and joining of protein-coding sequences (exons).

sales representatives: Salespersons in biotechnology companies; sales representatives are "people persons" who work closely with medical doctors, hospitals, and health care providers to promote a company's products.

scale-up processes: The industrial implementation of processes in which chemical or microbiological conversion of material takes place that behave differently on a small scale (in laboratories or pilot plants) and on a large scale (in production).

scaling-up: Process manufacturing modification from original research purification.

SDS-PAGE : Sodium Dodecyl Sulfate Polyacrylamide Gel Electrophoresis is a process where proteins are boiled in the presence of SDS to denature the protein and attach sulfate (negatively charged) groups to each amide link in a protein so that they can be separated by their number of charges by electrophoresis in a gel made of polyacrylimide.

secondary structure: a-Helix or b-sheet structure of proteins.

seeding: See *bioaugmentation*.

selection (antibiotic, blue-white screening): Laboratory technique used to identify bacteria containing recombinant DNA of interest; involves growing bacteria on media with antibiotic or other selection molecules.

selective breeding: Mating organisms with desired features to produce offspring with the same characteristics.

semiconservative replication (of DNA): Process by which DNA is copied; one original (parent) DNA molecule gives rise to two molecules, each of which has one original strand and one new strand.

senescence: Cellular aging process.

severe acute respiratory syndrome (SARS): Severe form of pneumonia caused by the SARS virus; can lead to death due to respiratory failure.

severe combined immunodeficiency (SCID): Genetic condition created by a mutation in the gene encoding the enzyme adenosine deaminase. Affected individuals have no functional immune system and are prone to death by common, normally minor infections. Commonly known as "boy in the bubble" condition because of need for SCID individuals to live in germ-free environments.

sex chromosomes: Contain majority of genes that determine an organism's sex; the X (female) and Y (male) chromosomes in humans.

short interfering RNA (siRNA): Small (21 or 22 nt) double-stranded pieces of nonprotein coding RNA, so named because they were shown to bind to mRNA and subsequently block or interfere with translation of bound mRNAs.

short tandem repeat (STR): One to six nucleotide repeats that are dispersed throughout chromosomes. Because these repeated regions can occur in many locations within the DNA, the probes used to identify them complement the DNA regions that surround the specific microsatellite being analyzed.

"shotgun" cloning: Random cloning of many fragments at once; no individual gene is specifically targeted for cloning.

silent mutation: Base-pair substitution that has no effect on the amino acid sequence of a protein.

single-locus probe: A probe that will bind to only complementary sequences of DNA, found in only one location in the genome.

single-nucleotide polymorphisms (SNPs): Single-nucleotide variations in the gene sequence, or a type of DNA mutation; the basis of genetic variation among humans.

single-strand binding proteins: Proteins that bind to unraveled single strands of parental DNA during DNA replication to prevent DNA from reforming double strands before being copied.

sister chromatids: Exact copies of double-stranded DNA molecules and proteins (chromatin) joined to form a chromosome.

site-directed mutagenesis: With this technique, mutations can be created in specific nucleotides of a cloned gene contained in a vector. The gene can then be expressed in cells, which results in translation of a mutated protein. This allows researchers to study the effects of particular mutations on protein structure and functions as a way to determine what nucleotides are important for specific functions of the protein.

size-exclusion chromatography (SEC): Separation based on molecular size.

sludge: A semisolid material produced from treated sewage waste; consists largely of small particles of feces, waste papers, and microorganisms.

slurry-phase bioremediation: Process in which contaminated soil is removed from a site and mixed with water and fertilizers (and often oxygen) in large drums or bioreactors to create a mixture (slurry) that stimulates bioremediation by microorganisms in the soil.

small interfering RNAs (siRNA): Double-stranded RNA molecules, usually 20-25 bp in length, which can be used for RNA interference (silencing) experiments. Naturally occurring in many species and contribute to gene expression regulation through RNA silencing.

solid-phase bioremediation: Ex situ soil clean-up strategies that primarily involve composting, landfarming, or biopiles.

somatic cell nuclear transfer: DNA transfer process that can be used for reproductive or therapeutic cloning purposes; involves removing DNA from one cell and inserting it into an egg cell that has had its DNA removed (enucleated).

somatic cells: All cell types in multicellular organisms except for gametes (sperm and egg cells).

Southern blot analysis: Laboratory technique invented by Ed Southern that involves transferring (blotting) DNA fragments onto a filter-paper blot for use in probe hybridization studies.

Southern blotting: See *Southern blot analysis*.

species integrity: Generally refers to maintaining the natural functions, abilities, and genetic constitution of a particular species; for example, creating recombinant animals or plants such as transgenic or polyploid organisms may change "species integrity" by making a species less well adapted to life in the wild.

specific utility: Requirement of the USPTO that a researcher must know exactly what the DNA sequence does to patent it.

spectral karyotyping: A technique involving probes specific for each chromosome that fluoresces different colors to identify chromosomes according to their color pattern.

sperm-mediated transfer: The DNA is injected or attached directly to the nucleus of the sperm before fertilization.

StarLink: A transgenic corn not approved for humans that was suspected of contaminating seed intended for food.

startup company: Formed by a small team of scientists who believe they may have a promising product to make (such as a recombinant protein to treat disease). The team must typically then seek investors to fund their company so they can buy or rent laboratory facilities, buy equipment and supplies, and continue the research and development necessary to make their product.

statistical probability: Using statistical measures to calculate the possibility (probability) that a particular event will happen.

stem cells (embryonic and adult-derived stem cells): Immature (undifferentiated) cells that are capable of forming all mature cell types in animals and that can be derived from embryos at several days of age or from adult tissues.

Stone Age genomics: A number of laboratories around the world are involved in analyzing "ancient" DNA. These studies are generating fascinating data from minuscule amounts of ancient DNA from bone and other tissues and fossil samples that are tens of thousands of years old.

substantial utility: Requirement of the USPTO that the product must have a real-world function before it can be patented.

substrate: Molecule or molecules on which an enzyme performs a reaction.

subtilisin: Protease derived from Bacillus subtilis, a valuable component of many laundry detergents, where it functions to degrade and remove protein stains from clothing. Several bacterial enzymes are also used to manufacture foods, such as carbohydrate-digesting enzymes called amylases that are used to degrade starches.

subunit vaccine: Vaccine created from components of a pathogen such as viral proteins or lipid molecules.

Superfund Program: Program established by the U.S. Congress in 1980 through the U.S. Environmental Protection Agency, designed to identify and clean up hazardous waste sites and protect citizens from harmful effects of waste sites.

synthetic biology: The use of manmade DNA sequences for creating modified cells or organisms.

synthetic genome approaches: See *synthetic biology*.

T lymphocytes (T cells): Play essential roles in helping B cells recognize and respond to antigens.

Taq DNA polymerase: DNA-synthesizing enzyme isolated from Thermus aquaticus, a thermophilic Archae that lives in hot springs; its ability to withstand high temperatures (thermostable) without denaturation makes it valuable for use in PCR experiments.

TATA box: Short nucleotide sequence (TATA) usually located approximately 20 to 30 base pairs "upstream" (in the 5′ direction) of the start site of many eukaryotic genes; part of promoter sequence bound by transcription factors used to stimulate RNA polymerase.

telomerase: Enzyme that fills in gaps of DNA nucleotides at the ends of chromosomes (telomeres) that remain after DNA polymerase has copied DNA.

telomere: The end structures of a eukaryotic chromosome; in humans, consists of specific repeating sequences of DNA (TTAGGG).

template strand: After RNA polymerase binds to a promoter, it unwinds a region of the DNA to separate the two strands. Only one of the strands, called the template strand (the opposite strand is called the coding strand), is copied by RNA polymerase.

teratogenic: a substance that creates abnormal developmental effects in a an embryo.

tertiary structure: The unique three-dimensional shape of a protein.

thalidomide: Compound used initially to combat morning sickness in pregnant women; certain chemical forms of thalidomide caused severe birth defects; thalidomide derivatives are still being investigated for their use in treating cancer, HIV, and other diseases.

therapeutic cloning: Using a patient's DNA to create (clone) an embryo as a source of stem cells that could be used in therapeutic applications to treat the patient.

thermophiles: Organisms with high optimal growth temperatures. For example, bacteria that live in hot springs are thermophiles.

thermostable enzyme: An enzyme that is capable of withstanding high temperatures and is isolated from theromophiles. For example, Taq DNA polymerase is a thermostable enzyme.

3′ end: The end of a strand of DNA or RNA in which the last nucleotide is not attached to another nucleotide by a phosphodiester bond involving the 3′ carbon of the pentose sugar.

three Rs of animal research: Reduce the number of higher species used; Replace animals with alternative models whenever possible; and Refine tests and experiments to ensure the most humane conditions possible.

thymine: Abbreviated T; pyrimidine base present in DNA nucleotides.

Ti vector: Plasmid DNA vector derived from soil bacterium that can be used to clone genes in plant cells and deliver genes into plants.

T lymphocyte: Type of white blood cell (leukocyte); T lymphocytes, also called "T cells" play an essential role in helping the immune system recognize and respond to foreign materials (antigens).

tissue engineering: Designing and growing tissues for use in regenerative medicine applications.

tobacco mosaic virus (TMV): A virus that naturally invades certain plants.

traits: Inherited features of an organism such as skin color and body shape.

transcription: The synthesis of RNA from DNA, which occurs in the nucleus of eukaryotic cells and the cytoplasm of prokaryotic cells.

transcription factors: DNA-binding proteins that bind promoter regions of a gene and stimulate transcription of a gene by RNA polymerase.

transcriptional regulation: Form of gene-expression regulation that involves controlling the process of transcription by controlling the amount of RNA produced by a cell.

transfection: Introducing DNA into animal or plant cells.

transfer RNA (tRNA): Small RNA molecules that transport amino acids to a ribosome during protein synthesis. tRNA binds to specific codons in mRNA sequences during translation.

transformable spongiform encephalitis (TSE): Prion disease that can occur in sheep and goats (scrapie) and cows (bovine spongiform encephalitis, or "mad cow" disease). Human forms of these brain-destroying diseases include kuru and TSE. All these diseases involve changes in the conformation of prion precursor protein, a protein normally found in mammalian neurons as a membrane glycoprotein.

transformation: The process by which bacteria take in DNA from the surroundings. Term also is used to define changes that cause a normal cell to become a cancer cell.

transgene: Gene from one organism introduced into another organism to create a transgenic; term usually applies to genes used to create transgenic animals and plants.

transgenic animals: Animals that contain genes from another source. For instance, human genes for clotting proteins can be introduced into cows for the production of these proteins in their milk.

translation: The synthesis of proteins from genetic information in messenger (mRNA) molecules. Translation occurs in the cytoplasm of all cells.

triploid: Three sets of chromosomes (3n); used to describe organisms with three sets of chromosomes.

tuberculosis (TB): Disease caused by the bacterium Mycobacterium tuberculosis, which grows slowly and can exist in a human for several years before the individual develops TB.

tumor-inducing (Ti) plasmid: A plasmid vector found in Agrobacter.

two-dimensional electrophoresis: Separation based on charge in two directions.

type I, or insulin-dependent, diabetes mellitus: Disease caused by a lack of the pancreatic hormone insulin, which is required for carbohydrate metabolism. Creates elevated blood sugar levels (hyperglycemia).

ultrafiltration: Separation of particles smaller than 20 μm.

upstream processing: Adjustments in the purification process based on changes in the biologic process, making processing more efficient.

uracil: Abbreviated U; pyrimidine base present in RNA nucleotides.

U. S. Department of Agriculture (USDA): Agency created in 1862 that has many functions related to the advancement and regulation of agriculture. Some of those functions include the regulation of plant pests, plants, and veterinary biologics.

U.S. Environmental Protection Agency (EPA): Agency whose primary purpose is to protect human health and to safeguard the natural environment (air, water, and land) by working with other federal agencies in the United States and state and local governments to develop and enforce regulations under existing environmental laws.

U. S. Patent and Trademark Office (USPTO): The office of the federal government that issues patents (including patents for gene sequences).

utilitarian approach: Line of ethical thought that states that actions are moral if the result produces the greatest good for the greatest number of humans; also referred to as consequential ethics because it focuses on results or consequences, not on intentions.

utility patent: An invention is "useful" if it provides some identifiable benefit and is capable of use (the doctrine prevents the patenting of fantastic or inoperative devices such as perpetual motion machines).

vaccination: The process of administering a vaccine to provide an organism with immunity to an infectious microorganism.

vaccines: A preparation of a microorganism or its components that is used to stimulate the production of antibodies (antibody-mediated immunity) in an organism.

variable number tandem repeats (VNTRs): The recognition that variable numbers of repeated nucleotides can be found in DNA and can be used for identification of individuals.

vectors: DNA (or viruses) that can be used to carry and replicate other pieces of DNA in molecular biology experiments; for example, plasmid DNA, viruses used for gene therapy; also refers to organisms that carry disease.

venture capital (VC): Funding for an enterprise provided by financiers who see promise in the enterprise and expect a substantial financial benefit for the high risk (or expect to take a loss).

Western blot analysis: Laboratory technique for separating protein molecules by gel electrophoresis and transferring (blotting) proteins onto a filter paper blot that is usually probed with antibodies to study protein structure and function.

Western blotting: See *Western blot analysis.*

whole-genome "shotgun" sequencing: DNA sequencing approaches in which an entire genome can be fragmented and its sequence determined. Bioinformatics methods are then used to assemble the sequenced fragments of a genome to assemble the entire genome sequence.

xenotransplantation: The transfer of tissues or organs from one species to another; for instance, transplanting a pig organ into a human.

x-ray crystallography: Identification of molecules based on the patterns produced after bombardment of their crystals with x-rays.

yeast: A unicellular fungus.

yeast artificial chromosome (YAC): Plasmid vectors grown in yeast cells that can replicate very large pieces of DNA; used to clone pieces of human chromosomes for the Human Genome Project.

yeast two-hybrid system: Laboratory technique involving the use of yeast to join together different proteins (creating a hybrid protein) as a way to study protein function.

zinc-finger nucleases (ZFNs): Artificial restriction enzymes generated by fusing a zinc finger DNA-binding domain to a DNA-cleavage domain.

zygote: Diploid cell formed when haploid gametes such as a sperm and egg cell unite.

Index

Note: Page numbers followed by f indicate figures; those followed by t indicate tables.